Audio/Video Protocol Handbook: Broadcast Standards and Reference Data

On-Line Updates

Additional updates relating to audio/video engineering in general, and this book in particular, can be found at the *Standard Handbook of Video and Television Engineering* web site:

www.tvhandbook.com

The tvhandbook.com web site supports the professional audio/video community with news, updates, and product information relating to the broadcast, post production, consumer, and business/industrial applications of digital audio and video.

Check the site regularly for news, updated chapters, and special events related to broadcast engineering. The technologies encompassed by *Audio/Video Protocol Handbook* are changing rapidly, with new developments each month. Changing market conditions and regulatory issues are adding to the rapid flow of news and information in this area.

Specific services found at **www.tvhandbook.com** include:

- **Technology News**. News reports and technical articles on the latest developments in digital broadcasting, both in the U.S. and around the world. Check in at least once a month to see what's happening in the fast-moving area of digital technology.
- **Resource Center**. Check for the latest information on professional and broadcast audio/video systems. The Resource Center provides updates on implementation and standardization efforts, plus links to related web sites.
- **tvhandbook.com Update Port**. Updated material for books in the McGraw-Hill Audio/Video Engineering Series is posted on the site regularly. Material available includes updated sections and chapters in areas of rapidly advancing technologies.
- **tvhandbook.com Book Store**. Check to find the latest books on digital audio and video technologies. Direct links to authors and publishers are provided. You can also place secure orders from our on-line bookstore.

In addition to the resources outlined above, detailed information is available on other books in the McGraw-Hill Audio/Video Series.

Audio/Video Protocol Handbook: Broadcast Standards and Reference Data

Jerry C. Whitaker, Editor

McGraw-Hill
New York Chicago San Francisco
Lisbon London Madrid Mexico City Milan
New Delhi San Juan Seoul Singapore
Sydney Toronto

Cataloging-in-Publication Data is on file with the Library of Congress.

McGraw-Hill

A Division of The McGraw-Hill Companies

1 2 3 4 5 6 7 8 9 0 DOC/DOC 0 9 8 7 6 5 4 3 2

P/N 140078-8

PART OF

ISBN 0-07-139643-8

The sponsoring editor for this book was Steve Chapman and the production supervisor was Pamela Pelton. The book was set in Times New Roman and Helvetica by Technical Press, Morgan Hill, CA.

Printed and bound by R. R. Donnelley & Sons Company.

This book is printed on recycled, acid-free paper containing a minimum of 50% recycled, de-inked fiber.

Contents

Contributors **xv**

Preface **xvii**

How to Use this Book xviii

Chapter 1: Systems Engineering **1**

Introduction 1
Systems Theory 1
Systems Engineering 2
Functional Analysis 4
Synthesis 5
Modeling 5
Dynamics 5
Optimization 5
Evaluation and Decision 5
Trade Studies 6
Description of System Elements 8
Phases of a Typical System Design Project 9
Design Development 9
Electronic System Design 10
Detailed Design 11
Customer Support 13
Budget Requirements Analysis 13
Feasibility Study and Technology Assessment 14
Planning and Control of Scheduling and Resources 14
Project Tracking and Control 15
Executive Management 16
Project Manager 17
Systems Engineer 20
References 21
Bibliography 21

Chapter 2: Engineering Documentation **23**

Introduction 23
Basic Concepts 23
The Manuals 24
Documentation Conventions 24
Self Documentation 25
Database Documentation 26
Graphic Documentation 27
Update Procedures 27
Equipment Documentation 27

Operator/User Documentation 28
The True Cost of Documentation 29
Bibliography 29

Chapter 3: Safety Issues **31**
Introduction 31
Electric Shock 31
Effects on the Human Body 31
Circuit Protection Hardware 33
Working with High Voltage 33
RF Considerations 36
First Aid Procedures 36
Operating Hazards 36
OSHA Safety Considerations 38
Protective Covers 38
Identification and Marking 38
Grounding 39
Beryllium Oxide Ceramics 40
Corrosive and Poisonous Compounds 42
FC-75 Toxic Vapor 42
Nonionizing Radiation 43
NEPA Mandate 43
Revised Guidelines 44
Multiple-user Sites 45
Operator Safety Considerations 46
X-Ray Radiation Hazard 46
Implosion Hazard 46
Hot Coolant and Surfaces 46
Polychlorinated Biphenyls 47
Governmental Action 48
PCB Components 48
PCB Liability Management 50
References 50
Bibliography 51

Chapter 4: Video Standards Overview **53**
Introduction 53
The History of Modern Standards 53
American National Standards Institute (ANSI) 54
Professional Society Engineering Committees 55
Advanced Television Systems Committee 55
Electronics Industries Alliance 56
Institute of Electrical and Electronic Engineers 56
Society of Motion Picture and Televisions Engineers 56
Audio Engineering Society 57
Advanced Television Systems Committee 57
ATSC Digital Television Standard, Document A/53B 57
Guide to the use of the ATSC Digital Television Standard, Document A/54 57
Digital Audio Compression (AC-3), Document A/52A 58

Standard for Coding 25/50 Hz Video, Document A/63 58
Transmission Measurement and Compliance Standard for DTV, Document A/64 Rev. A 58
PSIP for Terrestrial Broadcast and Cable, Document A/65 Rev A with Amendment No. 1 58
Use of ATSC A/65A PSIP Standard in Taiwan, Document A/68 58
Conditional Access For Terrestrial Broadcast, Document A/70 58
ATSC Recommended Practice for Developing DTV Field Test Plans, Document A/75 59
Modulation & Coding Requirements for DTV Applications Over Satellite, Document A/80 59
Data Broadcast, Document A/90 59
Implementation Guidelines for the ATSC Data Broadcast Standard, Document A/91 59
DTV Application Software Environment (DASE) 59
Interactive Services 60
When is a Standard Finished? 60
Standards Relating to Digital Video 60
Video 60
Audio 61
ATSC DTV Standard 61
Service Multiplex and Transport Systems 61
System Information Standard 61
Receiver Systems 62
Program Guide 62
Program/Episode/Version Identification 62
DVB 63
General 63
Multipoint Distribution Systems 63
Interactive Television 63
Conditional Access 63
Interfaces 64
Acquiring Reference Documents 64
Bibliography 64

Chapter 5: SMPTE Documents 65

Introduction 65
General Topics 65
Ancillary 65
Digital Control Interfaces 66
Edit Decision Lists 67
Image Areas 67
Interfaces and Signals 67
Bit-Parallel Interfaces 67
Bit-Serial Interfaces 67
Scanning Formats 68
Monitors 68
MPEG-2 68
Test Patterns 69
Video Recording and Reproduction 69
Time and Control Code 70
Tape Recording Formats 70
SMPTE Documents by Number 71
Scopes of SMPTE Standards 84

Scopes of SMPTE Recommended Practices 119
Scopes of SMPTE Engineering Guidelines 143

Chapter 6: CEA Standards 149

Introduction 149
Antennas 149
Television 150
Television Receivers 157
Home Networks 157
Audio 163
Radio Broadcast 163
Loudspeakers 164

Chapter 7: SCTE Standards 165

Introduction 165
Selected Abstracts of SCTE Standards 167
Selected DTV-Related DVS Documents 169

Chapter 8: AES Standards 175

Introduction 175
Informational Documents 175
Project Reports 176
Standards and Recommended Practices 177

Chapter 9: References in ATSC Standards and Recommended Practices 183

Introduction 183
A/52: Digital Audio Compression Standard (AC-3) 183
Normative References 184
Informative References 184
A/53B: ATSC Digital Television Standard (Revision B) 184
Normative References 184
Informative References 185
A/54: Guide to the Use of the ATSC Digital Television Standard 185
Normative References 185
Informative References 185
A/57: Program/Episode/Version Identification 186
Normative References 186
Informative References 186
A/58: Harmonization with DVB SI in the use of the ATSC Digital Television Standard 186
Normative Reference 186
Informative Reference 186
A/63: Standard for Coding 25/50 Hz Video 186
Normative References 186
Informative References 187
A/64: Transmission Measurement and Compliance For Digital Television 187
Normative References 187
Informative References 187
A/65A: Program and System Information Protocol for Terrestrial Broadcast and Cable 187

Normative References 188
Informative References 188
A/68: Use of ATSC A/65A PSIP Standard in Taiwan 189
Normative References 189
Informative References 189
A/70: Conditional Access System for Terrestrial Broadcast 189
Normative References 189
Informative References 190
A/75: ATSC Recommended Practice for Developing DTV Field Test Plans 190
Normative References 190
Informative References 191
A/80: Modulation & Coding Requirements For Digital TV Applications Over Satellite 191
Normative References 191
Informative References 191
A/90:ATSC Data Broadcast Standard 192
Normative References 192
Informative References 193
A/91 Implementation Guidelines for the ATSC Data Broadcast Standard 193
Normative References 193
Informative References 193
Master Listings of References 194
Normative References (alphabetical order) 194
Listing of Organizations Referenced 196
Informative References (alphabetical order) 197
Listing of Organizations Referenced 199

Chapter 10: The Electromagnetic Spectrum 201

Introduction 201
Spectral Sub-Regions 202
Optical Spectrum 202
Visible Light Band 202
IR Band 203
UV Band 204
DC to Light 204
Microwave Band 204
Radio Frequency (RF) Band 205
Power Frequency (PF)/Telephone Band 205
Frequency Band Designations 205
Light to Gamma Rays 208
X Ray Band 208
Gamma Ray Band 210
Bibliography 210

Chapter 11: Frequency Assignment and Allocations 213

Introduction 213
The International Telecommunication Union 214
Purposes of the Union 215
Structure of the Union 215
The Federal Communications Commission 215

National Table of Frequency Allocations 216
U.S. Government Table of Frequency Allocations 217

Chapter 12: Dictionary of Electronics Terms 309

Terms Relating to Digital Television 309
General Electronics Terms 314
B 321
C 324
D 330
E 335
F 339
G 343
H 345
I 347
J 352
K 352
L 353
M 355
N 359
O 361
P 363
Q 365
R 365
S 370
T 373
U 375
V 377
W 379
X 380
Y 380
References 380
Bibliography 381

Chapter 13: Acronyms and Abbreviations 383

Acronyms and Abbreviations Relating to Digital Television 383
General Electronics Acronyms and Abbreviations 385
A 385
B 388
C 389
D 392
E 395
F 397
G 399
H 400
I 400
J 402
K 402
L 403
M 404

N 406
O 408
P 409
Q 412
R 412
S 413
T 415
U 417
V 418
W 419
X 419
Z 420
References 420
Bibliography 420

Chapter 14: Reference Data and Tables **421**

Standard Units 421
Standard Prefixes 423
Common Standard Units 423
Conversion Reference Data 425
Reference Tables 450
Power Conversion Factors 450
Standing Wave Ratio 451
Specifications of Standard Copper Wire Sizes 452
Celcius-to-Fahrenheit Conversion Table 454
Inch-to-Millimeter Conversion Table 455
Conversion of Millimeters to Decimal Inches 455
Convertion of Common Fractions to Decimal and Millimeter Units 457
Decimal Equivalent Size of Drill Numbers 458
Decimal Equivalent Size of Drill Letters 459
Conversion Ratios for Length 459
Conversion Ratios for Area 460
Conversion Ratios for Mass 461
Conversion Ratios for Volume 461
Conversion Ratios for Cubic Measure 462
Conversion Ratios for Electrical Quantities 462

Chapter 15: Informative Documents by Subject **463**

Introduction 463
Audio 463
Principles and Sound and Hearing 463
The Audio Spectrum 466
Architectural Acoustic Principles and Design Techniques 468
Microphone Devices and Systems 469
Sound Reproduction Devices and Systems 471
Digital Coding of Audio Signals 472
Compression Technologies for Audio 473
Audio Networking 474
Audio Recording Systems 476

Audio Production Facility Design 478
Radio Broadcast Transmission Systems 478
Radio Receivers 481
Standards and Practices 482
Video 483
Light, Vision, and Photometry 483
Color Vision, Representation, and Reproduction 485
Optical Components and Systems 487
Digital Coding of Video Signals 487
Electron Optics and Deflection 488
Video Cameras 490
Monochrome and Color Image Display Devices 491
Video Recording Systems 499
Video Production Standards, Equipment, and System Design 501
Film for Video Applications 504
Compression Technologies for Video and Audio 505
Video Networking 507
Digital Television Transmission Systems 510
Frequency Bands and Propagation 515
Television Transmission Systems 517
Television Antenna Systems 520
Television Receivers and Cable/Satellite Distribution Systems 522
Video Signal Measurement and Analysis 527
Standards and Practices 529

Appendix A: A Brief History of Radio **533**
Introduction 533
In the Beginning 533
Sarnoff: The Visionary 534
Radio Central 535
Radio City at Rockefeller Center 535
Hanson's Vision 536
WLW: The Nation's Station 537
The FCC Enters the Picture 538
The Golden Era of Radio 538
The Origins of the Networks 539
FM Grows Up 540
Bibliography 541

Appendix B: A Brief History of Television **543**
Introduction 543
Television: A Revolution in Communications 543
The Nipkow Disc 543
Zworykin: The Brains of RCA 544
Farnsworth: The Boy Wonder 545
Other Experimenters 546
Pickup Tubes 547
Sold State Imaging 548
Image Reproduction 549

Who was First? 550
TV Grows Up 551
Transmission Standard Developed 552
Color Standard 552
UHF Comes of Age 554
Birth of the Klystron 554
Early Development of HDTV 555
1125/60 Equipment Development 556
The 1125/60 System 558
European HDTV Systems 559
A Perspective on HDTV Standardization 560
Digital Systems Emerge 560
HD-DIVINE 561
Eureka ADTT 562
Digital Video Broadcasting 562
Involvement of the Film Industry 563
Political Considerations 564
Terminology 565
Digital Television in the U.S. 565
The Process Begins 566
System Testing: Round 2 568
Formation of the Grand Alliance 570
Testing the Grand Alliance System 571
The Home Stretch for the Grand Alliance 573
Digital Broadcasting Begins 574
Continuing Work on the ATSC Standard 574
A Surprising Challenge 575
Petition Filed 577
Petition Rejected 577
FCC Reviews the Digital Television Conversion Process 578
ATSC Comments Filed 579
A Turning Point is Reached 580
Receiver Performance Issues 581
References 582
Bibliography 582

Subject Index **585**

About the Editor **589**

Contributors

Fred Baumgartner

Terrence M. Baun

Gene DeSantis

Donald C. McCroskey

John Norgard

Some material in this book has been adapted from the McGraw-Hill *Standard Handbook of Video and Television Engineering*, 3rd edition. Used with permission.

Preface

A widely used dictionary lists seven meanings for the word "standard," but only one of these, "anything authorized as a measurement of quantity and quality," seems to relate to the idea of a standard in the industrial area. The original idea of commercial standards was to be bound legally on units of weights and measure for fairness in trade. In the field of electronics, the standardization of electrical units of measurement would be the counterpart. Today, standards are essentially recommendations for users and/or manufacturers to adhere to basic specifications to allow operational interchangeability in the use of equipment and supplies.

Anyone concerned with interchangeability of equipment or product should be concerned with standards. A prospective user hesitates to purchase equipment that does not conform to recognized interface standards for connectors, input/output levels, control, timing, and test specifications. A manufacturer may find a limited market for a good product if it is not compatible with other equipment in common use.

To most video professionals, the term "standards" envisions a means of promoting interchange of basic hardware. To others, it evokes thoughts of a slowdown of progress, of maintaining a status quo—perhaps for the benefit of a particular group. Both camps can cite examples to support their viewpoint, but no one can seriously contend that we would better off without standards. Standards promote economies of scale that tend to produce more reliable products at a lower cost.

For most people, the question is: "How do standards affect my life? Do they stifle progress? Do they prevent products from appearing on the market in a timely fashion? Do they discourage alternate technologies that might be beneficial in the long run?" Some would respond affirmatively to one or more of these questions, but consider the upside. Standards ensure that the needs of the user are considered. Interconnection of equipment from different manufacturers is facilitated. The current rollout of digital television products at a record pace attests to the need for, and benefits of, standards. The progress made so far in the DTV era would have been wholly impossible without the considerable efforts of organizations such as the ATSC, SMPTE, SCTE, CES, and NAB.

Rapid improvements in technology tend to make many standards technically obsolete by the time they are adopted. But such is the nature of our rapidly expanding technology-based society. There is no need to apologize for this natural phenomena. A standard still provides a stable platform for manufacturers to market their product and assures the user of some degree of compatibility. Technical chaos are real possibilities if standards are not adopted in a timely manner. Only the strongest companies could be expected to survive in an atmosphere where standards are lacking. A successful standard promises a stable period of income to manufacturers while giving users assurance of multiple sources during the active life of the product.

Standardization usually starts within a company as a way to reduce costs associated with parts stocking, design drawings, training, and retraining of personnel. The next level might be a cooperative agreement between firms making similar equipment to use standardized dimensions, parts, and components. Competition, trade secrets, and the *NIH factor* (not invented here) often generate an atmosphere that prevents such an understanding. Enter the professional engineering society, which promises a forum for discussion between users and engineers while down playing the commercial and business aspects.

Of the many standards-setting organizations in the professional audio/video field, the most prominent are:

- Society of Motion Picture and Television Engineers
- Audio Engineering Society
- National Association of Broadcasters
- Consumer Electronics Association
- Advanced Television Systems Committee
- Society of Cable and Telecommunications Engineers

Standards, whether for a new television broadcast system or VTR connector pin assignments, are vital for the continued growth of the communications industry. Much of the material contained in this publication is provided courtesy of these organizations, which the editor gratefully acknowledges.

Web Resources

Several chapters in this book draw heavily upon material made available by leading standards organizations. Web site addresses are given where applicable, and readers are encouraged to explore these valuable resources. Standards documents can be downloaded or ordered on-line from most of the sites. Because of the rapidly changing nature of digital audio and video implementation, readers are encouraged to check in regularly.

Another valuable resource is the SMPTE Television Standards on CD-ROM. This product, available for purchase from the SMPTE, contains all existing and proposed Standards, Engineering Guidelines, and Recommended Practices for television work. In this era of DTV, this product is indispensable. Similar products are available from other standards organizations, either on a per-document basis or as a complete package, as in the SMPTE offering. Subscription services are also available and should be considered for organizations that require the latest Standards, Recommended Practices, Engineering Guidelines, and related documents on hand.

How to Use This Book

This is not a tutorial handbook in the conventional sense. Most of the information provided is intended to guide readers to sources of additional information. One of the challenges in keeping up with the myriad of standards documents applicable to professional audio/video engineering is determining how the various documents relate to the work at hand. It is the goal of this publication to make this task easier to accomplish by providing a one-stop comprehensive listing of standards, engineering guidelines, and recommended practices applicable to professional audio and video engineering.

Specific elements of the *Audio/Video Protocol Handbook* include the following:

- A primer on professional engineering concepts. Contained in Chapters 1–3, the important elements of systems engineering, engineering documentation, and safety issues are examined. These chapters set the stage for the material that follows.

- An encyclopedia of standards documents. Chapters 4–9 represent the most detailed, comprehensive listing of standards relating to professional audio and video technologies ever published. Specific standards organizations included in this document inventory are ATSC, SMPTE, SCTE, AES, and CEA. Document abstracts are provided for nearly all standards of importance to broadcast and related industries.
- A spectrum of frequencies and assignments. Chapters 10 and 11 detail the major bands of the spectrum, with particular emphasis on radio frequency services and applications. Included is an up-to-date 100 page table of frequency allocations.
- A comprehensive glossary of terms. A 100+ page dictionary of communications terms is provided in Chapter 12, followed by a 50 page listing of acronyms and abbreviations in Chapter 13.
- Reference data and tables. In Chapter 14, readers will find the most comprehensive units conversion table available anywhere.
- A detailed listing of informative documents. Listed by subject matter, Chapter 15 is a valuable resource for any reader seeking additional information on a specific area of audio or video. This chapter lists reference documents for all areas of broadcast communication—from the audio spectrum to video test and measurement.
- A perspective of history. No examination of standards would be complete without some perspective of the road that led to the development of the radio and television broadcast systems that we enjoy today. These two separate but related technologies have been the driving force behind much of the electronics business for decades. Appendix A gives a brief history of radio, and Appendix B looks at the development of television.

A CD-ROM is provided with this book that permits searching for key words, standards numbers, and subjects. On the CD-ROM is a copy of this book in the Adobe Acrobat format. It is intended to help readers easily find the information they want.

It is the goal of the *Audio/Video Protocol Handbook* to bring the diverse concepts and technologies of professional audio/video engineering together in an understandable and readily-usable form.

Jerry C. Whitaker

January, 2002

For

Jennifer and Andy

welcome home

Audio/Video Protocol Handbook: Broadcast Standards and Reference Data

Chapter

1

Systems Engineering

Gene DeSantis

1.1 Introduction[1]

Modern systems engineering emerged during World War II as—due to the degree of complexity in design, development, and deployment—weapons evolved into weapon systems. In the sixties, the complexities of the space program made a systems engineering approach to design and problem solving even more critical. Indeed, the Department of Defense and NASA are two of the staunchest practitioners. With the increase in size and complexity of television and nonbroadcast video systems during that same period, the need for a systems approach to planning, designing and building facilities gained increased attention.

Today, large engineering organizations utilize a systems engineering process. Much has been published about system engineering practices in the form of manuals, standards, specifications, and instruction. In 1969, MIL-STD-499 was published to help government and contractor personnel involved in support of defense acquisition programs. In 1974, this standard was updated to MIL-STD-499A, which specifies the application of system engineering principles to military development programs. Likewise, the builders of turnkey television systems and facilities have adopted their own unique systems engineering approaches to projects. The tools and techniques of this processes continue to evolve in order to do each job a little better, save time, and cut costs.

1.2 Systems Theory

Although there are other areas of application outside of the broadcast industry, we will be concerned with systems theory as it applies to television systems engineering. We will be concerned with audio, video, RF, control, time code, telecommunications, computer systems, and software. Systems theory can be applied to engineering of all of these elements. Building and vehicle systems—including space planning, power and lighting, environmental control, and safety sys-

1. This chapter is based on: DeSantis, Gene: "Systems Engineering," in *Standard Handbook of Video and Television Engineering*, 3rd ed., Jerry C. Whitaker (ed.), McGraw-Hill, New York, N.Y., 2000. Used with permission. All rights reserved.

tems—can all benefit from the systems engineering approach. These systems are made up of component elements that are interconnected and programmed to function together in a facility.

For the purpose of this discussion, a system is defined as a set of related elements that function together as a single entity.

Systems theory consists of a body of concepts and methods that guide the description, analysis, and design of complex entities.

Decomposition is an essential tool of systems theory. The systems approach attempts to apply an organized methodology to completing large complex projects by breaking them down into simpler, more manageable components. These elements are treated separately, analyzed separately, and designed separately. In the end, all of the components are recombined to build the whole.

Holism is an element of systems theory in that the end product is greater than the sum of its component elements. In systems theory, modeling and analytical methods enable all essential effects and interactions within a system and those between a system and its surroundings to be taken into account. Errors resulting from the idealization and approximation involved in treating parts of a system in isolation, or reducing consideration to a single aspect, are thus avoided.

Another holistic aspect of system theory describes *emergent properties*. Properties that result from the interaction of system components, properties that are not those of the components themselves, are referred to as emergent properties.

Although dealing with concrete systems, *abstraction* is an important feature of systems models. Components are described in terms of their function rather than in terms of their form. Graphical models such as block diagrams, flow diagrams, and timing diagrams are commonly used.

Mathematical models may also be employed. Systems theory shows that, when modeled in abstract formal language, apparently diverse kinds of systems show significant and useful *isomorphisms* of structure and function. Similar interconnection structures occur in different types of systems. Equations that describe the behavior of electrical, thermal, fluid, and mechanical systems are essentially identical in form.

Isomorphism of structure and function implies isomorphism of behavior of a system. Different types of systems exhibit similar dynamic behavior such as response to stimulation.

The concept of *hard* and *soft* systems appears in system theory. In hard systems, the components and their interactions can be described by mathematical models. Soft systems can not be described so easily. They are mostly human activity systems that imply unpredictable behavior and non uniformity. They introduce difficulties and uncertainties of conceptualization, description, and measurement. The kinds of system concepts and methodology described previously can not be applied.

1.2a Systems Engineering

Systems engineering depends on the use of a process methodology based on systems theory. In order to deal with the complexity of large projects, systems theory breaks down the process into logical steps.

Even though underlying requirements differ from program to program, there is a consistent, logical process that can best be used to accomplish system design tasks. The basic product development process is illustrated in Figure 1.1. The systems engineering starts at the beginning of this process to describe the product to be designed. It includes four activities:

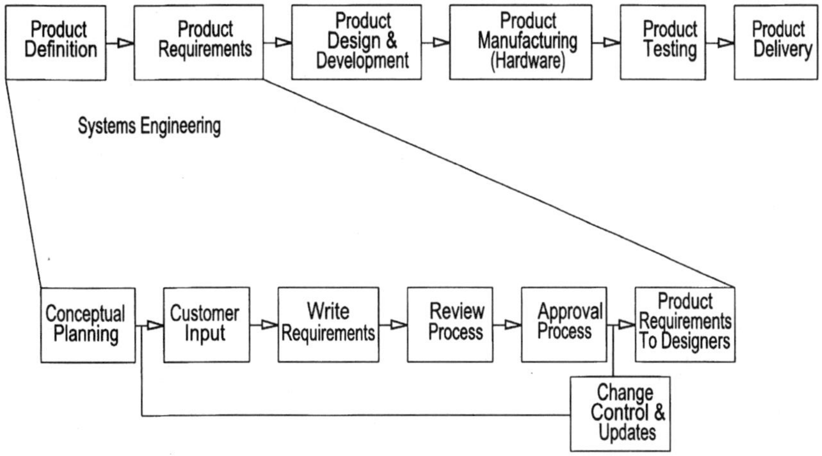

Figure 1.1 The product development and documentation process.

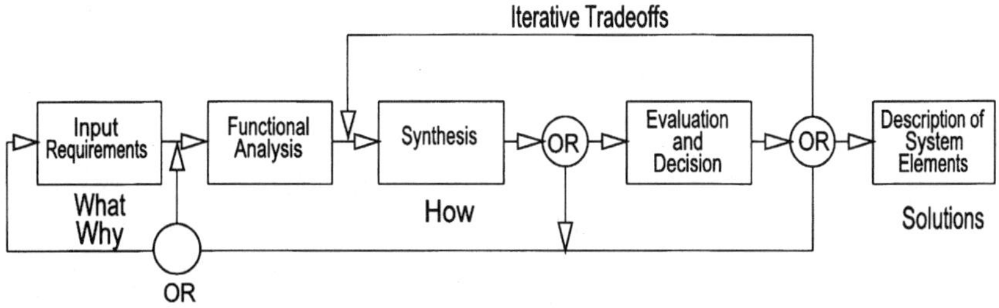

Figure 1.2 The systems engineering process. (*After* [1].)

- Functional analysis
- Synthesis
- Evaluation and decision
- Description of system elements

This process is illustrated in Figure 1.2. The process is iterative. That is, with each successive pass, the product element description becomes more detailed. At each stage in the process a decision is made whether to accept, make changes, or return to an earlier stage of the process and produce new documentation. The result of this activity is documentation that fully describes all system elements and which can be used to develop and produce the elements of the system. The systems engineering process does not produce the actual system itself.

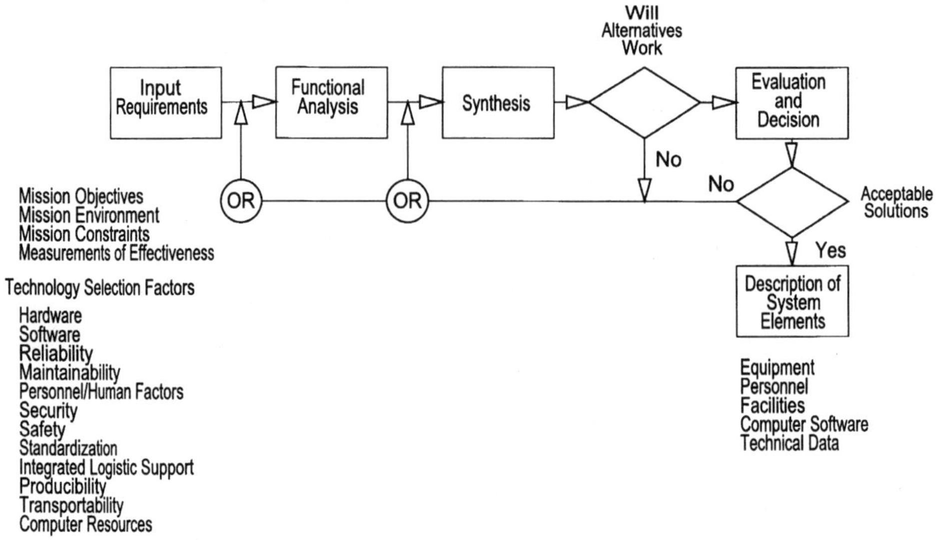

Figure 1.3 The systems engineering decision process. (*After* [1].)

1.2b Functional Analysis

A *systematic approach* to systems engineering will include elements of systems theory. (See Figure 1.3.) To design a product, hardware and software engineers need to develop a vision of the product—the product requirements. These requirements are usually based on customer needs researched by a marketing department. An organized process to identify and validate customer needs will help minimize false starts. System objectives are first defined. This may take the form of a mission statement that outlines the objectives, the constraints, the mission environment, and the means of measuring mission effectiveness.

The purpose of the system is defined, analysis is carried out to identify the requirements and what essential functions the system must perform, and why. The *functional flow block diagram* is a basic tool used to identify *functional needs*. It shows logical sequences and relationships of operational and support functions at the system level. Other functions such as maintenance, testing, logistics support, and productivity may also be required in the functional analysis. The functional requirements will be used during the *synthesis phase* to show the allocation of the functional performance requirements to individual system elements or groups of elements. Following evaluation and decision, the functional requirements provide the functionally oriented data required in the description of the system elements.

Analysis of time critical functions is also a part of this functional analysis process when functions have to take place sequentially, or concurrently, or on a particular schedule. Time line documents are used to support the development of requirements for the operation, testing, and maintenance functions.

1.2c Synthesis

Synthesis is the process by which concepts are developed to accomplish the functional requirements of a system. Performance requirements and constraints, as defined by the functional analysis, are applied to each individual element of the system, and a design approach is proposed for meeting the requirements. Conceptual schematic arrangements of system elements are developed to meet system requirements. These documents can be used to develop a description of the system elements and can be used during the *acquisition phase.*

1.2d Modeling

The concept of *modeling* is the starting point of synthesis. Because we must be able to weigh the effects of different design decisions in order to make choices between alternative concepts, modeling requires the determination of those quantitative features that describe the operation of the system. We would, of course, like a model with as much detail as possible describing the system. Reality and time constraints, however, dictate that the simplest possible model be selected in order to improve our chances of design success. The model itself is always a compromise. The model is restricted to those aspects that are important in the evaluation of system operation. A model might start off as a simple block diagram with more detail being added as the need becomes apparent.

1.2e Dynamics

Most system problems are dynamic in nature. The signals change over time and the components determine the *dynamic response* of the system. The system behavior depends on the signals at a given instant, as well as on the rates of change of the signals and their past values. The term "signals" can be replaced by substituting human factors, such as the number of users on a computer network for example.

1.2f Optimization

The last concept of synthesis is *optimization*. Every design project involves making a series of compromises and choices based on relative weighting of the merit of important aspects. The best candidate among several alternatives is selected. Decisions are often subjective when it comes to deciding the importance of various features.

1.2g Evaluation and Decision

Program costs are determined by the tradeoffs between operational requirements and engineering design. Throughout the design and development phase, decisions must be made based on evaluation of alternatives and their effects on cost. One approach attempts to correlate the characteristics of alternative solutions to the requirements and constraints that make up the selection criteria for a particular element. The rationale for alternative choices in the decision process are documented for review. Mathematical models or computer simulations can be employed to aid in this evaluation decision making process.

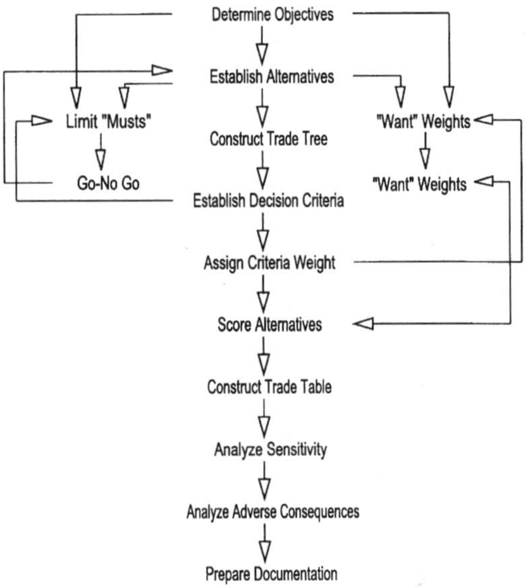

Figure 1.4 Trade study process flow. (*After* [2].)

1.2h Trade Studies

A structured approach is used in the *trade study process* to guide the selection of alternative configurations and insure that a logical and unbiased choice is made. Throughout development, trade studies are carried out to determine the best configuration that will meet the requirements of the program. In the concept exploration and demonstration phases, trade studies help define the system configuration. Trade studies are used as a detailed design analysis tool for individual system elements in the full scale development phase. During production, trade studies are used to select alternatives when it is determined that changes need to be made. Figure 1.4 illustrates the relationship of the various of elements that can be employed in a trade study. To provide a basis for the selection criteria, the objectives of the trade study must first be defined. Functional flow diagrams and system block diagrams are used to identify trade study areas that can satisfy certain requirements. Alternative approaches to achieving the defined objectives can then be established.

Complex approaches can be broken down into several simpler areas and a decision tree constructed to show the relationship and dependences at each level of the selection process. This *trade tree*, as it is called, is illustrated in Figure 1.5. Several trade study areas may be identified as possible candidates for accomplishing a given function. A trade tree is constructed to show relationships and the path through selected candidate trade areas at each level to arrive at a solution.

Several alternatives may be candidates for solutions in a given area. The selected candidates are then submitted to a systematic evaluation process intended to weed out unacceptable candidates. Criteria are determined that are intended to reflect the desirable characteristics. Undesirable characteristics may also be included to aid in the evaluation process. Weights are assigned to each criteria to reflect its value or impact on the selection process. This process is subjective. It

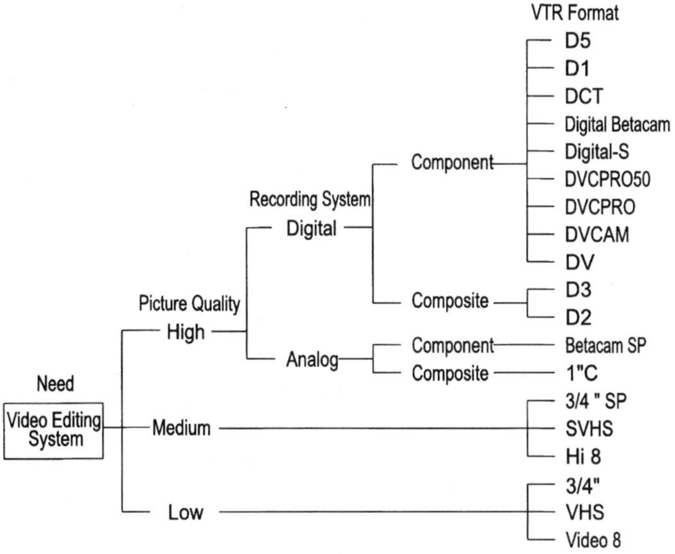

Figure 1.5 Trade tree example for a video project.

should also take into account cost, schedule, and hardware availability restraints that may limit the selection.

The criteria data on the candidates is collected and tabulated on a decision analysis work sheet. The attributes and limitations are listed in the first column and the data for each candidate listed in adjacent columns to the right. The performance data is available from vendor specification sheets or can require laboratory testing and analysis to determine. Each attribute is given a relative score from 1 to 10 based on its comparative performance relative to the other candidates. *Utility function graphs* can be used to assign logical scores for each attribute. The utility curve represents the advantage rating for a particular value of an attribute. A graph is made of ratings on the *y* axis versus an attribute value on the *x* axis. Specific scores can then be applied that correspond to particular performance values. The shape of the curve may take into account requirements, limitations, and any other factor that will influence its value regarding the particular criteria being evaluated. The limits to which the curves should be extended should run from the minimum value below which no further benefit will accrue to the maximum value above which no further benefit will accrue.

The scores are filled in on the decision analysis work sheet and multiplied by the weights to calculate the weighted score. The total of the weighted scores for each candidate then determines their ranking. As a general rule, at least a 10 percent difference in score is acceptable as "meaningful."

Further analysis can be applied in terms of evaluating the sensitivity of the decision to changes in the value of attributes, weights, subjective estimates, and cost. Scores should be checked to see if changes in weights or scores would reverse the choice.

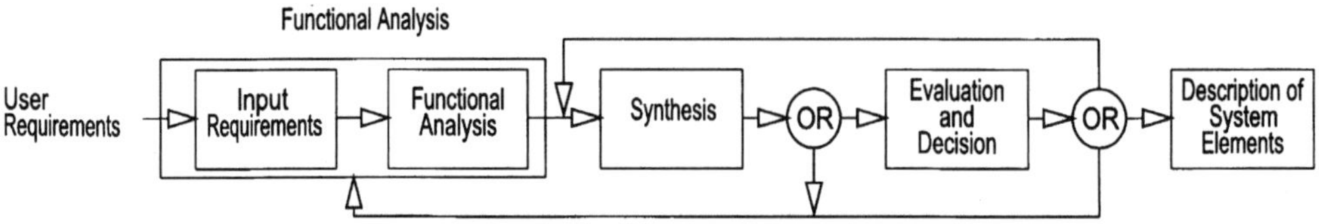

Figure 1.6 Basic documentation for systems engineering. (*After* [1].)

A *trade table* can be prepared to summarize the selection results. Pertinent criteria are listed for each alternative solution. The alternatives may be described in a quantitative manner such as high, medium, or low.

Finally, the results of the trade study are documented in the form of a report that discusses the reasons for the selections and may include the trade tree and the trade table.

There has to be a formal system of change control throughout the systems engineering process to prevent changes from being made without proper review and approval by all concerned parties, and to keep all parties informed. Change control also insures that all documentation is kept up to date and can help to eliminate redundant documents. Finally, change control helps to control project costs.

1.2i Description of System Elements

Five categories of interacting system elements can be defined: equipment (hardware), software, facilities, personnel, and procedural data. Performance, design, and test requirements must be specified and documented for equipment, components, and computer software elements of the system. It is necessary to specify environmental and interface design requirements that are necessary for proper functioning of system elements within a facility.

The documentation produced by the systems engineering process controls the evolutionary development of the system. Figure 1.6 illustrates the process documentation used by one organization in each step of the systems engineering effort.

The requirements are formalized in written specifications. In any organization, there should be clear standards for producing specifications. This can help reduce the variability of technical content and improve product quality as a result. It is also important to make the distinction here that the product should not be over specified to the point of describing the design or making it too costly. On the other hand, requirements should not be too general or so vague that the product would fail to meet the customer needs. In large departmentalized organizations, commitment to schedules can help assure that other members of the organization can coordinate their time.

The system engineering process does not actually design the system. The system engineering process produces the documentation necessary to define, design, build, and test the system. The technical integrity provided by this documentation ensures that the design requirements for the system elements reflect the functional performance requirements, that all functional performance requirements are satisfied by the combined system elements, and that such requirements are optimized with respect to system performance requirements and constraints.

1.3 Phases of a Typical System Design Project

The video industry has always been a very dynamic industry as a result of the rapid advancement of electronic technology. The design of a complex modern video facility can be used to illustrate the systems engineering approach.

1.3a Design Development

System design is carried out in a series of steps that lead to an operational facility. Appropriate research and preliminary design work is completed in the first phase of the project, the *design and development phase*. It is the intent of this phase to fully delineate all requirements of the project and to identify any constraints. Based on initial concepts and information, the design requirements are modified until all concerned parties are satisfied and approval is given for the final design work to proceed. The first objective of this phase is to answer the following questions.

- What are the functional requirements of the product of this work?
- What are the physical requirements of the product of this work?
- What are the performance requirements of the product of this work?
- Are there any constraints limiting design decisions?
- Will existing equipment be used?
- Is the existing equipment acceptable?
- Will this be a new facility or a renovation?
- Will this be a retrofit or upgrade to an existing system?
- Will this be a stand alone system?

Working closely with the customer's representatives, the equipment and functional requirements of each of the major technical areas of the facility are identified. In the case of facility renovation, the systems engineer's first order of business is to analyze existing equipment. A visit is made to the site to gather detailed information about the existing facility. Usually confronted with a mixture of acceptable and unacceptable equipment, the systems engineer must sort out those pieces of equipment that meet current standards, and determine which items should be replaced. Then, after soliciting input from the facility's technical and operational personnel, the systems engineer develops a list of needed equipment.

One of the systems engineer's most important contributions is the ability to identify and meet the needs of the customer, and do it within the project budget. Based on the customer's initial concepts and any subsequent equipment utilization research conducted by the systems engineer, the desired capabilities are identified as precisely as possible. Design parameters and objectives are defined and reviewed. Functional efficiency is maximized to allow operation by a minimum number of personnel. Future needs are also investigated at this time. Future technical systems expansion is considered.

After the customer approves the equipment list, preliminary system plans are drawn up for review and further development. If architectural drawings of the facility are available, they can be used as a starting point for laying out an equipment floor plan. The systems engineer uses this floor plan to be certain adequate space is provided for present and future equipment, as well as adequate clearance for maintenance and convenient operation. Equipment identification is then added to the architect's drawings.

Documentation should include, but not be limited to:

- Equipment list with prices
- Technical system functional block diagrams
- Custom item descriptions
- Rack and console elevations
- Equipment floor plans

The preliminary drawings and other supporting documents are prepared to record design decisions and to illustrate the design concepts to the customer. Renderings, scale models, or full size mock ups may also be needed to better illustrate, clarify, or test design ideas.

Ideas and concepts have to be exchanged and understood by all concerned parties. Good communication skills are essential for the team members. The bulk of the creative work is carried out in the design development phase. The physical layout—the look and feel—and the functionality of the facility will all have been decided and agreed upon by the completion of this phase of the project. If the design concepts appear feasible, and the cost is within the anticipated budget, management can authorize work to proceed on the final detailed design.

1.3b Electronic System Design

Performance standards and specifications have to be established up-front in a technical facility project. These will determine the performance level of equipment that will be acceptable for use in the system and affect the size of the budget. Signal quality, stability, reliability, and accuracy are examples of the kinds of parameters that have to be specified. Access and processor speeds are important parameters when dealing with computer driven products. The systems engineer has to confirm weather selected equipment conforms to the standards.

At this point it must be determined what functions each component in the system will be required to fulfill, and how each will function together with other components in the system. The management and operation staff usually know what they would like the system to do, and how they can best accomplish it. They have probably selected equipment that they think will do the job. With a familiarity of the capabilities of different equipment, the systems engineer should be able to contribute to this function-definition stage of the process. Questions that need to be answered include:

- What functions must be available to the operators?
- What functions are secondary and therefore not necessary?
- What level of automation should be required to perform a function?

- How accessible should the controls be?

Over-engineering or over-design should be avoided. This serious and costly mistake can be made by engineers and company staff when planning technical system requirements. A staff member may, for example, ask for a seemingly simple feature or capability without fully understanding its complexity or the additional cost burden it may impose on a project. Other portions of the system may have to be compromised in order to implement the additional feature. An experienced systems engineer will be able to spot this type of issue and determine whether the tradeoffs and added engineering time and cost are really justified.

When existing equipment is going to be used, it will be necessary to make an inventory list. This list will be the starting point for developing a final equipment list. Usually, confronted with a mixture of acceptable and unacceptable equipment, the systems engineer must sort out what meets current standards and what should be replaced. Then, after soliciting input from facility technical personnel, the systems engineer develops a summary of equipment needs, including future acquisitions. One of the systems engineer's most important contributions is the ability to identify and meet these needs within the facility budget.

A list of major equipment is prepared. The systems engineer selects the equipment based on experience with the products, and on customer preferences. Often some of the existing equipment may be reused. A number of considerations are discussed with the facility customer to arrive at the best product selection. Some of the major points include:

- Budget restrictions
- Space limitations
- Performance requirements
- Ease of operation
- Flexibility of use
- Functions and features
- Past performance history
- Manufacturer support

The goal here is the specification of equipment to meet the functional requirements of the project efficiently and economically. Simplified block diagrams for the video, audio, control, data, and communication systems are drawn. They are discussed with the customer and presented for approval.

1.3c Detailed Design

With the research and preliminary design development completed, the details of the design must now be concluded. The design engineer prepares complete detailed documentation and specifications necessary for the fabrication and installation of the technical systems, including all major and minor components. Drawings must show the final configuration and the relationship of each component to other elements of the system, as well as how they will interface with other building services, such as air conditioning and electrical power. This documentation must communicate the design requirements to the other design professionals, including the construction and installation contractors.

In this phase, the systems engineer develops final, detailed flow diagrams and schematics that show the interconnection of all equipment. Cable interconnection information for each type of signal is taken from the flow diagrams and recorded on the cable schedule. Cable paths are measured and timing calculations are made for signals requiring synchronization, such as video, synchronizing pulses, subcarrier, and digital audio and/or video (when required by the design). These timed cable lengths are entered onto the cable schedule.

The *flow diagram* is a schematic drawing used to show the interconnections between all equipment that will be installed. It is different from a block diagram in that it contains much more detail. Every wire and cable must be included on the drawings.

The starting point for preparing a flow diagram can vary depending upon the information available from the design development phase of the project, and on the similarity of the project to previous projects. If a similar system has been designed in the past, the diagrams from that project can be modified to include the equipment and functionality required for the new system. New models of the equipment can be shown in place of their counterparts on the diagram, and only minor wiring changes made to reflect the new equipment connections and changes in functional requirements. This method is efficient and easy to complete.

If the facility requirements do not fit any previously completed design, the block diagram and equipment list are used as a starting point. Essentially, the block diagram is expanded and details added to show all of the equipment and their interconnections, and to show any details necessary to describe the installation and wiring completely.

An additional design feature that might be desirable for specific applications is the ability to easily disconnect a rack assembly from the system and relocate it. This would be the case if the system where to be prebuilt at a systems integration facility and later moved and installed at the client's site. When this is a requirement, the interconnecting cable harnessing scheme must be well planned in advance and identified on the drawings and cable schedules.

Special custom items need to be defined and designed. Detailed schematics and assembly diagrams are drawn. Parts lists and specifications are finalized, and all necessary details worked out for these items. Mechanical fabrication drawings are prepared for consoles and other custom-built cabinetry.

The design engineer provides layouts of cable runs and connections to the architect. Such detailed documentation simplifies equipment installation and facilitates future changes in the system. During preparation of final construction documents, the architect and the design engineer can firm-up the layout of the technical equipment wire ways, including access to flooring, conduits, trenches, and overhead wire ways.

Dimensioned floor plans and elevation drawings are required to show placement of equipment, lighting, electrical cable ways, duct, conduit, and HVAC ducting. Requirements for special construction, electrical, lighting, HVAC, finishes, and acoustical treatments must be prepared and submitted to the architect for inclusion in the architectural drawings and specifications. This type of information, along with cooling and electrical power requirements, also must be provided to the mechanical and electrical engineering consultants (if used on the project) so they can begin their design calculations.

Equipment heat loads are calculated and submitted to the HVAC consultant. Steps are taken when locating equipment to avoid any excessive heat buildup within the equipment enclosures while maintaining a comfortable environment for the operators.

Electrical power loads are calculated and submitted to the electrical consultant and steps taken to provide for sufficient power, proper phase balance, and backup electricity as required.

1.3d Customer Support

The systems engineer can assist in purchasing equipment and help to coordinate the move to a new or renovated facility. This can be critical if a great deal of existing equipment is being relocated. In the case of new equipment, the customer will find the systems engineer's knowledge of prices, features, and delivery times to be an invaluable asset. A good systems engineer will see to it that equipment arrives in ample time to allow for sufficient testing and installation. A good working relationship with equipment manufacturers helps guarantee their support and speedy response to the customer's needs.

The systems engineer can also provide engineering management support during planning, construction, installation, and testing to help qualify and select contractors, resolve problems, explain design requirements, and assure quality workmanship by the contractors and the technical staff.

The procedures described in this section outline an ideal scenario. In reality, management may often try to bypass many of the foregoing steps to save money. This, the reasoning goes, will eliminate unnecessary engineering costs and allow construction to start right away. Utilizing in-house personnel, a small company may attempt to handle the job without professional help. This puts an added burden on the staff who are already working full time taking care of the daily operation of the facility. With inadequate design detail and planning, which can result when using unqualified people, the job of setting technical standards and making the system work then defaults to the construction contractors, in-house technical staff, or the installation contractor. This can result in costly and uncoordinated work-arounds and—of course—delays and added costs during construction, installation, and testing. It makes the project less manageable and less likely to be completed successfully.

The complexity of a project can be as simple as interconnecting a few pieces of equipment together to designing the software for an automated robotic storage system. The size of a technical facility can vary from a small one room operation to a large multimillion dollar facility. Where large amounts of money and other resources are going to be involved, management is well advised to recruit the services of qualified system engineers.

1.3e Budget Requirements Analysis

The need for a project may originate with customers, management, operations staff, technicians, or engineers. In any case, some sort of logical reasoning or a specific production requirement will justify the cost. On small projects, like the addition of a single piece of equipment, money only has to be available to make the purchase and cover installation costs. When the need may justify a large project, it is not always immediately apparent how much the project will cost to complete. The project has to be analyzed by dividing it up into its constituent elements. These elements include:

- Equipment
- Materials
- Resources (including money and man hours needed to complete the project)

An executive summary or capital project budget request containing a detailed breakdown of these elements can provide the information needed by management to determine the return on investment, and to make an informed decision on weather or not to authorize the project.

A *capital project budget request* containing the minimum information might consist of the following items:

- Project name. Use a name that describes the result of the project, such as "control room upgrade."
- Project number (if required). A large organization that does many projects will use a project numbering system of some kind, or may use a budget code assigned by the accounting department.
- Project description. A brief description of what the project will accomplish, such as "design the technical system upgrade for the renovation of production control room 2."
- Initiation date. The date the request will be submitted.
- Completion date. The date the project will be completed.
- Justification. The reason the project is needed.
- Material cost breakdown. A list of equipment, parts, and materials required for construction, fabrication, and installation of the equipment.
- Total material cost.
- Labor cost breakdown. A list of personnel required to complete the project, their hourly pay rates, the number of hours they will spend on the project, and the total cost for each.
- Total project cost. The sum of material and labor costs.
- Payment schedule. Estimation of individual amounts that will have to be paid out during the course of the project and the approximate dates each will be payable.
- Preparer's name and the date prepared.
- Approval signature(s) and date(s) approved.

More detailed analysis, such as return on investment, can be carried out by an engineer, but financial analysis should be left to the accountants who have access to company financial data.

Feasibility Study and Technology Assessment

In the case where it is required that an attempt be made to implement new technology, and where a determination must be made as to weather certain equipment can perform a desired function, it will be necessary to conduct a *feasibility study*. The systems engineer may be called upon to assess the state-of-the-art in order to develop a new application. An executive summary or a more detailed report of evaluation test results may be required, in addition to a budget estimate, in order to help management make their decision.

Planning and Control of Scheduling and Resources

Several planning tools have been developed for planning and tracking progress toward the completion of projects and scheduling and controlling resources. The most common graphical project management tools are the *Gantt Chart* and the *Critical Path Method* (CPM) utilizing the *Project Evaluation and Review* (PERT) technique. Computerized versions of these tools have greatly enhanced the ability of management to control large projects.

1.3f Project Tracking and Control

A project team member may be called upon by the project manager to report the status of work during the course of the project. A standardized project status report form can provide consistent and complete information to the project manager. The purpose is to supply information to the project manager regarding work completed, and money spent on resources and materials.

A project status report containing the minimum information might contain the following items:

- Project number (if required)
- Date prepared
- Project name
- Project description
- Start date
- Completion date (the date this part of the project was completed)
- Total material cost
- Labor cost breakdown
- Preparer's name
- Change Control

After part or all of a project design has been approved and money allocated to build it, any changes may increase or decrease the cost. Factors that effect the cost include:

- Components and materials
- Resources, such as labor and special tools or construction equipment
- Costs incurred because of manufacturing or construction delays

Management will want to know about such changes, and will want to control them. For this reason, a method of reporting changes to management and soliciting their approval to proceed with the change may have to be instituted. The best way to do this is with a *change order request* or change order. A change order includes a brief description of the change, the reason for the change, a summary of the effect it will have on costs, and what effect it will have on the project schedule.

Management will exercise its authority and approve or disapprove each change based upon its understanding of the cost and benefits, and their perceived need for the modification of the original plan. Therefore, it is important that the systems engineer provide as much information and explanation as may be necessary to make the change clear and understandable to management.

A change order form containing the minimum information might include the following items:

- Project number
- Date prepared
- Project name
- Labor cost breakdown

- Preparer's name
- Description of the change
- Reason for the change
- Equipment and materials to be added or deleted
- Material costs or savings
- Labor costs or savings
- Total cost of this change (increase or decrease)
- Impact on the schedule
- Program Management

Systems engineering is both a technical process and a management process. Both processes must be applied throughout a program if it is to be successful. The persons who plan and carry out a project constitute the *project team*. The makeup of a project team will vary depending on the size of the company and the complexity of the project. It is up to management to provide the necessary human resources to complete the project.

1.3g Executive Management

The executive manager is the person who can authorize that a project be undertaken. This person can allocate funds and delegate authority to others to accomplish the task. Motivation and commitment is toward the goals of the organization. The ultimate responsibility for a project's success is in the hands of the executive manager. This person's job is to get things done through other people by assigning group responsibilities, coordinating activities between groups, and resolving group conflicts. The executive manager establishes policy, provides broad guidelines, approves the project master plan, resolves conflicts, and assures project compliance with commitments.

Executive management delegates the project management functions and assigns authority to qualified professionals, allocates a capital budget for the project, supports the project team, and establishes and maintains a healthy relationship with project team members.

Management has the responsibility to provide clear information and goals—up front—based upon their needs and initial research. Before initiating a project, the company executive should be familiar with daily operation of the facility, analyze how the company works, how the staff does their jobs, and what tools they need to accomplish the work. Some points that may need to be considered by an executive before initiating a project include:

- What is the current capital budget for equipment?
- Why does the staff currently use specific equipment?
- What function of the equipment is the weakest within the organization?
- What functions are needed but cannot be accomplished with current equipment?
- Is the staff satisfied with current hardware?
- Are there any reliability problems or functional weaknesses?

- What is the maintenance budget, and is it expected to remain steady?
- How soon must the changes be implemented?
- What is expected from the project team?

Only after answering the appropriate questions will the executive manager be ready to bring in expert project management and engineering assistance. Unless the manager has made a systematic effort to evaluate all the obvious points about the facility requirements, the not-so-obvious points may be overlooked. Overall requirements must be broken down into their component parts. Do not try to tackle ideas that have to many branches. Keep the planning as basic as possible. If the company executive does not make a concerted effort to investigate the needs and problems of a facility thoroughly before consulting experts, the expert advice will be shallow and incomplete, no matter how good the engineer.

Engineers work with the information they are given. They put together plans, recommendations, budgets, schedules, purchases, hardware, and installation specifications based upon the information they receive from interviewing management and staff. If the management and staff have failed to go through the planning, reflection, and refinement cycle before those interviews, the company will likely waste time and money.

1.3h Project Manager

Project management is an outgrowth of the need to accomplish large complex projects in the shortest possible time, within the anticipated cost, and with the required performance and reliability. Project management is based upon the realization that modern organizations may be so complex as to preclude effective management using traditional organizational structures and relationships. Project management can be applied to any undertaking that has a specific end objective.

The project manager is the person who has the authority to carry out a project. This person has been given the legitimate right to direct the efforts of the project team members. The manager's power comes from the acceptance and respect accorded him or her by superiors and subordinates. The project manager has the power to act, and is committed to group goals.

The project manager is responsible for getting the project completed properly, on schedule and within budget, by utilizing whatever resources are necessary to accomplish the goal in the most efficient manner. The manager provides project schedule, financial, and technical requirement direction and evaluates and reports on project performance. This requires planning, organizing, staffing, directing, and controlling all aspects of the project.

In this leadership role, the project manager is required to perform many tasks including the following:

- Assemble the project organization.
- Develop the project plan.
- Publish the project plan.
- Set measurable and attainable project objectives.
- Set attainable performance standards.
- Determine which scheduling tools (PERT, CPM, and/ or GANTT) are right for the project.

- Using the scheduling tools, develop and coordinate the project plan, which includes the budget, resources, and the project schedule.
- Develop the project schedule.
- Develop the project budget.
- Manage the budget.
- Recruit personnel for the project.
- Select subcontractors.
- Assign work, responsibility, and authority so team members can make maximum use of their abilities.
- Estimate, allocate, coordinate, and control project resources.
- Deal with specifications and resource needs that are unrealistic.
- Decide upon the right level of administrative and computer support.
- Train project members on how to fulfill their duties and responsibilities.
- Supervise project members, giving them day-to-day instructions, guidance, and discipline as required to fulfill their duties and responsibilities.
- Design and implement reporting and briefing information systems or documents that respond to project needs.
- Control the project.

Some basic project management practices can improve the chances for success. Consider the following:

- Secure the necessary commitments from top management to make the project a success.
- Set up an action plan that will be easily adopted by management.
- Use a work breakdown structure that is comprehensive and easy to use.
- Establish accounting practices that help, not hinder, successful completion of the project.
- Prepare project team job descriptions properly up-front to eliminate conflict later on.
- Select project team members appropriately the first time.

After the project is under way, follow these steps:

- Manage the project, but make the oversight reasonable and predictable.
- Get team members to accept and participate in the plans.
- Motivate project team members for best performance.
- Coordinate activities so they are carried out in relation to their importance with a minimum of conflict.
- Monitor and minimize inter-departmental conflicts.
- Get the most out of project meetings without wasting the team's productive time. Develop an agenda for each meeting, and start on time. Conduct one piece of business at a time. Assign

responsibilities where appropriate. Agree on follow-up and accountability dates. Indicate the next step for the group. Set the time and place for the next meeting. End on time.

- Spot problems and take corrective action before it is too late.
- Discover the strengths and weaknesses in project team members and manage them to get the desired results.
- Help team members solve their own problems.
- Exchange information with subordinates, associates, superiors, and others about plans, progress, and problems.
- Make the best of available resources.
- Measure project performance.
- Determine, through formal and informal reports, the degree to which progress is being made.
- Determine causes of and possible ways to act upon significant deviations from planned performance.
- Take action to correct an unfavorable trend, or to take advantage of an unusually favorable trend.
- Look for areas where improvements can be made.
- Develop more effective and economical methods of managing.
- Remain flexible.
- Avoid "activity traps".
- Practice effective time management.

When dealing with subordinates, each person must:

- Know what they are supposed to do, preferably in terms of an end product.
- Have a clear understanding of what their authority is, and its limits.
- Know what their relationship with other people is.
- Know what constitutes a job well done in terms of specific results.
- Know when and what they are doing exceptionally well.
- Understand that there are just rewards for work well done, and for work exceptionally well done.
- Know where and when they are falling short of expectations.
- Be made aware of what can and should be done to correct unsatisfactory results.
- Feel that their superior has an interest in them as an individual.
- Feel that their superior believes in them and is anxious for them to succeed and progress.

By fostering a good relationship with associates, the manager will have less difficulty communicating with them. The fastest, most effective communication takes place among people with common points of view.

The competent project manager watches what is going on in great detail and can, therefore, perceive problems long before they flow through the paper system. Personal contact is faster than filing out formal forms. A project manager who spends most of his or her time in the management office instead of roaming through the places where the work is being done, is headed for catastrophe.

1.3i Systems Engineer

The term *systems engineer* means different things to different people. The systems engineer is distinguished from the engineering specialist, who is concerned with only one aspect of a well-defined engineering discipline in that he must be able to adapt to the requirements of almost any type of system. The systems engineer provides the employer with a wealth of experience gained from many successful approaches to technical problems developed through hands-on exposure to a variety of situations. This person is a professional with knowledge and experience, possessing skills in a specialized and learned field or fields. The systems engineer is an expert in these fields; highly trained in analyzing problems and developing solutions that satisfy management objectives.

A competent systems engineer has a wealth of technical information that can be used to speed up the design process and help in making cost effective decisions. The experienced systems engineer is familiar with proper fabrication, construction, installation, and wiring techniques and can spot and correct improper work.

Training in personnel relations, a part of the engineering curriculum, helps the systems engineer communicate and negotiate professionally with subordinates and management.

Small in-house projects can be completed on an informal basis and, indeed, this is probably the normal routine where the projects are simple and uncomplicated. In a large project, however, the systems engineer's involvement usually begins with preliminary planning and continues through fabrication, implementation, and testing. The degree to which program objectives are achieved is an important measure of the systems engineer's contribution.

During the design process the systems engineer:

- Concentrates on results and focuses work according to the management objectives.
- Receives input from management and staff.
- Researches the project and develops a workable design.
- Assures balanced influence of all required design specialties.
- Conducts design reviews.
- Performs trade-off analyses.
- Assists in verifying system performance.
- Resolves technical problems related to the design, interface between system components, and integration of the system into the facility.

Aside from designing a system, the systems engineer has to answer any questions and resolve problems that may arise during fabrication and installation of the hardware. This person must also monitor the quality and workmanship of the installation. The hardware and software will have to be tested and calibrated upon completion. This too is the concern of the systems engineer. During the production or fabrication phase, systems engineering is concerned with:

- Verifying system capability
- Verifying system performance
- Maintaining the system baseline
- Forming an analytical framework for further analysis

Depending on the complexity of the new installation, the systems engineer may have to provide orientation and operating instruction to the users. During the operational support phase, system engineers:

- Receive input from users
- Evaluate proposed changes to the system
- Establish their effectiveness
- Facilitates the effective incorporation of changes, modifications, and updates

Depending on the size of the project and the management organization, the systems engineer's duties will vary. In some cases the systems engineer may have to assume the responsibilities of planning and managing smaller projects.

1.4 References

1. Hoban, F. T., and W. M. Lawbaugh: *Readings In Systems Engineering*, NASA, Washington, D.C., 1993.
2. *System Engineering Management Guide,* Defense Systems Management College, Virginia, 1983.

1.5 Bibliography

Delatore, J. P., E. M. Prell, and M. K. Vora: "Translating Customer Needs Into Product Specifications", *Quality Progress*, January 1989.

DeSantis, Gene: "Systems Engineering Concepts," in *NAB Engineering Handbook*, 9th ed., Jerry C. Whitaker (ed.), National Association of Broadcasters, Washington, D.C., 1999.

DeSantis, Gene: "Systems Engineering," in *The Electronics Handbook*, Jerry C. Whitaker (ed.), CRC Press, Boca Raton, Fla., 1996.

Finkelstein, L.: "Systems Theory", *IEE Proceedings*, vol. 135, Part A, no. 6, July 1988.

Shinners, S. M.: *A Guide to Systems Engineering and Management*, Lexington, 1976.

Tuxal, J. G.: *Introductory System Engineering*, McGraw-Hill, New York, N.Y., 1972.

Chapter

2

Engineering Documentation

Fred Baumgartner, Terrence M. Baun

2.1 Introduction[1]

Video facilities are designed to have as little down-time as possible. Yet, inadequate documentation is a major contributor to the high cost of systems maintenance and the resulting widespread replacement of poorly documented facilities. The cost of neglecting *hours* of engineering documentation is paid in *weeks* of reconstruction.

Documentation is a management function every bit as important as project design, budgeting, planning, and quality control; it is often the difference between an efficient and reliable facility and a misadventure. If the broadcast engineer does not feel qualified to attempt documentation of a project, the engineer must at the very least oversee and approve the documentation developed by others.

Within the last few years the need for documentation has increased with the complexity of the broadcast systems. Fortunately, the power of documentation tools has kept pace.

2.2 Basic Concepts

The first consideration in the documentation process is the complexity of the installation. A basic video editing station may require almost no formal documentation, while a large satellite or network broadcast facility may require computerized databases and a full time staff doing nothing but documentation updates. Most facilities will fall somewhere in the middle of that spectrum.

A second concern is the need for flexibility at the facility. Seldom does a broadcast operation get "completely rewired" because the cabling wears our or fails. More often, it is the supporting documentation that has broken down, frustrating the maintainability of the system. Retroactive documentation is physically difficult and emotionally challenging, and seldom generates the level of commitment required to be entirely successful or accurate; hence, a total rebuild is often

1. This chapter is based on: Baumgartner, Fred, and Terrence M. Baun: "Engineering Documentation," in *Standard Handbook of Video and Television Engineering*, 3rd ed., Jerry C. Whitaker (ed.), McGraw-Hill, New York, N.Y., 2000. Used with permission. All rights reserved.

the preferred solution to documentation failure. Documentation must be considered a hedge against such unnecessary reconstruction.

Finally, consider efficiency and speed. Documentation is a prepayment of time. Repairs, rerouting, replacements, and reworking all go faster and smoother with proper documentation. If your installation is one in which any downtime or degradation of service is unacceptable, then budgeting sufficient time for the documentation process is critically important.

Because, in essence, documentation is education, knowing "how much is enough" is a difficult decision. We will begin by looking at the basics, and then expand our view of the documentation process to fit specialized situations.

2.2a The Manuals

Even if you never take a pen to paper, you do have one source of documentation to care for, since virtually every piece of commercial equipment comes with a manual.

Place those manuals in a centralized location, and arrange them in an order that seems appropriate for your station. Most engineering shops file manuals alphabetically, but some prefer a filing system based on equipment placement. (For example, production studio equipment manuals would be filed together under a "Production Studio" label and might even be physically located in the referenced studio.) But whatever you do, be consistent Few things are as frustrating to a technician as being unable to locate a manual when needing something as simple as a part number or the manufacturer's address.

Equipment manuals are the first line of documentation and deserve our attention and respect.

2.2b Documentation Conventions

The second essential item of documentation is the statement of "conventions." By this we mean a document containing basic information essential to an understanding of the facility, posted in an obvious location and available to all who maintain the plant. Consider the following examples of conventions:

- Where are the equipment manuals and how are they organized?
- What is the architecture of the ground system? Where is the central station ground and is it a star, grid, or other distribution pattern? Are there separate technical and power grounds? Are audio shield grounded at the source, termination, or both locations?
- What is the standard audio input/output architecture? Is this a +8 dBm, +4 dBm, or 0 dBm facility? Is equipment sourced at 600 Ω terminated at its destination, or left unterminated? Are unbalanced audio sources wired with the shield as ground or is the low side of a balanced pair used for that purpose? How are XLR-type connectors wired—pin 2 high or low?
- How can a technician disconnect utility power to service line voltage wiring within racks? Where are breakers located, and how are they marked? What equipment is on the UPS power, generator, or utility power? Whom do you call for power and building systems maintenance?
- Where are the keys to the transmitter? Is the site alarmed?

For such an essential information source, you will find it takes very little time to generate the conventions document. Keep it short—it is not meant to be a book. A page or two should be sufficient for most installations and, if located in a obvious place, this document will keep the technical staff on track and will save service personnel from stumbling around searching for basic information. This document is the key to preventing many avoidable embarrassments.

The next step beyond the conventions document is a documentation *system*. There are three primary methods:

- Self documentation
- Database documentation
- Graphic documentation

In most cases, a mixture of all three is necessary and appropriate. In addition to documenting the physical plant and its interconnections, each piece of equipment, whether commercially produced or custom made, must be documented in an organized manner.

Self Documentation

In situations where the facility is small and very routine, self documentation is possible. Self documentation relies on a set of standard practices that are repeated. Telephone installations, for example, appear as a mass of perplexing wires. In reality, the same simple circuits are simply repeated in an organized and universal manner. To the initiated, any telephone installation that follows the rules is easy to understand and repair or modify, no matter where or how large. Such a system is truly self documenting. Familiarity with telephone installations is particularly useful, because the telephone system was the first massive electronic installation. It is the telephone system that gave us "relay" racks, grounding plans, demarcation points, and virtually all of the other concepts that are part of today's electronic control or communications facility.

The organization, color codes, terminology, and layout of telephone systems is recorded in minute detail. Once a technician is familiar with the rules of telephone installations, drawings and written documentation are rarely required for routine expansion and repair. The same is true for many parts of other facilities. Certainly, much of the wiring in any given rack of equipment can be self documenting. For example, a video tape recorder will likely be mounted in a rack with a picture monitor, audio monitor, waveform monitor, and vectorscope. The wiring between each of these pieces of equipment is clearly visible, with all wires short and their purpose obvious to any technician familiar with the rules of video. Furthermore, each video cable will conform to the same standards of level or data configuration. Additional documentation, therefore, is largely unnecessary.

By convention, there are rules of grounding, power, and signal flow in all engineering facilities. In general, it can be assumed that in most communications facilities, the ground will be a star system, the power will be individual 20 amp feeds to each rack, and the signal flow within each rack will be from top to bottom. Rules that might vary from facility to facility include color coding, connector pin outs, and rules for shield and return grounding.

To be self documenting, the rules must be determined and all of the technicians working on the facility must know and follow the conventions. The larger the number of technicians, or the higher the rate of staff turn-over, the more important it is to have a readily available document that clearly covers the conventions in use.

One thing must be very clear: a facility that does not have written documentation is not automatically self documenting. Quite the contrary. A written set of conventions and unfaltering adherence to them are the trade marks of a self documenting facility.

While it is good engineering practice to design all facilities to be as self documenting as possible, there are limits to the power of self documentation. In the practical world, self documentation can greatly reduce the amount of written documentation required, but can seldom replace it entirely.

Database Documentation

As facilities expand in size and complexity, a set of conventions will longer answer all of the questions. At some point, a wire leaves an equipment rack and its destination is no longer obvious. Likewise, equipment will often require written documentation as to its configuration and purpose, especially if it is utilized in an uncharacteristic way.

Database documentation records the locations of both ends of a given circuit. For this, each cable must be identified individually. There are two common systems for numbering cables: *ascension numbers*, and *from-to coding*. In ascension numbering schemes, each wire or cable is numbered in increasing order, one, two, three, and so on. In from-to coding, the numbers on each cable represent the source location, the destination location, and normally some identification as to purpose and/or a unique identifier. For example, a cable labeled 31-35-B6 might indicated that cable went from a piece of equipment in rack 31 to another unit in rack 35 and carries black, it is also the sixth cable to follow the same route and carry the same class of signal.

Each method has its benefits. Ascension numbering is easier to assign, and commonly available preprinted wire labels can be used. On the other hand, ascension numbers contain no hints as to wire purpose or path, and for that reason "purpose codes" are often added to the markings.

From-to codes can contain a great deal more information without relying on the printed documentation records, but often space does not permit a full delineation on the tag itself. Here again, supplemental information may still be required in a separate document or database.

Whatever numbering system is used, a complete listing must be kept in a database of some type. In smaller installations, this might simply be a spiral notebook that contains a complete list of all cables, their source, destination, any demarcations, and signal parameters.

Because all cabling can be considered as a transmission line, all cabling involves issues of termination. In some data and analog video applications, it is common for a signal to "loop-through" several pieces of equipment. Breaking or tapping into the signal path often has consequences elsewhere, resulting in unterminated or double terminated lines. While more forgiving, analog audio has similar concerns. Therefore, documentation must include information on such termination.

Analog audio and balanced lines used in instrumentation have special concerns of their own. It is seldom desirable to ground both ends of a shielded cable. Again, the documentation must reflect which end(s) of a given shield is grounded.

In many cases, signal velocity is such that the length of the lines and the resulting propagation delay is critical. In such circumstances, this is significant information that should be retained. In cases where differing signal levels or configurations are used (typical in data and control systems), it is the documentor's obligation to record those circumstances as well.

However or wherever the database documentation is retained, it represents the basic information that defines the facility interconnections and must be available for updating and duplication as required.

Graphic Documentation

Electronics is largely a graphical language. Schematics and flow charts are more understandable than net lists or cable interconnection lists. Drawings, either by hand—done with the aid of drafting machines and tools—or accomplished on CAD (computer aided design) programs are highly useful in conveying overall facility design quickly and clearly. Normally, the wire numbering scheme will follow that used in the database documentation, so that the graphic and text documentation can be used together.

CAD drawings are easy to update and reprint. For this reason, documentation via CAD is becoming more popular, even in smaller installations. Because modern CAD programs not only draw but also store information, they can effectively serve as an electronic file cabinet for documentation. While there have been attempts to provide electronic/telecommunications engineering documentation "templates" and corresponding technical graphics packages for CAD programs, most of the work in this area has been done by engineers working independently to develop their own systems. Obviously, the enormous scope of electronic equipment and telecommunications systems make it impractical for a "standard" CAD package to suit every user.

2.2c Update Procedures

Because documentation is a dynamic tool, as the facility changes, so too must the documentation. It is common for a technician to "improve" conditions by reworking a circuit or two. Most often this fixes a problem that should be corrected as a maintenance item. But sometimes, it plants a "time bomb" wherein a future change, based on missing or incomplete documentation of the previous work, will cause problems.

It is essential that there be a means of consistently updating the documentation. The most common way of accomplishing this is the mandatory "change sheet." Here, whenever a technician makes a change it is reported back to those who keep the documentation. If the changes are extensive, the use of the "red-line" drawing and "edit sheet" come into play; the original drawing and database, respectively, are printed and corrected with a "red pen." This document then is used to update the original documentation.

In some cases, the updating process can be tied to the engineering reports or discrepancy process. Most facilities use some form of "trouble ticket" to track equipment and system performance and to report and track maintenance. This same form may be used to report changes required of the documentation or errors in existing documentation.

2.2d Equipment Documentation

Plant documentation does not end when all of the circuit paths in the facility are defined. Each circuit begins and ends at a piece of equipment, which can be modified, reconfigured, or removed from service. Keep in mind that unless the lead technician lives forever, never changes jobs, and never takes a day off, someone unfamiliar with the equipment will eventually be asked to return it to service. For this reason, a documentation file for each piece of equipment must be maintained.

Equipment documentation contains these key elements:

- The equipment manual
- Modification record

- Configuration information
- Maintenance record

The equipment manual is the manufacturer's original documentation. As mentioned previously, the manuals must be organized in such a manner that they can be easily located. Typically, the manuals are kept at the site where the equipment is installed (if practical). Remember that equipment with two "ends" such as STLs, RPUs, or remote control systems need manuals at *each* location!

Of course, if a piece of equipment is of custom construction, there must be particular attention paid to creating a manual. For this reason, a copy of the key schematics and documentation is often attached directly to the equipment. This "built-in manual" may be the only documentation to survive over the years.

Many pieces of equipment, over time, will require modification. Typically, the modification is recorded in three ways. First, internally to the equipment. A simple note glued into the chassis may be suitable, or a marker pen is used to write on a printed circuit board or other component. Second, the changes can be recorded in the manual, either inside the cover, or on the schematic or relevant pages. If the manual serves several machines, this may not be appropriate. If this is the case, a third option is to keep the modification information in a separate *equipment file.*

Equipment files are typically kept in standard file folders, and may be filed with the pertinent equipment manuals. Ideally, the equipment file is started when the equipment is purchased, and should contain purchase date, serial number, all modifications, equipment location(s), and a record of service.

The equipment file is the proper place to keep the configuration information. An increasingly large amount of equipment is microprocessor based or otherwise configurable for a specific mode of operation. Having a record of the machine's default configuration is extremely helpful when a power glitch (or an operator) reconfigures a machine unexpectedly.

Equipment files should also contain repair records. With most equipment, documenting failures, major part replacements, operating time, and other service-related events serves a valuable purpose. Nothing is more useful in troubleshooting than a record of previous failures, configuration, modifications, and—of course—a copy of the original manual.

2.2e Operator/User Documentation

User documentation provides, at its most basic level, instructions on how to use a system. While most equipment manufacturers provide reasonably good instruction and operations manuals for their products, when those products are integrated into a system another level of documentation may be required. Complex equipment may require interface components that need to be adjusted from time to time, or various machines may be incompatible in some data transmission modes—all of which is essential to the proper operation of the system.

Such information often resides only in the heads of certain key users and is passed on by word of mouth. This level of informality can be dangerous, especially when changes take place in the users or maintenance staff, resulting in differing interpretations between operators and maintenance people about how the system normally operates. Maintenance personnel will then spend considerable time tracking hypothetical errors reported by misinformed users.

A good solution is to have the operators write an operating manual, providing a copy to the maintenance department. Such documentation will go a long way toward improving inter-departmental communications and should result in more efficient maintenance as well.

2.2f The True Cost of Documentation

There is no question that documentation is expensive—in some cases it can equal the cost of installation. Still, both installation and documentation expenses pale in comparison to the cost of equipment and potential revenue losses resulting from system down-time.

Documentation must be seen as a management and personnel issue of the highest order. Any lapse in the documentation updating process can result in disaster. Procrastination and the resulting lack of follow through will destroy any documentation system and ultimately result in plant failures, extend down-time, and premature rebuilding of the facility.

Making a business case for documentation is similar to making any business case. Gather together all the costs in time, hardware, and software on one side of the equation, and balance this against the savings in time and lost revenue on the other side. Engineering managers are expected to project costs accurately, and the allocation of sufficient resources for documentation and its requisite updating is an essential part of that responsibility.

2.3 Bibliography

Baumgartner, Fred, and Terrence Baun: "Broadcast Engineering Documentation," in *NAB Engineering Handbook*, 9th ed., Jerry C. Whitaker (ed.), National Association of Broadcasters, Washington, D.C., 1999.

Baumgartner, Fred, and Terrence Baun: "Engineering Documentation," in *The Electronics Handbook*, Jerry C. Whitaker (ed.), CRC Press, Boca Raton, Fla., 1996.

Chapter

3

Safety Issues

Jerry C. Whitaker, editor

3.1 Introduction

Electrical safety is important when working with any type of electronic hardware. Because transmitters and many other systems operate at high voltages and currents, safety is doubly important. The primary areas of concern, from a safety standpoint, include:

- Electric shock
- Nonionizing radiation
- Beryllium oxide (BeO) ceramic dust
- Hot surfaces of vacuum tube devices
- Polychlorinated biphenyls (PCBs)

3.2 Electric Shock

Surprisingly little current is required to injure a person. Studies at Underwriters Laboratories (UL) show that the electrical resistance of the human body varies with the amount of moisture on the skin, the muscular structure of the body, and the applied voltage. The typical hand-to-hand resistance ranges between 500 Ω and 600 kΩ, depending on the conditions. Higher voltages have the capability to break down the outer layers of the skin, which can reduce the overall resistance value. UL uses the lower value, 500 Ω, as the standard resistance between major extremities, such as from the hand to the foot. This value is generally considered the minimum that would be encountered and, in fact, may not be unusual because wet conditions or a cut or other break in the skin significantly reduces human body resistance.

3.2a Effects on the Human Body

Table 3.1 lists some effects that typically result when a person is connected across a current source with a hand-to-hand resistance of 2.4 kΩ. The table shows that a current of approximately 50 mA will flow between the hands, if one hand is in contact with a 120 V ac source and the

Table 3.1 The Effects of Current on the Human Body

Current	Effect
1 mA or less	No sensation, not felt
More than 3 mA	Painful shock
More than 10 mA	Local muscle contractions, sufficient to cause "freezing" to the circuit for 2.5 percent of the population
More than 15 mA	Local muscle contractions, sufficient to cause "freezing" to the circuit for 50 percent of the population
More than 30 mA	Breathing is difficult, can cause unconsciousness
50 mA to 100 mA	Possible ventricular fibrillation
100 mA to 200 mA	Certain ventricular fibrillation
More than 200 mA	Severe burns and muscular contractions; heart more apt to stop than to go into fibrillation
More than a few amperes	Irreparable damage to body tissue

other hand is grounded. The table indicates that even the relatively small current of 50 mA can produce *ventricular fibrillation* of the heart, and perhaps death. Medical literature describes ventricular fibrillation as rapid, uncoordinated contractions of the ventricles of the heart, resulting in loss of synchronization between heartbeat and pulse beat. The electrocardiograms shown in Figure 3.1 compare a healthy heart rhythm with one in ventricular fibrillation. Unfortunately, once ventricular fibrillation occurs, it will continue. Barring resuscitation techniques, death will ensue within a few minutes.

The route taken by the current through the body has a significant effect on the degree of injury. Even a small current, passing from one extremity through the heart to another extremity, is dangerous and capable of causing severe injury or electrocution. There are cases where a person has contacted extremely high current levels and lived to tell about it. However, usually when this happens, the current passes only through a single limb and not through the body. In these instances, the limb is often lost, but the person survives.

Current is not the only factor in electrocution. Figure 3.2 summarizes the relationship between current and time on the human body. The graph shows that 100 mA flowing through a human adult body for 2 s will cause death by electrocution. An important factor in electrocution, the *let-go range*, also is shown on the graph. This range is described as the amount of current that causes "freezing", or the inability to let go of the conductor. At 10 mA, 2.5 percent of the population will be unable to let go of a "live" conductor. At 15 mA, 50 percent of the population will be unable to let go of an energized conductor. It is apparent from the graph that even a small amount of current can "freeze" someone to a conductor. The objective for those who must work around electric equipment is how to protect themselves from electric shock. Table 3.2 lists required precautions for personnel working around high voltages.

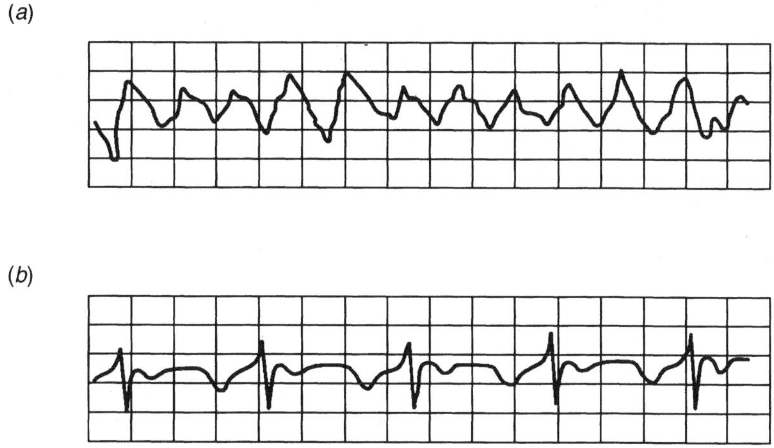

Figure 3.1 Electrocardiogram of a human heartbeat: (*a*) healthy rhythm, (*b*) ventricular fibrillation.

3.2b Circuit Protection Hardware

The typical primary panel or equipment circuit breaker or fuse will not protect a person from electrocution. In the time it takes a fuse or circuit breaker to blow, someone could die. However, there are protection devices that, properly used, may help prevent electrocution. The *ground-fault current interrupter* (GFCI), shown in Figure 3.3, works by monitoring the current being applied to the load. The GFI uses a differential transformer and looks for an imbalance in load current. If a current (5 mA, ±1 mA) begins to flow between the neutral and ground or between the hot and ground leads, the differential transformer detects the leakage and opens up the primary circuit within 2.5 ms.

GFIs will not protect a person from every type of electrocution. If the victim becomes connected to both the neutral and the hot wire, the GFI will not detect an imbalance.

3.2c Working with High Voltage

Rubber gloves are commonly used by engineers working on high-voltage equipment. These gloves are designed to provide protection from hazardous voltages or RF when the wearer is working on "hot" ac or RF circuits. Although the gloves may provide some protection from these hazards, placing too much reliance on them can have disastrous consequences. There are several reasons why gloves should be used with a great deal of caution and respect. A common mistake made by engineers is to assume that the gloves always provide complete protection. The gloves found in many facilities may be old or untested. Some may show signs of user repair, perhaps with electrical tape. Few tools could be more hazardous than such a pair of gloves.

Another mistake is not knowing the voltage rating of the gloves. Gloves are rated differently for both ac and dc voltages. For example, a *class 0* glove has a minimum dc breakdown voltage of 35 kV; the minimum ac breakdown voltage, however, is only 6 kV. Furthermore, high-voltage

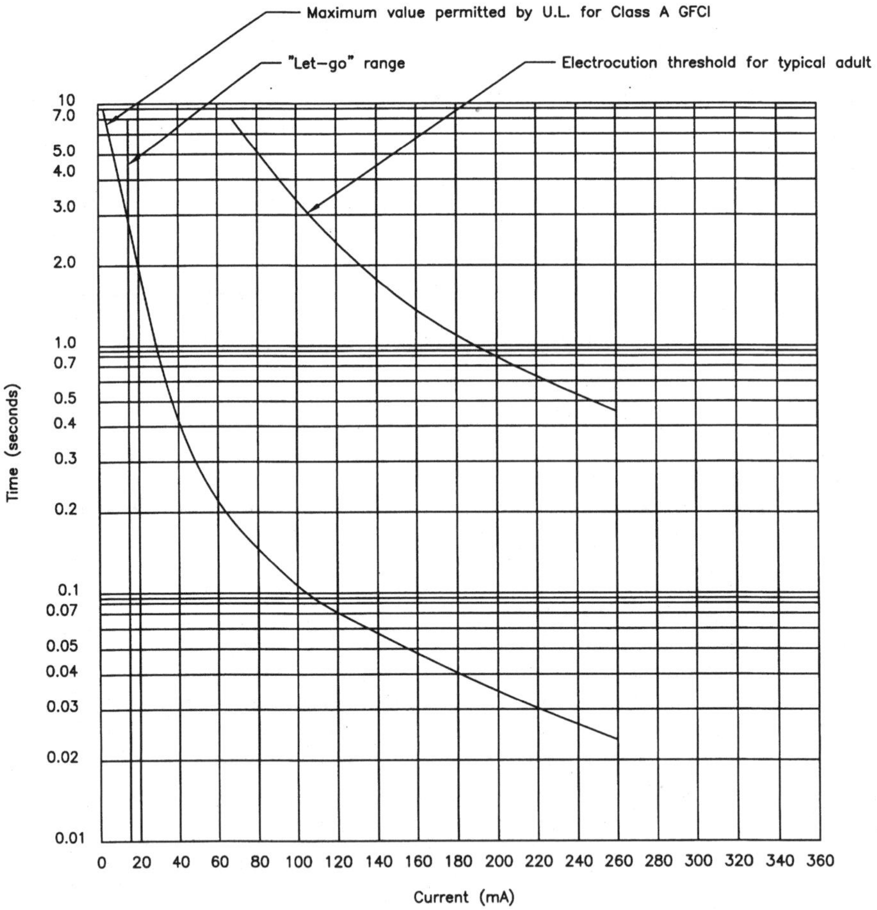

Figure 3.2 Effects of electric current and time on the human body. Note the "let-go" range.

rubber gloves are not usually tested at RF frequencies, and RF can burn a hole in the best of them. It is possible to develop dangerous working habits by assuming that gloves will offer the required protection.

Gloves alone may not be enough to protect an individual in certain situations. Recall the axiom of keeping one hand in a pocket while working around a device with current flowing? That advice is actually based on simple electricity. It is not the "hot" connection that causes the problem, but the ground connection that lets the current begin to flow. Studies have shown that more than 90 percent of electric equipment fatalities occurred when the grounded person contacted a live conductor. Line-to-line electrocution accounted for less than 10 percent of the deaths.

Table 3.2 Required Safety Practices for Engineers Working Around High-Voltage Equipment

High-Voltage Precautions	
✓	Remove all ac power from the equipment. Do not rely on internal contactors or SCRs to remove dangerous ac.
✓	Trip the appropriate power distribution circuit breakers at the main breaker panel.
✓	Place signs as needed to indicate that the circuit is being serviced.
✓	Switch the equipment being serviced to the local control mode as provided.
✓	Discharge all capacitors using the discharge stick provided by the manufacturer.
✓	Do not remove, short circuit, or tamper with interlock switches on access covers, doors, enclosures, gates, panels, or shields.
✓	Keep away from live circuits.

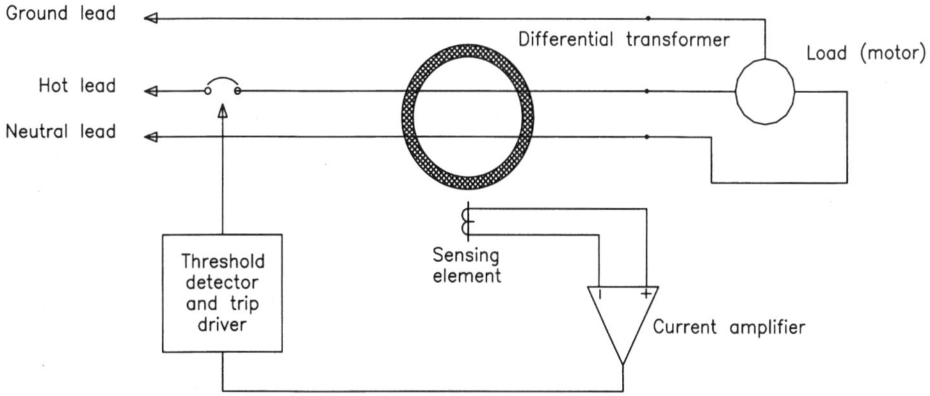

Figure 3.3 Basic design of a ground-fault interrupter (GFI).

When working around high voltages, always look for grounded surfaces. Keep hands, feet, and other parts of the body away from any grounded surface. Even concrete can act as a ground if the voltage is sufficiently high. If work must be performed in "live" cabinets, then consider using, in addition to rubber gloves, a rubber floor mat, rubber vest, and rubber sleeves. Although this may seem to be a lot of trouble, consider the consequences of making a mistake. Of course, the best troubleshooting methodology is never to work on any circuit without being certain that no hazardous voltages are present. In addition, any circuits or contactors that normally contain hazardous voltages should be firmly grounded before work begins.

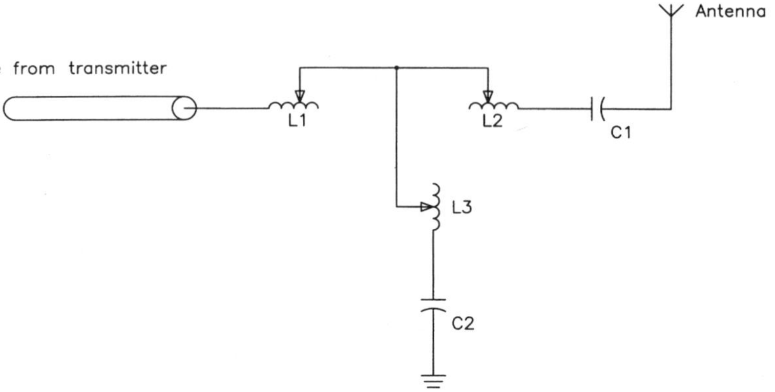

Figure 3.4 Example of how high voltages can be generated in an RF load matching network.

RF Considerations

Engineers often rely on electrical gloves when making adjustments to live RF circuits. This practice, however, can be extremely dangerous. Consider the typical load matching unit shown in Figure 3.4. In this configuration, disconnecting the coil from either L2 or L3 places the full RF output literally at the engineer's fingertips. Depending on the impedances involved, the voltages can become quite high, even in a circuit that normally is relatively tame.

In the Figure 3.4 example, assume that the load impedance is approximately $106 + j202\ \Omega$. With 1 kW feeding into the load, the rms voltage at the matching output will be approximately 700 V. The peak voltage (which determines insulating requirements) will be close to 1 kV, and perhaps more than twice that if the carrier is being amplitude-modulated. At the instant the output coil clip is disconnected, the current in the shunt leg will increase rapidly, and the voltage easily could more than double.

3.2d First Aid Procedures

All engineers working around high-voltage equipment should be familiar with first aid treatment for electric shock and burns. Always keep a first aid kit on hand at the facility. Figure 3.5 illustrates the basic treatment for victims of electric shock. Copy the information, and post it in a prominent location. Better yet, obtain more detailed information from the local heart association or Red Cross chapter. Personalized instruction on first aid usually is available locally.

3.2e Operating Hazards

A number of potential hazards exist in the operation and maintenance of high-power vacuum tube RF equipment. Maintenance personnel must exercise extreme care around such hardware. Consider the following guidelines:

If the victim is not responsive, follow the A–B–Cs of basic life support.

A AIRWAY: If the victim is unconscious, open airway.

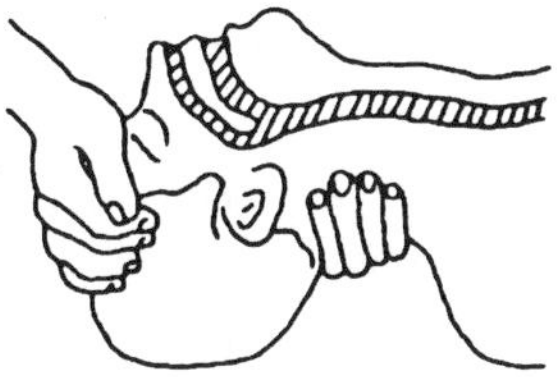

1. Lift up neck
2. Push forehead back
3. Clear out mouth if necessary
4. Observe for breathing

B BREATHING: If the victim is not breathing, begin artificial breathing.

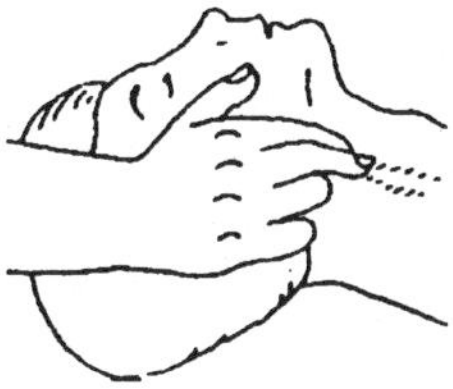

1. Tilt head
2. Pinch nostrils
3. Make airtight seal
4. Provide four quick full breaths

Check carotid pulse. If pulse is absent, begin artificial circulation. Remember that mouth–to–mouth resuscitation must be commenced as soon as possible.

C CIRCULATION: Depress the sternum 1.2 to 2 inches.

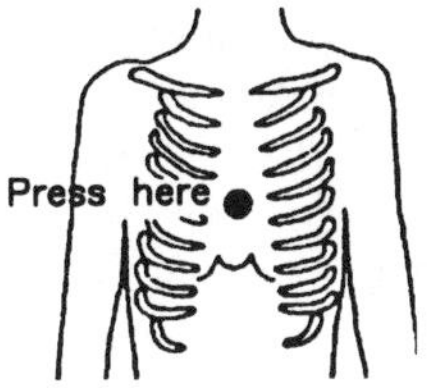

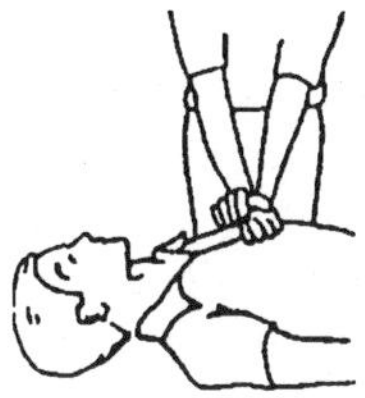

For situations in which there is one rescuer, provide 15 compressions and then 2 quick breaths. The approximate rate of compressions should be 80 per minute.
For situations in which there are two rescuers, provide 5 compressions and then 1 breath. The approximate rate of compressions should be 60 per minute.
Do not interrupt the rhythm of compressions when a second person is giving breaths.

If the victim is responsive, keep warm and quiet, loosen clothing, and place in a reclining position. Call for medical assistance as soon as possible.

Figure 3.5 Basic first aid treatment for electric shock.

- Use caution around the high-voltage stages of the equipment. Many power tubes operate at voltages high enough to kill through electrocution. Always break the primary ac circuit of the power supply, and discharge all high-voltage capacitors.
- Minimize exposure to RF radiation. Do not permit personnel to be in the vicinity of open, energized RF generating circuits, RF transmission systems (waveguides, cables, or connectors), or energized antennas. High levels of radiation can result in severe bodily injury, including blindness. Cardiac pacemakers may also be affected.
- Avoid contact with beryllium oxide (BeO) ceramic dust and fumes. BeO ceramic material may be used as a thermal link to carry heat from a tube to the heat sink. Do not perform any operation on any BeO ceramic that might produce dust or fumes, such as grinding, grit blasting, or acid cleaning. Beryllium oxide dust and fumes are highly toxic, and breathing them can result in serious injury or death. BeO ceramics must be disposed of as prescribed by the device manufacturer.
- Avoid contact with hot surfaces within the equipment. The anode portion of many power tubes is air-cooled. The external surface normally operates at a high temperature (up to 250°C). Other portions of the tube also may reach high temperatures, especially the cathode insulator and the cathode/heater surfaces. All hot surfaces may remain hot for an extended time after the tube is shut off. To prevent serious burns, avoid bodily contact with these surfaces during tube operation and for a reasonable cool-down period afterward. Table 3.3 lists basic first aid procedures for burns.

3.3 OSHA Safety Considerations

The U.S. government has taken a number of steps to help improve safety within the workplace under the auspices of the Occupational Safety and Health Administration (OSHA). The agency helps industries monitor and correct safety practices. OSHA has developed a number of guidelines designed to help prevent accidents. OSHA records show that electrical standards are among the most frequently violated of all safety standards. Table 3.4 lists 16 of the most common electrical violations, including exposure of live conductors, improperly labeled equipment, and faulty grounding.

3.3a Protective Covers

Exposure of live conductors is a common safety violation. All potentially dangerous electric conductors should be covered with protective panels. The danger is that someone may come into contact with the exposed current-carrying conductors. It is also possible for metallic objects such as ladders, cable, or tools to contact a hazardous voltage, creating a life-threatening condition. Open panels also present a fire hazard.

3.3b Identification and Marking

Circuit breakers and switch panels should be properly identified and labeled. Labels on breakers and equipment switches may be many years old and may no longer reflect the equipment actually in use. This is a safety hazard. Casualties or unnecessary damage can be the result of an improp-

Table 3.3 Basic First Aid Procedures for Burns (More detailed information can be obtained from any Red Cross office.)

	Extensively Burned and Broken Skin
✓	Cover affected area with a clean sheet or cloth.
✓	Do not break blisters, remove tissue, remove adhered particles of clothing, or apply any salve or ointment.
✓	Treat victim for shock as required.
✓	Arrange for transportation to a hospital as quickly as possible.
✓	If arms or legs are affected, keep them elevated.
✓	If medical help will not be available within an hour and the victim is conscious and not vomiting, prepare a weak solution of salt and soda: 1 level teaspoon of salt and level teaspoon of baking soda to each quart of tepid water. Allow the victim to sip slowly about 4 ounces (half a glass) over a period of 15 minutes. Discontinue fluid intake if vomiting occurs. (Do not offer alcohol.)
	Less Severe Burns (First and Second Degree)
✓	Apply cool (not ice-cold) compresses using the cleanest available cloth article.
✓	Do not break blisters, remove tissue, remove adhered particles of clothing, or apply salve or ointment.
✓	Apply clean, dry dressing if necessary.
✓	Treat victim for shock as required.
✓	Arrange for transportation to a hospital as quickly as possible.
✓	If arms or legs are affected, keep them elevated.

erly labeled circuit panel if no one who understands the system is available in an emergency. If a number of devices are connected to a single disconnect switch or breaker, a diagram should be provided for clarification. Label with brief phrases, and use clear, permanent, and legible markings.

Equipment marking is a closely related area of concern. This is not the same thing as equipment identification. Marking equipment means labeling the equipment breaker panels and ac disconnect switches according to device rating. Breaker boxes should contain a nameplate showing the manufacturer, rating, and other pertinent electrical factors. The intent is to prevent devices from being subjected to excessive loads or voltages.

3.3c Grounding

OSHA regulations describe two types of grounding: *system grounding* and *equipment grounding*. System grounding actually connects one of the current-carrying conductors (such as the terminals of a supply transformer) to ground. (See Figure 3.6.) Equipment grounding connects all of the noncurrent-carrying metal surfaces together and to ground. From a grounding standpoint, the only difference between a grounded electrical system and an ungrounded electrical system is that the *main bonding jumper* from the service equipment ground to a current-carrying conductor is omitted in the ungrounded system. The system ground performs two tasks:

- It provides the final connection from equipment-grounding conductors to the grounded circuit conductor, thus completing the ground-fault loop.

Table 3.4 Sixteen Common OSHA Violations (After [1].)

Fact Sheet	Subject	NEC[1] Reference
1	Guarding of live parts	110-17
2	Identification	110-22
3	Uses allowed for flexible cord	400-7
4	Prohibited uses of flexible cord	400-8
5	Pull at joints and terminals must be prevented	400-10
6.1	Effective grounding, Part 1	250-51
6.2	Effective grounding, Part 2	250-51
7	Grounding of fixed equipment, general	250-42
8	Grounding of fixed equipment, specific	250-43
9	Grounding of equipment connected by cord and plug	250-45
10	Methods of grounding, cord and plug-connected equipment	250-59
11	AC circuits and systems to be grounded	250-5
12	Location of overcurrent devices	240-24
13	Splices in flexible cords	400-9
14	Electrical connections	110-14
15	Marking equipment	110-21
16	Working clearances about electric equipment	110-16

[1] National Electrical Code

- It solidly ties the electrical system and its enclosures to their surroundings (usually earth, structural steel, and plumbing). This prevents voltages at any source from rising to harmfully high voltage-to-ground levels.

Note that equipment grounding—bonding all electric equipment to ground—is required whether or not the system is grounded. Equipment grounding serves two important tasks:

- It bonds all surfaces together so that there can be no voltage difference among them.
- It provides a ground-fault current path from a fault location back to the electrical source, so that if a fault current develops, it will rise to a level high enough to operate the breaker or fuse.

The National Electrical Code (NEC) is complex and contains numerous requirements concerning electrical safety. The fact sheets listed in Table 3.4 are available from OSHA.

3.4 Beryllium Oxide Ceramics

Some tubes, both power grid and microwave, contain beryllium oxide (BeO) ceramics, typically at the output waveguide window or around the cathode. Never perform any operations on BeO

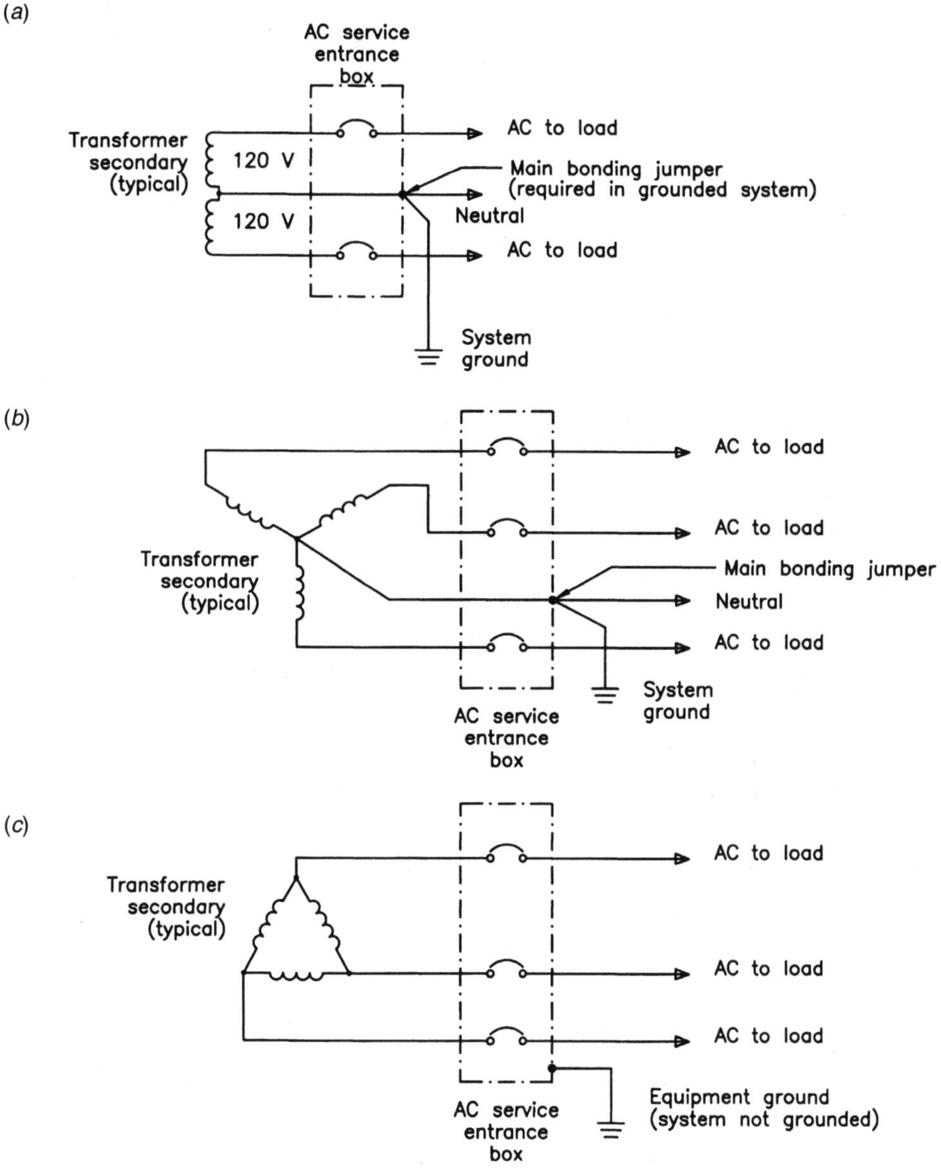

Figure 3.6 AC service entrance bonding requirements: (*a*) 120 V phase-to-neutral (240 V phase-to-phase), (*b*) 3-phase 208 V wye (120 V phase-to-neutral), (*c*) 3-phase 240 V (or 480 V) delta. Note that the main bonding jumper is required in only two of the designs.

ceramics that produce dust or fumes, such as grinding, grit blasting, or acid cleaning. Beryllium oxide dust and fumes are highly toxic, and breathing them can result in serious personal injury or death.

If a broken window is suspected on a microwave tube, carefully remove the device from its waveguide, and seal the output flange of the tube with tape. Because BeO warning labels may be obliterated or missing, maintenance personnel should contact the tube manufacturer before performing any work on the device. Some tubes have BeO internal to the vacuum envelope.

Take precautions to protect personnel working in the disposal or salvage of tubes containing BeO. All such personnel should be made aware of the deadly hazards involved and the necessity for great care and attention to safety precautions. Some tube manufacturers will dispose of tubes without charge, provided they are returned to the manufacturer prepaid, with a written request for disposal.

3.4a Corrosive and Poisonous Compounds

The external output waveguides and cathode high-voltage bushings of microwave tubes are sometimes operated in systems that use a dielectric gas to impede microwave or high-voltage breakdown. If breakdown does occur, the gas may decompose and combine with impurities, such as air or water vapor, to form highly toxic and corrosive compounds. Examples include Freon gas, which may form lethal *phosgene*, and sulfur hexafluoride (SF_6) gas, which may form highly toxic and corrosive sulfur or fluorine compounds such as *beryllium fluoride*. When breakdown does occur in the presence of these gases, proceed as follows:

- Ventilate the area to outside air
- Avoid breathing any fumes or touching any liquids that develop
- Take precautions appropriate for beryllium compounds and for other highly toxic and corrosive substances
- If a coolant other than pure water is used, follow the precautions supplied by the coolant manufacturer.

FC-75 Toxic Vapor

The decomposition products of FC-75 are highly toxic. Decomposition may occur as a result of any of the following:

- Exposure to temperatures above 200°C
- Exposure to liquid fluorine or alkali metals (lithium, potassium, or sodium)
- Exposure to ionizing radiation

Known thermal decomposition products include *perfluoroisobutylene* (PFIB; $[CF_3]_2\ C = CF_2$), which is highly toxic in small concentrations.

If FC-75 has been exposed to temperatures above 200°C through fire, electric heating, or prolonged electric arcs, or has been exposed to alkali metals or strong ionizing radiation, take the following steps:

- Strictly avoid breathing any fumes or vapors.

- Thoroughly ventilate the area.
- Strictly avoid any contact with the FC-75.
- Under such conditions, promptly replace the FC-75 and handle and dispose of the contaminated FC-75 as a toxic waste.

3.5 Nonionizing Radiation

Nonionizing radio frequency radiation (RFR) resulting from high-intensity RF fields is a growing concern to engineers who must work around high-power transmission equipment. The principal medical concern regarding nonionizing radiation involves heating of various body tissues, which can have serious effects, particularly if there is no mechanism for heat removal. Recent research has also noted, in some cases, subtle psychological and physiological changes at radiation levels below the threshold for heat-induced biological effects. However, the consensus is that most effects are thermal in nature.

High levels of RFR can affect one or more body systems or organs. Areas identified as potentially sensitive include the ocular (eye) system, reproductive system, and the immune system. Nonionizing radiation also is thought to be responsible for metabolic effects on the central nervous system and cardiac system.

In spite of these studies, many of which are ongoing, there is still no clear evidence in Western literature that exposure to medium-level nonionizing radiation results in detrimental effects. Russian findings, on the other hand, suggest that occupational exposure to RFR at power densities above 1.0 mW/cm^2 does result in symptoms, particularly in the central nervous system.

Clearly, the jury is still out as to the ultimate biological effects of RFR. Until the situation is better defined, however, the assumption must be made that potentially serious effects can result from excessive exposure. Compliance with existing standards should be the minimum goal, to protect members of the public as well as facility employees.

3.5a NEPA Mandate

The National Environmental Policy Act of 1969 required the Federal Communications Commission to place controls on nonionizing radiation. The purpose was to prevent possible harm to the public at large and to those who must work near sources of the radiation. Action was delayed because no hard and fast evidence existed that low- and medium-level RF energy is harmful to human life. Also, there was no evidence showing that radio waves from radio and TV stations did not constitute a health hazard.

During the delay, many studies were carried out in an attempt to identify those levels of radiation that might be harmful. From the research, suggested limits were developed by the American National Standards Institute (ANSI) and stated in the document known as ANSI C95.1-1982. The protection criteria outlined in the standard are shown in Figure 3.7.

The energy-level criteria were developed by representatives from a number of industries and educational institutions after performing research on the possible effects of nonionizing radiation. The projects focused on absorption of RF energy by the human body, based upon simulated human body models. In preparing the document, ANSI attempted to determine those levels of

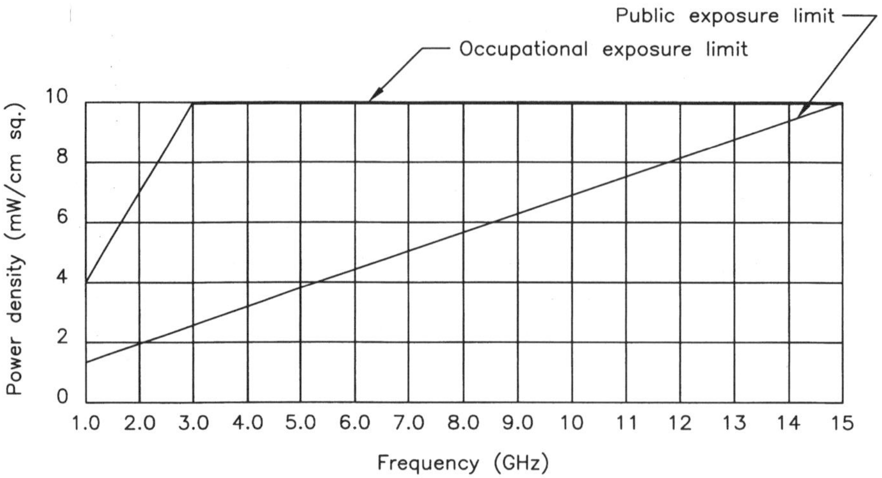

Figure 3.7 The power density limits for nonionizing radiation exposure for humans

incident radiation that would cause the body to absorb less than 0.4 W/kg of mass (averaged over the whole body) or peak absorption values of 8 W/kg over any 1 gram of body tissue.

From the data, the researchers found that energy would be absorbed more readily at some frequencies than at others. The absorption rates were found to be functions of the size of a specific individual and the frequency of the signal being evaluated. It was the result of these absorption rates that culminated in the shape of the *safe curve* shown in the figure. ANSI concluded that no harm would come to individuals exposed to radio energy fields, as long as specific values were not exceeded when averaged over a period of 0.1 hour. It was also concluded that higher values for a brief period would not pose difficulties if the levels shown in the standard document were not exceeded when averaged over the 0.1-hour time period.

The FCC adopted ANSI C95.1-1982 as a standard that would ensure adequate protection to the public and to industry personnel who are involved in working around RF equipment and antenna structures.

Revised Guidelines

The ANSI C95.1-1982 standard was intended to be reviewed at 5-year intervals. Accordingly, the 1982 standard was due for reaffirmation or revision in 1987. The process was indeed begun by ANSI, but was handed off to the Institute of Electrical and Electronics Engineers (IEEE) for completion. In 1991, the revised document was completed and submitted to ANSI for acceptance as ANSI/IEEE C95.1-1992.

The IEEE standard incorporated changes from the 1982 ANSI document in four major areas:

- An additional safety factor was provided in certain situations. The most significant change was the introduction of new *uncontrolled* (public) exposure guidelines, generally established

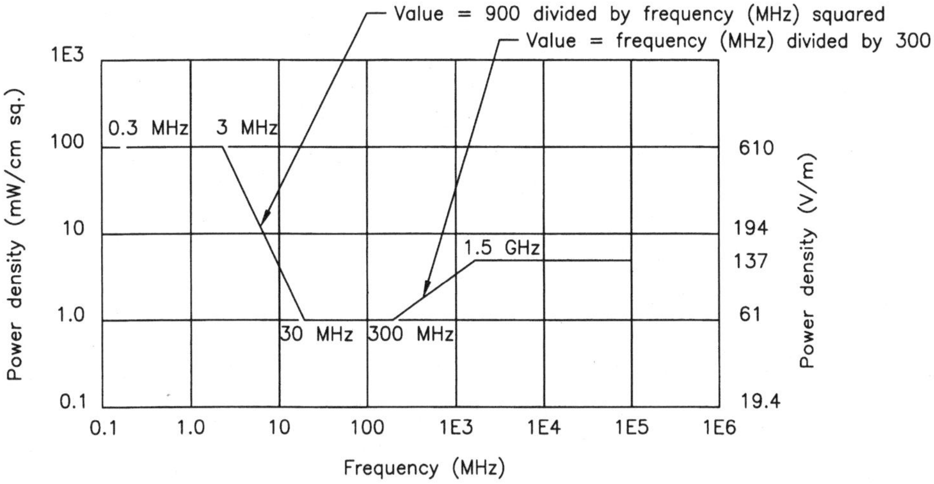

Figure 3.8 ANSI/IEEE exposure guidelines for microwave frequencies.

at one-fifth of the *controlled* (occupational) exposure guidelines. Figure 3.8 illustrates the concept for the microwave frequency band.

- For the first time, guidelines were included for body currents; examination of the electric and magnetic fields were determined to be insufficient to determine compliance.
- Minor adjustments were made to occupational guidelines, including relaxation of the guidelines at certain frequencies and the introduction of *breakpoints* at new frequencies.
- Measurement procedures were changed in several aspects, most notably with respect to spatial averaging and to minimum separation from reradiating objects and structures at the site.

The revised guidelines are complex and beyond the scope of this handbook. Refer to the ANSI/IEEE document for details.

3.5b Multiple-user Sites

At a multiple-user site, the responsibility for assessing the RFR situation—although officially triggered by either a new user or the license renewal of all site tenants—is, in reality, the joint responsibility of all the site tenants. In a multiple-user environment involving various frequencies, and various protection criteria, compliance is indicated when the fraction of the RFR limit within each pertinent frequency band is established and added to the sum of all the other fractional contributions. The sum must not be greater than 1.0. Evaluating the multiple-user environment is not a simple matter, and corrective actions, if indicated, may be quite complex.

Operator Safety Considerations

RF energy must be contained properly by shielding and transmission lines. All input and output RF connections, cables, flanges, and gaskets must be RF leakproof. The following guidelines should be followed at all times:

- Never operate a power tube without a properly matched RF energy absorbing load attached.
- Never look into or expose any part of the body to an antenna or open RF generating tube, circuit, or RF transmission system that is energized.
- Monitor the RF system for radiation leakage at regular intervals and after servicing.

3.6 X-Ray Radiation Hazard

The voltages typically used in microwave tubes are capable of producing dangerous X rays. As voltages increase beyond 15 kV, metal-body tubes are capable of producing progressively more dangerous radiation. Adequate X-ray shielding must be provided on all sides of such tubes, particularly at the cathode and collector ends, as well as at the modulator and pulse transformer tanks (as appropriate). High-voltage tubes never should be operated without adequate X-ray shielding in place. The X-ray radiation of the device should be checked at regular intervals and after servicing.

3.7 Implosion Hazard

Because of the high internal vacuum in power grid and microwave tubes, the glass or ceramic output window or envelope can shatter inward (implode) if struck with sufficient force or exposed to sufficient mechanical shock. Flying debris could result in bodily injury, including cuts and puncture wounds. If the device is made of beryllium oxide ceramic, implosion may produce highly toxic dust or fumes.

In the event of such an implosion, assume that toxic BeO ceramic is involved unless confirmed otherwise.

3.8 Hot Coolant and Surfaces

Extreme heat occurs in the electron collector of a microwave tube and the anode of a power grid tube during operation. Coolant channels used for water or vapor cooling also can reach high temperatures (boiling—100°C—and above), and the coolant is typically under pressure (as high as 100 psi). Some devices are cooled by boiling the coolant to form steam.

Contact with hot portions of the tube or its cooling system can scald or burn. Carefully check that all fittings and connections are secure, and monitor back pressure for changes in cooling system performance. If back pressure is increased above normal operating values, shut the system down and clear the restriction.

For a device whose anode or collector is air-cooled, the external surface normally operates at a temperature of 200 to 300°C. Other parts of the tube also may reach high temperatures, particularly the cathode insulator and the cathode/heater surfaces. All hot surfaces remain hot for an extended time after the tube is shut off. To prevent serious burns, take care to avoid bodily contact with these surfaces during operation and for a reasonable cool-down period afterward.

3.9 Polychlorinated Biphenyls

PCBs belong to a family of organic compounds known as *chlorinated hydrocarbons*. Virtually all PCBs in existence today have been synthetically manufactured. PCBs have a heavy oil-like consistency, high boiling point, a high degree of chemical stability, low flammability, and low electrical conductivity. These characteristics resulted in the widespread use of PCBs in high-voltage capacitors and transformers. Commercial products containing PCBs were widely distributed between 1957 and 1977 under several trade names including:

- Aroclor
- Pyroclor
- Sanotherm
- Pyranol
- Askarel

Askarel is also a generic name used for nonflammable dielectric fluids containing PCBs. Table 3.5 lists some common trade names used for Askarel. These trade names typically will be listed on the nameplate of a PCB transformer or capacitor.

PCBs are harmful because once they are released into the environment, they tend not to break apart into other substances. Instead, PCBs persist, taking several decades to slowly decompose. By remaining in the environment, they can be taken up and stored in the fatty tissues of all organisms, from which they are slowly released into the bloodstream. Therefore, because of the storage in fat, the concentration of PCBs in body tissues can increase with time, even though PCB exposure levels may be quite low. This process is called *bioaccumulation*. Furthermore, as PCBs accumulate in the tissues of simple organisms, and as they are consumed by progressively higher organisms, the concentration increases. This process is called *biomagnification*. These two factors are especially significant because PCBs are harmful even at low levels. Specifically, PCBs have been shown to cause chronic (long-term) toxic effects in some species of animals and aquatic life. Well-documented tests on laboratory animals show that various levels of PCBs can cause reproductive effects, gastric disorders, skin lesions, and cancerous tumors.

PCBs may enter the body through the lungs, the gastrointestinal tract, and the skin. After absorption, PCBs are circulated in the blood throughout the body and stored in fatty tissues and a variety of organs, including the liver, kidneys, lungs, adrenal glands, brain, heart, and skin.

The health risk from PCBs lies not only in the PCB itself, but also in the chemicals that develop when PCBs are heated. Laboratory studies have confirmed that PCB by-products, including *polychlorinated dibenzofurans* (PCDFs) and *polychlorinated dibenzo-p-dioxins* (PCDDs), are formed when PCBs or *chlorobenzenes* are heated to temperatures ranging from approximately 900 to 1300°F. Unfortunately, these products are more toxic than PCBs themselves.

Table 3.5 Commonly Used Trade Names for PCB Insulating Material

Apirolio	Abestol	Askarel	Aroclor B	Chlorexto	Chlophen
Chlorinol	Clorphon	Diaclor	DK	Dykanol	EEC-18
Elemex	Eucarel	Fenclor	Hyvol	Inclor	Inerteen
Kanechlor	No-Flamol	Phenodlor	Pydraul	Pyralene	Pyranol
Pyroclor	Sal-T-Kuhl	Santothern FR	Santovac	Solvol	Thermin

3.9a Governmental Action

The U.S. Congress took action to control PCBs in October 1975 by passing the Toxic Substances Control Act (TSCA). A section of this law specifically directed the EPA to regulate PCBs. Three years later the Environmental Protection Agency (EPA) issued regulations to implement the congressional ban on the manufacture, processing, distribution, and disposal of PCBs. Since that time, several revisions and updates have been issued by the EPA. One of these revisions, issued in 1982, specifically addressed the type of equipment used in industrial plants and transmitting stations. Failure to properly follow the rules regarding the use and disposal of PCBs has resulted in high fines and even jail sentences.

Although PCBs are no longer being produced for electrical products in the United States, there are thousands of PCB transformers and millions of small PCB capacitors still in use or in storage. The threat of widespread contamination from PCB fire-related incidents is one reason behind the EPA's efforts to reduce the number of PCB products in the environment. The users of high-power equipment are affected by the regulations primarily because of the widespread use of PCB transformers and capacitors. These components usually are located in older (pre-1979) systems, so this is the first place to look for them. However, some facilities also maintain their own primary power transformers. Unless these transformers are of recent vintage, it is quite likely that they too contain a PCB dielectric. Table 3.6 lists the primary classifications of PCB devices.

3.9b PCB Components

The two most common PCB components are transformers and capacitors. A PCB transformer is one containing at least 500 ppm (parts per million) PCBs in the dielectric fluid. An Askarel transformer generally has 600,000 ppm or more. A PCB transformer may be converted to a *PCB-contaminated device* (50 to 500 ppm) or a *non-PCB device* (less than 50 ppm) by having it drained, refilled, and tested. The testing must not take place until the transformer has been in service for a minimum of 90 days. Note that this is *not* something a maintenance technician can do. It is the exclusive domain of specialized remanufacturing companies.

PCB transformers must be inspected quarterly for leaks. If an impervious dike is built around the transformer sufficient to contain all of the liquid material, the inspections can be conducted yearly. Similarly, if the transformer is tested and found to contain less than 60,000 ppm, a yearly inspection is sufficient. Failed PCB transformers cannot be repaired; they must be properly disposed of.

If a leak develops, it must be contained and daily inspections begun. A cleanup must be initiated as soon as possible, but no later than 48 hours after the leak is discovered. Adequate records must be kept of all inspections, leaks, and actions taken for 3 years after disposal of the compo-

Table 3.6 Definition of PCB Terms as Identified by the EPA

Term	Definition	Examples
PCB	Any chemical substance that is limited to the biphenyl molecule that has been chlorinated to varying degrees, or any combination of substances that contain such substances.	PCB dielectric fluids, PCB heat-transfer fluids, PCB hydraulic fluids, 2,2',4-trichlorobiphenyl
PCB article	Any manufactured article, other than a PCB container, that contains PCBs and whose surface has been in direct contact with PCBs.	Capacitors, transformers, electric motors, pumps, pipes
PCB container	A device used to contain PCBs or PCB articles, and whose surface has been in direct contact with PCBs.	Packages, cans, bottles, bags, barrels, drums, tanks
PCB article container	A device used to contain PCB articles or equipment, and whose surface has not been in direct contact with PCBs.	Packages, cans, bottles, bags, barrels, drums, tanks
PCB equipment	Any manufactured item, other than a PCB container or PCB article container, that contains a PCB article or other PCB equipment.	Microwave systems, fluorescent light ballasts, electronic equipment
PCB item	Any PCB article, PCB article container, PCB container, or PCB equipment that deliberately or unintentionally contains, or has as a part of it, any PCBs.	
PCB transformer	Any transformer that contains PCBs in concentrations of 500 ppm or greater.	
PCB contaminated	Any electric equipment that contains more than 50, but less than 500 ppm of PCBs. (Oil-filled electric equipment other than circuit breakers, reclosers, and cable whose PCB concentration is unknown must be assumed to be PCB-contaminated electric equipment.)	Transformers, capacitors, contaminated circuit breakers, reclosers, voltage regulators, switches, cable, electromagnets

nent. Combustible materials must be kept a minimum of 5 m from a PCB transformer and its enclosure.

As of October 1, 1990, the use of PCB transformers (500 ppm or greater) was prohibited in or near commercial buildings when the secondary voltages are 480 V ac or higher.

The EPA regulations also require that the operator notify others of the possible dangers. All PCB transformers (including PCB transformers in storage for reuse) must be registered with the local fire department. The following information must be supplied:

- The location of the PCB transformer(s).
- Address(es) of the building(s) and, for outdoor PCB transformers, the location.
- Principal constituent of the dielectric fluid in the transformer(s).
- Name and telephone number of the contact person in the event of a fire involving the equipment.

Any PCB transformers used in a commercial building must be registered with the building owner. All building owners within 30 m of such PCB transformers also must be notified. In the event of a fire-related incident involving the release of PCBs, the Coast Guard National Spill Response Center (800-424-8802) must be notified immediately. Appropriate measures also must be taken to contain and control any possible PCB release into water.

Capacitors are divided into two size classes, *large* and *small*. A PCB small capacitor contains less than 1.36 kg (3 lbs) of dielectric fluid. A capacitor having less than 100 in^3 also is considered to contain less than 3 lb of dielectric fluid. A PCB large capacitor has a volume of more than 200 in^3 and is considered to contain more than 3 lb of dielectric fluid. Any capacitor having a volume between 100 and 200 in^3 is considered to contain 3 lb of dielectric, provided the total weight is less than 9 lb. A PCB *large high-voltage capacitor* contains 3 lb or more of dielectric fluid and operates at voltages of 2 kV or greater. A *large low-voltage capacitor* also contains 3 lb or more of dielectric fluid but operates below 2 kV.

The use and servicing of PCB small capacitors is not restricted by the EPA unless there is a leak. In that event, the leak must be repaired or the capacitor disposed of. Disposal may be handled by an approved incineration facility, or the component may be placed in a specified container and buried in an approved chemical waste landfill. Items such as capacitors that are leaking oil greater than 500 ppm PCBs should be taken to an EPA-approved PCB disposal facility.

3.9c PCB Liability Management

Properly managing the PCB risk is not particularly difficult; the keys are understanding the regulations and following them carefully. Any program should include the following steps:

- Locate and identify all PCB devices. Check all stored or spare devices.
- Properly label PCB transformers and capacitors according to EPA requirements.
- Perform the required inspections and maintain an accurate log of PCB items, their location, inspection results, and actions taken. These records must be maintained for 3 years after disposal of the PCB component.
- Complete the annual report of PCBs and PCB items by July 1 of each year. This report must be retained for 5 years.
- Arrange for necessary disposal through a company licensed to handle PCBs. If there are any doubts about the company's license, contact the EPA.
- Report the location of all PCB transformers to the local fire department and to the owners of any nearby buildings.

The importance of following the EPA regulations cannot be overstated.

3.10 References

1. National Electrical Code, NFPA #70.

3.11 Bibliography

"Current Intelligence Bulletin #45," National Institute for Occupational Safety and Health, Division of Standards Development and Technology Transfer, February 24, 1986.

Code of Federal Regulations, 40, Part 761.

"Electrical Standards Reference Manual," U.S. Department of Labor, Washington, D.C.

Hammett, William F.: "Meeting IEEE C95.1-1991 Requirements," *NAB 1993 Broadcast Engineering Conference Proceedings*, National Association of Broadcasters, Washington, D.C., pp. 471–476, April 1993.

Hammar, Willie: *Occupational Safety Management and Engineering*, Prentice Hall, New York, N.Y.

Markley, Donald: "Complying with RF Emission Standards," *Broadcast Engineering*, Intertec Publishing, Overland Park, Kan., May 1986.

"Occupational Injuries and Illnesses in the United States by Industry," OSHA Bulletin 2278, U.S. Department of Labor, Washington, D.C, 1985.

OSHA, "Electrical Hazard Fact Sheets," U.S. Department of Labor, Washington, D.C, January 1987.

OSHA, "Handbook for Small Business," U.S. Department of Labor, Washington, D.C.

Pfrimmer, Jack, "Identifying and Managing PCBs in Broadcast Facilities," *1987 NAB Engineering Conference Proceedings*, National Association of Broadcasters, Washington, D.C, 1987.

"Safety Precautions," Publication no. 3386A, Varian Associates, Palo Alto, Calif., March 1985.

Smith, Milford K., Jr., "RF Radiation Compliance," *Proceedings of the Broadcast Engineering Conference*, Society of Broadcast Engineers, Indianapolis, IN, 1989.

"Toxics Information Series," Office of Toxic Substances, July 1983.

Whitaker, Jerry C.: *AC Power Systems*, 2nd Ed., CRC Press, Boca Raton, Fla., 1998.

Whitaker, Jerry C.: G. DeSantis, and C. Paulson: *Interconnecting Electronic Systems*, CRC Press, Boca Raton, Fla., 1993.

Whitaker, Jerry C.: *Maintaining Electronic Systems*, CRC Press, Boca Raton, Fla. 1991.

Whitaker, Jerry C.: *Power Vacuum Tubes Handbook*, 2nd Ed., CRC Press, Boca Raton, Fla., 1999.

Whitaker, Jerry C.: *Radio Frequency Transmission Systems: Design and Operation*, McGraw-Hill, New York, N.Y., 1990.

Chapter

4

Video Standards Overview

Donald C. McCroskey

Jerry C. Whitaker, editor

4.1 Introduction

Standardization usually starts within a company as a way to reduce costs associated with parts stocking, design drawings, training, and retraining of personnel. The next level might be a cooperative agreement between firms making similar equipment to use standardized dimensions, parts, and components. Competition, trade secrets, and the *NIH factor* (not invented here) often generate an atmosphere that prevents such an understanding. Enter the professional engineering society, which promises a forum for discussion between users and engineers while down playing the commercial and business aspects.

4.2 The History of Modern Standards

In 1836, the U. S. Congress authorized the Office of Weights and Measures (OWM) for the primary purpose of ensuring uniformity in custom house dealings. The Treasury Department was charged with its operation. As advancements in science and technology fueled the industrial revolution, it was apparent that standardization of hardware and test methods was necessary to promote commercial development and to compete successfully with the rest of the world. The industrial revolution in the 1830s introduced the need for interchangeable parts and hardware. Economical manufacture of transportation equipment, tools, weapons, and other machinery were possible only with mechanical standardization.

By the late 1800's professional organizations of mechanical, electrical, chemical, and other engineers were founded with this aim in mind. The Institute of Electrical Engineers developed standards between 1890 and 1910 based on the practices of the major electrical manufacturers of the time. Such activities were not within the purview of the OWM, so there was no government involvement during this period. It took the pressures of war production in 1918 to cause the formation of the American Engineering Standards Committee (AESC) to coordinate the activities of various industry and engineering societies. This group became the American Standards Association (ASA) in 1928.

Parallel developments would occur worldwide. The International Bureau of Weights and Measures was founded in 1875, the International Electrotechnical Commission (IEC) in 1904, and the International Federation of Standardizing Bodies (ISA) in 1926. Following World War II (1946) this group was reorganized as the International Standards Organization (ISO) comprised of the ASA and the standardizing bodies of 25 other countries. Present participation is approximately 55 countries and 145 technical committees. The stated mission of the ISO is to *facilitate the internationalization and unification of industrial standards*.

The International Telecommunications Union (ITU) was founded in 1865 for the purpose of coordinating and interfacing telegraphic communications worldwide. Today, its member countries develop regulations and voluntary recommendations, and provide coordination of telecommunications development. A sub-group, the International Radio Consultative Committee (CCIR) (which no longer exists under this name), is concerned with certain transmission standards and the compatible use of the frequency spectrum, including geostationary satellite orbit assignments. Standardized transmission formats to allow interchange of communications over national boundaries are the purview of this committee. Because these standards involve international treaties, negotiations are channeled through the U. S. State Department.

4.2a American National Standards Institute (ANSI)

ANSI coordinates policies to promote procedures, guidelines, and the consistency of standards development. Due process procedures ensure that participation is open to all persons who are materially affected by the activities without domination by a particular group. Written procedures are available to ensure that consistent methods are used for standards developments and appeals. Today, there are more than 1000 members who support the U.S. voluntary standardization system as members of the ANSI federation. This support keeps the Institute financially sound and the system free of government control.

The functions of ANSI include: (1) serving as a clearinghouse on standards development and supplying standards-related publications and information, and (2) the following business development issues:

- Provides national and international standards information necessary to market products worldwide.
- Offers American National Standards that assist companies in reducing operating and purchasing costs, thereby assuring product quality and safety.
- Offers an opportunity to voice opinion through representation on numerous technical advisory groups, councils, and boards.
- Furnishes national and international recognition of standards for credibility and force in domestic commerce and world trade.
- Provides a path to influence and comment on the development of standards in the international arena.

Prospective standards must be submitted by an ANSI accredited standards developer. There are three methods which may be used:

- **Accredited organization method**. This approach is most often used by associations and societies having an interest in developing standards. Participation is open to all interested parties as well as members of the association or society. The standards developer must fashion its own operating procedures, which must meet the general requirements of the ANSI procedures.
- **Accredited standards committee method**. Standing committees of directly and materially affected interests develop documents and establish consensus in support of the document. This method is most often used when a standard affects a broad range of diverse interests or where multiple associations or societies with similar interests exist. These committees are administered by a *secretariat*, an organization that assumes the responsibility for providing compliance with the pertinent operating procedures. The committee can develop its own operating procedures consistent with ANSI requirements, or it can adopt standard ANSI procedures.
- **Accredited canvass method**. This approach is used by smaller trade association or societies that have documented current industry practices and desire that these standards be recognized nationally. Generally, these developers are responsible for less than five standards. The developer identifies those who are directly and materially affected by the activity in question and conducts a letter ballot *canvass* of those interests to determine consensus. Developers must use standard ANSI procedures.

Note that all methods must fulfill the basic requirements of public review, voting, consideration, and disposition of all views and objections, and an appeals mechanism.

The introduction of new technologies or changes in the direction of industry groups or engineering societies may require a mediating body to assign responsibility for a developing standard to the proper group. The Joint Committee for Intersociety Coordination (JCIC) operates under ANSI to fulfill this need.

4.2b Professional Society Engineering Committees

The engineering groups that collate and coordinate activities that are eventually presented to standardization bodies encourage participation from all concerned parties. Meetings are often scheduled in connection with technical conferences to promote greater participation. Other necessary meetings are usually scheduled in geographical locations of the greatest activity in the field. There are no charges or dues to be a member or to attend the meetings. An interest in these activities can still be served by reading the reports from these groups in the appropriate professional journals. These wheels may seem to grind exceedingly slowly at times, but the adoption of standards that may have to endure for 50 years or more should not be taken lightly.

Advanced Television Systems Committee

The Advanced Television Systems Committee (ATSC) is an international, non-profit organization developing voluntary standards for digital television. The ATSC has over 200 member organizations representing the broadcast, broadcast equipment, motion picture, consumer electronics, computer, cable, satellite, and semiconductor industries. The following is a partial list of ATSC Standards and technical activities. Additional information is available at www.atsc.org.

Electronics Industries Alliance

Many of the early standards relating to radio and television broadcasting were developed by equipment manufacturers under the banner of the Radio Manufacturers Association (RMA), later the RETMA (add Electronics and Television), then the Electronics Industries Association (EIA), and now the Electronic Industries Alliance (www.eia.org).

The EIA is a national trade organization made up of a number of product divisions. Some of the best known EIA standards activities are in the areas of data communications, instrumentation, broadcast transmitters, video transmission, video cameras, test charts, video monitors, and RF interference.

With the proliferation and expansion of electronics, the EIA is now divided into many sectors and groups. Of particular interest to the broadcaster are the Consumer Electronics Association (CEA), Electronic Components, Assemblies, Equipment, & Supplies Association (ECA), and Telecommunications Industry Association (TIA). For a comprehensive, searchable listing of all electronic international standards visit http://global.ihs.com/.

Institute of Electrical and Electronic Engineers

The IEEE has many branches (professional groups) that serve the standardization needs of the electrical, electronic, and computer industries. Presently available standards relate to definitions, measurement techniques, and test methods. The Institute of Radio Engineers (long since joined the IEEE) was responsible for measurement standards and techniques in televisions early years. The Standards Coordinating Committees publish books and documents covering definitions of electric and electronics terms, graphic symbols and reference designations for engineering drawings, and letter symbols for measurement units. Anyone concerned with power wiring and distribution should be interested in the National Electrical Safety Code books. Nearly all of these documents are available from IEEE (www.ieee.org) or ANSI.

Society of Motion Picture and Televisions Engineers

Organizations such as the SMPTE (www.smpte.org), composed primarily of users of equipment and processes, are able to accomplish what is nearly impossible in the manufacturing community. Namely, to provide a forum where users and manufacturers can distill the best of current technology to promote basic interchangeability in hardware and software. A chronology of the development of this engineering society provides insights as to how such organizations adapt to the needs of advancing technologies.

Around 1915 it became obvious that the rapidly expanding motion picture industry must standardize basic dimensions and tolerances of film stocks and transport mechanisms. After two unsuccessful attempts to form industry based standardizing committees, the Society of Motion Picture Engineers was formed. The founding goals were to standardize the nomenclature, equipment, and processes of the industry; to promote research and development in the industry's science and technology; and to remain independent of, while cooperating with, its business partners. It is this independent quality of a professional society that makes it possible to mediate strongly held opinions of business competitors.

By the late 1940's it was apparent that the future of motion pictures and television would involve sharing technology, techniques, and the market for visual education and entertainment. SMPE became SMPTE. In comparatively recent times the Society has been assigned more responsibility for television standards. The recording and reproduction of television signals has become the province of SMPTE standardization efforts. An index of the work of the engineering committees is published yearly. The basic SMPTE documents issued as a part of the organization's standardization efforts are:

- **Engineering Guidelines**—guidelines for the implementation of test materials and equipment operation.
- **Recommended Practices**—these include specifications for test materials, generic equipment setup and operating techniques, and mechanical dimensions involving operational procedures.
- **Standards**—mechanical specifications for film, tape, cassettes, and transport mechanisms; electrical recording and reproduction characteristics; and protocol and software issues for digital video systems.

Audio Engineering Society

The AES (www.aes.org) was organized in 1948 primarily to serve the needs of the high quality audio recording and reproduction community. The Society maintains a standards committee (AESSC) that supervises the work of subcommittees and working groups. Drafts of proposed standards are published in the *Journal of the Audio Engineering Society* (JAES) for review and comment by all interested parties. Any substantive comments (as opposed to editorial) are then considered by the committee before submitting the final document to a vote. Current AES standards address measurement methods, commercial loudspeaker specifications, and digital audio recording/transmission systems.

4.2c Advanced Television Systems Committee

The Advanced Television Systems Committee (ATSC) is an international, non-profit organization developing voluntary standards for digital television. The ATSC has nearly 200 member organizations representing the broadcast, broadcast equipment, motion picture, consumer electronics, computer, cable, satellite, and semiconductor industries. The following is a partial list of ATSC Standards and technical activities. Additional information is available at www.atsc.org.

ATSC Digital Television Standard, *Document A/53B*

The Digital Television Standard describes the system characteristics of the advanced television (ATV) system. The document and its normative annexes provide detailed specification of the parameters of the system including the video encoder input scanning formats and the pre-processing and compression parameters of the video encoder, the audio encoder input signal format and the pre-processing and compression parameters of the audio encoder, the service multiplex and transport layer characteristics and normative specifications, and the VSB RF/transmission subsystem. The system is modular in concept and the specifications for each of the modules are provided in the appropriate annex.

Guide to the use of the ATSC Digital Television Standard, *Document A/54*

This guide provides an overview and tutorial of the system characteristics of the advanced television (ATV) system defined by ATSC Standard A/53, *ATSC Digital Television.*

Digital Audio Compression (AC-3), *Document A/52A*

This document specifies coded representation of audio information and the decoding process, as well as information on the encoding process. The coded representation specified is suitable for use in digital audio transmission and storage applications, and may convey from 1 to 5 full bandwidth audio channels, along with a low frequency enhancement channel. A wide range of encoded bit-rates is supported by this specification. Typical applications of digital audio compression are in satellite or terrestrial audio broadcasting, delivery of audio over metallic or optical cables, or storage of audio on magnetic, optical, semiconductor, or other storage media.

Standard for Coding 25/50 Hz Video, *Document A/63*

This document describes the characteristics for the video subsystem of a digital television system operating at 25 Hz and 50 Hz frame rates.

Transmission Measurement and Compliance Standard for DTV, *Document A/64 Rev. A*

This document describes methods for testing, monitoring, and measurement of the transmission subsystem intended for use in the digital television (DTV) system, including specifications for maximum out-of-band emissions, parameters affecting the quality of the inband signal, symbol error tolerance, phase noise and jitter, power, power measure, frequency offset, and stability. In addition, it describes the condition of the RF symbol stream upon loss of MPEG packets. (The ATSC approved a revision to this document on May 30, 2000, that includes the revised FCC DTV emission mask.

PSIP for Terrestrial Broadcast and Cable, *Document A/65 Rev A with Amendment No. 1*

The Program and System Information Protocol Standard provides a methodology for transporting digital television system information and electronic program guide data. The standard includes an amendment that provides new functionality known as *Directed Channel Change* (DCC). This new feature will allow broadcasters to tailor programming or advertising based upon parameters defined by the viewer such as: postal, zip or location code, program identifier, demographic category, and content subject category. Potential applications include customized programming services, commercials based upon demographics, and localized weather and traffic reports.

Use of ATSC A/65A PSIP Standard in Taiwan, *Document A/68*

This standard specifies the use of character sets, rating regions, and major channel numbers for the implementation of ATSC A/65A *Program and System Information Protocol* (PSIP) in Taiwan.

Conditional Access For Terrestrial Broadcast, *Document A/70*

This document defines a standard for the Conditional Access system for ATSC terrestrial broadcasting to enable broadcasters to fully utilize the capabilities of digital broadcasting. This standard is based, whenever possible, on existing open standards and defines the building blocks necessary to ensure interoperability. The ATSC CA module is replaceable; to ensure that ATSC hosts are protected against obsolescence as security is upgraded. This standard applies to all CA vendors that supply CA service on behalf of an ATSC service provider. An overview of the CA standard is given in Annex C. (This document includes an Amendment that the ATSC approved on May 30, 2000.)

ATSC Recommended Practice for Developing DTV Field Test Plans, *Document A/75*

This document presents the objectives of, and general methodology for conducting field tests of over-the-air terrestrial digital television (DTV) systems. The scope of the work includes reception, demodulation, and recovery of the transmitted data. The scope of the work herein is not concerned with the decoded data or analog signals except when these signals are used as a means to determine that the data has been correctly recovered.

Modulation and Coding Requirements for DTV Applications Over Satellite, *Document A/80*

This document defines a standard for modulation and coding of data delivered over satellite for digital television contribution and distribution applications. The data can be a collection of program material including video, audio, data, multimedia, or other material. It includes the ability to handle multiplexed bit streams in accordance with the MPEG-2 systems layer, but it is not limited to this format and makes provision for arbitrary types of data as well. QPSK, 8PSK and 16 QAM modulation modes are included, as well as a range of forward error correction techniques.

Data Broadcast, *Document A/90*

The ATSC Data Broadcast Standard defines protocols for data transmission compatible with digital multiplex bit streams constructed in accordance with ISO/IEC 13818-1 (MPEG-2 systems). The standard supports data services that are both TV program related and non-program related. Applications may include enhanced television, webcasting, and streaming video services. Data broadcasting receivers may include PCs, televisions, set-top boxes, or other devices. The standard provides mechanisms for download of data, delivery of datagrams, and streaming data.

Implementation Guidelines for the ATSC Data Broadcast Standard, *Document A/91*

This document provides a set of guidelines for the use and implementation of the ATSC Data Broadcast Standard as described in ATSC Standard, A/90 (2000), "Data Broadcast Standard." As such, they facilitate the efficient and reliable implementation of data broadcast services.

DTV Application Software Environment (DASE)

The DASE standard will define a software layer (middleware) that allows programming content and applications to run on a common receiver platform. Interactive and enhanced applications need access to common receiver features in a platform-independent manner. The standard will provide enhanced and interactive content creators the specifications necessary to ensure that their applications and data will run uniformly on all brands and models of receivers. Manufacturers will be able to choose hardware platforms and operating systems for receivers, but provide the commonality necessary to support applications made by many content creators.

Interactive Services

This standard defines session level protocols carried over interaction channels associated with interactive services. The interaction channel may be one or two way and connects a user (operating through a DTV receiver) with some service provider. The ATSC Interactive Services Protocols are intended to operate on a variety of physical networks, by focusing on higher layer protocols but not addressing specific applications

4.2d When is a Standard Finished?

Rapid improvements in technology tend to make many standards technically obsolete by the time they are adopted. But such is the nature of our rapidly expanding technology-based society. There is no need to apologize for this natural phenomena. A standard still provides a stable platform for manufacturers to market their product and assures the user of some degree of compatibility. Technical chaos and anarchy are real possibilities if standards are not adopted in a timely manner. Only the strongest companies could be expected to survive in an atmosphere where standards are lacking. A successful standard promises a stable period of income to manufacturers while giving users assurance of multiple sources during the active life of the product.

Standards are seldom available at no cost. The number of man-hours that go into the development of a typical standard is mind-boggling. The participant's time and expenses are donated, but the standards developing organization has associated administrative expenses that must be recovered. The cost of a printed standard reflect these costs as distributed over the anticipated demand. Standards issuing organizations will supply lists of available standards on request. Internet web sites of the organizations also contain up-to-date listings, descriptions, and ordering information. Many organizations now issue sets of standards on a CD-ROM at considerable savings over the cost of printed versions. Proposed new and revised standards are published for comment in professional society journals, such as SMPTE and AES.

4.3 Standards Relating to Digital Video

The following references provide additional information on digital television in general, and the ATSC and DVB standards in particular.

4.3a Video

ISO/IEC IS 13818-1, International Standard (1994), MPEG-2 Systems

ISO/IEC IS 13818-2, International Standard (1994), MPEG-2 Video

ITU-R BT.601-4 (1994), Encoding Parameters of Digital Television for Studios

SMPTE 274M-1995, Standard for Television, 1920 × 1080 Scanning and Interface

SMPTE 293M-1996, Standard for Television, 720 × 483 Active Line at 59.94 Hz Progressive Scan Production, Digital Representation

SMPTE 294M-1997, Standard for Television, 720 × 483 Active Line at 59.94 Hz Progressive Scan Production, Bit-Serial Interfaces

SMPTE 295M-1997, Standard for Television, 1920 × 1080 50 Hz, Scanning and Interface

SMPTE 296M-1997, Standard for Television, 1280 × 720 Scanning, Analog and Digital Representation, and Analog Interface

4.3b Audio

ATSC Standard A/52 (1995), Digital Audio Compression (AC-3)

AES 3-1992 (ANSI S4.40-1992), AES Recommended Practice for digital audio engineering—Serial transmission format for two-channel linearly represented digital audio data

ANSI S1.4-1983, Specification for Sound Level Meters

IEC 651 (1979), Sound Level Meters

IEC 804 (1985), Amendment 1 (1989), Integrating/Averaging Sound Level Meters

4.4 ATSC DTV Standard

The following documents form the basis for the ATSC digital television standard.

4.4a Service Multiplex and Transport Systems

ATSC Standard A/52 (1995), Digital Audio Compression (AC-3)

ISO/IEC IS 13818-1, International Standard (1994), MPEG-2 Systems

ISO/IEC IS 13818-2, International Standard (1994), MPEG-2 Video

ISO/IEC CD 13818-4, MPEG Committee Draft (1994), MPEG-2 Compliance

4.4b System Information Standard

ATSC Standard A/52 (1995), Digital Audio Compression (AC-3)

ATSC Standard A/53 (1995), ATSC Digital Television Standard

ATSC Standard A/80 (1999), Modulation And Coding Requirements For Digital TV (DTV) Applications Over Satellite

ISO 639, Code for the Representation of Names of Languages, 1988

ISO CD 639.2, Code for the Representation of Names of Languages: alpha-3 code, Committee Draft, dated December 1994

ISO/IEC 10646-1:1993, Information technology—Universal Multiple-Octet Coded Character Set (UCS) — Part 1: Architecture and Basic Multilingual Plane

ISO/IEC 11172-1, Information Technology—Coding of moving pictures and associated audio for digital storage media at up to about 1.5 Mbit/s—Part 1: Systems

ISO/IEC 11172-2, Information Technology—Coding of moving pictures and associated audio for digital storage media at up to about 1.5 Mbit/s—Part 2: Video

ISO/IEC 11172-3, Information Technology—Coding of moving pictures and associated audio for digital storage media at up to about 1.5 Mbit/s—Part 3: Audio

ISO/IEC 13818-3:1994, Information Technology—Coding of moving pictures and associated audio—Part 3: Audio

ISO/CD 13522-2:1993, Information Technology—Coded representation of multimedia and hypermedia information objects—Part 1: Base notation

ISO/IEC 8859, Information Processing—8-bit Single-Octet Coded Character Sets, Parts 1 through 10

ITU-T Rec. H. 222.0 / ISO/IEC 13818-1:1994, Information Technology—Coding of moving pictures and associated audio—Part 1: Systems

ITU-T Rec. H. 262 / ISO/IEC 13818-2:1994, Information Technology—Coding of moving pictures and associated audio—Part 2: Video

ITU-T Rec. J.83:1995, Digital multi-programme systems for television, sound, and data services for cable distribution

ITU-R Rec. BO.1211:1995, Digital multi-programme emission systems for television, sound, and data services for satellites operating in the 11/12 GHz frequency range

4.4c Receiver Systems

47 CFR Part 15, FCC Rules

EIA IS-132, EIA Interim Standard for Channelization of Cable Television

EIA IS-23, EIA Interim Standard for RF Interface Specification for Television Receiving Devices and Cable Television Systems

EIA IS-105, EIA Interim Standard for a Decoder Interface Specification for Television Receiving Devices and Cable Television Decoders

4.4d Program Guide

ATSC Standard A/53 (1995), ATSC Digital Television Standard

ANSI/EIA-608-94 (1994), Recommended Practice for Line 21 Data Service

ISO/IEC IS 13818-1, International Standard (1994), MPEG-2 Systems

4.4e Program/Episode/Version Identification

ATSC Standard A/53 (1995), Digital Television Standard

ATSC Standard A/65 (1998), Program and System Information Protocol for Terrestrial Broadcast and Cable

ATSC Standard A/70 (1999), Conditional Access System for Terrestrial Broadcast

ISO/IEC IS 13818-1, International Standard (1994), MPEG-2 systems

4.5 DVB

The following documents form the basis of the DVB digital television standard. For additional information contact DVB at http://www.dvb.org.

4.5a General

Digital Satellite Transmission Systems, ETS 300 421

Digital Cable Delivery Systems, ETS 300 429

Digital Terrestrial Broadcasting Systems, ETS 300 744

Digital Satellite Master Antenna Television (SMATV) Distribution Systems, ETS 300 473

Specification for the Transmission of Data in DVB Bitstreams, TS/EN 301 192

Digital Broadcasting Systems for Television, Sound and Data Services; Subtitling Systems, ETS 300 743

Digital Broadcasting Systems for Television, Sound and Data Services; Allocation of Service Information (SI) Codes for Digital Video Broadcasting (DVB) Systems, ETR 162

4.5b Multipoint Distribution Systems

Digital Multipoint Distribution Systems at and Above 10 GHz, ETS 300 748

Digital Multipoint Distribution Systems at or Below 10 GHz, ETS 300 749

4.5c Interactive Television

Return Channels in CATV Systems (DVB-RCC), ETS 300 800

Network-Independent Interactive Protocols (DVB-NIP), ETS 300 801

Interaction Channel for Satellite Master Antenna TV (SMATV), ETS 300 803

Return Channels in PSTN/ISDN Systems (DVB-RCT), ETS 300 802

Interfacing to PDH Networks, ETS 300 813

Interfacing to SDH Networks, ETS 300 814

4.5d Conditional Access

Common Interface for Conditional Access and Other Applications, EN50221

Technical Specification of SimulCrypt in DVB Systems, TS101 197

4.5e Interfaces

DVB Interfaces to PDH Networks, prETS 300 813

DVB Interfaces to SDH Networks, prETS 300 814

4.6 Acquiring Reference Documents

ATSC Standards

Advanced Television Systems Committee (ATSC), 1750 K Street N.W., Suite 1200 Washington, D.C. 20006 USA; Phone 202.828.3130; Fax 202.828.3131; Internet http://www.atsc.org/

IEEE Standards

Institute of Electrical and Electronics Engineers, Inc. 445 Hoes Lane, PO Box 1331, Piscataway, N.J. 08855-1331, USA; Phone 800.678.IEEE (4333); Outside USA and Canada 732.981.9667; Internet http://www.ieee.org; Email customer-service@ieee.org

IETF Standards

Internet Engineering Task Force (IETF), c/o Corporation for National Research Initiatives, 1895 Preston White Drive, Suite 100, Reston, VA 20191-5434 USA; Phone 703.620.8990; Fax 703.758.5913; Internet: http://www.ietf.org/rfc

ISO Standards

Global Engineering Documents, World Headquarters, 15 Inverness Way East, Englewood, CO 80112-5776; Phone 800.854.7179; Fax 303.397.2750; Internet http://global.ihs.com

ISO Central Secretariat, 1, rud de Varembe, Case postale 56, CH-1211 Geneve 20, Switzerland; Phone +41 22 749 01 11; Fax + 41 22 733 34 30; Internet http://www.iso.ch; Email central@iso.ch

SCTE Standards

Society of Cable Telecommunications Engineers Inc., 140 Philips Road, Exton, PA 19341; Phone 610.363.6888; Fax 610.363.5898; Internet http://scte.org Email scte@scte.org

SMPTE Standards

Society of Motion Picture and Television Engineers, 595 W. Hartsdale Avenue, White Plains, NY 10607-1824; Phone 914.761.1100; Fax: 914.761.3115; Internet http://www.smpte.org

4.7 Bibliography

McCroskey, Donald C.: "Standardization: History and Purpose," in *The Electronics Handbook*, Jerry C. Whitaker (ed.), CRC Press, Boca Raton, Fla., 1996.

Chapter

5

SMPTE Documents

Jerry C. Whitaker, editor

5.1 Introduction

The following documents relating to digital television have been approved (or are pending at this writing) by the Society of Motion Picture and Television Engineers. For additional information, contact the SMPTE at http://www.smpte.org.

5.1a General Topics

AES/EBU Emphasis and Preferred Sampling Rate, EG 32

Alignment Color Bar Signal, EG 1

Audio: Linear PCM in MPEG-2 Transport Stream, SMPTE 302M

Camera Color Reference Signals, Derivation of, RP 176

Color, Equations, Derivation of, RP 177

Color, Reference Pattern, SMPTE 303M

Wide-Screen Scanning Structure, SMPTE RP 199

5.1b Ancillary

AES/EBU Audio and Auxiliary Data, SMPTE 272M

Camera Positioning by Data Packets, SMPTE 315M

Data Packet and Space Formatting, SMPTE 291M

DTV Closed-Caption Server to Encoder Interface, SMPTE 333M

Error Detection and Status Flags, RP 165

Format for Non-PCM Audio and Data in an AES3 Serial Digital Audio Interface, SMPTE 337M

Format for Non-PCM Audio and Data in an AES3 Serial Digital Audio Interface—ATSC A/52 (AC-3) Data Type, SMPTE 340M

Format for Non-PCM Audio and Data in an AES3 Serial Digital Audio Interface—Captioning Data Type, SMPTE 341M

Format for Non-PCM Audio and Data in an AES3 Serial Digital Audio Interface—Data Types, SMPTE 338M

Format for Non-PCM Audio and Data in an AES3 Serial Digital Audio Interface—Generic Data Types, SMPTE 339M

HDTV 24-bit Digital Audio, SMPTE 299M

LTC and VITC Data as HANC Packets, RP 196

Time and Control Code, RP 188

Transmission of Signals Over Coaxial Cable, SMPTE 276M

5.1c Digital Control Interfaces

Common Messages, RP 172

Control Message Architecture, RP 138

Electrical and Mechanical Characteristics, SMPTE 207M

ESlan Implementation Standards, EG 30

ESlan Remote Control System, SMPTE 275M

ESlan Virtual Machine Numbers, RP 182

Glossary, Electronic Production, EG 28

Remote Control of TV Equipment, EG 29

Status Monitoring and Diagnostics, Fault Reporting, SMPTE 269M

Status Monitoring and Diagnostics, Processors, RP 183

Status Monitoring and Diagnostics, Protocol, SMPTE 273M

Supervisory Protocol, RP 113

System Service Messages, RP 163

Tributary Interconnection, RP 139

Type-Specific Messages, ATR, RP 171

Type-Specific Messages, Routing Switcher, RP 191

Type-Specific Messages, VTR, RP 170

Universal Labels for Unique ID of Digital Data, SMPTE 298M

Video Images: Center, Aspect Ratio and Blanking, RP 187

Video Index: Information Coding, 525- and 625-Line, RP 186

5.1d Edit Decision Lists

Device Control Elements, SMPTE 258M

Storage, 3-1/2-in Disk, RP 162

Storage, 8-in Diskette, RP 132

Transfer, Film to Video, RP 197

5.1e Image Areas

8 mm Release Prints, TV Safe Areas, RP 56

16 mm and 35 mm Film and 2 × 2 slides, SMPTE 96

Review Rooms, SMPTE 148

Safe Areas, RP 27.3

5.1f Interfaces and Signals

12-Channel for Digital Audio and Auxiliary Data, SMPTE 324M

Checkfield, RP 178

Development of NTSC, EG 27

Key Signals, RP 157

NTSC Analog Component 4:2:2, SMPTE 253M

NTSC Analog Composite for Studios, SMPTE 170M

Pathological Conditions, EG 34

Bit-Parallel Interfaces

1125/60 Analog Component, RP 160

1125/60 Analog Composite, SMPTE 240M

1125/60 High-Definition Digital Component, SMPTE 260M

NTSC Digital Component, SMPTE 125M

NTSC Digital Component, 16 × 9 Aspect Ratio, SMPTE 267M

NTSC Digital Component 4:4:4:4 Dual Link, RP 175

NTSC Digital Component 4:4:4:4 Single Link, RP 174

NTSC Digital Composite, SMPTE 244M

Bit-Serial Interfaces

4:2:2p and 4:2:0p Bit Serial, SMPTE 294M

540 Mbits/s Serial Digital Interface, SMPTE 344M

Digital Component 4:2:2 AMI, SMPTE 261M

Digital Component S-NRZ, SMPTE 259M

Digital Composite AMI, SMPTE 261M

Digital Composite, Error Detection Checkwords/Status Flag, RP 165

Digital Composite, Fiber Transmission System, SMPTE 297M

Digital Composite, S-NRZ, SMPTE 259M

Element and Metadata Definitions for the SDTI-CP, SMPTE 331M

Encapsulation of Data Packet Streams over SDTI (SDTI-PF), SMPTE 332M

HDTV, SMPTE 292M

High Data-Rate Serial Data Transport Interface (HD-SDTI), SMSPTE 348M

HDTV, Checkfield, RP 198

Jitter in Bit Serial Systems, RP 184

Jitter Specification, Characteristics and Measurements, EG 33

Jitter Specification, Measurement, RP 192

SDTI Content Package Format (SDTI-CP), SMPTE 326M

Serial Data Transport Interface, SMPTE 305.2M

Time Division Multiplexing Video Signals and Generic Data over High-Definition Interfaces, SMPTE 346M

Vertical Ancillary Data Mapping for Bit-Serial Interface, SMPTE 334M

Scanning Formats

1280 × 720 Scanning, SMPTE 296M

1920 × 1080 Scanning, 60 Hz, SMPTE 274M

1920 × 1080 Scanning, 50 Hz, SMPTE 295M

720 × 483 Digital Representation, SMPTE 293M

5.1g Monitors

Alignment, RP 167

Colorimetry, RP 145

Critical Viewing Conditions, RP 166

Receiver Monitor Setup Tapes, RP 96

5.1h MPEG-2

4:2:2 Profile at High Level, SMPTE 308M

4:2:2 P@HL Synchronous Serial Interface, SMPTE 310M

Alignment for Coding, RP 202

MPEG-2 Video Elementary Stream Editing Information, SMPTE 328M

MPEG-2 Video Recoding Data Set, SMPTE 327M

MPEG-2 Video Recoding Data Set—Compressed Stream Format, SMPTE 329M

Opportunistic Data Broadcast Flow Control, SMPTE 325M

Splice Points for the Transport Stream, SMPTE 312M

Transport of MPEG-2 Recoding Information as Ancillary Data Packets, SMPTE 353M

Transporting MPEG-2 Recoding Information Through 4:2:2 Component Digital Interfaces, SMPTE 319M

Transporting MPEG-2 Recoding Information Through High-Definition Digital Interfaces, SMPTE 351M

Unique Material Identifier (UMID), SMPTE 330M

5.1i Test Patterns

Alignment Color Bars, EG 1

Camera Registration, RP 27.2

Deflection Linearity, RP 38.1

Mid-Frequency Response, RP 27.5

Operational Alignment, RP 27.1

Safe Areas, RP 27.3

Telecine Jitter, Weave, Ghost, RP 27.4

5.1j Video Recording and Reproduction

Audio Monitor System Response, SMPTE 222M

Channel Assignments, AES/EBU Inputs, EG 26

Channel Assignments and Magnetic Masters to Stereo Video, RP 150

Cassette Bar Code Readers, EG 31-1995

Data Structure for DV-Based Audio, Data, and Compressed Video, SMPTE 314M

Loudspeaker Placement, HDEP, RP 173

Relative Polarity of Stereo Audio Signals, RP 148

Tape Care, Handling, Storage, RP 103

Telecine Scanning Capabilities, EG 25

Time and Control Code

Binary Groups, Date and Time Zone Transmissions, SMPTE 309M

Binary Groups, Storage and Transmission, SMPTE 262M

Directory Index, Auxiliary Time Address Data, RP 169

Directory Index, Dialect Specification of Page-Line, RP 179

Specifications, TV, Audio, Film, SMPTE 12M

Time Address Clock Precision, EG 35

Vertical Interval, 4:2:2 Digital Component, SMPTE 266M

Vertical Interval, Encoding Film Transfer Information, 4:2:2 Digital, RP 201

Vertical Interval, Location, RP 164

Vertical Interval, Longitudinal Relationship, RP 159

Vertical Interval, Switching Point, RP 168

Tape Recording Formats

SMPTE Documents Relating to Tape Recording Formats (*Courtesy of SMPTE.*)

	B	C	D-1	D-2	D-3	D-5	D-6	D-7 (1)	D-9 (2)	E (3)	G (4)	H (5)	L (6)	M-2
Basic system parameters														
525/60	15M	18M	EG10	EG20	264M	279M	277M	306M	316M	21M			RP144	RP158
625/50					265M	279M	277M	306M	316M					
Record dimensions	16M	19M	224M	245M	264/5M	279M	277M			21M		32M	229M	249M
Characteristics														
Video signals	RP84	RP86								RP87		32M	230M	251M
Audio and control signals	17M	20M	RP155	RP155	264/5M	279M	278M			RP87		32M	230M	251M
Data and control record			227M	247M	264/5M	279M	278M							
Tracking control record	RP83	RP85				279M	277M							
Pulse code modulation audio														252M
Time and control recording	RP93		228M	248M	264/5M	279M	278M						230M	251M
Audio sector time code, equipment type information			RP181											
Nomenclature		18M	EG21	EG21						21M		32M		
Index of documents				EG22										
Stereo channels	RP142	RP142								RP142	RP142	RP142	RP142	
Relative polarity	RP148	RP148	RP148	RP148						RP148	RP148	RP148	RP148	
Tape	25M	25M	225M	246M	264/5M	279M	277M				35M	32M	238M	250M
Reels	24M	24M												
Cassettes			226M	226M	263M	263M	226M	307M	317M	22M	35M	32M	238M	250M
Small										31M				
Bar code labeling			RP156	RP156										
Dropout specifications	RP121	RP121												
Reference tape and recorder														
System parameters	29M													
Tape	26M	26M												
High-definition														
Encoding process/data format						342M								
Transmission over 360 Mb/s						RP209								

Notes:
1DVCPRO, 2 Digital S, 3 U-matic, 4 Beta, 5 VHS, 6 Betacam

5.2 SMPTE Documents by Number

Documents without year designation were at the proposal stage at printing.

SMPTE Standards, Recommended Practices, and Engineering Guidelines by Number

SMPTE Standards	
SMPTE 1-1996	Video Recording—2-in Magnetic Recording Tape
SMPTE 3-1998	Television Analog Recording—Frequency Response and Operating Level of Recorders and Reproducers—Audio 1 Record on 2-in Tape Operating at l5 and 7.5 in/s
SMPTE 4-1995	Television Analog Recording—2-in Magnetic Tape for Quadruplex Recording—Speed
SMPTE 5-1995	Television Analog Recording—2-in Reels
SMPTE 6-1998	Video Recording—2-in Quadruplex Tape—Video, Audio and Tracking-Control Records
SMPTE 7-1999	Motion-Picture Film (16-mm)—Camera Aperture Image and Usage
SMPTE 8-1995	Video Recording—Quadruplex Recorders Operating at 15 in/s—Audio Level and Multifrequency Test Tape
SMPTE 11-1995	Video Recording—Quadruplex Recorders Operating at 7.5 in/s—Audio Level and Multifrequency Test Tape
SMPTE 12M-1999	Television, Audio, and Film—Time and Control Code
SMPTE 15M-1998	Television Analog Recording—1-in Type B Helical Scan—Basic System Parameters
SMPTE 16M-1998	Television Analog Recording—1-in Type B Helical Scan—Records
SMPTE 17M-1998	Television Analog Recording—1-in Type B Helical Scan—Frequency Response and Operating Level
SMPTE 18M-1996	Television Analog Recording—1-in Type C- Basic System and Transport Geometry Parameters
SMPTE 19M-1996	Television Analog Recording—1-in Type C--Records
SMPTE 20M-1996	Television Analog Recording—1-in Type C Recorders and Reproducers—Longitudinal Audio Characteristics
SMPTE 21M-1997	Video Recording—3/4-in Type E Helical Scan—Records
SMPTE 22M-1997	Video Recording—3/4-in Type E Helical Scan—Cassette
SMPTE 24M-1996	Television Recording—1-in Reels
SMPTE 25M-1995	Video Recording—1-in Magnetic Recording Tape
SMPTE 26M-1995	Video Recording—1-in Helical-Scan Recorders—Raw Stock for Reference Tapes
SMPTE 29M-1995	Television Analog Recording—1-in Type B Reference Recorders—Basic System and Transport Geometry
SMPTE 30M-1995	Television Analog Recording—1-in Type B Reference Recorders—Records on Reference Tapes
SMPTE 31M-1995	Television Analog Recording—3/4-in Type E—Small Video Cassette
SMPTE 32M-1998	Video Recording—1/2-in Type H—Cassette, Tape, and Records
SMPTE 35M-1997	Television Analog Recording—1/2-in Type G—Cassette and Tape
SMPTE 37M-1994	Motion-Picture Equipment—Raw Stock Cores
SMPTE 40-1997	Motion-Picture Film (35-mm)—Release Prints—Photographic Audio Records
SMPTE 41-1999	Motion-Picture Film (16-mm)—Prints—Photographic Audio Records
SMPTE 48-1995	Motion-Picture Film (16-mm)—Picture and Sound Contact Printing—Printed Areas

SMPTE Standards, Recommended Practices, and Engineering Guidelines by Number

SMPTE 55-2000	Motion-Picture Film—35- and 16-mm Audio Release Prints—Leaders and Cue Marks
SMPTE 56-1996	Motion-Picture Film—Nomenclature for Studios and Processing Laboratories
SMPTE 59-1998	Motion-Picture Film (35-mm)—Camera Aperture Images and Usage
SMPTE 73-1998	Motion-Picture Film (32-mm)—35-mm Film Perforated 32-mm, 2R
SMPTE 74-1993	Motion-Picture Cameras (16- and 8-mm)—Zero Point for Focusing Scales (R1998)
SMPTE 75M-1994	Motion-Picture Film—Raw Stock—Designation of A and B Windings
SMPTE 76-1996	Motion-Picture Cameras—16-mm and 8-mm Threaded Lens Mounts
SMPTE 83-1996	Motion-Picture Film (16-mm)—Edge Numbers—Location and Spacing
SMPTE 86-1996	Motion-Picture Film—Magnetic Audio Records—Two, Three, Four, and Six Records on 35-mm and One Record on 17.5-mm Magnetic Film
SMPTE 87M-1996	Motion-Picture Film (16-mm)—100-Mil Magnetic Striping
SMPTE 93-1998	Motion-Picture Film (35-mm)—Perforated BH
SMPTE 96-1999	Television—35- and 16-mm Motion-Picture Film and 2 × 2-in Slides—Scanned Area and Photographic Image Area for 4:3 Aspect Ratio
SMPTE 97-1999	Motion-Picture Film (16-mm)—200-Mil Position-Edge Magnetic Audio Record
SMPTE 101-1998	Motion-Picture Film (16-mm)—Perforated 2R-3000—Magnetic Striping
SMPTE 102-1997	Motion-Picture Film (35-mm)—Perforated CS-1870
SMPTE 109-1998	Motion-Picture Film (16-mm)—Perforated 1R and 2R
SMPTE 111-1996	Motion-Picture Film (35-mm)—Prints Made on Continuous Contact Printers—Exposed Areas for Picture and Audio
SMPTE 112-1999	Motion-Picture Film (16-mm)—100-Mil Magnetic Audio Record
SMPTE 117M-1996	Motion-Picture Film—Photographic Audio Record—Spectral Diffuse Density
SMPTE 119-1999	Motion-Picture Film (70-mm)—Perforated 65-mm, KS-1870
SMPTE 125M-1995	Television—Component Video Signal 4:2:2—Bit-Parallel Digital Interface
SMPTE 127-1994	Motion-Picture Film (16-mm)—Magnetic Photographic Audio Records—Magnetic Striping of Prints
SMPTE 137-2000	Motion-Picture Film (35-mm)—Release Prints—Four Magnetic Audio Records
SMPTE 139-1996	Motion-Picture Film (35-mm)—Perforated KS
SMPTE 143-1994	Motion-Picture Film (8-mm Type R)—Length of Film on Camera Spools—25-Ft Capacity
SMPTE 145-1999	Motion-Picture Film (65-mm)—Perforated KS
SMPTE 146M-1996	Motion-Picture Film—16- and 8-mm Reversal Color Camera Films—Determination of Speed
SMPTE 148-1991	Motion-Picture Film—35- and 16-mm Prints for Television Transmission—Film Image Area for Review Room Viewing
SMPTE 149-1999	Motion-Picture Film (8-mm Type S)—Perforated 1R
SMPTE 151-1998	Motion-Picture Film (8-mm Type S)—16-mm Film Perforated 8-mm Type S, (1-3)
SMPTE 152-1994	Motion-Picture Film (70-mm)—Projectable Image Area
SMPTE 153-1996	Motion-Picture Film (8-mm Type S)—16-mm Film Perforated 8-mm Type S (1-4)—Printed Areas
SMPTE 154-1998	Motion-Picture Film (8-mm Type S)—Projectable Image Area and Projector Usage

SMPTE Standards, Recommended Practices, and Engineering Guidelines by Number

SMPTE 157-1999	Motion-Picture Film (8-mm Type S)—Camera Aperture Image and Usage
SMPTE 159.1-1996	Motion-Picture Film (8-mm Type S)—Model 1 Camera Cartridge, Cartridge-Camera Interface, and Take-Up Core Drive
SMPTE 159.2-1996	Motion-Picture Film (8-mm Type S)—Model 1 Camera Cartridge Aperture, Camera Aperture Profile, Film Position, Pressure Pad, and Flatness
SMPTE 160M-1995	Motion-Picture Equipment (8-mm Type S)—Projection Reels—100- to 312-mm Diameter
SMPTE 161-1998	Motion-Picture Film (8-mm Type S)—Magnetic Striping
SMPTE 162-1998	Motion-Picture Film (8-mm Type S)—16-mm Film Perforated 8-mm Type S, (1-4)—Magnetic Striping
SMPTE 163-1998	Motion-Picture Film (8-mm Type S)—35-mm Film Perforated 8-mm Type S, 5R—Magnetic Striping
SMPTE 164-1993	Motion-Picture Film (8-mm Type S)—Magnetic Audio Record—Position, Dimensions, and Reproducing Speed
SMPTE 165-1999	Motion-Picture Film (35-mm)—Perforated 8-mm Type S, 5R (1-3-5-7-0)
SMPTE 166-1999	Motion-Picture Film (8-mm Type S)—Exposure Control and Stock Identification—Sound and Silent Camera Cartridge Notches
SMPTE 168-1996	Motion-Picture Film (16-mm)—Perforated 8-mm Type S, (1-4)
SMPTE 169-1997	Motion-Picture Film (35-mm)—Perforated 8-mm Type S, 2R-1664 (1-0)
SMPTE 170M-1999	Television—Composite Analog Video Signal—NTSC for Studio Applications
SMPTE 171-1996	Motion-Picture Film (35-mm)—Perforated 16-mm, 3R (1-3-0)
SMPTE 173-1999	Motion-Picture Equipment (8-mm Type R)—Double 8-mm Camera Spools—100-Ft Capacity
SMPTE 174-1994	Motion-Picture Equipment (16-mm)—Camera Spools—50- to 400-ft Capacity
SMPTE 176-1999	Motion-Picture Film (8-mm Type S)—16-mm Film Perforated 2R-1667 (1-3)—Magnetic Striping
SMPTE 177-1995	Motion-Picture Film (35-mm)—Four-Track Magnetic Audio Release Prints—Magnetic Striping
SMPTE 179-1996	Motion-Picture Film (8-mm Type S)—35-mm Film Perforated 2R (1-0) and 5R (1-3-5-7-0)—Printed Areas
SMPTE 181-1996	Motion-Picture Film (8-mm Type S)—16-mm Film Perforated 8-mm Type S (1-3)—Printed Areas
SMPTE 183M-1996	Motion-Picture Film—Photographic Audio Level Test Films—Measurement of Photoelectric Output Factor
SMPTE 184M-1998	Motion-Picture Film—Raw Stock Identification and Labeling
SMPTE 185-1993	Motion-Picture Film (70-mm)—Six Magnetic Records on Release Prints—Position, Dimensions, Reproducing Speed, and Identity
SMPTE 188M-1994	Motion-Picture Equipment (8-mm Type S)—Model II Camera Cartridges (15-m Capacity)—Camera-Run Film Length
SMPTE 189M-1994	Motion-Picture Equipment (8-mm Type S)—Model II Camera Cartridges—Loaded Film Location
SMPTE 190M-1994	Motion-Picture Equipment (8-mm Type S)—Model II Camera Cartridges—Cartridge-Camera Fit and Core
SMPTE 191M-1994	Motion-Picture Equipment (8-mm Type S)—Model II Camera Cartridges—Slots, Projections, and Cartridge Hole
SMPTE 192-1997	Motion-Picture Equipment (35-mm)—Shipping Reels for Prints

SMPTE Standards, Recommended Practices, and Engineering Guidelines by Number

SMPTE 194-1997	Motion-Picture Film (35-mm)—Projector Usage—Release Prints Having Four Perforations per Frame
SMPTE 195-2000	Motion-Picture Film (35-mm)—Motion-Picture Prints—Projectable Image Area
SMPTE 196M-1995	Motion -Picture Film—Indoor Theater and Review Room Projection—Screen Luminance and Viewing Conditions
SMPTE 197-1998	Motion-Picture Film (8-mm Type S)—50-Ft Model 1 Sound Camera Cartridge—Cartridge, Cartridge-Camera Interface, and Take-Up Core
SMPTE 198-1998	Motion-Picture Film (8-mm Type S)—50-Ft Model 1 Sound Camera Cartridge—Aperture, Pressure Pad, and Film Position
SMPTE 199-1998	Motion-Picture Film (8-mm Type S)—50-Ft Model 1 Sound Camera Cartridge—Pressure Pad Flatness and Camera Aperture Profile
SMPTE 200M-1998	Motion-Picture Equipment (8-mm Type S)—Model 1 Camera Cartridge—Camera Run Length, Perforation Cutout, and End-of-Run Notch
SMPTE 201M-1996	Motion-Picture Film (16-mm)—Type W Camera Aperture Image
SMPTE 202M-1998	Motion-Pictures—B-Chain Electroacoustic Response—Dubbing Theaters, Review Rooms, and Indoor Theaters
SMPTE 203-1998	Motion-Picture Film (35-mm)—Prints—Two-Track Photographic Audio Records
SMPTE 205-1993	Motion-Picture Equipment (8-mm Type S)—Model 1 Camera Cartridge—Interface and Take-Up Core Drive (200-ft Capacity) (R1998)
SMPTE 206-1998	Motion-Picture Equipment (8-mm Type S)—Model 1 Sound Camera Cartridge—Aperture, Profile, Film Position, Pressure Pad, and Flatness (200-ft Capacity)
SMPTE 207M-1997	Television—Digital Control Interface—Electrical and Mechanical Characteristics
SMPTE 208M-1992	Motion-Picture Film—35- and 16-mm Magnetic Audio Records—Recorded Characteristics (R1998)
SMPTE 209M-1996	Motion-Picture Film (8-mm Type S)—Magnetic Audio Records—Recorded Characteristic
SMPTE 210M-1999	Motion-Picture Film (16-mm)—Magnetic Audio Records—Two Records on 16-mm Magnetic Film
SMPTE 211M-1996	Motion-Picture Film—16- and 35-mm Variable-Area Photographic Audio Records—Signal-to-Noise Ratio
SMPTE 212M-1995	Motion-Picture Equipment (8-mm Type S)—Projection Reels—75-mm Diameter
SMPTE 214M-1999	Motion-Picture Film (35-mm)—Photographic Audio Reproduction Characteristic
SMPTE 215-1995	Motion-Picture Film (65-mm)—Camera Aperture Image
SMPTE 216-1998	Motion-Picture Film (35-mm)—Recorded Characteristic of Magnetic Audio Records—Four Track Striped Release Prints
SMPTE 217-1998	Motion-Picture Film (70-mm)—Recorded Characteristic of Magnetic Audio Records—Striped Release Prints
SMPTE 218M-1996	Motion-Picture Film (16-mm)—200-mil Center-Position Magnetic Audio Record
SMPTE 220-1996	Motion-Picture and Television Equipment—Camera Mounting Connections—1/4-inch-20 Thread and 3/8-inch-16 Thread Tripod Screws
SMPTE 221-1998	Motion-Picture Film (70-mm)—Six-Track Audio Release Prints—Magnetic Striping
SMPTE 222M-1994	Television—Control and Review Rooms—Monitor System Electroacoustic Response
SMPTE 223M-1996	Motion-Picture Film—Safety Film

SMPTE Standards, Recommended Practices, and Engineering Guidelines by Number

SMPTE 224M-1996	Television Digital Component Recording—19-mm Type D-1—Tape Record
SMPTE 225M-1996	Television Digital Component Recording—19-mm Type D-1—Magnetic Tape
SMPTE 226M-1996	Television Digital Recording—19-mm Tape Cassettes
SMPTE 227M-1996	Television Digital Component Recording—19-mm Type D-1—Helical Data and Control Records
SMPTE 228M-1996	Television Digital Component Recording—19-mm Type D-1—Time and Control Code and Cue Records
SMPTE 229M-1996	Television Analog Recording—1/2-in Type L—Records
SMPTE 230M-1996	Television Analog Recording—1/2-in Type L—Electrical Parameters, Control Code, and Tracking Control
SMPTE 231-1999	Motion-Picture Film (8-mm Type R)—Camera Aperture Image and Usage
SMPTE 233-1998	Motion-Picture Film (16-mm)—Projectable Image Area and Projector Usage
SMPTE 234-1998	Motion-Picture Film (8-mm Type R)—Projectable Image Area and Projector Usage
SMPTE 235-1998	Motion-Picture Equipment (16-mm)—Projection Reels—200- to 2300-ft Capacity
SMPTE 236-1998	Motion-Picture Equipment (8-mm Type R)—Projection Reels
SMPTE 237-1998	Motion-Picture Film (35-mm)—Perforated DH-1870
SMPTE 238M-1998	Television Analog Recording—1/2-in Type L—Tapes and Cassettes
SMPTE 239-1999	Motion-Picture Film (16-mm)—Perforated 8-mm Type R, 2R
SMPTE 240M-1999	Television—Signal Parameters—1125-Line High-Definition Production Systems
SMPTE 241-1995	Motion-Picture Equipment—35- and 70-mm Projection Reels
SMPTE 242-1993	Motion-Picture Equipment (35-mm)—Universal Intermittent Sprockets
SMPTE 243M-1993	Motion-Picture Equipment—35- and 70-mm Projection Lenses and Mounts (R1998)
SMPTE 244M-1995	Television—System M/NTSC Composite Video Signals—Bit-Parallel Digital Interface
SMPTE 245M-1993	Television Digital Recording—19-mm Type D-2 Composite Format—Tape Record
SMPTE 246M-1993	Television Digital Recording—19-mm Type D-2 Composite Format—Magnetic Tape
SMPTE 247M-1993	Television Digital Recording—19-mm Type D-2 Composite Format—Helical Data and Control Records
SMPTE 248M-1993	Television Digital Recording—19-mm Type D-2 Composite Format—Cue Record, and Time and Control Code Record
SMPTE 249M-1996	Television Analog Recording—1/2-in Type M-2—Records
SMPTE 250M-1996	Television Analog Recording—1/2-in Type M-2—Tapes and Cassettes
SMPTE 251M-1996	Television Analog Recording—1/2-in Type M-2—Electrical Parameters of Video, Audio, Time and Control Code, and Tracking Control
SMPTE 252M-1996	Television Analog Recording—1/2-in Type M-2--Pulse Code Modulation Audio
SMPTE 253M-1998	Television—Three-Channel RGB Analog Video Interface
SMPTE 254-1998	Motion-Picture Film (35-mm)—Manufacturer-Printed, Latent Image Identification Information
SMPTE 256M-1996	Television—Specifications for Video Tape Leader
SMPTE 257-1998	Motion-Picture Film (35-mm)—Stereoscopic Prints with Vertically Positioned Subframes—Projectable Image Areas
SMPTE 258M-1993	Television—Transfer of Edit Decision Lists

SMPTE Standards, Recommended Practices, and Engineering Guidelines by Number

SMPTE 259M-1997	Television—10-Bit 4:2:2 Component and $4f_{sc}$ Composite Digital Signals—Serial Digital Interface
SMPTE 260M-1999	Television—1125/60 High-Definition Production System—Digital Representation and Bit-Parallel Interface
SMPTE 261M-1993	Television—10-Bit Serial Digital Television Signals: 4:2:2 Component and $4f_{sc}$ NTSC Composite—AMI Transmission Interface
SMPTE 262M-1995	Television, Audio, and Film—Binary Groups of Time and Control Codes—Storage and Transmission of Data
SMPTE 263M-1996	Television Digital Recording—1/2-in Type D-3 Composite and 1/2-in Type D-5 Component Formats—Tape Cassette
SMPTE 264M-1998	Television Digital Recording—1/2-in Type D-3 Composite Format—525/60
SMPTE 265M-1993	Television Digital Recording—1/2-in Type D-3 Composite Format—625/50
SMPTE 266M-1994	Television—4:2:2 Digital Component Systems—Digital Vertical Interval Time Code
SMPTE 267M-1995	Television—Bit-Parallel Digital Interface—Component Video Signal 4:2:2 16 × 9 Aspect Ratio
SMPTE 268M-1994	File Format for Digital Moving-Picture Exchange (DPX)
SMPTE 269M-1999	Television—Fault Reporting in Television Systems
SMPTE 270-1994	Motion-Picture Film (65-mm)—Manufacturer-Printed Latent Image Identification Information
SMPTE 271-1994	Motion-Picture Film (16-mm)—Manufacturer-Printed Latent Image Identification Information
SMPTE 272M-1994	Television—Formatting AES/EBU Audio and Auxiliary Data into Digital Video Ancillary Data Space
SMPTE 273M-1995	Television—Status Monitoring and Diagnostics Protocol
SMPTE 274M-1998	Television—1920 × 1080 Scanning and Analog and Parallel Digital Interfaces for Multiple Picture Rates
SMPTE 275M-1995	Television and Audio Equipment—Eslan-1 Remote Control System
SMPTE 276M-1995	Television—Transmission of AES-EBU Digital Audio Signals Over Coaxial Cable
SMPTE 277M-1996	Television Digital Recording—19-mm Type D-6—Helical Data, Longitudinal Index, Cue, and Control Records
SMPTE 278M-1996	Television Digital Recording—19-mm Type D-6—Content of Helical Data, and Time and Control Code Records
SMPTE 279M-1996	Digital Video Recording—1/2-in Type D-5 Component Format—525/60 and 625/50
SMPTE 291M-1998	Television—Ancillary Data Packet and Space Formatting
SMPTE 292M-1998	Television—Bit-Serial Digital Interface for High-Definition Television Systems
SMPTE 293M-1996	Television—720 × 483 Active Line at 59.94-Hz Progressive Scan Production—Digital Representation
SMPTE 294M-1997	Television—720 × 483 Active Line at 59.94-Hz Progressive Scan Production—Bit-Serial Interfaces
SMPTE 295M-1997	Television—1920 × 1080 50-Hz—Scanning and Interface
SMPTE 296M-2001	Television—1280 × 720 Scanning, Analog and Digital Representation, and Analog Interface
SMPTE 297M-2000	Television—Serial Digital Fiber Transmission System for ANSI/SMPTE 259M Signals
SMPTE 298M-1997	Television—Universal Labels for Unique Identification of Digital Data
SMPTE 299M-1997	Television—24-Bit Digital Audio Format for HDTV Bit-Serial Interface

SMPTE Standards, Recommended Practices, and Engineering Guidelines by Number

SMPTE 300-1997	Motion-Picture Color Print Film (35-mm)—Manufacturer-Printed Latent Image Identification Information
SMPTE 301M-1999	Motion-Picture Film—Theater Projection Leader, Trailer, and Cue Marks
SMPTE 302-1998	Television—Mapping of AES3 Data into MPEG-2 Transport Stream
SMPTE 303M	Television—Color Reference Pattern
SMPTE 304M-1998	Television—Broadcast Cameras—Hybrid Electrical and Fiber-Optic Connector
SMPTE 305M-2-2000	Television—Serial Data Transport Interface
SMPTE 306M-1998	Television Digital Recording—6.35-mm Type D-7 Component Format—Video Compression at 25 Mb/s—525/60 and 625/50
SMPTE 307M-1998	Television Digital Recording—6.35-mm Type D-7 Component Format—Tape Cassette
SMPTE 308M-1998	Television—MPEG-2 4:2:2 Profile at High Level
SMPTE 309M-1999	Television—Transmission of Date and Time Zone Information in Binary Groups of Time and Control Code
SMPTE 310M-1998	Television—Synchronous Serial Interface for MPEG-2 Digital Transport Streams
SMPTE 311M-1998	Television—Hybrid Electrical and Fiber Optical Camera Cable
SMPTE 312M-1999	Television—Splice Points for MPEG-2 Transport Streams
SMPTE 313-1999	Motion-Picture Film (65-mm)—Manufacturer-Printed Latent Image Identification Information—120 Perforation Repeat
SMPTE 314M-1999	Television—Data Structure for DV-Based Audio, Data, and Compressed Video—25 and 50 Mbit/s
SMPTE 315M-1999	Television—Camera Positioning Information Conveyed by Ancillary Data Packets
SMPTE 316M-1999	Television Digital Recording—12.65-mm Type D-9 Component Format—Video Compression—525/60 and 625/50
SMPTE 317M-1999	Television Digital Recording—12 .65-mm Type D-9 Component Format—Tape Cassette
SMPTE 318M-1999	Television and Audio—Reference Signals for the Synchronization of 59.94- or 50-Hz Related Video and Audio Systems in Analog and Digital Areas
SMPTE 319M-2000	Television—Transporting MPEG-2 Recoding Information through 4:2:2 Component Digital Interfaces
SMPTE 320M-1999	Television—Channel Assignments and Levels on Multichannel Audio Media
SMPTE 321M-1999	Television—Data Stream Format for the Exchange of DV-Based Audio, Data, and Compressed Video Over a Serial Data Transport Interface
SMPTE 322M-1999	Television—Format for Transmission of DV Compressed Video, Audio, and Data Over a Serial Data Transport Interface
SMPTE 323M-1999	Motion-Picture Film—Channel Assignments and Levels on Multichannel Audio Media
SMPTE 324M-2000	Television—12-Channel Serial Interface for Digital Audio and Auxiliary Data
SMPTE 325M-1999	Digital Television—Opportunistic Data Broadcast Flow Control
SMPTE 326M-2000	Television—SDTI Content Package Format (SDTI-CP)
SMPTE 327M-2000	Television—MPEG-2 Video Recoding Data Set
SMPTE 328M-2000	Television—MPEG-2 Video Elementary Stream Editing Information
SMPTE 329M-2000	Television—MPEG-2 Video Recoding Data Set—Compressed Stream Format

SMPTE Standards, Recommended Practices, and Engineering Guidelines by Number

SMPTE 330M-2000	Television—Unique Material Identifier (UMID)
SMPTE 331M-2000	Television—Element and Metadata Definitions for the SDTI-CP
SMPTE 332M-2000	Television—Encapsulation of Data Packet Streams over SDTI (SDTI-PF)
SMPTE 333M-1999	Television—DTV Closed-Caption Server to Encoder Interface
SMPTE 334M-2000	Television—Vertical Ancillary Data Mapping for Bit-Serial Interface
SMPTE 337M-2000	Television—Format for Non-PCM Audio and Data in an AES3 Serial Digital Audio Interface
SMPTE 338M-2000	Television—Format for Non-PCM Audio and Data in an AES3 Serial Digital Audio Interface—Data Types
SMPTE 339M-2000	Television—Format for Non-PCM Audio and Data in an AES3 Serial Digital Audio Interface—Generic Data Types
SMPTE 340M-2000	Television—Format for Non-PCM Audio and Data in an AES3 Serial Digital Audio Interface—ATSC A/52 (AC-3) Data Type
SMPTE 341M-2000	Television—Format for Non-PCM Audio and Data in an AES3 Serial Digital Audio Interface—Captioning Data Type
SMPTE 342M-2000	Television—HD-D5 Compressed Video 1080i and 720p Systems—Encoding Process and Data Format
SMPTE 344M-2000	Television—540 Mbits/s Serial Digital Interface
SMPTE 346M-2000	Television—Time Division Multiplexing Video Signals and Generic Data over High-Definition Interfaces
SMPTE 348M-2000	Television—High Data-Rate Serial Data Transport Interface (HD-SDTI)
SMPTE 351M-2000	Television—Transporting MPEG-2 Recoding Information through High-Definition Digital Interface
SMPTE 353M-2000	Transport of MPEG-2 Recoding Information as Ancillary Data Packets
SMPTE Recommended Practices	
RP 6-1994	Recorded Carrier Frequencies and Preemphasis Characteristics for 2-in Quadruplex Video Magnetic Tape Recording for 525-Line/60-Field Television Systems (R1999)
RP 9-1995	Dimensions of Double-Frame 35-mm 2 × 2 Slides for Precise Applications in Television
RP 11-1994	Tape Vacuum Guide Configuration and Position for Quadruplex Video Magnetic Tape Recording (R1999)
RP 12-1997	Screen Luminance for Drive-In Theaters
RP 14-1997	Plotting Data from Sensitometric Strips Exposed on Type Ib (Intensity Scale) Sensitometers
RP 15-1997	Calibration of Densitometers Used for Black-and-White Photographic Density Measurement
RP 16-1993	Specifications of Tracking Control Record for 2-in Quadruplex Video Magnetic Tape Recordings (R1997)
RP 17-1964	Photographic Recording Technique for Measuring High-Speed Camera Image Unsteadiness (R1997)
RP 18-1995	Specifications for Test Film for Subjective Checking of 16-mm Motion-Picture Audio Projectors
RP 19-1995	Specifications for 8-mm Type R Registration Test Film
RP 20-1995	Specifications for 16-mm Registration Test Film
RP 21-1997	Dimensions of 35- and 70-mm Motion-Picture Rewind Spindles
RP 24-1997	Dimensions for 16-mm Motion-Picture Camera Spindles

SMPTE Standards, Recommended Practices, and Engineering Guidelines by Number

RP 25-1995	Audio and Picture Synchronization on Motion-Picture Film Relative to the Universal Leader for Magnetic and Photographic Records
RP 27.1-1989	Specifications for Operational Alignment Test Pattern for Television
RP 27.2-1989	Specifications for Operational Registration Test Pattern for Multiple-Channel Television Cameras
RP 27.3-1989	Specifications for Safe Action and Safe Title Areas Test Pattern for Television Systems
RP 27.4-1994	Specifications for an Operational Test Pattern for Checking Jitter, Weave and Travel Ghost in Television Projectors
RP 27.5-1989	Specifications for Mid-Frequency Response Test Pattern for Television
RP 32-1995	Specifications for 8-mm Type S Test Film for Projectors and Printers
RP 34-1998	Dimensions for 16-mm Motion-Picture Projector Reel Spindles
RP 36-1999	Positioning the Headwheel and Adjacent Tape Guides for 2-in Quadruplex Video Magnetic Tape Recorders
RP 38.1-1989	Specifications for Deflection Linearity Test Pattern for Television
RP 39-1993	Specifications for Maintaining an Emulsion-In Orientation on Theatrical Release Prints (R1997)
RP 40-1995	Specifications for 35-mm Projector Alignment and Screen Image Quality Test Film
RP 45-1972	Use and Care of Sound Test Films (R1987)
RP 47-1999	Electronic Method of Dropout Detection and Counting
RP 48-1999	Lubrication of 16- and 8-mm Motion-Picture Prints
RP 49-1995	Leaders for 8-mm Type R and S Motion-Picture Release Prints Used in Continuous-Loop Cartridges
RP 50-1995	Dimensions for 8-mm Type S Motion-Picture Projector Reel Spindles
RP 51-1995	Screen Luminance and Viewing Conditions for 8-mm Review Rooms
RP 53-1993	Scene-Change Methods for Printing 35-mm, 16-mm and 8-mm Type S Motion-Picture Film (R1997)
RP 54-1999	Edge Numbering on 16-mm Release Prints
RP 55-1997	8-mm Type S Sprocket Design
RP 56-1995	Safe Action and Safe Title Areas for 8-mm Release Prints
RP 58-1995	Nomenclature for Devices Enclosing 8-mm Motion-Picture Film for Projection
RP 59-1999	Color and Luminance of Review Room Screens for Viewing Motion-Picture Materials Intended for Slides or Film Strips
RP 63-1997	Specifications for Sound-Focusing Test Film for 16-mm Audio Reproducers, Photographic Type
RP 64-1999	Specifications for Audio-Focusing Test Film for 35-mm Audio Reproducers, Photographic Type
RP 65-2000	Motion-Picture Enlargement/Reduction Ratios
RP 67-1997	Specifications for Buzz-Track Test Film for 16-mm Motion-Picture Audio Reproducers, Photographic Type
RP 68-1997	Specifications for Buzz-Track Test Film for 35-mm Motion-Picture Photographic Audio Reproducers
RP 69-1997	Specifications for Scanning-Beam Uniformity Test Film for 35-mm Motion-Picture Audio Reproducers
RP 70-1997	Specifications for Flutter Test Film for 16-mm Audio Reproducers, Photographic Type

SMPTE Standards, Recommended Practices, and Engineering Guidelines by Number

RP 73-1992	8-mm Type R (Regular 8) Sprocket Design (R1997)
RP 74-1992	16-mm Sprocket Design (R1997)
RP 75-1997	Specifications for Flutter Test Film for 35-mm Studio Audio Reproducers, Magnetic Type
RP 76-1997	Specifications for Flutter Test Film for 16-mm Audio Reproducers, Magnetic Type
RP 77-1994	Specifications for Azimuth Test Film for 35-mm Studio Audio Reproducers, Magnetic Type
RP 78-1997	Specifications for Azimuth Test Film for 16-mm Audio Projectors, Magnetic Type
RP 79-1999	Specifications for Flutter Test Film for 35-mm Four-Track Striped Release Print Audio Reproducers, Magnetic Type
RP 81-1999	Specifications for Scanning-Beam Uniformity Test Film for 16-mm Motion-Picture Photographic Audio Reproducers
RP 82-1995	Specifications for 16-mm Projector Alignment and Screen Image Quality Test Film
RP 83-1996	Specifications of Tracking-Control Record for 1-in Type B Helical-Scan Television Analog Recording
RP 84-1996	Reference Carrier Frequencies and Preemphasis Characteristics for 1-in Type B Helical-Scan Television Analog Recording
RP 85-1999	Tracking-Control Record for 1-in Type C Helical-Scan Television Tape Recording
RP 86-1991	Video Record Parameters for 1-in Type C Helical-Scan Television Tape Recording (R1995)
RP 87-1999	Reference Carrier Frequencies, Preemphasis Characteristic and Audio and Control Signals for 3/4-in Type E Helical-Scan Video Tape Cassette Recording
RP 90-1999	Specifications for Type U Audio Level and Multifrequency Test Film for 16-mm Audio Reproducers, Magnetic Type
RP 91-1997	Specifications for 70-mm Projector Alignment and Screen Image Quality Test Film
RP 92-1995	Specifications for Audio Level and Multifrequency Test Films for 8-mm Type S Audio Reproducers, Magnetic Type
RP 93-1999	Requirements for Recording American National Standard Time and Control Code for 1-in Type B Helical-Scan Video Tape Recorders
RP 94-2000	Gain Determination of Front Projection Screens
RP 95-1994	Installation of Gain Screens
RP 96-1993	Specifications for Subjective Reference Tapes for Helical-Scan Video Tape Reproducers for Checking Receiver/Monitor Setup
RP 97-1997	Specifications for Flutter Test Film for 35-mm Audio Reproducers, Photographic Type
RP 98-1995	Measurement of Screen Luminance in Theaters
RP 103-1995	Care, Storage, Operation, Handling and Shipping of Video Tape for Television
RP 104-1994	Cross-Modulation Tests for Variable-Area Photographic Audio Tracks
RP 105-1995	Method for Determining the Degree of Jump and Weave in 70-, 35,- and 16-mm Motion-Picture Projected Images
RP 106-1994	Film Tension in 35-mm Motion-Picture Systems Operating Under 0.9 m/s (180 ft/min)
RP 107-1995	Video and Audio Reference Tape for 1-in Type B Helical-Scan Format
RP 109-1994	Spectral Response of Photographic Audio Reproducers for 8-mm Type S Motion-Picture Film (R1999)

SMPTE Standards, Recommended Practices, and Engineering Guidelines by Number

RP 110-1992	Specifications for an Alignment Test Film for Anamorphic Attachments to 35-mm Motion-Picture Projectors
RP 111-1999	Dimensions for 70-, 65- and 35-mm Motion-Picture Film Splices
RP 113-1996	Supervisory Protocol for Digital Control Interface
RP 114-1994	Dimensions of Photographic Control and Data Record on 16-mm Motion-Picture Film (R1999)
RP 115-1997	Dimensions of Photographic Control and Data Record on 35-mm Motion-Picture Release Prints
RP 116-2000	Dimensions of Photographic Control and Data Record on 35-mm Motion-Picture Camera Negatives
RP 117-1994	Dimensions of Magnetic Control and Data Record on 8-mm Type S Motion-Picture Film
RP 120-1994	Measurement of Intermodulation Distortion in Motion-Picture Audio Systems
RP 121-1997	Tape Dropout Specifications for 1-in Types B and C Video Tape Recorders/Reproducers
RP 122-1993	Dimensions of Cemented Splices on 8-mm Type S Motion-Picture Film, Projection Type (R1997)
RP 123-1997	Dimensions of Tape Splices on 8-mm Type S Motion-Picture Film, Projection Type
RP 124-1998	Insertion Pivot for Studio Lighting Units and Mating Holders for Use with Standing and Hanging Support Systems
RP 127-1999	Specifications for Type U Audio Level and Multifrequency Test Film for 35-mm Studio Audio Reproducers, Magnetic Full-Coat Type
RP 128-1997	Specifications for Audio Level and Multifrequency Test Film for 70-mm Striped Six-Track Release Print Audio Reproducers, Magnetic Type
RP 129-1995	Requirements for 35-, 16- and 8-mm Type S Tape Splices on Magnetic Audio Recording Motion-Picture Film
RP 130-1995	Dimensions of Tape Splices on 16-mm and 8-mm Type R Motion-Picture Film, Projection Type
RP 131-1994	Storage of Motion-Picture Films
RP 132-1994	Storage of Edit Decision Lists on 8-in Flexible Diskette Media
RP 133-1999	Specifications for Medical Diagnostic Imaging Test Pattern for Television Monitors and Hard-Copy Recording Cameras (R1995)
RP 134-1994	Polarity for Analog Audio Magnetic Recording and Reproduction (R1999)
RP 135-1999	Use of Binary User Groups in Motion-Picture Time and Control Codes
RP 136-1999	Time and Control Codes for 24, 25, or 30 Frame-Per-Second Motion-Picture Systems
RP 138-1996	Control Message Architecture for Digital Control Interface
RP 139-1997	Tributary Interconnection for Digital Control Interface
RP 140-1995	Position of Photographic Audio Record for Routine Test Signals
RP 141-1995	Background Acoustic Noise Levels in Theaters and Review Rooms
RP 142-1997	Stereo Audio Track Allocations and Identification of Noise Reduction for Video Tape Recording
RP 143-1999	Specifications for Type U Audio Level and Multifrequency Test Film for 35-mm Striped Four-Track Release Print Audio Reproducers, Magnetic Type
RP 144-1999	Basic System and Transport Geometry Parameters for 1/2-in Type L Format
RP 145-1999	SMPTE C Color Monitor Colorimetry
RP 148-1987	Relative Polarity of Stereo Audio Signals (R1997)

SMPTE Standards, Recommended Practices, and Engineering Guidelines by Number

RP 149-1992	Dimensions of Transverse Cemented Splices on 16-mm and 8-mm Type R Motion-Picture Film (R1997)
RP 150-2000	Channel Assignments and Test Leader for Magnetic Film Masters Intended for Transfer to Video Media Having Stereo Audio
RP 151-1999	Lubrication of 35-mm Motion-Picture Prints for Projection
RP 152-1994	Edge Identification of Leader and Picture for 35-mm Release Prints (R1999)
RP 153-1999	Method for Measuring 35- and 70-mm Shutter Efficiency
RP 155-1997	Audio Levels for Digital Audio Records on Digital Television Tape Recorders
RP 156-1999	Bar Code Labeling for Type D-1 Component and Type D-2 Composite Cassette Identification
RP 157-1995	Key Signals
RP 158-1999	Basic System and Transport Geometry Parameters for 1/2-in Type M-2 Format
RP 159-1995	Vertical Interval Time Code and Longitudinal Time Code Relationship
RP 160-1997	Three-Channel Parallel Analog Component High-Definition Video Interface
RP 161-1999	Logic Design for Decoding Digital Audio Control Words in D-1 Helical Data and Control Records
RP 162-1993	Storage of Edit Decision Lists on 3-1/2 in Disks
RP 163-1992	Television—System Service Messages
RP 164-1996	Location of Vertical Interval Time Code
RP 165-1994	Error Detection Checkwords and Status Flags for Use in Bit-Serial Digital Interfaces for Television
RP 166-1995	Critical Viewing Conditions for Evaluation of Color Television Pictures
RP 167-1995	Alignment of NTSC Color Picture Monitors
RP 168-1993	Definition of Vertical Interval Switching Point for Synchronous Video Switching
RP 169-1995	Television, Audio and Film Time and Control Code—Auxiliary Time Address Data in Binary Groups—Dialect Specification of Directory Index Locations
RP 170-1993	Video Tape Recorder Type-Specific Messages for Digital Control Interface
RP 171-1993	Type-Specific Messages for Digital Control Interface of Analog Audio Tape Recorders
RP 172-1993	Common Messages for Digital Control Interface
RP 174-1993	Bit-Parallel Digital Interface for 4:4:4:4 Component Video Signal (Single Link)
RP 175-1997	Digital Interface for 4:4:4:4 Component Video Signals (Dual Link)
RP 176-1997	Derivation of Reference Signals for Television Camera Color Evaluation
RP 177-1993	Derivation of Basic Television Color Equations (R1997)
RP 178-1996	Serial Digital Interface Checkfield for 10-Bit 4:2:2 Component and $4f_{sc}$ Composite Digital Signals
RP 179-1994	Dialect Specification of Page-Line Directory Index for Television, Audio, and Film Time and Control Code for Video-Assisted Film Editing
RP 180-1999	Spectral Conditions Defining Printing Density in Motion-Picture Negative and Intermediate Films
RP 181-1999	Audio Sector Time Code and Equipment-Type Information for 19-mm Type D-1 Digital Component Recording
RP 182-1995	List of Virtual Machine Numbers for ESbus and ESlan Systems
RP 183-1995	Monitoring and Diagnostics Processors

SMPTE Standards, Recommended Practices, and Engineering Guidelines by Number

RP 184-1996	Specification of Jitter in Bit-Serial Digital Systems
RP 185-1995	Classification of Projection Depth of Focus
RP 186-1995	Video Index Information Coding for 525- and 625-Line Television Systems
RP 187-1995	Center, Aspect Ratio and Blanking of Video Images
RP 188-1999	Transmission of Time Code and Control Code in the Ancillary Data Space of a Digital Television Data Stream
RP 189-1996	Organization of DPX Files on TAR Tapes
RP 190-1996	Care and Preservation of Audio Magnetic Recordings
RP 191-1996	Routing Switcher Type-Specific Messages for Remote Control of Broadcast Equipment
RP 192-1996	Jitter Measurement Procedures in Bit-Serial Digital Interfaces
RP 193	Test Patterns and Test Images for DPX Leader
RP 194-1998	Film Negative Cutter's Conform List
RP 195-1998	Use of the Reference Mark in Manufacturer-Printed Latent Image Key Numbers for Unambiguous Film Frame Identification
RP 196-1997	Transmission of LTC and VITC Data as HANC Packets in Serial Digital Television Interfaces
RP 197-1998	Film to Video Transfer List
RP 198-1998	Bit-Serial Digital Checkfield for Use in High-Definition Interfaces
RP 199-1999	Mapping of Pictures in Wide-Screen (16:9) Scanning Structure to Retain Original Aspect Ratio of the Work
RP 200-1999	Relative and Absolute Sound Pressure Levels for Motion-Picture Multichannel Sound Systems
RP 201-1999	Encoding Film Transfer Information Using Vertical Interval Time Code
RP 202	Video Alignment for MPEG-2 Coding
RP 203-2000	Real Time Opportunistic Data Flow Control in an MPEG-2 Transport Emission Multiplex
RP 204-2000	SDTI-CP MPEG Decoder Templates
RP 205-2000	Application of Unique Material Identifiers in Production and Broadcast Environments
RP 206-1999	Opportunistic Data Flow Control Using Ethernet as a Control Channel in an MPEG-2 Transport Emissions Multiplex
RP 209-2000	Format for Transmission of HD-D5 Compressed Video and Audio Data over 360 Mb/s Serial Digital Interface
RP 211-2000	Implementation of 24P, 25P and 30P Segmented Frames for 1920 x 1080 Production Format
SMPTE Engineering Guidelines	
EG 1-1990	Alignment Color Bar Test Signal for Television Picture Monitors
EG 2-1999	Edge Identification of Motion-Picture Raw Stock Containers
EG 3-1994	Projection for Technical Conferences
EG 5-1994	Projected Image Quality of 70-, 35-, and 16-mm Motion-Picture Projection Systems
EG 7-1994	Audio Sync Pulse for 8-mm Type S Cameras, Magnetic Audio Recorders, and Rerecording Projectors (R1999)
EG 8-1993	Specifications for Motion-Picture Camera Equipment Used in Space Environment (R1997)

SMPTE Standards, Recommended Practices, and Engineering Guidelines by Number

EG 9-1995	Audio Recording Reference Level for Post-Production of Motion-Picture Related Materials
EG 10-1996	Tape Transport Geometry Parameters for 19-mm Type D-1 Television Digital Component Recording
EG 12-1994	Control of Basic Parameters in the Manufacture of SMPTE Photographic and Magnetic Audio Test Films (R1999)
EG 13-1986	Use of Audio Magnetic Test Films (R1997)
EG 14-1999	Acoustical Background Noise Levels in Dubbing Stages
EG 15-1987	Recording Level for Dialog in Motion-Picture Production
EG 16-1997	Measurement Methods for Motion-Picture Camera Acoustical Noise—Field Method
EG 17-1997	B-Chain Electroacoustical Response for Preparing Magnetic Masters for Transfer to 16-mm or 35/32-mm Monaural Photographic Film
EG 18-1994	Design of Effective Cine Theaters
EG 20-1997	Tape Transport Geometry Parameters for 19-mm Type D-2 Composite Format for Television Digital Recording
EG 21-1997	Nomenclature for Television Digital Recording of 19-mm Type D-1 Component and Type D-2 Composite Formats
EG 22-1997	Description and Index of Documents for 19-mm Type D-2 Composite Television Digital Recording
EG 23-1996	Transfer of Two-Channel Stereo Audio from Audio Magnetic Film or Tape to Video Tape
EG 24-1995	Video and Audio Alignment Tapes and Procedures for 1-in Type C Helical-Scan Television Analog Recorders
EG 25-1996	Telecine Scanning for Film Transfer to Television
EG 26-1995	Audio Channel Assignments for Digital Television Tape Recorders with AES/EBU Digital Audio Input
EG 27-1994	Supplemental Information for ANSI/SMPTE 170M and Background on the Development of NTSC Color Standards (R1999)
EG 28-1993	Annotated Glossary of Essential Terms for Electronic Production
EG 29-1993	Remote Control of Television Equipment
EG 30-1995	Implementation of ESlan Standards
EG 31-1995	Considerations for Cassette Bar Code Readers
EG 32-1996	Emphasis of AES/EBU Audio in Television Systems and Preferred Audio Sampling Rate
EG 33-1998	Jitter Characteristics and Measurements
EG 34-1999	Pathological Conditions in Serial Digital Video Systems
EG 35-1999	Time and Control Code Time Address Clock Precision for Television, Audio and Film
EG 36	Transformations Between Television Component Color Signals

5.3 Scopes of SMPTE Standards

SMPTE 1-1996: for Video Recording—2-in Magnetic Recording Tape

This standard specifies the dimensions for the width, thickness, and curvature of 2-in video magnetic recording tape.

SMPTE 3-1998: Frequency Response and Operating Level of Recorders and Reproducers, Audio 1 Record on 2-in Tape Operating at 15 and 7.5 in/s

This standard specifies the frequency response and operating level for recorders and reproducers for audio 1 record for 2-in quadruplex video magnetic tape recording at 15 in/s and 7.5 in/s (381 mm/s and 190.5 mm/s), as defined in ANSI/SMPTE 6-1988. It also specifies the field method of calibration of recorders and reproducers, utilizing the test tapes as defined in ANSI/SMPTE 8-1989 and ANSI/SMPTE 11-1989.

SMPTE 4-1995: Two Inch Magnetic Tape for Quadruplex Recording—Speed

This standard specifies the nominal rates of travel of 2-in wide magnetic tape for quadruplex video magnetic tape recording.

SMPTE 5-1995: Two Inch Reels

This standard specifies the dimensions of reels in maximum capacities of 750, 1650, 3600, 5540, and 7230 ft designed to accommodate the maximum thickness of 2-in wide magnetic tape for television recording, as specified in ANSI/SMPTE 1.

SMPTE 6-1998: Two Inch Quadruplex Tape—Video, Audio and Tracking-Control Records

This standard specifies both the locations for the edges of the video, audio, and tracking-control records, and the mechanical separation of the simultaneously recorded information of the video and audio records, as recorded at 15 and 7.5 in/s on 2-in quadruplex video magnetic tape.

SMPTE 7-1999: Camera Aperture Image and Usage

This standard specifies the dimensions of the camera aperture image and its relative position to the reference edge and the perforations of 16-mm motion-picture film. The location of the perforations is based on dimensions given in ANSI/SMPTE 109. This standard also specifies the position of the emulsion and the frame rate for 16-mm motion-picture film perforated one or two edges.

SMPTE 8-1995: Quadruplex Recorders Operating at 15 in/s—Audio Level and Multifrequency Test Tape

This standard specifies an audio frequency test tape to be used for adjusting the sensitivity and frequency response of audio 1 record (program audio track) and audio 2 record (cue track) of quadruplex video magnetic tape recorders operating at a tape speed of 15 in/s (381 mm/s). The tape shall be used on recorders operating in accordance with ANSI/SMPTE 3. The operating level and frequency response for audio 2 record are specified in SMPTE RP 102.

SMPTE 11-1995: Quadruplex Recorders Operating at 7.5 in/s—Audio Level and Multifrequency Test Tape

This standard specifies an audio frequency test tape to be used for adjusting the sensitivity and frequency response of audio 1 record (program audio track) and audio 2 record (cue track) of quadruplex video magnetic tape recorders operating at a tape speed of 7.5 in/s (190.5 mm/s). The tape shall be used on recorders operating in accordance with ANSI/SMPTE 3. The operating level and frequency response for audio 2 record are specified in SMPTE RP 102.

SMPTE 12M-1999: Time and Control Code

The standard specifies a digital time and control code for use in television, film, and accompanying audio systems operating at 30, 25, and 24 frames per second. Clauses 4, 5, and 6 specify the manner in which time is represented in frame-based systems. Clause 7 describes the structure of the time address and control bits of the code, and sets guidelines for storage of user data in the code. Clause 8 specifies the modulation method and interface characteristics of a linear time code (LTC) source. Clause 9 specifies the modulation method for inserting the code into the vertical interval of a television signal. Clause 10 summarizes the relationship between the two forms of time and control code.

SMPTE 15M-1998: One Inch Type B Helical Scan—Basic System Parameters

This standard specifies the basic system parameters; i.e., the positions of recording head gaps, the scanning configuration, the axis of rotation of the video head wheel, and the appropriate tape tension for 1-in type B helical-scan television tape recorders for 525/60 monochrome or NTSC color systems.

SMPTE 16M-1998: One Inch Type B Helical Scan—Records

This standard specifies the dimensions and location of the video, audio, and tracking-control records and the longitudinal separation of the simultaneously recorded information of the video and audio records, as recorded on 1-in type B helical-scan television tape recordings.

SMPTE 17M-1998: One Inch Type B Helical Scan—Frequency Response and Operating Level

This standard specifies the frequency response and operating level of recorders and reproducers for audio records for 1-in type B helical-scan video tape recording.

SMPTE 18M-1996: One Inch Type C—Basic System and Transport Geometry Parameters

This standard specifies the general video record system, video pole tip locations, scanner parameters, scanner-guide locations, tape tension, and test conditions for 1-in type C helical-scan television analog recorders operating on the 525/60 monochrome or NTSC color systems.

SMPTE 19M-1996: One Inch Type C—Records

This standard specifies the dimensions and location of recorded video, audio, and tracking-control records for 1-in type C helical-scan television analog recorders operating in the 525/60 monochrome or NTSC color systems.

SMPTE 20M-1996: One Inch Type C Recorders and Reproducers—Longitudinal Audio Characteristics

This standard specifies the frequency response and reference level of recorders and reproducers for audio and longitudinal time and control code records for 1-in type C helical-scan television analog recording.

SMPTE 21M-1997: 3/4-in Type E Helical Scan—Records

This standard specifies the location of the edges of the video, audio, and tracking-control records and the mechanical separation of the simultaneously recorded information of the video and audio records, as recorded on a 3/4-in Type E helical-scan video tape recording cassette system, operating at a tape speed of 95.3 mm/s (3.752 in/s).

SMPTE 22M-1997: 3/4-in Type E Helical Scan—Cassette

This standard specifies the dimensions of a video cassette for use with a 3/4-in Type E helical-scan video tape recording cassette system, operating at a tape speed of 95.3 mm/s (3.752 in/s).

SMPTE 24M-1996: One Inch Reels

This standard specifies the configuration and dimensions for reels intended for 1-in magnetic tape for television recording on helical-scan video recorders, as specified in ANSI/SMPTE 25M-1989.

SMPTE 25M-1995: One Inch Magnetic Recording Tape

This standard specifies the dimensions of 1-in video magnetic recording tape, and the direction of magnetic orientation.

SMPTE 26M-1995: One Inch Helical-Scan Recorders—Raw Stock for Reference Tapes

This standard specifies dimensions for the width, thickness, and width deviation of raw tape stock used to record reference tapes for 1-in helical-scan video tape recorders. Conditions before testing and testing environment are also specified.

SMPTE 29M-1995: One Inch Type B Reference Recorders—Basic System and Transport Geometry

This standard specifies test conditions, general video record system, video pole-tip locations, scanner parameters, scanner-guide locations, and the tape tension for 1-in type B helical-scan video tape reference recorders operating on the 525/60 monochrome or NTSC color systems.

SMPTE 30M-1995: One Inch Type B Reference Recorders—Records on Reference Tapes

This standard specifies the dimensions and location of video, audio, and tracking- control records on reference tapes for 1 in type B helical-scan video tape recorders, operating on the 525/60 monochrome or NTSC color system, as described in ANSI/SMPTE 29M.

SMPTE 31M-1995: 3/4-in Type E—Small Video Cassette

This standard specifies the dimensions of a video cassette for use with a 3/4-in type E helical-scan video tape recording cassette system, operating at a tape speed of 95.3 mm/s (3.75 in/s).

SMPTE 32M-1998: 1/2-in Type H—Cassette, Tape, and Records

This standard specifies the characteristics and parameters for type H, helical-scan, 1/2-in video tape recorders operating with video signals as defined by CCIR Recommendation 624-4, and having a typical scanning structure of 525 lines, 59.94 fields per second, and 2:1 interlace.

This standard also specifies the double record time mode (long play, LP) and the triple record time mode (extended play, EP), and optional frequency modulation (FM) audio recording. This standard further specifies the dimensions of the video tape and compact video cassette for use in a type H, helical-scan, 1/2-in video tape recording cassette system with the aid of a cassette adapter. Finally, this standard specifies the characteristics and parameters for high-performance type H, helical-scan, 1/2-in video tape recorders operating with video signals as defined by CCIR Recommendation 624-4, and having a typical scanning structure of 525 lines, 59.94 fields per second, and 2:1 interlace.

SMPTE 35M-1997: 1/2-in Type G—Cassette and Tape

This standard specifies the dimensions of a video cassette and a video magnetic tape intended for use with 1/2-in type G video systems operating at tape speeds of 40, 20, and/or 13.3 mm/s (1.57, 0.79, and/or 0.52 in/s).

SMPTE 37M-1994: Raw Stock Cores

This standard specifies the recommended sizes and dimensions of raw stock cores for 8-, 16-, 35-, 65-, and 70-mm motion-picture films.

SMPTE 40-1997: Photographic Audio Records—Release Prints

This standard specifies the position, dimensions, and reproducing speed of variable-area and variable-density photographic audio records on 35-mm motion-picture release prints. This standard also specifies the longitudinal picture-audio displacement.

SMPTE 41-1999: Prints—Photographic Audio Records

This standard specifies the lateral location, dimensions, and the reproducing speed of variable-area and variable-density photographic audio records on 16-mm motion-picture prints. This standard also specifies the longitudinal picture-audio displacement. The standard further specifies the width scanned in the audio reproducer.

SMPTE 48-1995: Picture and Sound Contact Printing—Printed Areas

This standard specifies the location and size of the printed picture and photographic sound track areas for both negative/positive and reversal contact printing operations. An opaque line should appear between picture and sound in the finished print. The dimensions given for picture-printed area and for the two widths of sound-printed area provide for an overlap of picture- and sound-printed areas, or for a gap between them. The specific aperture to be used for printing sound will be chosen to provide the desired black line, as dictated by the printing materials and processing conditions in use.

SMPTE 55-2000: 35- and 16-mm Audio Release Prints—Leaders and Cue Marks

This standard specifies the make-up or assembly of leaders and cue marks for 35- and 16-mm audio motion-picture release prints for use in both motion-picture theaters and television studios.

SMPTE 56-1996: Nomenclature for Studios and Processing Laboratories

This standard defines terms used in motion-picture studios and processing laboratories.

SMPTE 59-1998: Camera Aperture Images and Usage

This standard specifies the dimensions of the camera aperture images and the relative positions of the vertical and horizontal centerlines of the intended image area with respect to the reference edge and the perforations of the camera negative film for 35- mm motion-picture cameras.

Motion-picture cameras used for different purposes require different aperture sizes. This standard specifies the image dimensions resulting from three styles of apertures used for the following purposes:

- Style A: Nonanamorphic sound motion pictures
- Style B: Anamorphic sound motion pictures
- Style C: Instrumentation photography and special processes

This standard also specifies the position of the photographic emulsion and the frame rate for 35-mm motion-picture cameras.

SMPTE 73-1998: 35-mm Film Perforated 32-mm, 2R

This standard specifies the cutting and perforating dimensions for 35-mm motion- picture film having two rows of 16-mm type perforations, one row near each edge of the 35-mm film, and a perforation pitch of either 0.2994 in or 0.3000 in (7.605 mm or 7.620 mm). The width of the 16-mm strip after processing and slitting is also specified.

SMPTE 74-1993: Zero Point for Focusing Scales (R1998)

This standard specifies a mark and its location to indicate the film plane on 16- and 8-mm cameras having lenses which can be focused for various object distances.

SMPTE 75M-1994: Raw Stock—Designation of A and B Windings

This standard specifies a method for designating the type of winding for rolls of single-row perforated and multiple-row, nonsymmetrically perforated motion-picture raw stock film in terms of the reference edge.

SMPTE 76-1996: Threaded Lens Mounts

This standard specifies the dimensions required for mechanical and optical interchangeability of lenses for 16-mm and 8-mm motion-picture cameras. For 16-mm cameras with threaded lens mounts, threads having a nominal major diameter of 1-in are often specified. Similarly, for 8-mm motion-picture cameras, threads having a nominal major diameter of 5/8 in are in common use. This standard does not apply to continuous type motion-picture cameras because of the type of optical system employed in those cameras.

SMPTE 83-1996: Edge Numbers—Location and Spacing

This standard defines the location within which edge numbers will appear on 16-mm motion-picture film. The maximum interval between successive numbers is also specified.

SMPTE 86-1996: Magnetic Audio Records—Two, Three, Four, and Six Records on 35-mm and One Record on 17.5-mm Magnetic Film

This standard specifies the position, dimensions, reproducing speed, and identity of the two-, three-, four-, or six-track magnetic audio records on 35-mm magnetic film, and one single-track record on 17.5-mm magnetic film. It also specifies the assignment of records to the various tracks on the magnetic coating on the film in relation to the direction of film travel.

SMPTE 87M-1996: 100-Mil Magnetic Striping

This standard specifies the location and dimensions of the magnetic striping material applied to 16-mm motion-picture film, which is used for both picture and sound.

SMPTE 93-1998: Perforated BH

This standard specifies the cutting and perforating dimensions for 35-mm motion-picture film with a BH-type perforation and a perforation pitch of either 0.1866 in or 0.1870 in (4.740 mm or 4.750 mm).

SMPTE 96-1999: 35- and 16-mm Motion-Picture Film and 2x2-in Slides—Scanned Area and Photographic Image Area for 4:3 Aspect Ratio

This standard specifies the size and location of that portion of 35- and 16-mm motion-picture film and 2x2-in slides to be reproduced by a 4:3 aspect ratio television film chain (telecine) and the size and location of the image area recorded on 35- and 16-mm motion-picture film in television-to-film recording equipment with a 4:3 aspect ratio.

SMPTE 97-1999: 200-Mil Edge Position—Magnetic Audio Record

This standard specifies the position, dimensions, and reproducing speed of the nominal 0.200-in (5.08-mm) magnetic audio record on 16-mm motion-picture film.

SMPTE 101-1998: Perforated 2R-3000—Magnetic Striping

This standard specifies the location and dimensions of the magnetic striping material applied to 16-mm motion-picture film with perforations along both edges to be used for both picture and audio.

SMPTE 102-1997: Perforated CS-1870

This standard specifies the cutting and perforating dimensions for 35-mm motion-picture film with a CS type perforation and a perforation pitch of 0.1870 in (4.750 mm).

SMPTE 109-1998: Perforated 1R and 2R

This standard specifies the cutting and perforating dimensions for 16-mm motion-picture film with perforations along one or both edges and a perforation pitch of either 0.2994 in or 0.3000 in (7.605 mm or 7.620 mm) for the following two categories: a) 16-mm motion-picture films; b) manufacturer-designated 16-mm professional motion-picture camera films with tighter tolerances.

SMPTE 111-1996: Prints Made on Continuous Contact Printers—Exposed Areas for Picture and Audio

This standard specifies the location and width dimensions of the exposed areas for picture and photographic audio on 35-mm motion-picture prints made on continuous contact printers. This standard is applicable to the printing of motion-picture raw stock which is cut and perforated in accordance with ANSI/SMPTE 139 or ANSI/SMPTE 237. This standard refers to the adjustment of the printer, and is in accordance with ANSI/SMPTE 40.

SMPTE 112-1999: 100-Mil Magnetic Audio Record

This standard specifies the position, dimensions and reproducing speed of the nominal 100-mil (2.54-mm) magnetic audio record on 16-mm motion-picture film. This standard also specifies the longitudinal picture-audio displacement on the film.

SMPTE 117M-1996: Photographic Audio Record Spectral Diffuse Density

This standard supplements American National Standards ANSI/NAPM 2.18 and ANSI IT2.19 by specifying spectral conditions suitable for determining the sensitometric characteristics of photographic audio records on three-component subtractive color films having records made up of dye images plus silver or a metallic salt. It does not apply to the density measurement of records composed of dyes only. The conditions of this standard are applicable to systems of audio reproduction using the S-1 photosurface. It is recognized that there are other types of photosurfaces used for photographic audio reproduction that do not fall within the scope of this standard. This standard defines a practical condition by means of which it is expected that most density measurements will be made.

SMPTE 119-1999: Perforated 65-mm, KS-1870

This standard specifies the cutting and perforating dimensions for 70-mm motion-picture film perforated 65-mm, with a KS-type perforation and a perforation pitch of 0.1870 in (4.750 mm).

SMPTE 125M-1995: Component Video Signal 4:2:2—Bit-Parallel Digital Interface

This standard defines an interface for system M (525/60) digital television equipment based on ITU-R BT.601. The standard has application in the television studio over distances up to 300 m (1000 ft).

SMPTE 127-1994: Magnetic-Photographic Audio Records—Magnetic Striping of Prints

This standard specifies the location and dimensions of the magnetic striping material applied to 16-mm motion-picture prints, containing a picture and photographic audio record, for the purpose of employing both a magnetic and the existing photographic audio record.

SMPTE 137-2000: Release Prints—Four Magnetic Audio Records

This standard specifies the position, dimensions, reproducing speed, identity, and use of the four magnetic audio records on 35-mm motion-picture release prints. The standard also specifies the longitudinal picture-audio displacement on the film.

SMPTE 139-1996: for Motion-Picture Film (35-mm) - Perforated KS

This standard specifies the cutting and perforating dimensions for 35-mm motion-picture film with a KS-type perforation and a perforation pitch of either 0.1866 or 0.1870 in (4.740 or 4.750 mm).

SMPTE 143-1994: Length of Film on Camera Spools—25-Ft Capacity

This standard describes the total length and the photographically useful length of raw film supplied on an 8-mm type R motion-picture camera spool of 25-ft (7.6-m) nominal capacity.

SMPTE 145-1999: Perforated KS

This standard specifies the cutting and perforating dimensions for 65-mm motion- picture film with a KS-type perforation and a perforation pitch of either 0.1866 in or 0.1870 in (4.740 mm or 4.750 mm).

SMPTE 146M-1996: 16- and 8-mm Reversal Color Camera Films—Determination of Speed

This standard specifies a method for the determination of American National Standard speed of 16-mm, 8-mm Type R, and 8-mm Type S reversal color camera films intended for direct projection in motion-picture photography.

SMPTE 148-1991: 35- and 16-mm Prints for Television Transmission—Film Image Area for Review Room Viewing

This standard specifies the dimensions of that part of the film image area used for review room viewing of 35- and 16-mm motion-picture prints intended for television transmission, and the placement of this area.

SMPTE 149-1999: Perforated 1R

This standard specifies the cutting and perforating dimensions for 8-mm motion-picture film with 8-mm type S perforations along one edge and a perforation pitch of either 0.1664 in or 0.1667 in (4.227 mm or 4.234 mm).

SMPTE 151-1998: 16-mm Film Perforated 8-mm Type S, (1, 3)

This standard specifies the cutting and perforating dimensions for 16-mm motion- picture film with 8-mm type S perforations in positions 1 and 3 and a perforation pitch of either 0.1664 in or 0.1667 in (4.227 mm or 4.234 mm). The width of the 8-mm strip after processing and slitting is also specified.

SMPTE 152-1994: Projectable Image Area

This standard specifies the maximum dimensions of the film image area intended for projection from a 70-mm motion-picture film, and the placement of this area relative to the perforations and the reference edge of the film.

SMPTE 153-1996: 16-mm Film Perforated 8-mm Type S (1, 4)—Printed Areas

This standard specifies the location and size of the printed picture area for negative/positive and reversal printing operations on 16-mm motion-picture film perforated 8-mm type S, 2R-1664 or 2R-1667, in row positions 1 and 4.

SMPTE 154-1998: Projectable Image Area and Projector Usage

This standard specifies the maximum dimensions of the film image area intended for projection and its relative position to the reference edge and the perforations of 8-mm type S motion-picture film, as specified in ANSI/SMPTE 149.This standard also specifies the projection frame rate for 8-mm type S motion-picture film.

SMPTE 157-1999: Camera Aperture Image and Usage

This standard specifies the dimensions of the camera aperture image and its relative position to the reference edge and the perforations of 8-mm type S motion-picture film, as specified in ANSI/SMPTE 149.This standard also specifies the position of the emulsion and the frame rate for 8-mm type S motion-picture film.

SMPTE 159.1-1996: Model 1 Camera Cartridge—Cartridge-Camera Interface and Take-Up Core Drive

This standard specifies the dimensions of the 8-mm type S motion-picture film camera cartridge and cartridge-camera interface. Also specified are the dimensions of the take-up core drive opening and critical dimensions of the take-up core as well as the driving force, direction of drive, and recommended drive ratio.

SMPTE 159.2-1996: Model 1 Camera Cartridge—Aperture, Profile, Film Position, Pressure Pad, and Flatness

This standard specifies the dimensions and location of the cartridge aperture, pressure pad, and characteristics essential to the appropriate flatness of the cartridge pressure pad. Also specified are the position of the 8-mm type S motion-picture film and its required clearances in the cartridge aperture.

SMPTE 160M-1995: Projection Reels—100- to 312-mm Diameter

This standard specifies the dimensions of 8-mm Type S motion-picture projection reels of 100- to 312-mm (3.94- to 12.28-in) diameter with nominal film capacities up to 360 m (1200 ft).

SMPTE 161-1998: Magnetic Striping

This standard specifies the location and dimensions of the magnetic recording stripe and the balance stripe applied to 8-mm motion-picture film with one row of 8-mm type S perforations.

SMPTE 162-1998: 16-mm Film Perforated 8-mm Type S (1, 4)—Magnetic Striping

This standard specifies the location and dimensions of the magnetic recording stripes and the balance stripes applied to 16-mm motion-picture film with two rows of 8-mm type S perforations in positions 1 and 4.

SMPTE 163-1998: 35-mm Film Perforated 8-mm Type S, 5R—Magnetic Striping

This standard specifies the location and dimensions of the magnetic recording stripes and the balance stripes applied to 35-mm motion-picture film with four rows of 8-mm type S perforations and one row of special perforations.

SMPTE 164-1993: Magnetic Audio Record—Position, Dimensions, and Reproducing Speed

This standard specifies the position, dimensions, and reproducing speed of the magnetic audio record on 8-mm type S motion-picture film having a nominal 0.027 in (0.69-mm) width magnetic stripe. This standard also specifies the longitudinal picture-audio displacement on the film.

SMPTE 165-1999: Perforated 8-mm Type S, 5R (1-3-5-7-0)

This standard specifies the cutting and perforating dimensions for 35-mm motion-picture film with four rows of 8-mm type S perforations and one row of special perforations having a perforation pitch of either 0.1664 in or 0.1667 in (4.227 mm or 4.234 mm). The film stock described in this standard is intended for the production of prints. The width of the 8-mm strip after processing and slitting is also specified.

SMPTE 166-1999: Exposure Control and Stock Identification—Sound and Silent Camera Cartridge Notches

This standard specifies the dimensions and location of 8-mm type S sound and silent motion-picture film camera cartridge notches intended to preset exposure devices automatically with respect to the film speed and color-balancing filter. This standard also specifies the dimensions and location of cartridge notches intended for identification of the motion-picture film inside the cartridge.

SMPTE 168-1996: Perforated 8-mm Type S (1, 4)

This standard specifies the cutting and perforating dimensions for 16-mm motion-picture film with 8-mm type S perforations in positions 1 and 4 and a perforation pitch of either 0.1664 in or 0.1667 in. The width of the 8-mm strip after processing and slitting is also specified.

SMPTE 169-1997: Perforated 8-mm Type S, 2R-1664 (1, 0)

This standard specifies the cutting and perforating dimensions for 35-mm motion-picture film with one row of 8-mm type S perforations and one row of special perforations having a perforation pitch of 0.1664 in (4.227 mm). The film stock described in this standard is intended for use as an intermediate film in the production of prints.

SMPTE 170M-1999: Composite Analog Video Signal—NTSC for Studio Applications

This standard describes the composite analog color video signal for studio applications: NTSC, 525 lines, 59.94 fields, 2:1 interlace with an aspect ratio of 4:3.This standard specifies the interface for analog interconnection and serves as the basis for the digital coding necessary for digital interconnection of NTSC equipment.

SMPTE 171-1996: Perforated 16-mm, 3R (1-3-0)

This standard specifies the cutting and perforating dimensions for 35-mm motion-picture film with 16-mm perforations in positions 1-3-0 and a perforation pitch of either 0.2994 in or 0.3000 in (7.605 mm or 7.620 mm). The width of the 16-mm strip after processing and slitting is also specified.

SMPTE 173-1999: Double 8-mm Camera Spools—100-Ft Capacity

The dimensions shown in this standard are for double 8-mm type R motion-picture film spools with a nominal capacity of 100 ft (30 m). These spools are used in cameras of the type in which each roll of film is passed through the camera twice for exposure in accordance with ANSI/SMPTE 231. The spindle holes in the spool are shown with splines which are intended to assist in assuring correct orientation of the spool in the camera.

SMPTE 174-1994: Camera Spools—50- to 400-ft Capacity

This standard specifies the dimensions for 16-mm motion-picture camera spools having capacities from 50 ft to 400 ft (15 m to 120 m) of film. This standard further specifies the configuration of the spindle holes in the two flanges.

SMPTE 176-1999: 16-mm Film Perforated 2R-1667 (1, 3)—Magnetic Striping

This standard specifies the location and dimensions of recording stripes and balance stripes applied to 16-mm motion-picture film with two rows of 8-mm type S perforations in positions 1 and 3.

SMPTE 177-1995: Four-Track Magnetic Audio Release Prints—Magnetic Striping

This standard specifies the location and dimensions of the magnetic recording stripes on 35-mm motion-picture film used for four-track magnetic audio release prints having an anamorphic-type picture image.

SMPTE 179-1996: 35-mm Film Perforated 2R (1, 0) and 5R (1-3-5-7-0)—Printed Areas

This standard specifies the location and size of the 8-mm type S printed picture areas for negative and intermediate optical reduction printing on 35-mm motion-picture film perforated 2R-1664 in row positions 1 and 0 and for print films derived by optical or contact printing on 35-mm film perforated 5R-1667 in row positions 1, 3, 5, 7, and 0.

SMPTE 181-1996: 16-mm Film Perforated 8-mm Type S (1, 3)—Printed Areas

This standard specifies the location and size of the 8-mm type S printed picture areas for negative/positive and reversal printing on 16-mm motion-picture film perforated 8- mm type S, 2R-1667 and 2R-1664 in row positions 1 and 3.

SMPTE 183M-1996: Photographic Audio Level Test Films—Measurement of Photoelectric Output Factor

This standard specifies the method of measurement of the photoelectric output factor of single-track photographic audio level test films in all film gauges, using a calibrating audio reproducer. It is applicable to both variable-area and variable-density audio records. The standard also specifies the intended performance of a calibrating audio reproducer. Calibrated audio-level test films are employed to measure the precise output level of photographic audio reproducers and the photoelectric output factor of different audio records, and to establish a reference level on a standard program-level meter appropriately chosen for the installation.

SMPTE 184M-1998: Raw Stock Identification and Labeling

This standard specifies the information to be included by the manufacturer covering the physical specifications and certain packaging characteristics of motion-picture raw stock. The suggested location of this information on the manufacturer's label is also specified.

SMPTE 185-1993: Six Magnetic Records on Release Prints—Position, Dimensions, Reproducing Speed, and Identity

This standard specifies the position, dimensions, reproducing speed, identity, and use of the six magnetic audio records on 70-mm motion-picture release prints. The standard also specifies the longitudinal picture-audio displacement on the film.

SMPTE 188M-1994: Model II Camera Cartridges (15-m Capacity)—Camera Run Film Length

This standard describes the camera run length of film supplied in 8-mm type S model II motion-picture film camera cartridges of 15-m (50-ft) nominal capacity and the length of film returned to the customer.

SMPTE 189M-1994: Model II Camera Cartridges—Loaded Film Location

This standard specifies the location of the film loaded in 8-mm type S model II motion-picture camera cartridges.

SMPTE 190M-1994: Model II Camera Cartridges—Cartridge-Camera Fit and Core

This standard specifies the external dimensions for the cartridge-camera fit and core specifications for 8-mm type S model II motion-picture film camera cartridges.

SMPTE 191M-1994: Model II Camera Cartridges—Slots, Projections and Cartridge Hole

This standard specifies the dimensions and location of cartridge slots, projections, and a hole for the 8-mm type S model II motion-picture film camera cartridge to preset cameras in accordance with the effective film speed and insert or exclude a color-balancing filter. This standard also describes the area available for visible film identification.

SMPTE 192-1997: Shipping Reels for Prints

This standard specifies the dimensions of shipping reels for 35-mm motion-picture prints having a nominal film capacity of 2000 ft (610 m). The use of 1000-ft capacity shipping reels is not recommended for the reasons specified in annex A.1.

SMPTE 194-1997: Projector Usage—Release Prints Having Four Perforations per Frame

This standard specifies the position of the emulsion for 35-mm motion-picture release prints having four perforations per frame and the position of the magnetic striping relative to the projector lens. The standard also specifies the rate of projection for defined systems and the relevant standards on location of the picture and audio records.

SMPTE 195-2000: Motion-Picture Prints—Projectable Image Area

This standard specifies the maximum dimensions of the film image area intended for projection from a 35-mm motion-picture film and the placement of this area relative to the perforations and the reference edge of the film. This standard specifies three types of image areas intended for theatrical projection:

- Style A: General theatrical release prints commonly referred to as nonanamorphic or wide screen
- Style B: Theatrical release prints with an anamorphic image
- Style C: Classic theatrical prints

SMPTE 196M-1995: Indoor Theater and Review Room Projection—Screen Luminance and Viewing Conditions

This standard specifies the screen luminance level, luminance distribution, and spectral distribution (color temperature) of the projection light for theatrical, review- room, and nontheatrical presentation of 16-, 35-, and 70-mm motion-picture prints intended for projection at 24 frames per second. This standard also specifies review-room viewing conditions. It is the purpose of these specifications to achieve the tone scale, contrast, and pictorial quality of the projected print that will be of the quality intended during its production.

SMPTE 197-1998: 50-ft Model I Sound Camera Cartridge—Cartridge, Cartridge-Camera Interface, and Take-Up Core

This standard specifies the dimensions of the 8-mm type S 50-ft model I sound motion-picture film camera cartridge and cartridge-camera interface. Also specified are the dimensions of the take-up core drive opening and critical dimensions of the take-up core as well as the driving force, direction of drive, and recommended drive ratio. An optional means of retaining the film supply scroll configuration until the cartridge is placed in the camera is also described.

SMPTE 198-1998: 50-ft Model I Sound Camera Cartridge—Aperture, Pressure Pad, and Film Position

This standard specifies the dimensions and location of the cartridge aperture and pressure pad as well as the position of the film in the aperture of 8-mm type S 50-ft model I sound motion-picture film camera cartridges.

SMPTE 199-1998: 50-ft Model I Sound Camera Cartridge—Pressure Pad Flatness and Camera Aperture Profile

This standard specifies the dimensions and characteristics necessary for the appropriate flatness of the cartridge pressure pads as well as the required clearances for the film in the aperture area in 8-mm type S 50 ft model I sound motion-picture film camera cartridges.

SMPTE 200M-1998: Model I Camera Cartridge—Camera Run Length, Perforation Cutout, and End-of-Run Notch

This standard describes the camera run length, perforation cutout notch, and end-of-run notch of film supplied in 8-mm type S model I motion-picture film camera cartridges of 15 m and 60 m (50 ft and 200 ft) nominal capacity, and the length of film returned to the customer.

SMPTE 201M-1996: Type W Camera Aperture Image

This standard specifies the dimensions of the image area produced by type W camera aperture on 16-mm motion-picture film intended for enlargement to nonanamorphic 35-mm motion- picture film with an aspect ratio of 1.66:1 or greater. It also specifies the position of the image relative to the reference edge of the film and to the perforations. This standard further specifies the dimensions and location of the enlarged image area on 35-mm internegatives or duplicate negatives and the enlargement ratio in optical printing from 16-mm type W originals.

SMPTE 202M-1998: B-Chain Electroacoustic Response—Dubbing Theaters, Review Rooms, and Indoor Theaters

This standard specifies the measurement methods and characteristic electroacoustic frequency response of the B-chain of motion-picture dubbing theaters, review rooms, and indoor theaters whose room volume exceeds 150 m^3 (5297 ft^3). It is intended to assist in standardization of reproduction of motion-picture sound in such rooms. It does not apply where the recorded sound is intended for reproduction under domestic listening conditions; i.e., for radio broadcasting, television broadcasting, tape, or disc. This standard does not cover that part of the motion-picture sound system from the transducer to the input terminals of the main fader, nor does it cover the electroacoustic response characteristic of motion-picture theater subbass loudspeakers (subwoofers).

SMPTE 203-1998: Prints—Two-Track Photographic Audio Records

This standard specifies the lateral location and dimensions of two-track variable-area audio records on 35-mm motion-picture prints. This standard also specifies the area scanned in the audio reproducer.

SMPTE 205-1993: Model I Camera Cartridge—Interface and Take-Up Core Drive, 200-ft Capacity (R1998)

This standard specifies the dimensions of the 8-mm type S 200 ft (60 m) capacity motion-picture camera cartridge and cartridge-camera interface. Also specified are the dimensions of the take-up core drive opening and critical dimensions of the take-up core as well as the driving force, direction of drive, and recommended drive ratio. An optional means of retaining the film supply until the cartridge is placed in the camera is described.

SMPTE 206-1998: Model I Sound Camera Cartridge—Aperture, Profile, Film Position, Pressure Pad, and Flatness, 200-ft Capacity

This standard specifies the dimensions and location of the cartridge aperture, pressure pad, and characteristics necessary for its appropriate flatness, clearance, and location of film in the camera aperture of 200-ft (60-m) capacity 8-mm type S model I sound motion-picture film camera cartridges.

SMPTE 207M-1997: Digital Control Interface—Electrical and Mechanical Characteristics

This standard defines the electrical and mechanical characteristics of an interface system comprised of a general-purpose communication channel and interface devices used for the transfer of data and digital control signals between equipment utilized in the production, post-production, and/or transmission of visual and aural information. It is intended that the communication channel and device(s) described in this standard be part of an overall equipment interface, allowing interconnection of programmable and nonprogrammable control and accessory equipment as required to configure an operational system with a defined function. The standard is also intended to allow rapid reconfiguration of a system providing more than one defined function utilizing a given group of equipment.

SMPTE 208M-1992: 35- and 16-mm Magnetic Audio Records—Recorded Characteristics (R1998)

This standard specifies the recorded characteristics of magnetic records on 35-mm motion-picture film intended for reproduction at 24 frames per second, and on 16-mm motion-picture film intended for reproduction at 24 frames per second.

SMPTE 209M-1996: Magnetic Audio Records—Recorded Characteristic

This standard specifies the recorded characteristic of magnetic audio records on 8-mm type S motion-picture prints and full-coat motion-picture magnetic film conforming to ANSI/SMPTE 149, running at the nominal speed of 24 frames (102 mm [4.0 in]) per second or 25 frames (106 mm [4.2 in]) per second.

SMPTE 210M-1999: Magnetic Audio Records—Two Records on 16-mm Magnetic Film

This standard specifies the lateral positions and width dimensions of two 3.81 mm (0.150 in) magnetic audio records and a control track on 16-mm single-perforated magnetic recording film. This standard also specifies the reproducing velocity of the film travel.

SMPTE 211M-1996: 16- and 35-mm Variable-Area Photographic Audio Records—Signal-to-Noise Ratio

This standard specifies a method for measuring the signal-to-noise ratio of 16- and 35-mm variable-area photographic audio records.

SMPTE 212M-1995: Projection Reels—75-mm Diameter

This standard specifies the dimensions of 8-mm type S motion-picture projection reels of 75-mm (2.95-in) diameter with a nominal film capacity of 15 m (50 ft) generally used for returning the film from the processing laboratory. The reels fit reel-to-reel equipment and are interchangeable in projection cassettes.

SMPTE 214M-1999: Photographic Audio Reproduction Characteristics

This standard specifies the electrical frequency response characteristics for photographic audio reproduction in motion-picture control rooms and indoor theaters. It is intended to assist in standardization of recording monitor and reproduction characteristics of motion-picture audio in studio dubbing theaters, review rooms, and indoor theaters. The standard covers that part of the motion-picture audio system from the transducer to the input terminals of the main fader.

SMPTE 215-1995: Camera Aperture Image

This standard specifies the dimensions of the camera aperture image and the relative positions of the vertical and horizontal centerlines of the intended image area with respect to the reference edge and the perforations of the camera negative film for 65-mm motion-picture cameras.

SMPTE 216-1998: Recorded Characteristic of Magnetic Audio Records—Four-track striped release prints

This standard specifies the recorded characteristic of magnetic audio records on 35-mm four-track striped motion-picture release prints, when reproduced at 24 frames per second or approximately 90 ft (27 m) per minute.

SMPTE 217-1998: Recorded Characteristic of Magnetic Audio Records—Striped Release Prints

This standard specifies the recorded characteristic of magnetic audio records on 70-mm striped motion-picture release prints, when reproduced at 120 perforations per second (approximately 112 ft [34 m] per minute or 22.4 in [569 mm] per second) which is 24 frames (5 perforations each) per second.

SMPTE 218M-1996: 200-mil Center-Position Magnetic Audio Record

This standard specifies the position, dimensions, and reproducing speed of the nominal 5.08-mm (0.200-in) center-position magnetic audio record on 16-mm motion-picture film.

SMPTE 220-1996: Camera Mounting Connections—1/4-Inch-20 Thread and 3/8-inch-16 Thread Tripod Screws

This standard specifies the location of the head mounting sockets in the mounting plates of motion-picture and television cameras intended to mate with 1/4-inch-20 thread and 3/8-inch-16 thread locking screws. The standard is not intended to prescribe design except for the dimensions affecting interchangeability; for this reason, the socket figures indicate two of many possible designs.

SMPTE 221-1998: Six-Track Audio Release Prints—Magnetic Striping

This standard specifies the location and dimensions of the magnetic recording stripes on 70-mm motion-picture film used for six-track magnetic audio release prints having a flat picture image with a maximum projection aperture of 1.914 in x 0.870 in (48.62 mm x 22.10 mm).

SMPTE 222M-1994: Control and Review Rooms—Monitor System Electroacoustic Response

This standard specifies the method of measurement and characteristics for the monitor chain electroacoustic response of television control and review rooms with volumes of 150 m^3 (5300 ft^3) and smaller. It is intended to assist in standardization of reproduction of program sound in television control and review rooms. It does not apply where the recorded sound is intended for reproduction under theater listening conditions; i.e., to motion-picture or public address systems. This standard does not cover equalization standards for other parts of the system, such as fixed or variable equalization applied for noise reduction around tape recorders, or the like.

SMPTE 223M-1996: Safety Film

This standard defines and specifies safety film for motion-picture use.

SMPTE 224M-1996: 19-mm Type D-1—Tape Record

This standard specifies the dimensions and location of the audio, video, and auxiliary data, analog cue track, time code, and control track records for 19-mm type D-1 television digital component recording, operating on the 525/60 television system encoded according to ANSI/SMPTE 125M.

SMPTE 225M-1996: 19-mm Type D-1—Magnetic Tape

This standard specifies the principal properties of the magnetic tape used for 19-mm type D-1 television digital component recording.

SMPTE 226M-1996: 19-mm Tape Cassettes

This standard specifies dimensions for three sizes of cassettes (S, M, and L) for use with 19-mm television digital recorders.

SMPTE 227M-1996: 19-mm Type D-1—Helical Data and Control Records

This standard specifies the content, format, and recording method of the data blocks forming the helical records on the tape containing video, audio, and associated data in 19-mm type D-1 television digital component recording. In addition, clause 6 of this document specifies the content, format, and recording method of the longitudinal record containing tracking information for the scanning head associated with the helical records. Track dimensions and locations are specified in SMPTE 224M.Thc standard applies to recorders operating in the 525-line television system with a frame frequency of 29.97 Hz nominal and in accord with ITU-R BT.601. One video channel and four independent audio channels are recorded. Audio channels operate in accord with ANSI S4.40 at a 48-kHz sampling frequency.

SMPTE 228M-1996: 19-mm Type D-1—Time and Control Code and Cue Records

This standard specifies the content, format, and modulation method of the longitudinal records contained in the cue track and the time-code track in 19-mm type D- 1 television digital component recording. Track dimensions and locations are specified in SMPTE 224M. The document applies to recorders operating in the 525-line television system with a frame frequency of 29.97 Hz.

SMPTE 229M-1996: 1/2-in Type L—Records

This standard specifies the dimensions and location of the video, audio, time code, and tracking-control records, as recorded by 1/2-in type L helical-scan video tape recorders operating with video signals having a typical scanning structure of 525 lines, 59.94 fields per second, and 2:1 interlace. Use is made of the video cassette and tape specified in ANSI/SMPTE 35M for type G format or SMPTE 238M for type L format.

SMPTE 230M-1996: 1/2-in Type L—Electrical Parameters, Control Code, and Tracking Control

This standard specifies the electrical parameters of video, audio, time and control code, and tracking-control signals for 1/2-in type L helical-scan video tape recorders operating with video signals having a typical scanning structure of 525 lines, 59.94 fields per second, and 2:1 interlace. This standard specifies two recording modes: mode 1 uses oxide-particle tape, and mode 2 uses metal-particle tape and permits audio frequency modulation (AFM) signals to be recorded.

SMPTE 231-1999: Camera Aperture Image and Usage

This standard specifies the dimensions of the camera aperture image and its relative position to the reference edge and the perforations of 8-mm type R motion-picture film. The location of the perforations is based on dimensions given in ANSI/SMPTE 239.This standard also specifies the position of the emulsion, the frame rate, and the orientation of the area being exposed for 8-mm type R film as used in a motion-picture camera.

SMPTE 233-1998: Projectable Image Area and Projector Usage

This standard specifies the maximum dimensions of the film image area intended for projection from a 16-mm motion-picture film perforated one or two edges as specified in SMPTE 109, and the placement of this area relative to the perforations and the reference edge of the film. This standard also specifies the position of the emulsion and the rate of projection for 16-mm motion-picture film perforated one or two edges, and the projector thread-up distance between audio and picture for 16-mm motion picture film with audio.

SMPTE 234-1998: Projectable Image Area and Projector Usage

This standard specifies the maximum dimensions of the film image area intended for projection from an 8-mm type R motion-picture film, and the placement of this area relative to the perforations and the reference edge of the film. This standard also specifies the position of the emulsion, the rate of projection, and the orientation of the image area for 8-mm type R motion-picture film as used in a motion-picture projector.

SMPTE 235-1998: Projection Reels—200- to 2300-ft Capacity

This standard specifies the dimensions for 16-mm motion-picture projection reels having capacities from 200 to 2300 ft (60 to 700 m) of film inclusive.

SMPTE 236-1998: Projection Reels

This standard specifies the dimensions for 8-mm type R motion-picture reels used for projection having film capacities of 50, 100, 200, 400, 600, 800, and 1200 ft (15, 30, 60, 120, 180, 240, and 360 m).

SMPTE 237-1998: Perforated DH-1870

This standard specifies the cutting and perforating dimensions for 35-mm motion-picture film with a DH-type perforation and a perforation pitch of 0.1870 in (4.750 mm).

SMPTE 238M-1998: 1/2-in Type L—Tapes and Cassettes

This standard specifies the magnetic tapes and video cassettes for the 1/2-in type L helical-scan video tape recorder system.

SMPTE 239-1999: Perforated 8-mm Type R, 2R

This standard specifies the cutting and perforating dimensions for 16-mm motion-picture film with two rows of 8-mm type R perforations and a perforation pitch of either 0.1500 in or 0.1497 in (3.810 mm or 3.802 mm). The width of the 8-mm strip after processing and slitting is also specified.

SMPTE 240M-1999: Signal Parameters—1125-Line High-Definition Production Systems

This standard defines the basic characteristics of the analog video signals associated with origination equipment operating in 1125-line high-definition television production systems. This standard defines systems operating at 60.00 Hz and 59.94 Hz field rates. The digital representation of the signals described in this standard may be found in SMPTE 260M. These two documents define between them both digital and analog implementations of 1125-line HDTV production systems.

SMPTE 241-1995: 35- and 70-mm Projection Reels

This standard specifies the dimensions of 35-mm projection reels for motion-picture and television applications and 35- and 70-mm projection reels intended for use on combination 70/35-mm projectors and rewinds. This standard does not apply to shipping reels as specified in ANSI/SMPTE 192.

SMPTE 242-1993: Universal Intermittent Sprockets

This standard specifies the dimensions of two types of 16-tooth intermittent sprockets for 35-mm motion-picture projectors. Other dimensions and definitions are given in annex A. This standard is applicable to sprockets used in conjunction with film perforated in accordance with ANSI/SMPTE 139 (0.1870 in pitch) or ANSI/SMPTE 102.

SMPTE 243M-1993: 35- and 70-mm Projection Lenses and Mounts (R1998)

This standard specifies the lens markings, focal length tolerances, mounting diameters, mechanical factors in mounting additional lens adapters to lenses, and the preferred value steps in focal lengths for lenses used in 35- and 70-mm motion-picture projectors . (Focal length referred to in this standard is the equivalent focal length, commonly known as EFL.)This standard also specifies the limiting or maximum available space for projector lens mounts where lenses of varying focal lengths and designs and attachments thereto are used.

SMPTE 244M-1995: System M/NTSC Composite Video Signals—Bit-Parallel Digital Interface

This standard describes a bit-parallel composite video digital interface for systems operating according to the 525-line, 59.94-Hz NTSC standard, as described by ANSI/SMPTE 170M, sampled at four times color subcarrier frequency. Sampling parameters for the digital representation of encoded video signals, the relationship between sampling phase and color subcarrier, and the digital levels of the video signal are defined. This standard has application for use with shielded twisted 12-pair cable of conventional design over distances up to 50 m, without transmission equalization or any special equalization of the receiver. Longer cable lengths may be used, but with rapidly increasing requirement for care in the cable selection and possible receiver equalization or the use of active repeaters or both. Digital composite video signals, as defined by this standard, are the signals conveyed by the composite implementation of the serial digital interface.

It should be noted that additional information to that described by this standard is also carried by the serial digital interface. The serial digital interface is the preferred method for the interconnection of composite digital equipments when cable lengths exceed 50 m.

SMPTE 245M-1993: 19-mm Type D-2 Composite Format—Tape Record

This standard specifies the dimensions and location of the audio, video, ancillary data, cue track, time code, and control-track records for 19-mm type D-2 helical-scan composite digital cassette television tape recorders operating on the 525/60 television system encoded according to ANSI/SMPTE 244M.

SMPTE 246M-1993: 19-mm Type D-2 Composite Format—Magnetic Tape

This standard specifies the principal properties of the magnetic tape used for the 19-mm type D-2 composite digital television format.

SMPTE 247M-1993: 19-mm Type D-2 Composite Format—Helical Data and Control Records

This standard specifies the content, format, and recording method of the data blocks forming the helical records on the tape containing video, audio, and ancillary data in the 19-mm type D-2 helical-scan television recorder. In addition, clause 4 of this standard specifies the content, format, and recording method of the longitudinal record containing tracking information for the scanning head associated with the helical records. Track dimensions and locations are specified in ANSI/SMPTE 245M.The standard applies to recorders operating in the 525-line television system with a frame frequency of 29.97 Hz nominal and in accord with ANSI/SMPTE 244M. One video channel and four independent audio channels are recorded. Audio channels operate in accord with ANSI S4.40 at a nominal 48-kHz sampling frequency.

SMPTE 248M-1993: 19-mm Type D-2 Composite Format—Cue Record and Time and Control Code Record

This standard specifies the content, format, and modulation method of the longitudinal records contained in the cue track and the time-code track in 19-mm type D- 2 helical-scan cassette video recorders. Track dimensions and locations are specified in ANSI/SMPTE 245M. The document applies to recorders operating in the 525-line television system with a frame frequency of 29.97 Hz.

SMPTE 249M-1996: 1/2-in Type M-2—Records

This standard specifies the dimensions and locations of the video, audio, time code, and tracking-control records, as recorded by 1/2-in type M-2 helical-scan video tape recorders operating with video signals having a typical scanning structure of 525 lines, 59.94 fields/s, 2:1 interlace, and utilizing the video cassettes specified in ANSI/SMPTE 250M. This standard also specifies the records for two different audio recording modes: common audio mode and pulse code modulation (PCM) audio mode.

SMPTE 250M-1996: 1/2-in Type M-2—Tapes and Cassettes

This standard specifies tapes and cassettes for the 1/2-in type M-2 helical-scan video tape recording system.

SMPTE 251M-1996: 1/2-in Type M-2—Electrical Parameters of Video, Audio, Time and Control Code, and Tracking Control

This standard specifies the recording system for the video, audio, time and control code, and tracking control signals for 1/2-in type M-2 helical-scan video tape recorders operating with video signals having a typical scanning structure of 525 lines, 59.94 fields/s, 2:1 interlace, and utilizing the video cassettes specified in ANSI/SMPTE 250M.The audio frequency modulation (AFM) recording shown in this standard is optional. Pulse code modulation (PCM) audio recording mode with limited interchangeability, as defined in ANSI/SMPTE 249M, is a secondary audio recording mode which is specified in ANSI/SMPTE 252M.

SMPTE 252M-1996: 1/2-in Type M-2—Pulse Code Modulation Audio

This standard specifies the pulse code modulation (PCM) audio mode of encoding and recording system utilizing a 1/2-in type M-2 helical-scan video tape recorder operating with video signals having a typical scanning structure of 525 lines, 59.94 fields/s, 2:1 interlace, and the cassettes specified in ANSI/SMPTE 250M.

SMPTE 253M-1998: Three-Channel RGB Analog Video Interface

This standard defines the component analog video interface for studio applications using three primary color signals carried on parallel channels for the interconnection of television equipment. The signals carried across the interface have a typical scanning structure of 525 lines, 59.94 fields/s, 2:1 interlace and 4.3 or 16.9 aspect ratio. The signals have a vertical blanking interval that is divided into an active line period and a horizontal blanking interval. Signal characteristics are defined by a gamma-corrected set of red, green, and blue R'G'B' primary video signals. In addition to general interconnection of television equipment at the component analog level of operation, the signals defined by this standard are suitable as inputs to analog-to-digital conversion systems in compliance with ANSI/SMPTE 125M and ANSI/SMPTE 267M or as inputs to NTSC composite encoders in compliance with ANSI/SMPTE 170M.

The three-channel interface defined by this standard (sync on green) is the preferred implementation. Some applications may require a four-channel interface (with sync carried separately). Such an implementation is described in annex A.

SMPTE 254-1998: Manufacturer-Printed, Latent Image Identification Information

This standard specifies the position and dimensions of machine-readable identification numbers. These numbers are intended to be a machine-readable version of the latent image key number. This standard also specifies the encoding format to be used for these machine readable numbers, as well as the area scanned and the spectral characteristics of the scanner. This standard also specifies the position, dimensions, and content of human-readable identification (key) numbers for use on 35-mm motion-picture films intended for original photography or intermediate printing which also include a machine-readable key number. These numbers will normally be exposed onto the film at the time of manufacture. This standard further specifies an area that may be used for optional manufacturer-specific film-type identification information. This standard also specifies an area on the film which is not to be exposed by the film manufacturer, thus leaving it available for customer data recording. Finally, this standard specifies an optional frameline index mark.

SMPTE 256M-1996: Specifications for Video Tape Leader

This standard specifies the minimum requirements for the content and duration of signals recorded prior to the start of the recorded program material to permit setup and adjustment of equipment for optimum performance during reproduction. The standard also specifies a visual and aural countdown sequence to facilitate program cueing and specifies the duration of video tape that precedes and follows the recorded material to provide the minimum lengths of tape required to ensure proper threading in video tape systems which do not employ tape cassettes.

SMPTE 257-1998: Stereoscopic Prints with Vertically Positioned Subframes—Projectable Image Areas

This standard specifies the maximum dimensions of the film image area intended for projection from a 35-mm motion-picture film using vertically positioned subframes for stereoscopic projection, and the placement of this area relative to the perforations and the reference edge of the film.

SMPTE 258M-1993: Transfer of Edit Decision Lists

This standard describes the data format for the interchange of a list of audio/video content decisions which specify an audio/video product. Exchange of this information allows the audio/video product to be reproduced on a compatible editing system. Such a list is commonly referred to as an edit decision list. The transfer medium and storage format are not specified in this standard.

SMPTE 259M-1997: 10-Bit 4:2:2 Component and $4f_{sc}$ NTSC Component Digital Signals—Serial Digital Interface

This standard describes a serial digital interface system M (525/60) digital television equipment operating with either 4:2:2 component signals or $4f_{sc}$ NTSC composite digital signals. (For 625-line PAL composite implementation, see annex E.) This standard has application when the signal loss at 70 MHz ($4f_{sc}$) or 135 MHz (4:2:2) due to coaxial cable characteristics does not exceed approximately 30 dB.

SMPTE 260M-1999: Digital Representation and Bit-Parallel Interface—1125/60 High-Definition Production System

This standard specifies the digital representation of the signal parameters of the 1125/60 high-definition production system as given in their analog form by SMPTE 240M.This standard also specifies the signal format and the mechanical and electrical characteristics of the bit-parallel digital interface for the interconnection of digital television equipment operating in the 1125/60 high-definition production system.

SMPTE 261M-1993: 10-Bit Serial Digital Television Signals—4:2:2 Component and $4f_{sc}$ NTSC Composite, AMI Transmission Interface

This standard describes an interface which transmits serial digital video data coded in scrambled bipolar NRZI per ANSI/SMPTE 259M, via alternate-mark-inversion (AMI) communication channels.

SMPTE 262M-1995: Binary Groups of Time and Control Codes—Storage and Transmission of Data

This standard specifies a directory index to classify various types of data to be recorded into the binary groups (user bits) of the SMPTE time and control code. The directory index, located in two binary groups of the time and control code frame, fully specifies the type of data stored in the remaining binary groups of that frame. This standard also specifies the group assignments of timing, application, and control data types to subsets of the directory index. It applies to both linear and vertical interval time code applications.

SMPTE 263M-1996: 1/2-in Type D-3 Composite and Type D-5 Component Format—Tape Cassette

This standard specifies dimensions for three sizes of cassettes (S, M, and L) for use with 1/2-in type D-3 composite and type D-5 component television digital recorders.

SMPTE 264M-1998: 1/2-in Type D-3 Composite Format—525/60

This standard specifies the content, format. and recording method of the data blocks containing video, audio, and associated data which form the helical records on 12.65-mm (0.5-in) tape in cassettes as specified in ANSI/SMPTE 263M. In addition, this standard specifies the content, format, and recording method of the longitudinal record containing tracking information for the scanning head associated with the helical records, and also the longitudinal cue audio and time code tracks. One video channel and four independent audio channels are recorded in the digital format. Each of these channels is designed to be capable of independent editing. The video channel records and reproduces a composite television signal in the 525-line system with a frame frequency of 29.97 Hz.

SMPTE 265M-1998: 1/2-in Type D-3 Composite Format—625/50

This standard specifies the content, format, and recording method of the data blocks containing video, audio, and associated data which form the helical records on 12.65-mm (0.5-in) tape in cassettes as specified in ANSI/SMPTE 263M. In addition, this standard specifies the content, format, and recording method of the longitudinal record containing tracking information for the scanning head associated with the helical records, and also the longitudinal cue audio and time code tracks. One video channel and four independent audio channels are recorded in the digital format. Each of these channels is designed to be capable of independent editing. The video channel records and reproduces a composite television signal in the 625-line system with a frame frequency of 25 Hz.

SMPTE 266M-1994: 4:2:2 Digital Component Systems—Digital Vertical Interval Time Code

This standard describes the signal format of a digital vertical interval time code suitable for use with the digital coding given in ANSI/SMPTE 125M (for 525-line, 59.94-Hz field rate, 4:2:2 component digital signals) or ITU-R BT.601 (for 625-line, 50-Hz field rate, 4:2:2 component digital signals).

SMPTE 267M-1995: Bit-Parallel Digital Interface—Component Video Signal 4:2:2 16x9 Aspect Ratio

This standard defines an interface for system M (525/59.94) wide screen, 16 × 9 aspect ratio, digital television equipment based on ITU-R 601-3. Two luminance sampling rates are provided, 13.5-MHz sampling providing full-signal compatibility with equipment operating in compliance with ANSI/SMPTE 125M, and 18-MHz sampling providing equivalent horizontal resolution for the 16 × 9 aspect ratio of this standard as compared to the 4 × 3 aspect ratio of ANSI/SMPTE 125M. Use of the 18-MHz sampling method also provides 16 × 9 to the 4 × 3 aspect ratio translation by sample selection rather than sample interpolation as would be required with 13.5-MHz sampling. The standard has application in the television studio over distances up to 300 m (1000 ft) for 13.5-MHz sampling and 225 m (750 ft) for 18-MHz sampling.

SMPTE 268M-1994: File Format for Digital Moving-Picture Exchange (DPX)

This standard defines a file format for the exchange of digital moving pictures on a variety of media between computer-based systems. It does not define the characteristics of input or output devices or displays. This format is known as the SMPTE digital picture exchange format version 1.0, or DPX in short form. The file extension is .dpx.

SMPTE 269M-1999: Fault Reporting in Television Systems

This standard describes a simple interface over which television equipment can report the occurrence of internal failures and faults in incoming signals. It is intended for use in all television equipment, from the simplest active devices to the most complex. The interface consists of an isolated closure which can assume one of three states: open, closed, or pulsing. These respectively signal that the reporting device is okay, has detected an internal fault, or is detecting incoming signal faults. Fault occurrence data may be collected from equipment complying with this standard by several means, ranging from simple "follow the lights to the trouble" summary alarm schemes to computerized logging systems. While full specification of such systems is beyond the scope of this standard, a general outline of one possible implementation is given in annex A.

SMPTE 270-1994: Manufacturer-Printed Latent Image Identification Information

This standard specifies the position and dimensions of machine-readable identification numbers. These numbers are intended to be a machine-readable version of the latent image key number. This standard also specifies the encoding format to be used for these machine-readable numbers, as well as the area scanned and the spectral characteristics of the scanner. This standard also specifies the position, dimensions, and content of human-readable identification (key) numbers for use on 65-mm motion-picture films intended for original photography or intermediate printing which also include a machine-readable key number. These numbers normally will be exposed onto the film at the time of manufacture.

This standard further specifies an area that may be used for optional manufacturer-specific film-type identification information. This standard also specifies an area on the film which is not to be exposed by the film manufacturer, thus leaving it available for customer data recording. Finally, this standard specifies an optional frame line index mark.

SMPTE 271-1994: Manufacturer-Printed Latent Image Identification Information

This standard specifies the position and dimensions of machine-readable identification numbers. These numbers are intended to be a machine-readable version of the latent image key number. This standard also specifies the encoding format to be used for these machine readable numbers, as well as the area scanned and the spectral characteristics of the scanner. This standard also specifies the position, dimensions and content of human-readable identification (key) numbers for use on 16-mm motion-picture films intended for original photography or intermediate printing which also include a machine-readable key number. These numbers normally will be exposed onto the film at the time of manufacture.

This standard further specifies an area that may be used for optional manufacturer-specific film-type identification information. This standard also specifies an area on the film which is not to be exposed by the film manufacturer, thus leaving it available for customer data recording.

SMPTE 272M-1994: Formatting AES/EBU Audio and Auxiliary Data into Digital Video Ancillary Data Space

This standard defines the mapping of AES digital audio data, AES auxiliary data, and associated control information into the ancillary data space of serial digital video conforming to ANSI/SMPTE 259M. The audio data and auxiliary data are derived from ANSI S4.40, generally referred to as AES audio.

This standard provides a minimum of two audio channels and a maximum of 16 audio channels based on available ancillary data space in a given format (four channels maximum for composite digital). Audio channels are transmitted in pairs combined, where appropriate, into groups of four. Each group is identified by a unique ancillary data ID. Several modes of operation are defined and letter suffixes are applied to the nomenclature for this standard to facilitate convenient identification of interoperation between equipment with various capabilities. The default form of operation is 48-kHz synchronous audio sampling carrying 20 bits of AES audio data and defined in a manner to ensure reception by all equipment conforming to this standard.

SMPTE 273M-1995: Status Monitoring and Diagnostics Protocol

This standard defines the status monitoring and diagnostics protocol (SMDP) used with a general-purpose communication link that connects to equipment used in the production, post-production, and/or transmission of visual and audio information. The communication link is separate and distinct from the supervisory interface for digital control (ANSI/SMPTE 207M). SMDP may be used when querying equipment for status and diagnostics information. The primary intent of this standard is to establish the protocol between a supervisory controller and associated video/audio equipment. In addition, it seeks to establish a common set of commands that should be used in SMDP-based systems. A provision allows for manufacturer-specific commands in addition to the common set.

SMPTE 274M-1998: 1920 x 1080 Scanning and Interface

This standard defines a family of raster-scanning systems for the representation of stationary or moving two-dimensional images sampled temporally at a constant frame rate and having an image format of 1920 x 1080 and an aspect ratio of 16:9. This standard specifies:

- R'G'B' color encoding
- R'G'B' analog and digital interfaces
- $Y'P'_BP'_R$ color encoding and analog interface
- $Y'C'_BC'_R$ color encoding and digital interface

An auxiliary component *A* may optionally accompany $Y'C'_BC'_R$; this interface is denoted $Y'C'_BC'_RA$.

SMPTE 275M-1995: ESlan-1 Remote Control System

This standard defines the services and protocols contained within the physical, data link, network, transport, and session layers of ESlan-1, a control and data network for use in television and audio program production, post-production, and distribution equipment. ESlan-1 is intended for application in small- to moderate-sized facilities requiring modest levels of performance. A study to determine limiting parameters on the size of ESlan-1 installations is currently being undertaken by the SMPTE. This standard is to be read in conjunction with SMPTE EG 29, SMPTE EG 30, and with other documents listed in clause 2 and annex C.

SMPTE 276M-1995: Transmission of AES/EBU Digital Audio Signals Over Coaxial Cable

This standard describes a point-to-point coaxial cable interface for the transmission of AES/EBU digital audio signals throughout television production and broadcast facilities. The purpose of this standard is to ensure that a level of compatibility exists between signals generated to this standard and analog video equipment, such as nonclamping distribution amplifiers, switchers, cables, and connectors, as normally used in television applications.

SMPTE 277M-1996: 19-mm Type D-6—Helical Data, Longitudinal Index, Cue, and Control Records

This standard specifies the format and recording method of the data blocks and associated data which form the helical records on 19-mm tape in cassettes as specified in SMPTE 226M. The data recorded may be digital video and audio data of various image standards up to approximately 1 Gbit/s as specified in SMPTE 278M. Also specified are the content, format, and recording method of the longitudinal record containing tracking information for the scanning heads associated with the helical records, and the longitudinal index and cue tracks. In addition, this standard specifies the principal properties of the magnetic tape used for 19-mm type D-6 digital recording.

SMPTE 278M-1996: 19-mm Type D-6—Content of Helical Data and Time and Control Code Records

This standard specifies the content of the data blocks which form the helical records as specified in ANSI/SMPTE 277M on 19-mm tape in cassettes as specified in ANSI/SMPTE 226M. Part of this standard is the specification of the time and control code record, which forms the longitudinal index track as specified in ANSI/SMPTE 277M. Digital video and audio data derived from various image standards are recorded with a data rate of approximately 1 Gbit/s. All image standards recorded by this format employ the identical track pattern, inner and outer block structure, and modulation code.

SMPTE 279M-1996: 1/2-in Type D-5—Component Format, 525/60 and 625/50

This standard specifies the content, format, and recording method of the data blocks containing video, audio, and associated data which form the helical records on 12.65-mm (0.5-in) tape in cassettes as specified in ANSI/SMPTE 263M. In addition, this standard specifies the content, format, and recording method of the longitudinal record containing tracking information for the scanning head associated with the helical records, and also the longitudinal cue audio, and time and control code tracks.

SMPTE 291M-1998: Ancillary Data Packet and Space Formatting

This standard specifies the basic formatting structure of the ancillary data space in the digital video data steam in the form of 10-bit words. Application of this standard includes 525-line, 625-line, component or composite, and high-definition digital television interfaces which provide 8- or 10-bit data ancillary data space. Space available for ancillary data packets is defined in the document specifying the connecting interface.

SMPTE 292M-1998: Bit-Serial Digital Interface for High-Definition Television Systems

This standard defines a bit-serial digital coaxial and fiber-optic interface for HDTV component signals operating at data rates in the range of 1.3 Gbits/s to 1.5 Gbits/s. Bit-parallel data derived from a specified source format are multiplexed and serialized to form the serial data stream. A common data format and channel coding are used based on modifications, if necessary, to the source format parallel data for a given high-definition television system. Coaxial cable interfaces are suitable for application where the signal loss does not exceed an amount specified by the receiver manufacturer. Typical loss amounts would be in the range of up to 20 dB at one-half the clock frequency. Fiber optic interfaces are suitable for application at up to 2 km of distance using single-mode fiber.

SMPTE 293M-1996: 720 x 483 Active Line at 59.94-Hz Progressive Scan Production—Digital Representation

This standard defines the digital representation of stationary or moving two-dimensional images for television production. The representation is sampled linearly in the spatial domain and sampled temporally at a constant frame rate. The scanned image has an aspect ratio of 16:9. This standard includes both R',G',B' and Y',C'_B,C'_R expressions for the signal representation.

SMPTE 294M-1997: 720 x 483 Active Line at 59.94-Hz Progressive Scan Production—Bit-Serial Interfaces

This standard defines two alternatives for bit-serial interfaces for the 720 × 483 active line at 59.94-Hz progressive scan digital signal for production, defined in ANSI/SMPTE 293M. Interfaces for coaxial cable are defined, each having a high degree of commonality with interfaces operating in accordance with ANSI/SMPTE 259M.

SMPTE 295M-1997: 1920 x 1080 50 Hz—Scanning and Interfaces

This standard defines a family of raster scanning systems for the representation of stationary or moving two-dimensional images sampled temporally at a constant frame rate and having an image format of 1920 x 1080 and an aspect ratio of 16:9 as given in table 1. This standard specifies:

- R'G'B' color encoding
- R'G'B' analog and digital interfaces
- $Y'P'_BP'_R$ color encoding and analog interface
- $Y'C'_BC'_R$ color encoding and digital interface

An auxiliary component *A* may optionally accompany $Y'C'_BC'_R$; this interface is denoted $Y'C'_BC'_RA$.

SMPTE 296M-1997: 1280 x 720 Scanning, Analog and Digital Representation and Analog Interface

This standard defines a family of raster scanning systems for the representation of stationary or moving two-dimensional images sampled temporally at a constant frame rate and having an image format of 1280 × 720 and an aspect ratio of 16:9. This standard specifies:

- R'G'B' color encoding
- R'G'B' analog and digital representation
- $Y'P'_BP'_R$ color encoding, analog representation and analog interface
- $Y'C'_BC'_R$ color encoding and digital representation

An auxiliary component *A* may optionally accompany $Y'C'_BC'_R$; this representation is denoted $Y'C'_BC'_RA$. A bit-parallel digital interface is incorporated by reference in clause 12.

SMPTE 297M-1997: Serial Digital Fiber Transmission System for ANSI/SMPTE 259M Signals

This standard defines an optical fiber system for transmitting bit-serial digital signals. It is specifically intended for transmitting ANSI/SMPTE 259M serial signals (143 through 360 Mbits/s). Its optical interface specifications and end-to-end system performance parameters are otherwise compatible with ANSI/SMPTE 292M, which covers transmission rates of 1.3 through 1.5 Gbits/sec.

SMPTE 298M-1997: Universal Labels for Unique Identification of Digital Data

This standard defines universal labels, a universal labeling mechanism to be used in identifying the type and encoding of data within a general-purpose data stream. The labeling mechanism is intended to function across all types of digital communications protocols and message structures, allowing the intermixture of data of any sort. Labels created using the mechanism specified are intended to be attached to the data they identify and to travel together with them through communications channels. This standard defines universal labels that can be used by any organization that wishes to label data in a manner that is universally unambiguous, globally unique, and traceable to the authorizing organization.

SMPTE 299M-1997: 24-Bit Digital Audio Format for HDTV Bit-Serial Interface

This standard defines the mapping of 24-bit AES digital audio data and associated control information into the ancillary data space of a serial digital video conforming to ANSI/SMPTE 292M. The audio data are derived from ANSI S4.40, hereafter referred to as AES audio. Audio signal, sampled at a clock frequency of 48 kHz locked (synchronous) to video, is the preferred implementation for intrastudio applications. As an option, this standard supports AES audio at synchronous or asynchronous sampling rates from 32 kHz to 48 kHz.

SMPTE 300-1997: Manufacturer-Printed Latent Image Identification Information

This standard specifies the position, dimensions, content, and exposure of human-readable, latent image information applied onto 35-mm color print film. This information is normally exposed onto the film at the time of manufacture. This standard also specifies spectral densities and a film area which is not to be exposed by the film manufacturer, thus leaving it available for subsequent customer data recording such as sound track recording.

SMPTE 301M-1999: Theater Projection Leader, Trailer, and Cue Marks

This standard specifies the make-up of assembly of leaders and cue marks for 70-, 35-, and 16-mm motion-picture release prints for use in motion-picture theaters and screening rooms.

SMPTE 302M-1998: Linear PCM Digital Audio in an MPEG-2 Transport Stream

This standard specifies the transport of uncompressed (linear PCM) digital audio in an MPEG-2 transport system. Some applications may require linear PCM (pulse code modulated) digital audio in conjunction with compressed video specified in the MPEG-2 4:2:2 profile. The MPEG audio standard defines compressed audio, but does not define uncompressed audio for carriage in an MPEG-2 transport system. This standard augments the MPEG standards to address the requirement for linear PCM digital audio.

SMPTE 303M: Color Reference Pattern

This standard defines the electrical and physical representation of a television color reference pattern. It also specifies colorimetry, geometry, and related parameters.

SMPTE 304M-1998: Broadcast Cameras—Hybrid Electrical and Fiber-Optic Connector

This standard defines a connector primarily intended for use in television broadcasting and video equipment, such as camera head to camera control-unit connections. It defines hybrid connectors, which contain a combination of electrical contacts and fiber-optic contacts for single-mode fibers. It also contains dimensional tolerances which ensure nondestructive mating of the electrical and optical interfaces, and functional operability of the electrical interface. Functional operability of the optical interface is dependent upon fiber preparation and termination and is, therefore, not guaranteed by this standard.

SMPTE 305M-2000: Serial Data Transport Interface

This standard defines a data stream used to transport packetized data within a studio/production center environment. The data packets and synchronizing signals are compatible with ANSI/ SMPTE 259M.

SMPTE 306M-1998: 6.35-mm Type D-7 Component Format—Video Compression at 25 Mbits/s, 525/60 and 625/50

This standard specifies the content, format, and recording method of the data blocks containing video, audio, and associated data which form the helical records on 6.35-mm tape in cassettes as specified in SMPTE 307M. In addition, this standard specifies the content, format, and recording method for longitudinal cue and control tracks.

SMPTE 307M-1998: 6.35-mm Type D-7 Component Format—Tape Cassette

This standard specifies the dimensions for two sizes of cassettes (M and L) for use with 6.35-mm type D-7 component television digital recorders.

SMPTE 308M-1998: MPEG-2 4:2:2 Profile at High Level

ISO/IEC 13818-2, commonly known as MPEG-2 video, includes specification of the MPEG-2 4:2:2 profile. Based on ISO/IEC 13818-2, this standard provides additional specification for the MPEG-2 4:2:2 profile at high level. It is intended for use in high-definition television production, contribution, and distribution applications. As in ISO/IEC 13818-2, this standard defines bit-streams, including their syntax and semantics, together with the requirements for a compliant decoder for 4:2:2 profile at high level, but does not specify particular encoder operating parameters.

SMPTE 309M-1999: Transmission of Date and Time Zone Information in Binary Groups of Time and Control Code

This standard specifies a coding technique for the transmission of date and time zone information in the user groups of a time and control code signal. A two-digit hexadecimal code in a pair of binary groups specifies the time zone and the format for the date encoding in the remaining six binary groups. Date information is encoded either as six decimal digits to display the date in the YYMMDD format or as six decimal digits in the modified Julian date (MJD) format.

SMPTE 310M-1998: Synchronous Serial Interface for MPEG-2 Digital Transport Stream

This standard describes the physical interface and modulation characteristics for a synchronous serial interface to carry MPEG-2 transport bit streams at rates up to 40 Mbits/s. It is a point-to-point interface intended for use in a low-noise environment. The low-noise environment is defined as a noise level that would corrupt no more than one MPEG-2 data packet per day at the transport clock rate. When other transmission systems (e.g., studio-to-transmitter microwave links, etc.) are interposed between devices employing this interface, higher noise levels may be encountered. In such cases, it is recommended that appropriate error correcting methods by used.

SMPTE 311M-1998: Hybrid Electrical and Fiber-Optic Camera Cable

This standard describes the minimum performance for a hybrid cable containing single-mode optical fibers and electrical conductors to convey signal and control in a variety of environments where moisture, weather, and ozone resistance are required. This document is not intended to be a cable manufacturing design standard. The cable described in this standard is intended to be used to interconnect cameras and base stations in conjunction with the connector interface standard.

SMPTE 312M-1999: Splice Points for MPEG-2 Transport Streams

This standard defines constraints on the encoding of and syntax for MPEG-2 transport streams such that they may be spliced without modifying the PES packet payload. Generic MPEG-2 transport streams, which do not comply with the constraints in this standard, may require more sophisticated techniques for splicing.

SMPTE 313-1999: Manufacturer-Printed Latent Image Identification Information—120 Perforation Repeat

This standard specifies the position and dimensions of 120 perforation repeat machine-readable identification numbers. These numbers are intended to be a machine-readable version of the latent image key number. This standard also specifies the encoding format to be used for these machine-readable numbers, as well as the area scanned and the spectral characteristics of the scanner.

SMPTE 314M-1999: Data Structure for DV-Based Audio, Data and Compressed Video—25 and 50 Mb/s

This standard defines the DV-based data structure for the interface of digital audio, subcode data, and compressed video with the following parameters: 525/60 system, 4:1:1 image sampling structure, 25 Mbits/s data rate; 525/60 system, 4:2:2 image sampling structure, 50 Mbits/s data rate; 625/50 system, 4:1:1 image sampling structure, 25 Mbits/s data rate; 625/50 system, 4:2:2 image sampling structure, 50 Mbits/s data rate.

SMPTE 315M-1999: Camera Positioning Information Conveyed by Ancillary Data Packets

This standard provides a method for multiplexing camera positioning information into the ancillary data space described in SMPTE 291M. Applications of the standard include the 525-line, 625-line, component or composite, and high-definition digital television interfaces which provide 10-bit ancillary data space. Two types of camera positioning information are defined in this standard: binary and ASCII.

SMPTE 316M-1999: 12.65-mm Type D-9 Component Format—Video Compression, 525/60 and 625/50

Intraframe bit-rate reduction is applied to video data prior to recording. This standard specifies the content, format, and recording method of the data blocks containing video, audio, and associated data that form the helical records on 12.65-mm tape in cassettes. In addition, it specifies the content, format, and recording method of the longitudinal record containing tracking information for the rotating head associated with the helical records, cue audio, and control tracks.

SMPTE 317M-1999: 12.65-mm Type D-9 Component Format—Tape Cassette

This standard specifies dimensions for the video tape cassette used with the 12.65-mm type D-9 digital component television recorder.

SMPTE 318M-1999: Reference Signals for the Synchronization of 59.94- or 50-Hz Related Video and Audio Systems in Analog and Digital Areas

This standard specifies the use of a derivative of a color black signal as a reference for the synchronization of all forms of composite or component, digital, or analog equipment using a system standard related to 59.94 Hz (60/1.001) or 50 Hz. It also provides the option for the reference signal to carry VITC. This will allow the reference to distribute local or UTC time data.

SMPTE 319M-2000: Transporting MPEG-2 Recoding Information through 4:2:2 Component Digital Interfaces

This standard specifies an embedded transport mechanism for the MPEG-2 recoding data set as defined in SMPTE 327M for the representation of MPEG-2 recoding information in ITU-R BT.656, 4:2:2 component digital interfaces.

SMPTE 320M-1999: Channel Assignments and Levels on Multichannel Audio Media

This standard specifies the audio channel assignment, and the relative levels of the audio channels, for recordings of audio programs containing between three and six audio channels, onto storage media for television sound. This standard may also be applied to other types of media (such as transmission) where a sequence of audio channels is available to carry a multichannel audio program. This standard is not intended for application in the area of film sound.

SMPTE 321M-1999: Data Stream Format for the Exchange of DV-Based Audio, Data and Compressed Video Over a Serial Data Transport Interface

This standard defines the format of the data stream for the synchronous exchange of DV-based audio, data, and compressed video (whose data structure is defined in SMPTE 314M) over the interface defined in SMPTE 305M (SDTI). It covers the transmission of audio, subcode data, and compressed video packets associated with DV-based 25 and 50 Mbits/s data structure for 525/60 and 625/50 systems, including faster-than-real-time transmission.

SMPTE 322M-1999: Format for Transmission of DV Compressed Video, Audio and Data Over a Serial Data Transport Interface

This standard specifies the data structure and the transmission format of DV compressed video, audio, and data over a serial data transport interface (SDTI, SMPTE 305M). The standard is a combination of video, audio, subcode, and control data optimized for the connection between DV-compliant VCRs and disk systems. It ensures high-speed data stream transfer up to five times faster than real time. The video, audio, and subcode data comply with IEC 61834-2 for both 525/60 and 625/50 systems.

SMPTE 323M-1999: Channel Assignments and Levels on Multichannel Audio Media

This standard specifies the audio channel assignment, and the relative levels of the audio channels, for recordings of audio programs containing six audio channels, onto storage media for film sound. Programs which are placed on media according to this standard are intended to reproduction in the cinema where the SW channel is always reproduced. This standard is not intended for application in the area of sound intended for reproduction in the domestic consumer environment.

SMPTE 324M: 12-Channel Serial Interface for Digital Audio and Auxiliary Data

This standard defines a synchronous, self-clocking serial interface for up to 12 channels of linearly encoded audio and auxiliary data. The interface is designed to allow multiplexing of six two-channel streams compliant with AES3.

SMPTE 325M-1999: Opportunistic Data Broadcast Flow Control

This standard defines the flow control protocol to be used between an emission multiplexer and data server for opportunistic data broadcast. Opportunistic data broadcast inserts data packets into the output multiplex to fill any available free bandwidth. The emission mutliplexer maintains a buffer from which it draws data to be inserted. The multiplexer will request additional MPEG-2 transport packets from the data server as its buffer becomes depleted. The number of packets requested depends upon the implementation, with the most stringent requirement being requesting a single MPEG-2 transport packet where the request and delivery can occur in less than the emission time of an MPEG-2 transport packet from the multiplexer. This protocol is designed to be extensible and provide a basis for low-latency, real-time backchannel communications from the emission multiplexer.

SMPTE 326M-2000: SDTI Content Package Format (SDTI-CP)

This standard specifies the format for the transport of content packages (CP) on the serial digital transport interface (SDTI) This format is abbreviated to the term SDTI-CP. This standard defines the structure of the content package mapped onto the SDTI transport. All element and metadata formats are defined by SMPTE 331M.

SMPTE 327M-2000: MPEG-2 Video Recoding Data Set

This standard specifies the content of the picture related recoding data set for the representation of ISO/IEC 13818-2 MPEG coding information for the purpose of optimally cascading decoders and recoders at any bit rate or GOP structure. The coding information is as derived from an ISO/IEC 13818 compliant MPEG bit stream during the picture decoding process, as described in ISO/IEC 13818-2.

SMPTE 328M-2000: MPEG-2 Video Elementary Stream Editing Information

This standard defines the MPEG video elementary stream (ES) information to facilitate seamless edits under defined circumstances. The video ES, as defined by the MPEG standards, are supplemented with additional information for professional studio applications. Supplementary information will be carried within the sequence header and the user data area of the video ES. This standard defines the data to be carried and the location of the data.

SMPTE 329M-2000: MPEG-2 Video Recoding Data Set—Compressed Stream Format

This standard specifies the stream format of the MPEG-2 recoding data set for the representation of compressed ISO/IEC 13818-2 MPEG coding information, as used in applications requiring transport systems of reduced data capacity.

SMPTE 330M-2000: Unique Material Identifier (UMID)

This standard specifies the format of the unique material identifier (UMID). The UMID is a unique identifier for picture, audio, and data material which is locally created but globally unique. It differs from many unique identifiers in that the number does not depend wholly upon a registration process, but can be generated automatically at the point of creation without reference to a central database.

SMPTE 331M-2000: Element and Metadata Definitions for the SDTI-CP

This standard specifies the formats of the elements and metadata used by the SDTI content package format standard (SDTI-CP), SMPTE 326M. This standard defines element and metadata formats where they are simply specified or where a publicly available reference is accessible. It is not intended that this standard provide detailed specifications for complex formats which may have broader application.

SMPTE 332M-2000: Encapsulation of Data Packet Streams Over SDTI (SDTI-PF)

This standard specifies an open framework for encapsulating data packet streams and associated control metadata over the SDTI transport (SMPTE 305M). Encapsulating data packet streams on SDTI allows them to be routed through conventional SDI (SMPTE 259M) equipment This standard specifies a range of packet types which may be carried over SDTI. The standard also offers a limited capability for metadata to be added providing packet control information to aid the successful transfer of packets.

SMPTE 333M-1999: DTV Closed-Caption Server to Encoder Interface

This standard defines a standard for interoperation of digital television closed-caption (DTVCC) data server devices and video encoders. The caption data server devices provide partially-formatted EIA 708 data to the video encoders using the request/response protocol and interface defined in this standard. The video encoder completes the formatting and includes the EIA 708 data in the video elementary stream picture-level user_data field. This standard describes an interface for transmission of DTVCC data from a caption server to video encoder.

SMPTE 334M-2000: Ancillary Data Mapping for Bit-Serial Interface

This standard defines a method of coding which allows data services to be carried in the vertical ancillary data space of a bit-serial component television signal conforming with SMPTE 292M or ANSI/SMPTE 259M.

SMPTE 337M-2000: Format for Non-PCM Audio and Data in an AES3 Serial Digital Audio Interface

This standard specifies an interface format for the transport of non-PCM audio and data in professional applications using the AES3 serial digital audio interface. This standard includes both physical and logical specifications, based on the existing AES3 format, to allow exchange of non-PCM data between different devices. The standard accommodates multiple non-PCM audio and data formats and allows carriage of multiple data streams within a single interface. This standard provides means for carrying time code or time alignment information so that the information conveyed over this interface may be synchronized with information content delivered over other interfaces.

SMPTE 338M-2000: Format for Non-PCM Audio and Data in an AES3—Data Types

This standard describes the data_type field defined in SMPTE 337M. This field describes data types that may be carried in an AES3 digital audio interface according to SMPTE 337M. This standard defines supported data types, but does not cover formatting that may be required for each data type. References are included for additional standards that describe data type specific formatting requirements.

SMPTE 339M-2000: Format for Non-PCM Audio and Data in an AES3—Generic Data Types

This standard specifies data type specific format requirements for several types of data bursts that may be carried within an AES3 interface according to SMPTE 337M. Included are descriptions of the data type, the format of the burst_payload for the data type, the coding of data type dependent fields in the burst_preamble, and additional data burst and bit stream formatting requirements not defined in SMPTE 337M. This includes specific synchronization methods which may affect formatting.

SMPTE 340M-2000: Format for Non-PCM Audio and Data in an AES3—ATSC A/52 (AC-3) Data Type

This standard specifies data type specific format requirements for AC-3 data bursts carried within an AES3 interface according to SMPTE 337M.

SMPTE 341M-2000: Format for Non-PCM Audio and Data in an AES3—Captioning Data Type

This standard specifies data type specific format requirements for caption data bursts carried within an AES3 interface according to SMPTE 337M.

SMPTE 342M-2000: HD-D5 Compressed Video 1080i and 720p Systems—Encoding Process and Data Format

This standard defines the encoding process of the HD-D5 video compression and its data format for the 1080i/59.94 system and the 720p/59.94 system.

SMPTE 344M: 540 Mbits/s Serial Digital Interface

This standard specifies a serial digital interface that operates at a nominal rate of 540 Mbits /s. This standard has application in the television studio over lengths of coaxial cable where the signal loss does not exceed an amount specified by the receiver manufacturer. Typical loss amounts would be in the range of 20 dB to 30 dB at one-half the clock frequency with appropriate receiver equalization. Receivers designed to work with lesser signal attenuation are acceptable. Separate SMPTE documents specify the mapping of source image formats onto the special 540 Mb/s serial interface.

SMPTE 346M: Time Division Multiplexing Video Signals and Generic Data Over High-Definition Interfaces

This standard defines the time division multiplexing (TDM) of various standard-definition digital video and generic 8-bit data signals over high-definition serial digital interfaces (SMPTE 292M).

SMPTE 348M: High Data-Rate Serial Data Transport Interface (HD-SDTI)

This standard provides the mechanisms necessary to facilitate the transport of packetized data over a synchronous data carrier. The HD-SDTI data packets and synchronizing signals provide a data transport interface which is compatible with SMPTE 292M (HD-SDI) such that it can be readily used by the infrastructure provided by this standard.

5.4 Scopes of SMPTE Recommended Practices

RP 6-1994: Recorded Carrier Frequencies and Preemphasis Characteristics for 2-in Quadruplex Video Magnetic Tape Recording for 525-Line 60-Field Television Systems

This practice specifies parameters of the recorded information essential to the interchange of 2-in quadruplex video magnetic tape recording of monochrome and NTSC color signals for 525-line/60-field television systems. The parameters include video preemphasis characteristics and recorded carrier frequencies for all recording practices and video pilot specifications for practice SHBP. Practices defined are:

- Practice SHBP: This practice is suitable for color and monochrome signals. A video pilot signal is added to the recorded information to be used as a playback reference.
- Practice HB: This practice is suitable for color and monochrome signals.
- Practice LBM: This practice is suitable only for monochrome signals. (It is considered to be obsolescent and is included for reference purposes only.)
- Practice LBC: This practice is suitable for color and monochrome signals. It is considered to be obsolescent and is included for reference purposes only.)

RP 9-1995: Dimensions of Double-Frame 35-mm 2x2 Slides for Precise Applications in Television

This practice specifies dimensions and tolerances for a double-frame 35-mm film clip and an associated 2x2-in mount which are intended to ensure that picture information is accurately and consistently positioned in a suitable slide projector. The slide mount described in clause 4 represents one suitable method for attaining accurate and consistent positioning of picture information in a suitable slide projector. The use of alternate methods of mounting the film clip to within the same accuracy shall be considered as meeting the requirements of this practice. This practice is not intended to replace or to void ANSI/SMPTE 96-1992 or ANSI PH3.43-1977 (R1991).

RP 11-1994: Tape Vacuum Guide Configuration and Position for Quadruplex Video Magnetic Tape Recording

This practice specifies the tape vacuum guide configuration and position for quadruplex video recordings on 2-in magnetic tape, and the test conditions for verifying these parameters.

RP 12-1997: Screen Luminance for Drive-in Theaters

This practice specifies the luminance (measured brightness) of the projection screens for drive-in theaters intended for the projection of motion-picture film at 24 frames/sec. The practice defines luminance ratios among portions of the total screen area, and defines the acceptable variations as viewed from positions within the audience area. The practice applies to both diffusing and directional screens.

RP 14-1997: Plotting Data from Sensitometric Strips Exposed on Type Ib (Intensity Scale) Sensitometers

The purpose of this practice is to specify the relationship of the spacings of the exposure scale (horizontal coordinate) of graph paper on which sensitometric data are plotted and the corresponding increments of the logarithm of exposure in the sensitometer when the exposure modulator is a step tablet.

RP 15-1997: Calibration of Densitometers Used for Black-and- White Photographic Density Measurement

The purpose of this practice is to specify the means to be employed in the calibration of densitometers utilized in the measurement of diffuse transmission densities. This practice applies to densitometers utilized for the measurement of processed black-and-white photographic films and plates or cast colloidal carbon tablets.

RP 16-1993: Specifications of Tracking-Control Record for 2-in Quadruplex Video Magnetic Tape Recordings

This practice specifies the recorded dimensional relationships among: a) tracking-control signal, b) frame pulse signal, and c) vertical synchronizing signal for 2-in (50.8 mm) quadruplex video magnetic tape recordings.

RP 17-1964: Photographic Recording Technique for Measuring High-Speed Camera Image Unsteadiness

This practice specifies a photographic method of recording and measuring high-speed motion-picture image unsteadiness. Existing image-steadiness methods are applicable to professional motion-picture cameras showing vertical image-steadiness accuracy within 0.0004 in. The procedure for measuring image unsteadiness, described in this practice, is far more practical for high-speed, rotating-prism motion-picture cameras usually showing unsteadiness characteristics greater than 0.0004 in.

RP 18-1995: Specifications for Test Film for Subjective Checking of 16-mm Motion-Picture Audio Projectors

This practice describes a test film and a method for subjective checking and demonstrating 16-mm motion-picture projection and audio performance.

RP 19-1995: Specifications for 8-mm Type R Registration Test Film

This practice specifies the subject material and the dimensions and location of the subject material for an 8-mm type R test film of high accuracy to assist the user in achieving several quantitative visual tests. The film can be used to test motion-picture projectors and printers.

RP 20-1995: Specifications for 16-mm Registration Test Film

This practice specifies the subject material and the dimensions and location of the subject material for a 16-mm test film of high accuracy to assist the user in achieving several quantitative visual tests. The film can be used to test motion-picture projectors and printers.

RP 21-1997: Dimensions of 35- and 70-mm Motion-Picture Rewind Spindles

This practice specifies the dimensions of 5/16-in and 1/2-in spindles for both 35-mm and 70-mm motion-picture rewinds.

RP 24-1997: Dimensions for 16-mm Motion-Picture Camera Spindles

This practice specifies the dimensions for 16-mm motion-picture camera spindles.

RP 25-1995: Audio and Picture Synchronization on Motion-Picture Film Relative to the Universal Leader for Magnetic and Photographic Records

It is the purpose of this practice to standardize the photographic and magnetic synchronizing signals and their position relative to the SMPTE Universal Leader, as specified in ANSI/SMPTE 55-1992.

RP 27.1-1989: Specification for Operational Alignment Test Pattern for Television

This practice describes the format, dimensions, and optical densities for a test pattern transparency to be used as an operational alignment tool for television systems.

RP 27.2-1989: Specifications for Operational Registration Test Pattern for Multiple-Channel Television Cameras

Format, dimensions, and optical densities are specified for a test pattern transparency to be used as an operational alignment tool for multiple channel color television cameras.

RP 27.3-1989: Specifications for Safe Action and Safe Title Areas Test Pattern for Television Systems

This practice specifies the format, dimensions, and optical densities for a test pattern for safe action and safe title areas for television systems.

RP 27.4-1994: Specifications for an Operational Test Pattern for Checking Jitter, Weave and Travel Ghost in Television Projectors

This practice specifies the format, dimensions, and optical densities for a test pattern transparency to be used as an operational tool for measurement of television film projector image stability.

RP 27.5-1989: Specifications for Mid-Frequency Response Test Pattern for Television

This practice specifies the format, dimensions. and optical densities for a test pattern to be used as an operational check of the mid-frequency response of a television system.

RP 32-1995: Specifications for 8-mm Type S Test Film for Projectors and Printers

This practice specifies the content and dimensions of an 8-mm type S test film useful in checking the performance of motion-picture projectors and printers. Its use is described in annex A.

RP 34-1998: Dimensions for 16-mm Motion-Picture Projector Reel Spindles

This practice specifies the dimensions for 16-mm motion-picture projector reel spindles. This practice specifies a test film for determining the presence of flutter in 35-mm motion-picture photographic audio reproducers operating at 90 ft (27.4 m) per minute.

RP 36-1999: Positioning the Headwheel and Adjacent Tape Guides for 2-in Quadruplex Video Magnetic Tape Recorders

This practice establishes the relative locations of critical elements in the path of the tape between the input and output guides for 2-in (51-mm) quadruplex video magnetic tape recorders operating at 15 in/s and 7.5 in/s (381 mm/s and 190.5 mm/s).

RP 38.1-1989: Specifications for Deflection Linearity Test Pattern for Television

Format, dimensions. and optical densities are specified for a test pattern transparency to be used in the measurement of geometric distortion of television systems.

RP 39-1993: Specifications for Maintaining an Emulsion-in Orientation on Theatrical Release Prints

This practice specifies the necessary handling changes in the laboratory, film exchange, and projection room to achieve the emulsion-in orientation of theatrical release prints. The practice also describes the advantages to be gained by the change to emulsion-in orientation of theatrical release prints. The practice further discusses the consequences of returning to the emulsion-out orientation during the exhibition life of theatrical release prints.

The practice suggests, in the annex, the various minor modifications that might be necessary in equipment used for projection, film rewind, and film inspection.

RP 40-1995: Specifications for 35-mm Projector Alignment and Screen Image Quality Test Film

This practice describes the content and dimensions of the image on a 35-mm motion-picture test film intended for alignment and evaluation of 35-mm motion-picture projection. This practice also gives procedures for the use of various elements of the test film image.

RP 45-1972: Use and Care of Sound Test Films

This recommended practice describes the proper method for use of sound test films, the reliability of the conclusions to be drawn from their output, the precautions necessary for reasonable accuracy, and the proper storage for preservation of calibration. This recommended practice is intended to apply to all sound test films, some of which bear a photographic recording and some a magnetic recording.

RP 47-1999: Electronic Method of Dropout Detection and Counting

This practice specifies the method of electronic dropout detection and counting for 2-in quadruplex video magnetic tape recordings made in accordance with practice HB of SMPTE RP 6.

RP 48-1999: Lubrication of 16- and 8-mm Motion-Picture Prints

This practice recognizes that surface treatment of 16- and 8-mm motion-picture prints to reduce the film surface friction coefficient is needed to promote good projection performance. The use of such treatment should result in increased steadiness, reduction of noise in the projector gate, and less tendency toward perforation damage during projection.

RP 49-1995: Leaders for 8-mm Type R and S Motion-Picture Release Prints Used in Continuous-Loop Cartridges

This practice provides guidelines for head and tail leaders on 8-mm type R and 8-mm type S motion-picture release prints intended for use in continuous loop cartridges.

RP 50-1995: Dimensions for 8-mm Type S Motion-Picture Projector Reel Spindles

This practice specifies the dimensions for 8-mm type S motion-picture projector reel spindles.

RP 51-1995: Screen Luminance and Viewing Conditions for 8-mm Review Rooms

This practice specifies the screen luminance level and characteristics of the projection screen and the viewing conditions for 8-mm review rooms.

RP 53-1993: Scene-Change Methods for Printing 35-mm, 16-mm and 8-mm Type S Motion-Picture Film

This practice specifies the dimensions and location of a scene-change notch or cueing spot for actuating the printer light-change mechanism when printing 35-mm, 16-mm, and 8-mm type S motion-picture films. (See annex A.1 for a frame-count cueing method which is more commonly utilized). Although this practice specifies a notch or a cueing spot for actuating the printer light mechanism, at no time should a film contain both.

RP 54-1999: Edge Numbering on 16-mm Release Prints

This practice covers the size, type, and frequency of numbers placed on 16-mm release prints. The purpose of these numbers is usually to determine specific locations on multiple-copy release prints of the same subjects.

RP 55-1997: 8-mm Type S Sprocket Design

This practice provides the dimensions and specifications for the design and maximum pitch of sprockets used with 8-mm type S motion-picture raw stock or processed film.

RP 56-1995: Safe Action and Safe Title Areas for 8-mm Release Prints

This practice specifies the dimensions of the nominal safe action and safe title image areas for 8-

mm release prints. The information provided applies to the use of these prints by either direct front or rear screen projection. The specifications are also applicable to reversal original photography.

RP 58-1995: Nomenclature for Devices Enclosing 8-mm Motion-Picture Film for Projection

This practice defines the terms cartridge and cassette when applied to devices enclosing 8-mm motion-picture film intended for projection. It does not apply to enclosures or containers for unexposed 8-mm camera film.

RP 59-1999: Color and Luminance of Review Room Screens for Viewing Motion-Picture Materials Intended for Slides or Film Strips

This practice specifies the luminance (photometric brightness) and color quality of projection illumination in review rooms for prints on motion-picture film intended for ultimate use as slides or film strips.

RP 63-1997: Specifications for Sound-Focusing Test Film for 16-mm Audio Reproducers, Photographic Type

This practice specifies a test film for use in focusing the scanning beam of 16-mm motion-picture photographic audio reproducers operating at 36 ft (11 m) per minute.

RP 64-1999: Specifications for Audio-Focusing Test Film for 35-mm Audio Reproducers, Photographic Type

This practice specifies a test film for use in focusing the scanning beam of 35-mm motion-picture photographic audio reproducers operating at 90 ft (27.4 m) per minute.

RP 65-2000: Motion-Picture Enlargement/Reduction Ratios

This practice specifies the enlargement/reduction ratios to be used in copying motion pictures from one film size to another while maintaining the aspect ratio and composition of the original film. This practice also specifies the dimensions of the projectable image area on the resulting copy, and gives the dimensions for a matte which produces an opaque border on the final projection copy. Also specified are dimensions for camera viewfinder marks which can be used in composing the original photography for the copy format.

RP 67-1997: Specifications for Buzz-Track Test Film for 16-mm Motion-Picture Audio Reproducers, Photographic Type

This practice specifies a test film for checking the lateral position of the sound scanning beam in 16-mm motion-picture photographic audio reproducers.

RP 68-1997: Specifications for Buzz-Track Test Film for 35-mm Motion-Picture Photographic Audio Reproducers

This practice specifies a test film for checking the lateral position of the audio scanning beam in 35-mm motion-picture photographic audio reproducers.

RP 69-1997: Specifications for Scanning-Beam Uniformity Test Film for 35-mm Motion-Picture Audio Reproducers

This practice describes a test film, the use of which is limited to the determination of the uniformity of scanning-beam illumination in 35-mm motion-picture audio reproducers. This test film is not intended to be used for the determination of the correct position of the scanning beam with respect to the reference edge of the film.

RP 70-1997: Specifications for Flutter Test Film for 16-mm Audio Reproducers, Photographic Type

This practice specifies a test film for determining the presence of flutter in 16-mm motion-picture photographic audio reproducers operating at 36 ft (11 m) per minute.

RP 73-1992: 8-mm Type R (Regular 8) Sprocket Design

This practice provides dimensions and specifications for the design of sprockets used with 8-mm type R (regular 8) motion-picture raw stock or processed film.

RP 74-1992: 16-mm Sprocket Design

This practice provides dimensions and specifications for the design of sprockets used with 16-mm motion-picture raw stock or processed film.

RP 75-1997: Specifications for Flutter Test Film for 35-mm Studio Audio Reproducers, Magnetic Type

This practice specifies a test film for determining the presence of flutter in 35-mm motion-picture studio magnetic audio reproducers operating at 96 perforations per second or approximately 90 ft (27 m) per minute for use with one-, three-, four-, and six-track audio systems.

RP 76-1997: Specifications for Flutter Test Film for 16-mm Audio Reproducers, Magnetic Type

This practice specifies a test film for determining the presence of flutter in 16-mm motion-picture magnetic audio reproducers operating at approximately 36 ft (11 m) per minute.

RP 77-1994: Specifications for Azimuth Test Film for 35-mm Studio Audio Reproducers, Magnetic Type

This practice specifies a test film for use in aligning the azimuth of magnetic head gaps in 35-mm motion-picture audio reproducers operating at 90 ft (27 m) per minute and designed for one, three-, four-, and six-track audio systems.

RP 78-1997: Specifications for Azimuth Test Film for 16-mm Audio Projectors, Magnetic Type

This practice specifies a test film for use in aligning the azimuth of magnetic head gaps in 16-mm motion-picture audio projectors operating at approximately 36 ft (11 m) per minute.

RP 79-1999: Specifications for Flutter Test Film for 35-mm Four-Track Striped Release Print Audio Reproducers, Magnetic Type

This practice specifies a test film for determining the presence of flutter in 35-mm motion-picture magnetic audio reproducers operating at 96 perforations per second or approximately 90 ft (27 m) per minute designed for four-track magnetic audio release prints.

RP 81-1999: Specifications for Scanning-Beam Uniformity Test Film for 16-mm Motion-Picture Photographic Audio Reproducers

This practice describes a test film to determine the uniformity of scanning-beam illumination in 16-mm motion-picture photographic audio reproducers. (This test film is not intended to be used for the determination of the correct position of the scanning beam with respect to the reference edge of the film.)

RP 82-1995: Specifications for 16-mm Projector Alignment and Screen Image Quality Test Film

This practice describes the artwork and dimensions for constructing a test chart to be used as the original subject for the manufacture of the test film. The practice also describes the types of photographic materials and densitometry necessary to manufacture the film.

RP 83-1996: Specifications of Tracking Control Record for 1-in Type B Helical-Scan Television Analog Recording

This practice specifies the recorded relationships among the tracking control signal, the edit pulse signal, and the video signal for 1-in type B helical-scan video tape recordings.

RP 84-1996: Reference Carrier Frequencies and Preemphasis Characteristics for 1 -in Type B Helical-Scan Television Analog Recording

This practice specifies the video reference frequencies to which the carrier is deviated and the associated video preemphasis for 1-in type B helical-scan television tape recording. (The video deemphasis to be used in reproduction is specified indirectly by requiring a flat input-to-output video response along with a specified preemphasis in recording.)

RP 85-1999: Tracking-Control Record for 1-in Type C Helical-Scan Television Tape Recording

This practice specifies the characteristics of the tracking-control record and the relationship between the recorded video and tracking-control signal for 1-in type C helical-scan television tape recorders operating on 525/60 monochrome or NTSC color systems.

RP 86-1991: Video Record Parameters for 1-in Type C Helical-Scan Television Tape Recording

This practice specifies parameters of the recorded information essential to the interchange of 1-in type C helical-scan television tape recordings of the 525/60 monochrome or NTSC color systems. The parameters include video preemphasis characteristics, recorded carrier frequencies, and record-current frequency response.

RP 87-1999: Reference Carrier Frequencies, Preemphasis Characteristic and Audio and Control Signals for 3/4-in Type E Helical-Scan Video Tape Cassette Recording

This practice specifies the reference frequencies for deviation of the frequency modulated carrier and associated video preemphasis characteristic for 3/4-in type E helical-scan video tape cassette recording of 525-line monochrome and NTSC color television signals at a tape speed of 95.3 mm/s (3.752 in/s). In addition, the characteristics of the audio and control signals are specified.

RP 90-1999: Specifications for Type U Audio Level and Multifrequency Test Film for 16-mm Audio Reproducers, Magnetic Type

This practice specifies a type U audio frequency test film to be used for adjusting the mechanical and electrical parameters of 16-mm motion-picture magnetic audio reproducers operating at 24 frames per second.

RP 91-1997: Specifications for 70-mm Projector Alignment and Screen Image Quality Test Film

This practice specifies a test film for quantitative measurements of 70-mm projector alignment and screen image quality. This practice also describes the artwork and dimensions for a test chart to be used as the original subject for the manufacture of a master negative.

RP 92-1995: Specifications for Audio Level and Multifrequency Test Films for 8-mm Type S Audio Reproducers, Magnetic Type

This practice specifies two audio frequency test films to be used for adjusting the sensitivity and frequency response of 8-mm type S motion-picture magnetic audio reproducers; one operating at 24 and the other at 18 frames per second.

RP 93-1999: Requirements for Recording American National Standard Time and Control Code for 1-in Type B Helical-Scan Video Tape Recorders

This practice specifies the recorded signal and the conditions for recording the time and control code on 1-in type B helical-scan video tape recorders as specified in ANSI/SMPTE 12M-1995.

RP 94-2000: Gain Determination of Front Projection Screens

This practice species a method for measurement of screen gain.

RP 95-1994: Installation of Gain Screens

This practice specifies the optimum installation parameters for gain screens used in motion-picture theaters.

RP 96-1993: Specifications for Subjective Reference Tapes for Helical-Scan Video Tape Reproducers for Checking Receiver/Monitor Setup

This practice specifies magnetic video reference tapes for subjective evaluation of receiver or monitor setup and overall performance of video and audio derived from 3/4-in type E and 1/2-in types G and H magnetic helical-scan tape reproducers. No test instruments are required.

RP 97-1997: Specifications for Flutter Test Film for 35-mm Audio Reproducers, Photographic Type

RP 98-1995: Measurement of Screen Luminance in Theaters

This practice specifies the procedure for a complete set of screen luminance measurements in theaters, intended to promote measured luminance uniformity that is widely acceptable to the audience.

RP 103-1995: Care, Storage, Operation, Handling and Shipping of Magnetic Recording Tape for Television

This practice provides guidance to technical managers, archivists, and technicians for the care, storage, operation, handling, and shipping conditions that help maximize life expectancy and interchange performance for television (video) magnetic recording tape.

RP 104-1994: Cross-Modulation Tests for Variable-Area Photographic Audio Tracks

This practice describes the cross-modulation method of measuring high-frequency distortion introduced during the production of variable-area audio motion-picture release prints.

RP 105-1995: Method for Determining the Degree of Jump and Weave in 70-, 35- and 16-mm Motion-Picture Projected Images

This practice identifies image motion, classifies the practical limits of acceptability of film jump and weave, and recommends a method of measurement for projection of70-, 35-, and 16-mm motion-picture prints.

RP 106-1994: Film Tension in 35-mm Motion-Picture Systems Operating Under 0.9 m/s (180 ft/min)

This practice specifies the film tension needed to transport 35-mm motion-picture film through a film-handling system operating under 0.9 m/s (180 ft/min) while minimizing conditions that contribute to film damage. This practice also recommends methods for testing film tension.

RP 107-1995: Video and Audio Reference Tape for 1-in Type B Helical-Scan Format

This practice specifies a video and audio reference tape to be used with l-in Type B helical-scan video tape recorders as defined in ANSI/SMPTE 15M-1992. It is to be used for: indication of video frequency response characteristics for both main and sync channels of the reproducing system; adjustment of gain of the video reproducing system; comparison of carrier frequencies of the video recording system; verification of level and phase of the control track recording system; adjustment of the gain of the program audio reproducing system; indication of the audio frequency response of the audio reproducing system; comparison of the audio recording gain and frequency response characteristics of the audio recording system; and verification of levels and timings of time code information recorded on audio 3 record.

RP 109-1994: Spectral Response of Photographic Audio Reproducers for 8-mm Type S Motion-Picture Film

This practice specifies the spectral response of the photographic audio reproducer light source and receptor as a unit, including any optical filtering that may be interposed.

RP 110-1992: Specifications for an Alignment Test Film for Anamorphic Attachments to 35-mm Motion-Picture Projectors

This practice specifies a test film for the alignment of anamorphic attachments or lenses for 35-mm motion-picture projectors.

RP 111-1999: Dimensions for 70-, 65- and 35-mm Motion-Picture Film Splices

This practice specifies the significant dimensions of splices for 70- , 65-, and 35-mm motion-picture film intended for projection and exhibition or for laboratory printing.

RP 113-1996: Supervisory Protocol for Digital Control Interface

This practice defines the supervisory protocol used within a general purpose communication channel of an interface system which transports data and digital control signals between equipment utilized in the production, post-production, and/or transmission of visual and aural information. It is intended that the supervisory protocol described in this practice be part of an overall system, allowing interconnection of programmable and nonprogrammable equipment as required to configure an operational system with a defined function, and to allow rapid reconfiguration of a system to provide more than one defined function utilizing a given group of equipment.

RP 114-1994: Dimensions of Photographic Control and Data Record on 16-mm Motion-Picture Film

This practice specifies the lateral location and dimensions of a photographic control and data record on 16-mm motion-picture originals, intermediates, and prints, the width scanned by the control and data reproducer and the reproducer spectral sensitivity.

RP 115-1997: Dimensions of Photographic Control and Data Record on 35-mm Motion-Picture Release Prints

This practice specifies the lateral location and dimensions of a photographic control and data record on 35-mm motion-picture release prints, the width scanned by the control and data reproducer, and the reproducer spectral sensitivity.

RP 116-2000: Dimensions of Photographic Control and Data Record on 35-mm Motion-Picture Camera Negatives

This practice specifies the lateral location and dimensions of a photographic control and data record on 35-mm motion-picture camera negatives, the width scanned by the control and data recorder and reproducer, the camera aperture, and the reproducer spectral sensitivity.

RP 117-1994: Dimensions of Magnetic Control and Data Record on 8-mm Type S Motion-Picture Film

This practice specifies the lateral location and dimensions of a magnetic control and data record on 8-mm type S motion-picture film.

RP 120-1994: Measurement of Intermodulation Distortion in Motion-Picture Audio Systems

This practice specifies the technique of measuring, by the intermodulation method, the signal distortion introduced by motion-picture audio systems.

RP 121-1997: Tape Dropout Specifications for 1-in Types B and C Video Tape Recorders/Reproducers

This practice defines the parameters for tape dropouts encountered in the reproduced FM signal of 1-in types B and C video magnetic recorders/reproducers.

RP 122-1993: Dimensions of Cemented Splices on 8-mm Type S Motion-Picture Film, Projection Type

This practice specifies the dimensions of cemented splices on 8-mm type S motion-picture film primarily intended for projection.

RP 123-1997: Dimensions of Tape Splices on 8-mm Type S Motion-Picture Film, Projection Type

This practice specifies the dimensions of mated cut splices on 8-mm type S motion-picture film made with an adhesive tape and intended only for projection.

RP 124-1998: Insertion Pivot for Studio Lighting Units and Mating Holders for Use with Standing and Hanging Support Systems

This practice specifies the dimensions for an Insertion pivot used for mounting studio lighting units on standing or hanging support devices, the mating holders for the Insertion pivot, and the mating devices for holding or hanging stage lighting devices weighing less than 22 lbs (10 kg).

RP 127-1999: Specifications for Type U Audio Level and Multifrequency Test Film for 35-mm Studio Audio Reproducers, Magnetic Full-Coat Type

This practice specifies a type U audio frequency test film to be used for adjusting the sensitivity and frequency response of 35-mm motion-picture magnetic studio audio reproducers operating at 96 perforations per second or approximately 90 ft (27 m) per minute for use with one-, three-, four-, and six-track audio systems.

RP 128-1997: Specifications for Audio Level and Multifrequency Test Film for 70-mm Striped Six-Track Release Print Audio Reproducers, Magnetic Type

This practice specifies an audio frequency test film to be used for adjusting the sensitivity and frequency response of 70-mm striped six-track motion-picture magnetic audio reproducers intended for release prints operating at 120 perforations per second or approximately 112 ft (34 m) per minute.

RP 129-1995: Requirements for 35-mm, 16-mm and 8-mm Type S Tape Splices on Magnetic Audio Recording Motion-Picture Film

This practice specifies the significant requirements for tape splices for 35-mm, 16-mm, and 8-mm Type S magnetic motion-picture film intended for audio recording.

RP 130-1995: Dimensions of Tape Splices on 16-mm and 8-mm Type R Motion Picture Film, Projection Type

This practice specifies the significant dimensions of mated cut splices for 16-mm and 8-mm type R motion-picture film made with an adhesive tape and intended for projection and exhibition.

RP 131-1994: Storage of Motion-Picture Films

This practice defines terms, classifications, and conditions for storage of motion-picture materials.

RP 132-1994: Storage of Edit Decision Lists on 8-in Flexible Diskette Media

This practice specifies the file and directory structure of an 8-in flexible diskette used for the interchange of edit decision lists (EDL).

RP 133-1999: Specifications for Medical Diagnostic Imaging Test Pattern for Television Monitors and Hard-Copy Recording Cameras

This practice describes the format, dimensions, and contrast required to make diagnostically significant measurements of the display and camera system resolution for both digital and analog monochrome signal sources. The practice provides users of medical diagnostic imaging systems with a comprehensive test pattern for day-to-day operational checks and adjustments of focus, brightness and contrast, resolution response, mid-band streaking, uniformity, and linearity of viewing monitors and hard copy recordings. This practice is not intended to create a standard for image characteristics such as resolution, geometry and linearity, uniformity, phosphor defects, etc. However, use of the pattern is encouraged as an appropriate tool for evaluating the measurement and specification of such image characteristics.

RP 134-1994: Polarity for Analog Audio Magnetic Recording and Reproduction

This practice specifies the polarity of the signal on the pin connections from a microphone presented with a positive pressure on the diaphragm. It also specifies the resulting positive magnetization when the positive microphone signal is recorded on any magnetic media.

This practice also specifies how this positive magnetization should be played back and fed through a reproducing system to provide a positive sound pressure from the loudspeaker.

RP 135-1999: Use of Binary User Groups in Motion-Picture Time and Control Codes

This practice specifies a method of coding data into the binary user groups of time and control codes for motion-picture systems. The type of data recorded is useful in the production of motion pictures. This practice also specifies a directory system to accommodate the various types of data that may need to be recorded. Whether or not to use a particular type of data (and, if used, the repetition frequency) is left to the discretion of the equipment manufacturer and/or the user. This practice also specifies the use of a checksum in one of the binary user groups.

RP 136-1999: Time and Control Codes for 245 25 or 30 Frame-Per-Second Motion-Picture Systems

This practice specifies digital code formats and modulation methods for motion-picture film to be used for timing, control, editing, and synchronization purposes. This practice also specifies the relationship of the code to the motion-picture frame. The codes described in this practice are similar to the continuous code described in ANSI/SMPTE 12M-1995.

There are two types of codes described in this practice. The first type, type C, is a continuous code which is very similar to the continuous code specified in ANSI/SMPTE 12M. This type of code can be used in situations where the film is moving continuously at the time of both recording and reproduction. The second type of code, type B, is a noncontinuous, block-type code, composed of blocks of data, each complete in itself, with gaps between the blocks. It is designed so that the code may be recorded and played back on equipment with intermittent film motion but still be decoded with the same type of electronic equipment used to read the type C or continuous time code.

The codes described in this practice can be used at various frame rates, the ones currently of interest being 24, 25, and 30 frames per second.

RP 138-1996: Control Message Architecture

This practice defines the architecture of the control message language used within a general-purpose communications channel of an interface system which transports data and control signals between equipment utilized in the production, post-production, and/or transmission of visual and aural information.

RP 139-1997: Tributary Interconnection

This practice describes the mechanism for the transfer of control messages between tributaries used within a general-purpose communications channel of an interface system which transports data and digital control signals between equipment utilized in the production, post-production, and/or transmission of visual and aural information. It is intended that the mechanism described in this practice be utilized when transferring control messages between tributaries used as a part of an overall system. The tributaries may be located either within a local network or on separate local networks which are interconnected by means of gateways and an interconnection bus.

It is further intended that this mechanism, when used as part of an overall system, shall allow the interconnection of programmable and nonprogrammable equipment as required to configure an operational system with defined functions, and will allow rapid reconfiguration of a system to provide more than one defined function utilizing a given group of equipment.

RP 140-1995: Position of Photographic Audio Record for Routine Test Signals

This practice reserves the section of the audio record available for routine audio tests on negative photographic audio records. The position specified is for audio records in the editorial or parallel sync position before the record has been advanced.

RP 141-1995: Background Acoustic Noise Levels in Theaters and Review Rooms

This practice provides measurement methods and recommended maximum levels for indoor background sound pressure levels in theaters and review rooms. The practice is limited to the noise of heating, ventilating, and air conditioning systems, intrusive noise from the projectors associated with the theater, and noise from any other mechanical or electrical equipment in the theater building. The practice is intended for application when the background noise is essentially a solid-state sound, without strong time-varying components.

The practice does not cover intrusive noise from other sources outside the theater, such as airplanes, highway traffic, adjacent theaters, or the like. The practice also does not cover noise resulting from the operation of the sound system in the theater. The practice does not cover vibration of the theater; i.e., movement of the building below 20 Hz.

RP 142-1997: Stereo Audio Track Allocations and Identification of Noise Reduction for Video Tape Recording

This practice recommends the allocation of stereo audio tracks on quadruplex television tape recorders employing a split audio head, 1-in types B and C, 3/4-in type E, and 1/2-in types H, L, and M, for the exchange of stereo program material between television organizations. It also recommends an identification procedure if noise reduction is employed, and identification and marking of the tape reel or cassette and the associated box.

RP 143-1999: Specifications for Type U Audio Level and Multifrequency Test Film for 35-mm Striped Four-Track Release Print Audio Reproducers, Magnetic Type

This practice specifies a type U audio frequency test film to be used for adjusting the mechanical and electrical parameters of 35-mm motion-picture magnetic audio reproducers intended for striped release prints perforated CS, operating at 24 frames (96 perforations) per second (approximately 90 ft [27 m] per minute [18 in or 457 mm per second]) for proper playback.

The International Organization for Standardization recognizes two reference levels for test films: Type U with a reference level of 185 nWb/m and type E with a reference level of 320 nWb/m, in order to account for differing meter types in common use in the United States and Europe.

RP 144-1999: Basic System and Transport Geometry Parameters for 1/2-in Type L Format

This practice specifies the tape speed, scanner parameters, tape tension, and test conditions for achieving the record dimensions specified in ANSI/SMPTE 229M-1991. The parameters are for reference purposes only and should not be interpreted as the only method available to attain the specifications in ANSI/SMPTE 230M-1991.

RP 145-1999: SMPTE C Color Monitor Colorimetry

This practice specifies the chromaticity values of the red, green, and blue visible radiation emitted by the primaries and the chromaticity of the white point for professional monitors used in systems based on SMPTE C colorimetry.

RP 148-1987: Relative Polarity of Stereo Audio Signals

This practice recommends the relative polarity of the signals of the two audio channels of stereo television magnetic tape recorders for the purpose of stereo program exchange.

RP 149-1992: Dimensions of Transverse Cemented Splices on 16-mm and 8-mm Type R Motion-Picture Film

This practice specifies the dimensions of transverse cemented splices on 16-mm and 8-mm type R motion-picture film. Two types of splices are specified: a laboratory splice for professional applications and a projection splice for release prints and consumer or amateur reversal films.

RP 150-2000: Channel Assignments and Test Leader for Magnetic Film Masters Intended for Transfer to Video Media Having Stereo Audio

This practice specifies the left- and right-channel assignments for stereo usage on magnetic film masters intended for transfer to video media. It also gives recommended test signals for use on the head leader of the magnetic master.

RP 151-1999: Lubrication of 35-mm Motion-Picture Prints for Projection

This practice recognizes that proper lubrication of 35-mm motion-picture prints is needed to promote good projection performance. Proper lubrication will result in improved steadiness, reduction of noise in the projector gate, reduced perforation damage, and increased projection life.

RP 152-1994: Edge Identification of Leader and Picture for 35-mm Release Prints

This practice specifies the content and location of information to appear on the edge of 35-mm release prints. As continuous platters become more prevalent, projectionists sometimes fail to resplice the leader on the appropriate reel, or resplice it on the incorrect reel. This practice provides enough information in the program area for independent reel identification and to identify the type of sound track on the print.

RP 153-1999: Method for Measuring 35- and 70-mm Shutter Efficiency

This practice specifies the method and factors to be considered when measuring and reporting the comparative shutter efficiency (remaining useful light) of 35- and 70-mm projectors. The purpose of this practice is to provide guidelines for the evaluation of projector shutter performance, either in an operating movie theater or in a test laboratory.

RP 155-1997: Audio Levels and Indicators for Digital Audio Records on Digital Television Tape Recorders

This practice specifies a reference amplitude to be used for the calibration of audio level indicators and to be recorded on the digital audio records of reference tapes intended for digital television tape recorders to facilitate the interchange of digital television tape recordings.

RP 156-1999: Bar Code Labeling for Type D-1 Component and Type D-2 Composite Cassette Identification

This practice describes the requirements for the generation of bar code labels for the automatic identification of type D-1 component and type D-2 composite cassettes. Dimensions and tolerances of the printed bar code symbols are specified. The symbol encoding, data structure, and formatting of the label information are also specified. Lastly, the label characteristics, size, orientation, and placement are specified. This practice includes both rear and side labels.

RP 157-1995: Key Signals

This practice describes the format of a key video signal which is used to control the contribution of an associated fill video signal into a composite of two or more signals. Such signals are commonly referred to simply as key signals. This description is given for composite and component analog and digital signals.

RP 158-1999: Basic System and Transport Geometry Parameters for 1/2-in Type M-2 Format

This practice specifies the video recording system, scanner parameters, and test conditions for 1/2-in type M-2 helical-scan video tape recorders operating with video signals having a typical scanning structure of 525 lines, 59.94 fields/sec, 2:1 interlace, and utilizing the cassettes specified in ANSI/SMPTE 250M-1991. The parameters described in this practice are for reference purposes only and should not be taken as the only method available to attain the specifications as defined in ANSI/SMPTE 249M-1991.

RP 159-1995: Vertical Interval Time Code and Longitudinal Time Code Relationship

This practice specifies the relationship between vertical interval time code (VITC) and longitudinal time code (LTC) when recorded on television tapes for use in the 525/60 television system.

RP 160-1997: Three-Channel Parallel Analog Component High-Definition Video Interface

This practice defines the physical characteristics of an interface using three parallel channels for the interconnection of equipment operating with analog component HDTV signals. For SMPTE 240M-1995, the signals carried across this interface have a scanning structure of 1125 lines, 60.00 fields per second, 16:9 aspect ratio, and 2:1 interlace. This interface is also appropriate for HDTV signals having other scanning structures. The intended uses of this interface are:

- To interconnect the elements of parallel analog HDTV video subsystems which use the same component sets within larger component islands or plants. Component HDTV editing and post-production suites are examples of such subsystems.
- To interconnect equipment into complete, self-contained HDTV analog component systems of relatively small size.

This practice applies to signals carried on the connectors described in the document and may not apply to component signals carried on other types of connectors. The practice also defines the preferred component video signals across the interface, including their waveform structure and levels.

RP 161-1999: Logic Design for Decoding Digital Audio Control Words in D-1 Helical Data and Control Records

This practice describes a way of decoding the digital audio control words embedded in audio sectors of the D-1 recorded tape. By the method described, recordings made by different interpretations of the early proposed format for D-1 data and control records can be retrieved with the least level of conflict.

RP 162-1993: Storage of Edit Decision Lists on 3-1/2 in Disks

This practice specifies the file and directory structure of 3-1/2 in floppy disks used for storage and interchange of edit decision lists (EDLs). The specification is by reference to International Standards for 3-1/2 in disk formats, more commonly known as the "generic DOS format." This practice does not specify the contents of an EDL.

RP 163-1992: System Service Messages

This practice details and defines the control message subset for the system service level. System service messages are used to perform system functions within a general-purpose communications channel of an interface system. This interface system shall transport data and digital control signals between equipment utilized in the production, post-production, and/or transmission of visual and aural information.

RP 164-1996: Location of Vertical Interval Time Code

The purpose of this practice is to define the preferred location of the vertical interval time code (VITC) at equipment interfaces, and its location on recorded media.

RP 165-1994: Error Detection Check words and Status Flags for Use in Bit-Serial Digital Interfaces for Television

This practice, also known as the error detection and handling (EDH) system, describes the generation of error detection check words and related status flags to be used optionally in conjunction with the serial digital interface for system M (525/59.94) and systems B, G, H, and I (625/50) digital television equipment operating with either 13.5-MHz or 18-MHz sampled 4:2:2 component digital signals or $4f_{sc}$ composite digital signals. Although it is preferred that this error-checking method be used in all serial transmitters and receivers, it is recognized that some equipment must minimize complexity. Additionally, there is nothing in this practice which should preclude its use in a parallel digital interface for 4:2:2 component digital signals.

RP 166-1995: Critical Viewing Conditions for Evaluation of Color Television Pictures

This practice specifies the environmental and surround conditions that are required in television or video program review areas for the consistent and critical evaluation of 525-line, 59.94-field television signals and other video program material at different technical facilities on properly aligned color picture monitors. This practice also is designed to provide for repeatable color grading or correction and the rendering of subjective assessments when used with RP 167-1995.

RP 167-1995: Alignment of NTSC Color Picture Monitors

This practice describes an alignment procedure for the consistent and repeatable alignment of television color picture monitors. For critical evaluation of picture program material, the aligned monitor shall be used in an environment such as that described in SMPTE RP 166-1995.

RP 168-1993: Definition of Vertical Interval Switching Point for Synchronous Video Switching

This practice defines an area of the vertical blanking interval to be used for switching of synchronous video signals. (The practice defines a switching area for both 525/60 and 625/50 signals. Values quoted for 625/50 systems should be regarded as tentative pending review by appropriate organizations. All such values are enclosed in brackets.)

The switching area is defined for 525/60 and 625/50 systems operating with composite or component, analog or digital signals. Digital signals may be in either the bit-parallel or bit-serial digital interface form.

The switching area is defined so as to be compatible with the error detection check word system used in bit-serial interfaces (see SMPTE RP 165-1994).

This practice is intended for guidance in new system design. It should be recognized that existing equipment may not switch within the defined area.

RP 169-1995: Television, Audio and Film Time and Control Cod—Auxiliary Time Address Data in Binary Groups—Dialect Specification of Directory Index Locations

This practice specifies a method of coding an auxiliary time address into the binary groups of SMPTE time and control codes, thus providing a second time address storage and storage location.

RP 170-1993: Video Tape Recorder Type-Specific Messages for Digital Control Interface

This practice describes the type-specific messages used for the control of video tape recorder devices. Video tape recorder type-specific messages are used to provide the means for the control and monitoring of a specific type of virtual machine within a general-purpose communications channel of an interface system. This interface system transports data and digital control signals between equipment utilized in the production, post-production, and/or transmission of visual and aural information.

RP 171-1993: Type-Specific Messages for Digital Control Interface of Analog Audio Tape Recorders

This practice details and defines the type-specific messages used for the control of analog audio tape recorder devices. These type-specific messages provide the means for the control and monitoring of a specific type of virtual machine and are implemented within an architecture of a greater control message language.

RP 172-1993: Common Messages for Digital Control Interface

This practice details and defines the control message subset common messages. Common messages are used to perform certain functions common to all equipment types within a general-purpose communications channel of an interface system. This interface system shall transport data and digital control signals between equipment utilized in the production, post-production, and/or transmission of visual and aural information.

RP 174-1993: Bit-Parallel Digital Interface for 4:4:4:4 Component Video Signal (Single Link)

This practice describes the means of interconnecting digital video equipment operating in system M (525/60) and complying with the 4:4:4 encoding parameters as defined in CCIR Recommendation 601-2, annex 1 with a nominal sampling frequency of 13.5 MHz. Provision is made to convey signal at 10-bit precision and to carry a fourth, auxiliary, channel as part of the signal multiplex (yielding 4:4:4:4 or 4x4 overall). The practice has application in the television studio over distances up to 100 m (320 ft).

RP 175-1997: Digital Interface for 4:4:4:4 Component Video Signals (Dual Link)

This practice describes a means of interconnecting digital video equipment operating in system M (525/60) and complying with the 4:4:4 sampling and encoding parameters defined in CCIR Recommendation 601-2, annex 1, with a nominal sampling frequency of 13.5 MHz. Provision is made to carry a fourth, auxiliary, channel as part of the signal multiplex, yielding 4:4:4:4 (or 4x4) overall.

The interface is primarily defined to convey signals having luminance, color-difference, and auxiliary components. Signals having green, red, blue, and auxiliary components may alternatively be conveyed.

This is a 10-bit interface; however, provision has been made to interconnect all signals with 8- or 10-bit precision.

RP 176-1997: Derivation of Reference Signals for Television Camera Color Evaluation

This practice is intended to define the numerical procedure for deriving reference video signals for television camera evaluation. A standard reflection test pattern is assumed and the signal levels computed by this procedure are the signals to be expected from an ideal camera as defined by the television system color specifications.

RP 177-1993: Derivation of Basic Television Color Evaluation

This practice is intended to define the numerical procedures for deriving basic color equations for color television and other systems using additive display devices. These equations are: 1) the normalized reference primary matrix which defines the relationship between RGB signals and CIE tristimulus values XYZ; 2) the system luminance equation; and 3) the color primary transformation matrix for transforming signals from one set of reference primaries to another set of reference primaries or to a set of display primaries.

RP 178-1996: Serial Digital Interface Checkfield for 10-Bit 4:2:2 Component and 4fsc Composite Digital Signals

This practice specifies digital test signals suitable for evaluating the low-frequency response of equipment handling serial digital video signals as defined by SMPTE 259M-1993. These test signals are fully valid digital component video as defined in ANSI/SMPTE 125M-1995. They are also useful digital composite video signals suitable for testing serial equipment in an out-of-service mode. Although a range of signals will produce the desired low-frequency effects, two specific signals are defined to test cable equalization and phase locked loop (PLL) lock-in, respectively.

RP 179-1994: Dialect Specification of Page-Line Directory Index for Television, Audio and Film Time and Control Code for Video-Assisted Film Editing

This practice specifies a method of coding film position, film frame to video frame phase relationship, transfer sync points, and production related data into binary groups of SMPTE time and control code. The dialect specified defines the structure of directory page-lines 0Ah to 0Fh (page 0, lines 10 through 15) 1Ah to 1Fh (page 1, lines 10 through 15), 2Fh (page 2, line 15), and D0h to EFh (pages 13 and 14, lines 0 through 15). This conforms to the multiplexing structure specified in SMPTE 262M-1995.

The page-line directory index, located in binary groups 8 and 7 of each time code frame, fully specifies the type of data stored in the remaining binary groups of that frame. The encoding method is suitable to both linear time code (LTC) and to vertical interval time code (VITC).

RP 180-1999: Spectral Conditions Defining Printing Density in Motion-Picture Negative and Intermediate Films

This practice defines the spectral conditions defining the printing density gammas of motion-picture color negative and intermediate materials. It is not intended as a replacement for the status M density spectral conditions given in ANSI/ISO 5/3 and commonly used for the evaluation of color photographic materials used for printing.

RP 181-1999: Audio Sector Time Code and Equipment Type Information for 19-mm Type D-1 Digital Component Recording

This practice is intended to provide an audio sector time code (ASTC) and an equipment type information (ETI). The ASTC can assist the D-1 DTTR to accurately read time code, especially under still-frame conditions. The ETI is designed to provide identification of the record DTTR. This practice applies to recorders operating on the 525-line television system with a frame frequency of 29.97 Hz and the 625-line television system with a frame frequency of 25 Hz.

RP 182-1995: List of Virtual Machine Numbers for ESbus and ESlan Systems

This practice defines the allocation of virtual machine type numbers to machines operating within the ESbus or ESlan remote control protocols. This practice is to be used in conjunction with SMPTE EG 29-1993, and other documents specified in annex A.

RP 183-1995: Monitoring and Diagnostics Processors

This practice defines the type-specific messages that enable the transport of status monitoring and diagnostics protocol (SMDP) data as defined in SMPTE 273M-1995 bidirectionally over ESbus and ESlan. It defines the diagnostics processor as a distinct virtual machine type. It is intended for use when it is desired to pipe transport) SMDP information over ESbus or ESlan as an alternative to the use of the dedicated SMDP transport mechanism.

RP 184-1995: Measurement of Jitter in Bit-Serial Digital Interfaces

This practice describes techniques for specifying and measuring output jitter from self-clocking bit-serial digital sources. It is specifically intended for, but not limited to, ANSI/SMPTE 259M-1993 serial systems.

RP 185-1995: Classification of Projection Depth of Focus

This practice specifies a test procedure for classification of the depth of focus of a projection system.

RP 186-1995: Video Index Information Coding for 525- and 625-Line Television Systems

This practice is intended to provide a method of coding the video index information data structure in 525-line and 625-line component digital video signals so that various picture and program related source data can be carried in conjunction with a video signal. Specific details of transporting the data structure through various video interconnection systems are not included in this practice. For 525-line systems, these may be found in ANSI/SMPTE 125M-1995 and ANSI/SMPTE 267M-1995 (625-line documentation is in preparation in ITU-R).

RP 187-1995: Center, Aspect Ratio and Blanking of Video Images

This practice defines picture center and aspect ratio for a number of existing video standards, and provides an extensible technique that may be used to define center and aspect ratio for future standards. Test charts and test patterns are described which permit image generation and display devices to be calibrated to the recommended geometry. Recommendations are made for blanking widths at various stages of the production/post-production process. Additionally, a definition is provided for a set of recommended screen units permitting simple cross reference between standards with aspect ratios of 4×3 and 16×9.

The practice is intended to be used for calibration of image generation and display devices. It is also intended as a reference for designers of such equipment (particularly graphics devices), and for designers of processing equipment such as image-manipulation devices.

RP 188-1999: Transmission of Time Code and Control Code in the Ancillary Data Space of a Digital Television Data Stream

This practice defines a transmission format for conveyance of linear (LTC) or vertical (VITC) time code data formatted according to ANSI/SMPTE 12M-1995 in 8- or 10-bit digital television data interfaces. Time code information is transmitted in the ancillary data space as defined in ANSI/SMPTE 291M-1996. Multiple codes can be transmitted within a single digital video data stream. Other time information, such as real time clock, DTTR tape timer information, and other user-defined information, may also be carried in the ancillary time code packet instead of time code. The actual information transmitted through the interface is identified by the coding of a distributed binary bit.

RP 189-1996: Organization of DPX Files on TAR Tapes

This practice describes a method (TAR) by which DPX files are organized on tape storage devices (and any other type of sequential block-oriented device). This practice describes the tape file format basic concept, physical and logical layout, the directory file format, file naming conventions, and examples of usage.

This practice describes a recommended method for tape labels that are attached to the tape shell or reel. This practice is intended for interchange only. It has not been optimized for other uses.

RP 190-1996: Care and Preservation of Audio Magnetic Recordings

This practice specifies the desirable storage conditions for audio magnetic recordings of continuing value, as they may remain in library or vault storage between periods of intermittent reproduction or duplication. Two categories of storage time are defined. The categories are in agreement with those defined by ANSI/NAPM IT9.11-1993:

- Medium-term storage conditions: Storage conditions suitable for the preservation of recorded information for a minimum of ten years.
- Extended-term storage conditions: Storage conditions suitable for the preservation of recorded information having permanent value.

RP 191-1996: Routing Switcher Type-Specific Messages for Remote Control of Broadcast Equipment

This dialect is intended to be used for the external control of routing switcher systems.

RP 192-1996: Jitter Measurement Procedures in Bit-Serial Digital Interfaces

This practice describes methods for measuring jitter performance in bit-serial digital interfaces. The techniques are specifically suited for jitter specifications that follow the form described in SMPTE RP 184.

RP 194-1998: Film Negative Cutter's Conform List

This practice describes the data format of a list generated by electronic film editing systems to be used as the master reference for a film negative cutter when conforming the original negative to the electronic work tape.

RP 195-1998: Use of the Reference Mark in Manufacturer-Printed Latent Image Key Numbers for Unambiguous Film Frame Identification

This practice specifies the procedure for unambiguously identifying film frames which have been exposed on film stock carrying latent image key numbers of an incompatible repetition rate.

RP 196-1997: Transmission of LTC and VITC Data as HANC Packets in Serial Digital Television Interfaces

This practice describes a transmission format for the transporting of linear time code (LTC) and vertical interval time code (VITC) over the SMPTE serial digital interface (ANSI/SMPTE 259M). The data packets will be transmitted in the horizontal ancillary data space (HANC). Mapping of data is for 10-bit interfaces only.

RP 197-1998: Film-to-Video Transfer List

This practice describes the data format of a list of film-to-video transfers. This list may be stored and referenced to document the process of transferring pictures and sound from film-related media to video media. The information documented in this list may include general production and post-production information (date, facility, film type). The list also contains the specific time (synchronization) relationships between source and destination elements. The storage medium for this list is not specified by this practice.

RP 198-1998: Bit-Serial Digital Checkfield for Use in High-Definition Interfaces

This practice specifies digital test signals suitable for evaluating the low-frequency response of equipment handling high-definition serial digital video signals as defined in ANSI/SMPTE 292M. Although a range of signals will produce the desired low-frequency effects, two specific signals are defined to test cable equalization and phase locked loop (PLL) lock-in, respectively. In the past, these two signals have been colloquially called "pathological signals."

RP 199-1999: Mapping of Pictures in Wide-Screen (16:9) Scanning Structure to Retain Original Aspect Ratio of the Work

This practice describes a method of mapping images originating in aspect ratios different from 16:9 into a 16:9 scanning structure in a manner that retains the original aspect ratio of the work.

RP 200-1999: Relative and Absolute Sound Pressure Levels for Motion-Picture Multi-channel Sound Systems

This practice specifies the measurement methods and wide-band sound pressure levels for review rooms and indoor theaters. Together with ANSI/SMPTE 202M, B-chain electroacoustic response, it is intended to assist in standardization of reproduction of motion-picture sound in such rooms.

RP 201-1999: Encoding Film Transfer Information Using Vertical Interval Time Code

This practice specifies a method of encoding video tape time code, film edge numbers, and production time code into three vertical interval time code lines. This practice is intended for use in post-production as a means of conveying the essential address elements that define the film to tape transfer. Normally this information will not be in the final program version. This practice defines the encoding in two parts; the first part specifies the data that will be encoded and the second part specifies specific methods of encoding the data into three vertical interval lines for analog and digital video signals, and into digital video ancillary data time code.

RP 202: Video Alignment for MPEG-2 Coding

Equipment conforming to this practice will minimize artifacts in multiple generations of encoding and decoding by optimizing macroblock alignment. As MPEG-2 becomes pervasive in emission, contribution, and distribution of video content, multiple compression and decompression (codec) cycles will be required. Concatenation of codecs may be needed for production, post-production, transcoding, or format conversion. Any time video transitions to or from the coefficient domain of MPEG-2 are performed, care must be exercised in alignment of the video both horizontally and vertically as it is coded from the raster format or decoded and placed in the raster format.

RP 203-2000: Real Time Opportunistic Data Flow Control in an MPEG-2 Transport Emission Multiplex

This practice defines the means of implementing opportunistic data flow control in a DTV MPEG-2 transport broadcast according to flow control messages defined in SMPTE 325M. An emissions multiplexer requests opportunistic data packets as the need for them arises and a data server responds by forwarding data already inserted into MPEG-2 transport stream packets. The control protocol that allows this transfer of asynchronous data is extensible in a backward compatible manner to allow for more advanced control as may be necessary in the future.

RP 204-2000: SDTI-CP MPEG Decoder Templates

This practice defines decoder templates for the encoding of SDTI content packages (SDTI-CP) with MPEG coded picture streams.

RP 205-2000: Application of Unique Material Identifiers in Production and Broadcast Environments

This practice defines the use in service of unique material identifiers (UMIDs) in production and broadcast environments. The UMID is one of a number of unique identifiers defined in the forthcoming SMPTE metadata dictionary, SMPTE RP 210. The UMID separately identifies picture, audio, data, and other essence.

RP 206-1999: Opportunistic Data Flow Control Using Ethernet as a Control Channel in an MPEG-2 Transport Emissions Multiplex

This practice proposes a means of implementing opportunistic data flow control in a DTV MPEG-2 transport broadcast using the flow control messages defined in SMPTE 325M. This practice defines an implementation that uses ethernet as the control channel from the emissions multiplexer to the data server. The data server's data are delivered to the emissions multiplexer via either an ASI, SDTI, or Ethernet point-to-point connection.

RP 209-2000: Format for Transmission of HD-D5 Compressed Video and Audio Data Over 360 Mb/s Serial Digital Interface

This practice defines the data stream used for synchronous transmission of HD-D5 compressed video and audio data over 360 Mb/s serial digital interface (SDI) for the 525/60 system as defined in ANSI/SMPTE 259M. This practice does not define data stream structure applicable for transmission over the serial data transport interface (SDTI), SMPTE 305M.

5.5 Scopes of SMPTE Engineering Guidelines

EG 1-1990: Alignment Color Bar Test Signal for Television Picture Monitors

This guideline specifies the purpose, format, and usage of a television picture monitor alignment color bar test signal with chroma set and black set signals.

EG 2-1999: Edge Identification of Motion-Picture Raw Stock Containers

This guideline specifies the system and items of raw stock identification to be used by the manufacturer when the edge of raw stock containers carries such identification. Recommendations for the physical size and location of identifying information are also specified.

EG 3-1994: Projection for Technical Conferences

This guideline specifies the minimum condition and parameters for effective presentation of papers at technical conferences.

EG 5-1994: Projected Image Quality of 70-, 35- and 16-mm Motion-Picture Projection Systems

This guideline specifies the conditions for the determination of image sharpness of 70-, 35-, and 16-mm motion-picture projection systems. It also classifies the practical limits of acceptability of image sharpness when using projector alignment test films.

EG 7-1994: Audio Sync Pulse for 8-mm Type S Cameras, Magnetic Audio Recorders and Rerecording Projectors

This guideline specifies the characteristics of the audio sync pulse for 8-mm type S cameras and film recorders and for magnetic tape recorders and audio projectors used for transfer of audio records from magnetic tape and film to a magnetic edge stripe on an 8-mm type S print.

EG 8-1993: Specifications for Motion-Picture Camera Equipment Used in Space Environment

This guideline specifies the technical and operational requirements for the use of documentary and theatrical motion-picture cameras aboard a space shuttle.

EG 9-1995: Audio Recording Reference Level for Post-Production of Motion-Picture Related Materials

This guideline specifies the audio recording reference level for intra- and inter-studio operations concerned with motion-picture post-production audio recording on both sprocketed and non-sprocketed analog and digital magnetic and photographic materials.

EG 10 (proposed): Tape Transport Geometry Parameters for 19-mm Type D-1 Television Digital Component Recording

This guideline describes three feasible examples of mechanical designs and test conditions for achieving the record dimensions specified in SMPTE 224M. The parameters are for reference purposes only.

EG 12-1994: Control of Basic Parameters in the Manufacture of SMPTE Photographic and Magnetic Audio Test Films

This guideline specifies and describes the basic control parameters to be followed in the production of SMPTE-distributed audio test films.

EG 13-1986: Use of Audio Magnetic Test Films

This guideline is intended to describe the usage and practical limitations of audio magnetic test films made to SMPTE specifications.

EG 14-1999: Acoustical Background Noise Levels in Dubbing Stages

This guideline provides measurement methods and recommended levels for background noise sound pressure levels in dubbing stages. The purpose of adding intentional background noise to dubbing stages is to match the environment of the dubbing stage to the average theater environment, for best translation of the program material from dubbing stage to theater. The practice is intended for application when the background noise is essentially a steady state sound, without strong time-varying components.

The background noise may be produced by air-handling systems, by adding noise to the monitoring system, or by a combination of these. An advantage to adding noise electronically is the ability to switch off the noise, for reproduction of the fullest possible dynamic range of the program material, exceeding that of most theaters, so long as the air-handling system note of the dubbing stage is quieter than most theaters.

EG 15-1987: Recording Level for Dialog in Motion-Picture Production

This guideline provides specifications for recording dialog under conditions normally encountered in motion-picture production. The use of a typical portable analog 1/4-in tape recorder operating at 7.5 in/s is assumed. Variations on typical uses are described in the appendix.

EG 16-1992: Measurement Methods for Motion-Picture Camera Acoustical Noise—Field Method

This guideline provides a simple method for measuring the acoustical noise output of motion-picture cameras in use on the set of a production. The guideline applies to noise occurring in only one circumstance: in front of a given camera in a specific acoustical environment. Thus, the measurements given by this guideline are not comparable with others made in different situations.

This guideline also gives limits on acceptability of measured camera noise due to the combined effects of the camera and its environment. Methods for reducing camera noise which are practicable on the set are included.

EG 17-1997: B-Chain Electroacoustical Response for Preparing Magnetic Masters for Transfer to 16-mm or 35/32-mm Monaural Photographic Film

This guideline specifies the electroacoustical frequency response characteristic of the monitor system when making magnetic masters intended for transfer to 16-mm photographic negative tracks.

EG 18-1994: Design of Effective Cine Theaters

Good design has often been compromised by practical solutions and the belief that effective cinemas are not economically feasible. Contrary to this belief, many effective cine theaters have been designed for museums, universities, visitor centers, and world's fairs by creative teams of architects, acousticians, and motion-picture engineers at reasonable cost. Similar results could be achieved for the film exhibitor by architects and their consultants based on the parameters and criteria contained in this guideline.

The effective cine theater is a place in which everyone can see and hear well. The summary lists the architectural parameters and the criteria which must be addressed by the designer.

EG 20-1997: Tape Transport Geometry Parameters for 19-mm Type D-2 Composite Format for Television Digital Recording

This guideline describes two feasible examples of mechanical design and test conditions for achieving the record dimensions specified in ANSI/SMPTE 245M-1993. The parameters are for reference purposes only.

EG 21-1997: Nomenclature for Television Digital Recording of 19-mm Type D-1 Component and Type D-2 Composite Formats

This guideline explains terms as used in the documents defining the D-1 and D-2 digital television recording formats.

EG 22-1997: Description and Index of Documents for 19-mm Type D-2 Composite Television Digital Recording

This guideline contains the index for the referenced documents describing the 19-mm type D-2 composite television recording format.

EG 23-1996: Transfer of Two-Channel Stereo Audio from Audio Magnetic Film or Tape to Video Tape

This guideline specifies the handling of volume range issues when transferring two-channel stereo audio from 35-mm or 16-mm magnetic film, or other audio tape formats utilizing time-code based methods of synchronization, to any video recording media. The guideline also specifies certain labeling requirements for audio masters.

EG 24-1995: Video and Audio Alignment Tapes and Procedures for 1-in Type C Helical-Scan Television Analog Recorders

This guideline describes the use of a manufacturer's alignment tape(s) intended for aligning type C television analog recorders to SMPTE specifications.

EG 25-1996: Telecine Scanning for Film Transfer to Television

This guideline specifies the maximum film image area and minimum image size range that a telecine should be capable of scanning to transfer motion-picture images to television.

EG 26-1995: Audio Channel Assignments for Digital Television Tape Recorders with AES/EBU Digital Audio Inputs

This guideline specifies the allocation of input audio signals to digital audio channels on digital television tape recorders (DTTRs) when the inputs are connected through AES/EBU digital interfaces. This guideline also specifies preferred assignments of programs to audio recording channels, on the basis of program type, for purposes of program exchange.

EG 27-1994: Supplemental Information for ANSI/SMPTE 170M and Background on the Development of NTSC Color Standards

This guideline is intended to provide supplemental and background information for ANSI/SMPTE 170M-1994. This guideline also provides information on the development of the National Television System Committee 1953 Recommendations for Transmission Standards for Color Television. Use of this guideline will aid in the understanding and implementation of modern NTSC television signals, equipment, and practices.

EG 28-1993: Annotated Glossary of Essential Terms for Electronic Production

This glossary brings together many of the terms that are essential in the production process. It draws upon the existing data bases when appropriate, noting the specific definitions and interpretations that apply to program production. Its intent is to include only those terms which impact upon production's basic needs and procedures, and whose clear interpretation in this environment is therefore essential.

EG 29-1993: Remote Control of Television Equipment

This guideline provides a guide to the architecture of the SMPTE/EBU ESbus digital control interface and related interfaces, which were developed for the purpose of standardizing the control of television equipment. The digital control interface was developed jointly by the SMPTE and the European Broadcasting Union (EBU).

The referenced documents define the technical specification and system characteristics required to allow the control of television production and distribution equipment.

EG 30-1995: Implementation of ESlan Standards

This guideline defines the architecture, the component layers, and the relationships governing the ESlan-1 and ESlan-2 suite of control and data networks to be used for audio and television program production, post-production, and distribution equipment. It should be read in conjunction with SMPTE EG 29-1993, which describes the basic control system; with SMPTE 275M-1995, which defines the services and protocols contained within the lower layers of the network model for moderate-scale systems; and with other documents as listed in annex A.

EG 31-1995: Considerations for Cassette Bar Code Readers

This guideline illustrates examples of possible manufacturing implementations of the cassette label recess area which may impact the reading of bar code labels. It applies to all SMPTE standardized cassettes which employ a label area utilized for bar code identification.

EG 32-1996: Emphasis of AES/EBU Audio in Television Systems and Preferred Audio Sampling Rate

This guideline defines the use of emphasis and sample rate in digital audio systems within television facilities. The AES/EBU digital audio standard has four conditions of emphasis and three basic sample rates. Although any or all of the possible emphasis and sample rate combinations can be used, it is good engineering practice to use a common emphasis and a single sample rate within a television facility.

EG 33-1998: Jitter Characteristics and Measurements

This guideline presents some of the special aspects of working with digital video signals that have been serialized in accordance with ANSI/SMPTE 259M, ANSI/SMPTE 292M, or ANSI/SMPTE 294M.

EG 34-1999: Pathological Conditions in Serial Digital Video Systems

This guideline examines the types of jitter in directly transmitted data signals, the methods for measuring each one, and some of the impacts they can have on system operation.

EG 35-1999: Time and Control Code Time Address Clock Precision for Television, Audio and Film

This guideline is specifically applicable to video systems operating at a frame rate of 29.97 Hz although its principles may be applicable to other systems.

EG 36-2000: Transformation Between Television Component Color Signals

This guideline describes the derivation of the transformation matrices and lists example transformations between color component signals adopting ITU-R BT.601 and ITU-R BT.709 luma/chroma equations for both the digital and analog component signal sets.

Chapter

6

CEA Standards

Jerry C. Whitaker, editor

6.1 Introduction

The following documents relating to television broadcasting have been adopted by the Consumer Electronics Association. For additional information contact the CEA at www.ce.org.

6.2 Antennas

CEB6-C: TV Receiving Antenna Manufacturers Guide to Categorizing Antennas for Use With the CEA TV Antenna Sector Map Program, Antenna Types and Characteristics, Minimum Performance Requirements, Packaging, and Marking Specifications

This bulletin was prepared to provide manufacturers of television receiver antennas with appropriate guidance on determining antenna categories and minimum performance requirements to comply with the CEA TV Antenna Selector Map program. Essential elements of this program include color coding of various television reception environments in a market, with corresponding color-coded marking of antennas to provide reception. See EIA-774 for test and measurement procedures. The CEA copyrighted logo for marking antennas and logo use guidelines are available from CEA.

CEB7: TV Receiving Antenna Manufacturers Guide to Indoor Antennas for Use with the CEA Indoor TV Antenna Certification Program, Indoor Antenna Characteristics, Packaging and Marking Specifications, and Minimum Performance Requirements"

This bulletin was prepared by the CEA R-5 Antennas Committee to provide manufacturers of indoor television receiver antennas with appropriate guidance on determining antenna characteristics and minimum performance requirements to comply with the CEA Indoor TV Antenna Certification program. Essential elements of this program include indoor TV antenna technical specifications and the use of a CEA copyrighted certification logo. Test and measurement procedures are contained in the EIA-774 standard, and the CEA copyrighted logo for marking antennas and logo use guidelines are available from CEA.

EIA-774: TV Receiving Antenna Performance Presentation and Measurement

This standard is intended to provide television receive antenna manufacturers with appropriate test and measurement procedures to examine antenna performance parameters necessary to comply with elements of the CEMA TV Antenna Selector Map program, specifically EIA CEB-6. Essential elements include procedures to determine antenna gain, front-to-back ratio, directivity, and distortion performance of active antennas with integrated amplifiers.

6.3 Television

CEB1-A: Recommended Practice for Content Advisories

This bulletin provided guidance to receiver manufacturers and designers concerning expected receiver response to content advisory information transported in the EIA-744-A U.S. or Canadian content advisory (program rating) packet. EIA/CEA-608-B incorporates this information.

CEB2: Recommended Practice for Expansion of Extended Data Service (XDS) to Include Cable Channel Mapping System Information

This bulletin provided guidance to receiver designers and manufacturers concerning expected receiver response to channel allocation data packets. EIA/CEA-608-B incorporates this information. EIA-CEB-2 is withdrawn. See EIA/CEA-608-B.

CEB4: Recommended Practice for VCR Specifications

The purpose of this recommended practice is to: 1) include essential information for the VHS VCR user, and 2) standardize the format for the presentation of the information.

CEB5: Recommended Practice for DTV Receiver Monitor Mode Capability

CEB5 is intended to provide recommendations to digital television (DTV) designers/manufacturers concerning a "monitor" mode capability. See EIA-762 for minimum specifications for a DTV remodulator.

CPEB1: Standard Method of Measurement of Ionizing Radiation from Television Receivers for Factory Quality Assurance

Establishes a standard method of obtaining ionizing radiation characteristics from production television receivers.

CPEB2: Definition of Normal Operating Conditions for Television Receivers

Specifies normal operating conditions of broadcast television receivers for the purpose of X-radiation measurements.

CPEB3: Measurement Instrumentation for X-Radiation from Television Receivers

Specifies instrumentation for the measurement of X-radiation emission from television receivers.

EIA/CEA-608-B: Line 21 Data Services

Serves as a technical guide for those providing encoding equipment and/or decoding equipment to produce material with encoded data embedded in Line 21 of the vertical blanking interval of the NTSC video signal. It is also a usage guide for those who will produce material using such equipment. Revision incorporates content advisory.

EIA/CEA-775.1: Web Enhanced DTV 1394 Interface Specification

This standard includes mechanisms to allow a source of MPEG service to utilize the MPEG decoding and display capabilities in a DTV.

EIA/CEA-775.2: Service Selection Information for Digital Storage Media Interoperability

A digital storage device such as a D-VHS or hard disk digital recorder may be used by the DTV or by another source device such as a cable set-top box to record or time-shift digital television signals. This standard supports the use of such storage devices by defining Service Selection Information (SSI), methods for managing discontinuities that occur during recording and playback, and rules for management of partial transport streams.

EIA/CEA-805: Data Services on the Component Video Interfaces

This standard specifies how data services are carried on component video interfaces (CVI), as described in EIA-770.1-A (for 2H 480p signals only), EIA-770.2-A (for 2H 480p signals only) and EIA-770.3-A. This standard applies to all CE devices carrying data services on the CVI vertical blanking interval (VBI). This standard does not apply to signals which originate in 1H 480I, as defined in EIA-770.1-A and EIA-770.2-A. The first data service defined is Copy Generation Management System (CGMS) information, including signal format and data structure when carried by the VBI of standard definition progressive and high definition YP_bP_r type component video signals. It is also intended to be usable when the YP_bP_r signal is converted into other component video interfaces including RGB and VGA.

EIA/CEA-818-B: Cable Compatibility Requirements

This standard defines the minimum requirements that shall be met by digital cable TV systems and digital TV receivers such that the receivers may be connected directly to the RF output of the cable system to provide selected baseline services.

EIA/CEA-819: Cable Compatibility Requirements for Two-Way Digital Cable TV Systems

This standard defines the minimum requirements that shall be met by two-way digital cable TV systems and two-way digital TV receivers such that the receivers may be connected directly to the RF output and input of the two-way cable system to provide specified services.

EIA/CEA-849: Application Profiles for EIA-775A Compliant DTVs

This standard specifies profiles for various applications of the EIA-775A standard. The application areas covered here include digital streams compliant with ATSC terrestrial broadcast, direct-broadcast satellite (DBS), OpenCable™, and standard definition Digital Video (DV) camcorders.

EIA/CEA-863: Connection Color Codes for Home Theater Systems

This standard defines the colors for marking connections commonly used for electronic devices in a home theater system. This standard adds continuity to installation information and assures consistency of information to installers.

EIA/CEA-CEB-8: Consideration of EIA-608-B Data Within the DTV Closed Captioning (EIA-708-B) Construct

EIA/CEA CEB-8 provides guidance on the use and processing of the EIA/CEA-608-B data stream embedded within the ATSC MPEG-2 video elementary transport stream. EIA/CEA CEB-8 augments EIA-708-B (addressing DTV Closed Captioning) Sections 4.3 and 9.23.

EIA/IS-105.1: Decoder Interface Standard

Specifies an interconnection method for attaching a cable decoder to a piece of consumer electronics equipment such as a TV or a VCR. Two ports—an audio/video/control port and an IF port—are covered under this interim standard. This specifies the physical characteristics of the interface between the decoder and the receiver.

EIA/IS-16-A: Immunity of Television Receivers and Video Cassette Recorders (VCRs) to Direct Radiation from Radio Transmissions, 0.5 to 30 MHz

This Interim Standard establishes as a performance guideline for the immunity of TV receivers and VCR's to direct radiation from radio transmissions below 30 MHz. It provides the recommended procedure for measuring the immunity of TV receivers and VCRs over the frequency range of 0.5 to 30 MHz.

EIA/IS-31: Recommended Design Guideline, Rejection of Educational Interference to Ch. 6 Television Reception

Establishes the design guideline for a color television receiver or VCR in the play-through mode to provide rejection of educational (and non-commercial) FM interference to Ch. 6 television reception equal to or exceeding that of the FM median receiver. It defines the signal and operation of the TV product, and the evaluation criteria to be used in implementing the interference immunity guideline. It also establishes guidelines for signal levels at which TV products should coexist with other RF sources, without material performance degradation.

EIA/IS-702: Copy Generation Management System (Analog)

This standard included packet description data relating to the Copy Generation Management System (Analog) (CGMS-A). EIA/CEA-608-B incorporates this information. EIA-702 is withdrawn. See EIA/CEA-608-B.

EIA/TIA-250-C: Electrical Performance for Television Transmission Systems (ANSI/EIA/TIA-250-C-89)

This Standard specifies the minimal transmission performance characteristics, consistent with good engineering practice, of television transmission of 525-line NTSC color or monochrome video and associated audio signals suitable for television broadcasting or for similar application. These limits are used for the acceptance of new systems or restoration of existing systems after maintenance. It should be noted that transmission systems utilize analog, digital, or a mixture of analog and digital techniques. Definitions, standards, and methods of measurement are given for both the video and related audio signals being carried from a few hundred feet to thousands of miles, including satellite transmission.

EIA-105.2: Decoder Interface Control Standard

The EIA/IS-105 specification is designed to allow consumer electronic devices such as television, VCRs, and other receiving devices to be connected to various external devices via the Decoder Interface Multi-pin connector.

EIA-170: Electrical Performance Standards Monochrome Television Studio Facilities

Establishes definitions, minimum standards, and methods of measurement for the electrical performance for monochrome television studio facilities. It is intended to apply only to locally generated signals; that is, signals generated in the studio itself or at a nearby point where control can be exercised over picture quality.

EIA-189-A: Encoded Color Bar Signal

The EIA Standard Color Bar Signal is intended for use as a test signal for the following principal reasons: a) adjustment of color monitors, b) adjustment of color encoders, and c) rapid checks of color television transmission systems.

EIA-23: RF Interface Specification for Television Receiving and Cable Television Systems

This specification is intended to apply to all cable systems and to all receiving devices which may be directly connected to a cable system residential outlet, including, but not limited to, television sets, video cassette recorders, and converters (whether furnished by cable operators or independently acquired by subscribers).

EIA-378: Measurement of Spurious Radiation from FM and TV Broadcast Receivers in the Frequency Range of 100 to 1000 MHz, Using the EIA Laurel Broadband Antenna

Describes potential sources of spurious radiation from frequency modulation and television broadcast receivers and establishes measurement methods whereby the strength of some radiations may be determined.

EIA-462: Electrical Performance Standards for Television Broadcast Demodulators

This Standard includes aural and visual performance standards for various functions and operating modes which may or may not be offered by a particular demodulator manufacturer, or may be offered by a manufacturer not in one instrument, but in two or more separate instruments.

EIA-544: Immunity of TV and VCR Tuners to Internally Generated Harmonic Interference from Signals in the Band 535 kHz to 30 MHz (ANSI/EIA-544-88)

This Standard establishes performance guidelines for rejection of interference by television receivers, video cassette recorders, and tuners. It details a measurement procedure which determines the level of interfering signal which will generate harmonics in the tuner, causing interference 40 dB below a desired signal at the intermediate frequency (IF) output of a tuner under test. This standard covers interference immunity to CB, amateur radio, and other transmissions.

EIA-679-B: National Renewable Security Standard (NRSS)

NRSS provides two physical designs. Part A defines a removable and renewable security element form factor that is an extension of the ISO-7816 standard. Part B defines a removable and renewable security element based on the PCMCIA ("C Card") form factor. The common attributes allow either an NRSS-A or NRSS-B device to provide security for applications involving pay and subscription cable or satellite television services, telephone, and all forms of electronic commerce.

EIA-693: Audio/Video Bus (AVBus) Physical Layer and Media Specification (ANSI/EIA-693-97)

This standard contains the performance specifications necessary to implement an Audio/Video (AV) Bus to carry baseband audio, video and control signals for limited distances between consumer audio and video equipment in the home. The purpose of this standard is to present all the information necessary for the development of an AV physical network and devices to communicate and share information over the network.

EIA-708-B: Digital Television (DTV) Closed Captioning

This document is intended as a definition of DTV Closed Captioning (DTVCC) and provides specifications and/or guidelines for caption service providers, DTVCC decoder and encoder manufacturers, DTV receiver manufacturers, and DTV signal processing equipment manufacturers. This specification includes: a) a description of the transport method of DTVCC data in the DTV signal, b) a description of DTVCC specific data packets and structures, c) a specification of how DTVCC information is to be processed, d) a list of minimum implementation recommendations for DTVCC receiver manufacturers, and e) a set of recommended practices for DTV encoder and decoder manufacturers.

EIA-744-A: Transport of Content Advisory Information Using Extended Data Service (XDS)

The XDS Content Advisory (program rating) packet transports essential content advisory information viewers need to block selected programming. EIA/CEA-608-B incorporates this information. EIA-744-A is withdrawn. See EIA/CEA-608-B.

EIA-745: Transport of Cable Channel Mapping System Information Using Extended Data Service (XDS)

This standard defined three new Miscellaneous Class data packets within XDS that transport information necessary to permit viewers to select a desired channel based on its broadcast channel assignment. EIA/CEA-608-B incorporates this information. EIA-745 is withdrawn. See EIA/CEA-608-B.

EIA-746-A: Transport of Internet Uniform Resource Locator (URL) Information Using Text-2 (T-2) Service

This document was a proposed amendment to EIA-608 to insert Internet Uniform Resource Locators (URLs) within the line-21 data system using the Text-2 (T-2) service. EIA/CEA-608-B incorporates this information. EIA-746-A is withdrawn. See EIA/CEA-608-B.

EIA-752: Transport of Transmission Signal Identifier (TSID) Using Extended Data Service (XDS)

This document was a proposed amendment to EIA-608 to include a unique 16-bit Transmission Signal Identifier in a new Extended Data Service (XDS) packet. EIA/CEA-608-B incorporates this information. EIA-752 is withdrawn. See EIA/CEA-608-B.

EIA-761-A: DTV Remodulator Specification with Enhanced OSD Capability

This standard defines minimum specifications for a one-way data path utilizing an 8 VSB trellis or a 16 VSB remodulator in compliance with ATSC Standard A/53, Annex D. This standard also defines on-screen display (OSD) capabilities. This standard applies to any type of device used to connect to an ATSC compliant digital television receiver (DTV) receiver. Devices meeting this standard should interoperate with any ATSC compliant receiver that also supports "monitor mode." This standard addresses required RF output specifications, on-screen display (OSD) capabilities, and capability profiles for a DTV remodulator and recommendations concerning input to the remodulator. This standard does not address 8-VSB without OSD. For information concerning 8 VSB without OSD, see also EIA-762 and EIA-799.

EIA-762: DTV Remodulator Specification

This standard defines a minimum specification for a one-way data path utilizing an 8-VSB trellis remodulator in compliance with ATSC A/53, Annex D. This standard applies to any type of device used to connect to an ATSC compliant digital television receiver (DTV) receiver. Devices meeting this standard should interoperate with any ATSC compliant receiver that also supports "monitor mode" (see EIA CEB-5.) This standard addresses both required RF output specifications for a DTV remodulator and recommendations concerning input to the remodulator.

EIA-766-A: U.S. and Canadian Region Rating Tables (RRT) and Content Advisory Descriptors for Transport of Content Advisory Information using ATSC A/65-A Program and System Information Protocol (PSIP)

This standard augments ATSC Standard A/65-A and SCTE DVS-097 Rev. 7, both titled Program and System Information Protocol for Terrestrial Broadcast and Cable (PSIP). Along with the above two standards, this standard designates the RRT which provides the receiver with the definition of the rating system and the Content Advisory Descriptor which provides the receiver with the specific program rating for each program. Specifically, this standard specifies the exact syntax to be used to define the U.S. and Canadian Rating Region Tables (RRT) in accordance with A/65-A Section 6.4 as well as the exact syntax to be used in the Content Advisory Descriptors that convey the rating information for each program in accordance with A/65-A Section 6.7.4. Thus DTV receivers may block unwanted programs as determined by the user.

EIA-770.1-A: Analog 525 Line Component Video Interface—Three Channels

This standard defines the physical characteristics of an interface and the parameters of the signals carried across the interface, using three parallel channels for the interconnection of equipment operating with analog component video signals. This standard includes specifications for two scanning structures: 1H, having 525 lines, 59.94 fields/second, 2:1 interlaced, and a horizontal scanning rate of 15.734 kHz; and 2H, for doubled scanned interfaces having 525 lines, 59.94 frames/second, progressively scanned, and having a horizontal scanning rate of 31.47 kHz. Both interfaces shall be capable of either 4:3 or 16:9 aspect ratios.

EIA-770.2-A: Standard Definition TV Analog Component Video Interface

This standard defines the physical characteristics of an interface and the parameters of the signals carried across that interface, using three parallel channels for the interconnection of equipment operating with analog component video signals. The standard includes specifications for: 1) 480i video format defined by 480 active lines, 525 total lines, 2:1 interlaced at 59.94 or 60 fields/second; and 2) 480p video format defined by 480 active lines, 525 total lines, progressively scanned at 59.94 or 60 frames/second. Both video formats shall be capable of either 4:3 or 16:9 aspect ratios.

EIA-770.3-A: High Definition TV Analog Component Video Interface

This standard defines two raster-scanning systems for the representation of stationary or moving two-dimensional images sampled temporally at a constant frame rate. The first image format specified is 1920 × 1080 samples (pixels) inside a total raster of 1125 lines. The second image format specified is 1280 × 720 samples (pixels) inside a total raster of 750 lines. Both image formats shall have an aspect ratio of 16:9.

EIA-775-A: DTV 1394 Interface Specification

This standard defines a specification for a baseband digital interface to a DTV using the IEEE-1394 bus and provides a level of functionality that is similar to the analog system. It is designed to enable interoperability between a DTV compliant with this standard and various types of consumer digital audio/video sources including digital set-top boxes (STBs) and analog/digital hard disk or videocassette recorders (VCRs).

EIA-796: NRSS Copy Protection Systems

The copy protection systems that have been included in EIA-796 are itemized for the purpose of identification. The systems outlined in EIA-796 all support the copy protection frameworks described in EIA-679-B, Parts A and B.

EIA-799: On-Screen Display Specification

This standard specifies syntax semantics for bitmapped graphics data typically used for on-screen display (OSD). The standard is applicable whenever it is necessary to specify a standard method for delivery of bitmapped graphics data. The pixel formats include optional alpha-blend and transparency attributes to support composition of graphics over analog or digitally decoded video within the display.

EIA-806: Carriage of DTV PSIP Information Using the XDS Method

This standard addresses transmission of DTV PSIP data using the XDS data carriage method. EIA/CEA-608-B incorporates this information. EIA-806 is withdrawn. See EIA/CEA-608-B.

IETNTS1: Industrial Electronic Tentative Standard No. 1 (IETNTS1), Color Television Studio Picture Line Amplifier Output Drawing

Specifies timing parameters within the horizontal blanking interval of a color television signal. This document is based on EIA-170.

TVSB5: Multichannel TV Sound System BTSC System Recommended Practices

Specifies the transmission of multichannel television sound (MTS) in accordance with the BTSC system and the FCC rules governing its use. This document is intended for both manufacturers and broadcasters.

6.4 Television Receivers

EIA-336: Color Coding for Chassis Wiring (ANSI/EIA-336-68) (R73) (R79)

The colors described in this Standard are as defined in EIA-359, "EIA Standard Colors for Color Identification and Coding."

6.5 Home Networks

ANSI/EIA-600.10: Introduction to the CEBus Standard

The EIA-600 Specification covers the overall topology of the EIA-600 network and the detailed topology for each individual medium used; the electrical and physical specifications for the media usable by EIA-600; the physical interface from a device to the medium and the signaling method specifications to be used on the medium; the protocol to be used for network access and the description of the control message format; and a command language that allows all devices to communicate a common set of functions to be performed. Aspects of the overall EIA-600 network that are not addressed in this specification are operation and maintenance of the network. This standard establishes a minimal set of rules for compliance. It does not rule out extended services that may be provided, as long as the rules of this standard are adhered to within the system. It is, in fact, the intention of the standards to permit extended services (defined by users) to exist.

ANSI/EIA-600.31: Power Line Physical Layer and Medium Specification (ANSI/EIA-600.31-97)

This document is the preliminary specification for the CEBus Power Line (PL) Physical Layer and Media portion of the Physical Layer and Media Specifications of EIA-600. Its purpose is to present the information necessary for the development of a PL physical network and devices to communicate and share information over the network. This is one of a series of documents covering the various media that comprise the CEBus standard.

ANSI/EIA-600.32: Twisted Pair Physical Layer and Medium Specification (ANSI/EIA-600.32-97)

This document is the specification for the CEBus Twisted Pair (TP) Physical Layer and Medium. Its purpose is to present all the information necessary for the development of a TP physical network and devices to communicate and share information over that network in an orderly manner. This is one of a series of documents covering the various media that comprise the CEBus standard.

ANSI/EIA-600.33: Coax Cable Physical Layer and Medium Specification (ANSI/EIA-600.33-97)

This document is the preliminary specification for the CEBus Coax (CX) Physical Layer and Medium. Its purpose is to present all the information necessary for the development of a CX physical network and devices to communicate and share information over that network in an orderly manner. This is one of a series of documents covering the various media that comprise the CEBus standard.

ANSI/EIA-600.34: IR Physical Layer and Medium Specification (ANSI/EIA-600.34-97)

This document is a preliminary specification for the CEBus Infrared (IR) Physical Layer and Medium portion of the Physical Layer and Medium specifications of EIA-600. Its purpose is to present all the information necessary for the development of a IR physical network and devices to communicate and share information over that network to and from IR and other CEBus media in an orderly manner. This is one of a series of documents covering the various media that comprise the CEBus standard.

ANSI/EIA-600.35: RF Physical Layer and Medium Specification (ANSI/EIA-600.35-97)

This document is the preliminary specification for the CEBus Radio Frequency (RF) Physical Layer and Medium portion of the Physical Layer and Medium specifications of EIA-600. Its purpose is to present all of the information necessary for the development of a RF physical layer for the CEBus device. This is one of a series of documents covering various media that comprise the CEBus standard.

ANSI/EIA-600.37: Symbol-Encoding Sublayer

This document describes the portion of the Node Physical Layer that interfaces to the Medium Access Control (MAC) Sublayer and to Layer System Management (LSM). This sublayer is called the Symbol Encoding (SE) Sublayer.

ANSI/EIA-600.38: Power Line/Radio Frequency Symbol Encoding Sublayer

This document describes the portion of the Power Line or RF Physical Layer that interfaces to the Medium Access Control (MAC) Sublayer and to Layer System Management (LSM). This sublayer is called the Power Line/RF Symbol Encoding (PL/RF SE) Sublayer.

ANSI/EIA-600.41: Description of the Data Link Layer (ANSI/EIA-600.41-97)

This document provides a prose description of the Data Link Layer Design for the CEBus Network. The intent of this document is to be descriptive, rather than provide a formal specification, and contains a discussion of the Data Link Layer interfaces to the Network Layer and Physical Layer, as well as a functional description of the Data Link Layer.

ANSI/EIA-600.42: Node Medium Access Control Sublayer (ANSI/EIA-600.42-97)

This part of the CEBus standard is a technical specification of the services and protocol for the Node Medium Access Control Sublayer.

ANSI/EIA-600.43: Node Logical Link Control Sublayer (ANSI/EIA-600.43-97)

This part of the CEBus standard is a technical specification of the services and protocol for the Node Logical Link Control Sublayer.

ANSI/EIA-600.45: Node Network Layer Specification

This document is the CEBus Node Network Layer part of EIA-600.

ANSI/EIA-600.46: Node Application Layer Specification

This document is the CEBus Node Application Layer part of EIA-600.

ANSI/EIA-600.81: Common Application Language (CAL) Specification (ANSI/EIA-600.81-97)

This document describes the basic framework of CAL. It is intended as an introduction to CAL operation and syntax that stresses the object-oriented aspects of CAL. It is believed that the object-oriented methodology offers the best means of understanding the complex interaction between devices, controls, and controllers present in the CEBus environment.

ANSI/EIA-600.82: CAL Context Description (ANSI/EIA-600.82-97)

This document describes the contexts, or main subsystems within a device, supported by the Common Application Language (CAL).

ANSI/EIA-633.10: Introduction to EIA-600 Conformance Specification

This standard is concerned with conformance of an implementation to the associated protocol specifications contained in EIA-600. A dual audience is expected for this standard. The first is laboratories that may be interested in acting as conformance testing agencies. Such agencies are tasked with converting these requirements into a hardware/software test system. The second audience is the set of designers of EIA-600 compatible products. Any designer of compatible products should understand the importance of the tests described in this standard, as they relate to implementing EIA-600. The second group should read this standard while considering whether its hardware and/or software implementation is likely to pass the stated tests.

ANSI/EIA-633.31: Power Line Physical Layer Conformance Specification

This portion of the conformance standard specifies tests to determine conformance of a Node's Power Line (PL) PL Physical Layer to IS-60. Part one of this standard provides an overview of the conformance philosophy. The reader is urged to review that material before attempting to use the details provided in this part.

ANSI/EIA-633.32: Twisted Pair Physical Layer Conformance (ANSI/EIA-633.32-97)

This standard specifies tests to determine conformance of a device's Twisted Pair Physical Layer to EIA-600.

ANSI/EIA-633.34: Infrared Physical Layer Conformance (ANSI/EIA-633.34-97)

This standard specifies tests to determine conformance of a Node's IR Physical Layer to EIA-600.

ANSI/EIA-633.37: Symbol Encoding Sublayer Physical Layer Conformance (ANSI/EIA-633.37-97)

This standard specifies tests to determine conformance of a Node's Symbol Encoding Sublayer to EIA-600.

ANSI/EIA-633.38: PL and RF Symbol Encoding Physical Layer Conformance (ANSI/EIA-633.38-97)

This standard specifies tests to determine conformance of a Node's Power Line or RF Symbol Encoding Sublayer to EIA-600.

ANSI/EIA-633.46: Node Application Layer Conformance Specification

This portion of the conformance standard specifies tests to determine conformance of a Node's Application Layer to EIA-600.

ANSI/EIA-633.81: CAL Conformance Specification

This portion of the conformance standard specifies tests to determine conformance of a Node's CAL to EIA-600.81. Part one of this standard provides an overview of the conformance philosophy. The reader is urged to review that material before attempting to use the details provided in this part.

ANSI/EIA-709.1-A: Control Network Protocol Specification (ANSI/EIA-709.1-A-99)

This specification applies to a communication protocol for networked control systems. The protocol provides peer-to-peer communication for networked control and is suitable for implementing both peer-to-peer and master-slave control strategies. This specification describes services in layers 2-7. In the layer 2 specification, it also describes the data link layer and the MAC sub-layer interface to the physical layer. The physical layer provides a choice of transmission media. The interface described in this specification supports multiple transmission media at the physical layer.

ANSI/EIA-709.2-A: Control Network Power Line (PL) Channel Specification (ANSI/EIA-709.2-A-99)

This document specifies the Control Network Power Line (PL) Channel and serves as a companion document to the EIA-709.1 Control Network Protocol Specification. Its purpose is to present the information necessary for the development of a PL physical network and nodes to communicate the share information over the network. This is one of a series of documents covering the various media that comprise the EIA-709 Standard.

ANSI/EIA-709.3: Free-Topology Twisted-Pair Channel Specification

This document specifies the EIA-709.3 free-topology twisted-pair channel and serves as a companion document to the EIA-709.1 Control Network Protocol Specification. The channel supports communication at 78.125 kbits/s between multiple nodes, each of which consists of a transceiver, a protocol processor, and application processor, a power supply, and application electronics.

ANSI/EIA-709.4: Fiber-Optic Channel Specification

In conjunction with ANSI/EIA-709.1-A Control Network Protocol Specification, EIA-709.4 defines a complete 7-layer protocol stack for communications on an EIA-709.4 single-fiber (half-duplex) fiber-optic channel. EIA-709.4 specifies the physical layer (OSI Layer 1) requirements for the EIA-709.4 fiber-optic channel which encompasses the interface to the Media Access Control (MAC) layer and the interface to the medium. The single-fiber channel implemented as specified in EIA-709.4 allows two nodes to communicate bidirectionally across a single piece of fiber cable.

ANSI/EIA-721.1: Generic Common Application Language (Generic CAL) Specification

This document describes the basic framework of Generic CAL. It is intended as an introduction to Generic CAL operation and syntax that stresses the object-oriented aspects of Generic CAL. It is believed that the object-oriented methodology offers the best means of understanding the complex interaction between devices, controls, and controllers present in a Generic Network environment.

ANSI/EIA-721.2: Generic CAL Context Description

This document describes the contexts, or main subsystems within a device, supported by the Generic Common Application Language (Generic CAL).

ANSI/EIA-721.3: Node Application Layer Specification

This Application Layer consists of four main elements. The application Process is the interface to the Application Layer. Services are provided by the Generic Common Application Language (Generic CAL) Element to the User Element of the Application Process. Generic CAL is the language framework through which Resource Allocation and Control functions are executed. Services are provided by the Message Transfer Element to the Generic CAL Element. The Message Transfer Element interfaces to the lower layers of the Generic Network either directly or through the Association Control Element. The lower layers are representative of some home automation networks. Additional OSI layers may be included. An adaptation layer may be required between the Generic CAL Application Layer and the Generic Network lower layers.

ANSI/EIA-721.4: Generic Common Application Language Quality of Service

This specification for Generic CAL consists of an Application Layer containing a command language and a Message Transfer Service Element. The specifications of the lower OSI layers are not within the scope of this standard. However, the services provided by the lower layers affect the performance and composition of messages issued from the Application Layer. These lower layer service options are collectively called the Quality-of-Service (QOS) available from the communications protocol. This portion of EIA-721 standard describes the lower layer QOS options that may impact the Application Layer. Recommended capabilities are specified. Also, a mechanism to convey these options to the Generic CAL Application Layer using Layer System Management functions is presented.

ANSI/EIA-776.1: CEBus-EIB Router Communications Protocol-Description of the CEBus-EIB Router

This document describes the operation of a CEBus-EIB Router. This document is not intended to define how a router should operate, but to provide an overview of the operation and the coordination of various router elements.

ANSI/EIA-776.2: CEBus-EIB Router Communications Protocol—CEBus-EIB Router Medium Access Control Sublayer

The CEBus-EIB Router Medium Control (MAC) Sublayer is almost identical to the CEBus or EIB Node MAC Sublayer corresponding to the "CEBus Side" or the Router. The differences are in the way the Router does address matching on a received packet and on the information exchanged in some of the service primitives. Rather than copy the Node MAC specification here and make minor changes, the Router MAC is specified by exception to the Node MAC for both the CEBus and EIB Specifications.

ANSI/EIA-776.3: CEBus-EIB Router Communications Protocol—CEBus-EIB Router Logical Link Control Sublayer

This section specifies the CEBus-EIB Router Logical Link Control Sublayer interfaces to the Router Network Layer and to the Layer System Management. The interfaces are described in terms of service primitives which are abstract interfaces across a layer boundary. A service primitive represents an exchange of information into or out of a layer. Although service primitives are defined using a format similar to that of programming language procedure calls, no implementation technique is implied.

ANSI/EIA-776.4: "CEBus-EIB Router Communications Protocol—CEBus-EIB Router Network Layer

The CEBus-EIB Router Network Layer is conceptually divided into several elements, each performing distinct well-defined services. Each element may be thought of as an independent process that communicates with the other elements and protocol layers through specified interfaces.

ANSI/EIA-776.5: CEBus-EIB Router Communications Protocol—The EIB Communications Protocol

EIB is a control system for related applications in homes and buildings. The EIB system offers standardized basic and system components, e.g., Bus Coupling Units (BCU), Power Supply Units (PSU), Bus Interface Modules (BIM), Routers, and RS-232 data interfaces. EIB offers the capability of constructing devices in a modular form using system devices like BCU or BIM that support communications-specific functions. A standardized interface called Physical External Interface (PEI) reduces the expense of developing EIB devices and allows them to be exchanged.

EIA/CEA-844: XML Encoding of Generic Common Application Language

This standard specifies the encoding of Generic Common Application Language (CAL) into XML. It is based on ANSI/EIA-721 and EIA-851.

EIA/CEA-851: VHN Home Network

The R-7.4 VHN Home Network Standard defines a flexible and open network architecture and communications protocol specification for digital devices in the home.

EIA-600 CEBus SET: EIA Home Automation System (CEBus)

Provides the necessary specifications for the Consumer Electronic Bus (CEBus), a local communications and control network designed specifically for the home. The CEBus network provides a standardized communication facility for exchange of control information as data among devices and services in the home. The major motivations for its development were: a) to develop a universal low cost method for devices in the home to communicate regardless of manufacturer; b) to allow the introduction of new products and services to the home at minimal cost and confusion to consumers; c) to meet the majority of anticipated home control requirements with a single multi-media network standard; and d) to minimize the redundancy of control and operation methods among devices and equipment in the home. CEBus is intended to handle existing and anticipated control communication requirements at minimum practical costs consistent with a broad spectrum of residential applications. It is intended for such functions as remote control, status indication, remote instrumentation, energy management, security systems, and entertainment device coordination. These situations require economical connection to a shared local communication network carrying relatively short digital messages. A major objective of this specification is compatibility. It is intended that every implementation of CEBus be able to co-exist with every other implementation; that every device that meets this specification can communicate with all other CEBus devices; and the language used for control functions will be understood by all devices. This version includes portions of IS-60 that have not been revised as well as new ANSI approved updates.

6.6 Audio

CPEB6-A: Preferred Voltage and Impedance Values for the Interconnection of Audio Products

Establishes preferred voltage and impedance values for inputs and outputs of generally available, mass produced, audio products and accessories. These values guide manufacturers in the product design in order to facilitate the interconnection of products of different manufacturers and permit the addition of other products or accessories to integrated systems which have input and output connections.

EIA-221-A: Polarity or Phase of Microphones for Broadcasting, Recording, and Sound Reinforcement

This standard deals with connections and methods of testing which will result in correct phasing of microphones for use in broadcasting, recording, and sound reinforcement. It is well-known that correct phase may be important to the operation of any system employing, simultaneously, more than one microphone. This is especially true when two similar microphones are placed in symmetrical relation to a performer.

EIA-490: Standard Test Methods of Measurement for Audio Amplifiers

Establishes methods to specify performance characteristics under both dynamic and static conditions for audio amplifiers.

EIA-560: Standard Method of Measurement for Compact Disc Players (ANSI/EIA-560-90)

Defines measurement methods and the form of disclosure for performance characteristics of consumer compact disc (CD) players. It applies to domestic reproducing equipment for CDs.

6.6a Radio Broadcast

EIA/IS-80: Audio Bandwidth and Distortion Recommendations for AM Broadcast Receivers

Specifies audio bandwidth and distortion recommendations for AM broadcast radio receivers. Applies to both AM monophonic and AM stereophonic receivers, as well as to receivers of single, multiple, or variable reception bandwidths.

EIA-549: NRSC AM Preemphasis/Deemphasis and Broadcast Audio Transmission Bandwidth Specifications (ANSI/EIA-549-88)

Describes specifications for the preemphasis of AM broadcasts, the preemphasis of AM receivers, and the audio bandwidth of AM stations prior to modulation. EIA-549 specifies the audio input to the AM broadcast transmitter (10 kHz audio, 75 μs maximum preemphasis) and deemphasis in the receiver. It applies to AM monophonic and AM stereo L + R transmissions, and to dual bandwidth and single bandwidth AM receivers.

6.6b Loudspeakers

ANSI/EIA-426-B: Loudspeakers, Optimum Amplifier Power (ANSI/EIA-426-B-98)

This standard recommends the maximum power rating for an amplifier to be connected to the speaker. This standard defines acceptable performance limits in the categories of power compression, distortion, and accelerated life testing.

ANSI/EIA-636: Recommended Loudspeaker Safety Practices (ANSI/EIA-636-96)

Provides guidance, including specifications and tests, for determining potential flammability, electric shock, and mechanical safety hazards associated with loudspeaker systems and assemblies Also provides guidance on product markings in the areas of quality control, customer service, product traceability, and proper product usage. This document addresses specific safety issues only, and should be used in conjunction with existing manufacturers' specifications and tests or as the basis of a new safety and testing program. Supersedes IS-33.

EIA-276-A: Acceptance Testing of Dynamic Loudspeakers

Specifies the acceptance testing of all types of dynamic loudspeakers (full range, woofers, horns, with or without baffle).

EIA-278-B: Mounting Dimensions for Loudspeakers

Specifies the equipment and techniques for measuring stiffness, as indicated by the deflection distance when the spider is loaded at its center by a specified mass.

EIA-299-A: Loudspeakers, Dynamic, Magnetic Structures and Impedance

Specifies nominal dimensions and weights for nine Alnico 5 magnets.

EIA-438: Loudspeaker Spiders, Test for Measuring Stiffness

Specifies the test method, equipment, and technique for measuring loudspeaker stiffness, as indicated by the deflection distance when the spider is loaded at its center by a specified mass.

Chapter

7

SCTE Standards

Jerry C. Whitaker, editor

7.1 Introduction

The standards listed in this section have been approved by the SCTE Engineering Committee as SCTE standards. Some have additionally been approved by the American National Standards Institute (ANSI) and are so designated. Note that SCTE standards numbering has been changed; the previous number is shown along with the new one.

Copies of standards which have not been approved by the Engineering Committee are no longer available outside the SCTE Standards Program because their accuracy and currency cannot be guaranteed. These include preliminary and draft standards, and standards listed as "adopted" by the relevant Subcommittee. Anyone with a need for one of these documents should contact the SCTE standards staff. SCTE can be reached on the Web at http://www.scte.org.

ANSI/SCTE 01 1996 (formerly IPS SP 400), *F Port (Female Outdoor) Physical Dimensions*

ANSI/SCTE 02 1997 (formerly IPS SP 406), *F Port (Female Indoor) Physical Dimensions*

ANSI/SCTE 03 1997 (formerly IPS TP 007), *Structural Return Loss*

ANSI/SCTE 04 1997 (formerly IPS TP 407), *F Connector Return Loss*

ANSI/SCTE 05 1999 (formerly IPS TP 408), *F Connector Return Loss In-line Pair*

ANSI/SCTE 06 1999 (formerly IPS TP 206), *Composite Distortion Measurements* (CTB, CSO)

ANSI/SCTE 07 2000 (formerly DVS 031), *Digital Video Transmission Standard for Cable Television*

SCTE 08 1996 (formerly DVS 011), *Cable and Satellite Extensions to ATSC System Information Standards*

SCTE 09 2001 (formerly IPS TP 001), *Test Method for Cold Bend*

SCTE 10 2001 (formerly IPS TP 002), *Test Method for Flexible Coaxial Cable Impact Test*

SCTE 11 2001 (formerly IPS TP 014), *Test Method for Aerial Cable Corrosion Protection Flow*

SCTE 12 2001 (formerly IPS TP 102), *Test Method for Center Conductor Bond to Dielectric for Trunk, Feeder, and Distribution Coaxial Cables*

SCTE 13 2001 (formerly IPS TP 103), *Dielectric Air Leakage Test Method for Trunk, Feeder and Distribution Coaxial Cable*

SCTE 14 2001 (formerly IPS TP 700), *Test Method for Hex Crimp Tool Verification/Calibration*

SCTE 15 2001 (formerly IPS SP 100), *Specification for Trunk, Feeder, and Distribution Coaxial*

SCTE 16 2001 (formerly IPS TP 204), *Hum Modulation*

SCTE 17 2001 (formerly IPS TP 216), *Carrier to Noise* (C/N, CCN, CIN, CTN)

SCTE 18 2001 (formerly DVS 208), *Emergency Alert Message for Cable*

SCTE 19 2001 (formerly DVS 132), *Standard Methods for Isochronous Data Services Transport*

SCTE 20 2001 (formerly DVS 157), *Standard Methods for Carriage of Closed Captions and Non-Real Time Sampled Video*

SCTE 21 2001 (formerly DVS 053), *VBI Extension for ATSC Digital Television Standards*

SCTE 22-1 1997 (formerly DSS 97-2), *Data-Over-Cable 1.0 Part 1: Service Interface Specification*

SCTE 23-2 2001 (formerly DSS 00-09), *Data-over Cable 1.1 Part 2: Baseline Privacy Interface Plus Specification*

SCTE 24-1 2001 (formerly DSS 00-02), *IPCablecom Part 1: Architecture Framework for the Delivery of Time-Critical Services Over Cable Television Networks Using Cable Modems*

SCTE 24-2 2001 (formerly DSS 00-01), *IPCablecom Part 2: Audio Codec Requirements for the Provision of Bi-directional Audio Service Over Cable Television Networks Using Cable Modems*

SCTE 24-3 2001 (formerly DSS 00-03), *IPCablecom Part 3: Network Call Signaling Protocol for the Delivery of Time Critical Services over Cable Television Networks Using Data Modems*

SCTE 24-4 2001 (formerly DSS 00-04), *IPCablecom Part 4: Dynamic Quality of Service for the Provision of Real-Time Services over Cable Television Networks Using Data Modems*

SCTE 24-5 2001 (formerly DSS 00-10), *IPCablecom Part 5: Media Terminal Adapter (MTA) Device Provisioning Requirements for the Delivery of Real-Time Services Over Cable Television Networks Using Cable Modems*

SCTE 24-6 2001 (formerly DSS 00-11), *IPCablecom Part 6: Management Information Base (MIB) Framework*

SCTE 24-7 2001 (formerly DSS 00-12), *IPCablecom Part 7: Media Terminal Adapter (MTA) Management Information Base (MIB) Requirements*

SCTE 24-8 2001 (formerly DSS 00-13), *IPCablecom Part 8: Network Call Signaling Management Information Base (MIB) Requirements*

SCTE 24-9 2001 (formerly DSS 00-14), *IPCablecom Part 9: Event Message Requirements*

SCTE 24-10 2001 (formerly DSS 00-15), *IPCablecom Part 10: Security Specification*

SCTE 24-11 2001 (formerly DSS 00-16), *IPCablecom Part 11: Internet Signaling Transport Protocol (ISTP)*

SCTE 24-12 2001 (formerly DSS 00-17), *IPCablecom Part 12: Trunking Gateway Control Protocol (TGCP)*

SCTE 24-13 2001 (formerly DSS 00-08), *IPCablecom Part 13: Electronic Surveillance Standard*

SCTE 27 1996 (formerly DVS 026), *SCTE Method—Subtitling Methods for Broadcast Cable*

SCTE 28 1998 (formerly DVS 097), *Integrated Program and System Information for Terrestrial Broadcast and Cable*

7.2 Selected Abstracts of SCTE Standards

ANSI/SCTE 07 2000 (DVS 031): Digital Video Transmission Standard for Cable Television

This standard describes the framing structure, channel coding, and channel modulation for a digital multi-service television distribution system that is specific to a cable channel. The system can be used transparently with the distribution from a satellite channel, as many cable systems are fed directly from the satellite links. The cable channel, including optical fiber, is primarily regarded as a bandwidth-limited linear channel, with a balanced combination of white noise, interference, and multi-path distortion. The quadrature amplitude modulation (QAM) technique used, together with adaptive equalization and concatenated coding, is well-suited to this application and channel. The specification covers both the 64 and 256 QAM. Most features of both modulation schemes are the same. Where there are differences, the specific details for each modulation scheme are covered in the standard. The design of the modulation, interleaving and coding is based upon testing and characterization of cable systems in North America. The modulation is QAM with a 64-point signal constellation (64 QAM) and with a 256-point signal constellation (256 QAM), transfer selectable. The forward error correction (FEC) is based on a concatenated coding approach that produces high-coding gain at moderate complexity and overhead. Concatenated coding offers improved performance over a block code, at a similar overall complexity. The system FEC is optimized for quasi-error free operation at a threshold output error event rate of one error per 15 minutes. The data format input to the modulation and coding is assumed to be a Moving Picture Experts Group (MPEG-2) transport. However, the method used for MPEG synchronization is de-coupled from FEC synchronization. For example, this enables the system to carry Asynchronous Transfer Mode (ATM) packets easily without interfering with ATM synchronization. There are two modes supported: Mode 1 has a symbol rate of 5.057 Ms/s and Mode 2 has a symbol rate of 5.361 Ms/s. Typically, Mode 1 will be used for 64 QAM and the Mode 2 will be used for 256 QAM. The system will be compatible with future implementations of higher data rate schemes employing high-order QAM extensions. ITU-T/SCTE.

SCTE 08 1996 (DVS 011): Cable and Satellite Extensions to ATSC System Information Standards

In conjunction with the Advanced Television Systems Committee document, ATSC A/56, Systems Information for Digital Television, this document defines a proposed Standard for System Information (SI) compatible with Moving Picture Experts Group (MPEG-2) compliant digital multiplex bitstreams constructed in accordance with ISO/IEC 13818-1 (MPEG-2) and transmitted on cable and satellite, and cable or satellite. The SI specifications here provides specific definitions and syntax for non-broadcast applications, extending the SI protocol by defining additional tables and descriptors applicable to cable and satellite, and cable or satellite transmission.

SCTE 20 2001 (DVS 157): Standard Methods for Carriage of Closed Captions and Non-Real Time Sampled Video

This document defines a standard for the carriage of vertical blanking interval (VBI) services in MPEG-2 compliant bitstreams constructed in accordance with ISO/IEC 13818-2.

SCTE 21 2000 (DVS 053): VBI Extension for ATSC Digital Television Standards

This document defines a standard for the carriage of Vertical Blanking Interval (VBI) services in MPEG-2 compliant bitstream constructed in accordance with ISO/IEC 13818-2. The approach builds upon a data structure defined in the ATSC A/53 Digital Television Standards, and is designed to be backward-compatible with the method.

SCTE 22-1 1997 (DSS 97-02): Data-Over-Cable Radio Frequency (RF) Interface Specification

This document defines the radio-frequency interface specifications for high speed data-over-cable systems that are developed by the Multimedia Cable Network System (MCNS). The intended service will allow transparent bi-directional transfer of Internet Protocol (IP) traffic, between the cable system headend and customer locations, over a hybrid fiber/coax (HFC) cable television system. The transmission path over the cable system is realized at he headend by a Cable Modem Termination System (CMTS), and at each customer location by a cable modem (CM). At he headend or hub, the interface to the data-over-cable system is the Cable Modem Termination System-Network-Side Interface (CMTS-NSI). At customer locations, it is the cable-modem-to-customer-premise-equipment interface (CMCI). The intent is for the MCNS operators to transparently transfer IP traffic between these interfaces, including but not limited to datagrams, Dynamic Host Configuration Protocol (DHCP), Internet Group Messaging Protocol (IGMP), and IP Group addressing (broadcast and multicast). ITU-T/SCTE.

SCTE 27 1996 (DVS 026): Subtitling Method for Broadcast Cable

This document defines a standard for a transmission protocol supporting multilingual subtitling services to augment video and audio within MPEG-2 multiplexes. This document is revised from a General Instrument standard entitled "Subtitles and Emergency Messages." Document Number STD-095-002, Revision 1.3, dated July 6, 1995.

SCTE 28 1998 (DVS 097): Program and System Information Protocol for Terrestrial Broadcast and Cable

This document defines a Standard for System Information (SI) and Program Guide (PG) data compatible with digital multiplex bit streams constructed in accordance with ISO/IEC 13818-1 (MPEG-2 Systems). The document defines the standard protocol for transmission of the relevant data tables contained within packets carried in the Transport Stream multiplex. The protocol defined herein will be referred to as Program and System Information Protocol (PSIP). Tables described in this document shall be applicable to terrestrial (over-the-air) and cable signals.

7.3 Selected DTV-Related DVS Documents

DVS 018: ATSC Digital Television Standard

This standard was prepared by the Advanced Television Systems Committee (ATSC). The Digital Television Standard describes the system characteristics of the U.S. advanced television (ATV) system. The document and its normative annex as provide detailed specifications and the parameters of the system including the video encoder, the audio encoder input signal format and the pre-processing and compression parameters of the audio encoders, the service multiplex and transport layer characteristics and normative specifications, and the VSB RF/transmission subsystem.

DVS 019: Digital Audio Compression (AC-3) Standard

The normative portions of this standard specify a coded representation of audio information and specify the decoding process. Informative information on the encoding process is included. The coded representation specified herein is suitable for use in the digital audio transmission and storage applications. The coded representation may convey from 1 to 5 full bandwidth audio channels, along with a low frequency enhancement channel.

DVS 020: Guide to the Use of the ATSC Digital Television Standard

This document provides a guide to understanding the system characteristics of the U.S. advanced television system as documented in the ATSC Digital Television Standard (DVS 018).

DVS 022: System Information for Digital Information

This document defines a Standard for System Information (SI) compatible with digital multiplex bit streams constructed in accordance with ISO/IEC 13818 (MPEG-2). It defines the standard protocol that carries relevant System Information tables contained within packets carried in the transport multiplex. This document describes tables that are applicable to terrestrial (over-the-air), cable, SMATV, MMDS, and satellite broadcast signals. Only certain messages are applicable in each of the domains. All tables and messages defined in this document are carried in the Network Packet Identifier (PID).

DVS 033: SCTE Video Compression Formats

The standard format consists of four tables: Table 1, Standardized Video Input Formats; Table 2, Compression Format constraints—Tier 1; Table 3, Compression Format Constraints—Tier 2 (Additional Modes); and Table 4, Legend for MPEG-2 codes values in Table 3.

DVS 046: Specifications for Digital Transmission Technologies

It is the goal of this document to promote the widespread availability of cost effective products and services, and the growth of digital video services without limiting technological change or innovation. This document provides the specification of digital video and related audio technologies in use or expected to be in use by the Cable TV industry in the U.S. In summary, it provides specifications to Channelization, Return Channel, Modulation, Forward Error Correction, Network, Transport, Conditional Access Interface, Video, Audio, Interactivity, and Cluster Interface.

DVS 064: National Renewable Security Standards (NRSS) Part A and Part B

NRSS provides a means for security for digital television receivers and digital VCR receivers. It has two physical designs, Part A and Part B. Both are removable and renewable security elements. Part A was from the ISO-7816 standard. It has 8 electrical contacts using a limited number of PC cards for security and in serial communication. This device costs less and can be smaller. Part B is more of interest to the cable television industry. It has 68 electrical contacts using parallel communications (one host and as many modules as one wants). It is more extensible and robust.

DVS 159: Optional Extensions for Carriage of NTSC VBI Data in Cable Digital Transport Streams

This document defines a standard for the carriage of Vertical Blanking Interval (VBI) services in MPEG-2 compliant bit streams constructed in accordance with ISO/IEC 13818-2. The approach builds upon a data structure defined in the ATSC A/53 Digital Television Standard, and is designed to be backwards-compatible with this method.

DVS 161: ATSC Data Broadcast Specification

This document defines a standard for data transmission compatible with digital multiplex bit streams constructed in accordance with ISO/IEC 13818-1 (MPEG-2 Systems).

DVS 165: DTV Interface Specification

This standard defines a minimum specification for an analog and baseband digital interface to a digital TV receiver. This specification is designed to enable interoperability between a DTV compliant and various types of consumer audio/video devices including digital set-top boxes (STBs) and analog/digital hard disk or video cassette recorders (VCRs).

DVS 167: Digital Broadband Delivery System: Out Of Band Transport—Mode B

This Physical Layer Interface supports transmission over radio frequency coax (up to 1 GHz bandwidth). It is referred to as the *bi-directional QPSK-link* on HFC (Hybrid Fiber Coax). The interface describes the complete physical layer structure, i.e. framing structure, channel coding, and modulation for each direction—downstream and upstream.

DVS 168: Emergency Alert System Interface to Digital Cable Network

The purpose of this document is to describe the interface between an Emergency Alert Receiver Server (EARS) and an Emergency Alert Controller (EAC).

DVS 178: Digital Broadband Delivery System: Out Of Band Transport—Mode A

This document specifies the Physical Layer and the Data Link Layer (including the MAC Layer) of the out-of-band cable system transport. Included are descriptions of the Physical Layer protocol and the Data Link Layer protocol.

DVS-181: Service Protocol

This document defines a standard wire line protocol that gives a cable operator a method for advertising services that do no conform to traditional broadcast content oriented services. These services may include e-mail, web browsing, e-commerce, and games. SP supports the notion of a "service" that is defined as an application and parameter data that is passed to the application upon activation. This expands on the existing content-centric definition of "service."

DVS 194: Home Digital Network Interface Specification with Copy Protection

The need for interfaces between cable set top boxes and digital television (DTV) receivers is one element of a general movement to interconnect multiple audio/visual (A/V) devices on a common bus or network. The IEEE 1394 interface has emerged as the preferred tool to accomplish this goal. DVS 194 contains requirements and options for an IEEE 1394 digital interface between a cable TV set top box (called a Host Device in this standard because it "hosts" a removable security module), and a DTV receiver.

IEEE 1394, which covers the physical interface, has been extended by EIA-775-A and EIA-799 which cover the command language and on-screen graphics display. This standard extends these to address the needs of cable set-top boxes. In addition, the Digital Transmission Content Protection specification governs copy protection of digital content on this interface.

One of the key advantages of a Host Device architecture is that all of the cable-specific protocols and interfaces are in the Host Device or its security module. This permits the cable operator to support proprietary legacy systems, as well as promote special cable-based applications without changes to the consumer's basic A/V equipment. In particular, the conditional access, network management, cable channel lineups, transport modulation, Internet transport technology, and other cable-specific technology are supported in the Host Device.

The functional partitioning defined in this standard locates the MPEG signal processing into the DTV receiver. This has the potential to produce a Host Device that has a lower cost than the equivalent digital set-top that decodes MPEG and produces analog output for an NTSC TV.

DVS 208: Emergency Alert Message for Cable

This Standard defines an Emergency Alert (EA) signaling method for use by cable TV systems to signal emergencies to digital "cable-ready" devices. Use of the EA signaling method defined in this Standard is designed for cable systems that support devices offered for retail sale and certified as "cable-ready". Such devices include digital set-top boxes that are sold to consumers at retail, cable-ready digital TV receivers, and cable-ready digital VCRs. Cable terminals owned by cable operators may use this or other proprietary methods for EA signaling.

The EA signaling scheme defined in this Standard allows a cable operator to disseminate emergency alert information related to state and local-level emergencies and warnings in a cost-effective and efficient way, while minimizing disruption to programming. While it is possible for a cable operator to comply with EAS requirements by simply replacing the source signal for all programs with an emergency information channel, such switching is disruptive to viewing, is overly intrusive for many kinds of local warnings, and is expensive and complex for the cable operator to implement in a digital cable environment where each transport stream may carry many programs that would have to be individually interrupted. To handle the rare case of a national-level EAS event, the EA message instructs a cable-ready device to force-tune to a designated emergency broadcast channel.

DVS-213: Copy Protection for POD Module Interface

The content protection mechanism presented in this document is intended to protect the interface between the Point of Deployment (POD) Removable Security Module and the host device. National control centers and/or cable system head-ends have the ability to communicate with individual addressable POD Modules for authorizing and de-authorizing of digital services on a channel-by-channel basis. Secured video services are encrypted throughout the entire delivery chain from originator to subscriber. Copy protection is used on the outputs of the host device.

DVS 234: Service Information Delivered Out-of-Band for Digital Cable Television

This document defines a standard for Service Information (SI) delivered out-of-band on cable. This standard is designed to support "navigation devices" on cable. The current specification defines the syntax and semantics for a standard set of tables providing the data necessary for such a device to discover and access digital and analog services offered on cable.

This specification defines SI tables delivered via an out-of-band path to support service selection and navigation by digital cable set-top boxes and other "digital cable-ready" devices. The SI tables defined in this standard are formatted in accordance with the Program Specific Information (PSI) data structures defined in MPEG-2 Systems.

This specification does not address the Electronic Program Guide application itself or any user interface which might deal with the presentation and application of the Service Information.

DVS 241: Digital Video Service Multiplex and Transport System Standard for Cable Television

This document describes the transport layer characteristics and normative specifications of the in-band Service Multiplex and Transport System Standard for Cable Television. The transport format and protocol for the Service Multiplex and Transport System Standard for Cable Television is a compatible subset of the MPEG-2 Systems specification defined in ISO/IEC 13818-1. It is based on a fixed-length packet Transport Stream approach which has been defined and optimized for digital television delivery applications.

DVS/258: Digital Video Systems Characteristics Standard for Cable Television

This document describes the characteristics and normative specifications for the Video Subsystem Standard for Cable Television.

DVS/295: HOST-POD Interface

This standard defines the characteristics and normative specifications for the interface between Point of Deployment (POD) security modules owned and distributed by cable operators, and commercially available consumer receivers and set-top terminals (Host devices) that are used to access multi-channel television programming carried on North American cable systems. These Host devices may also be supplied by the cable operators. The combination of a properly-authorized POD module and a Host device permits the unscrambled display of cable programming that is otherwise protected by a conditional access scrambling system.

This standard applies extensions, modifications, and constraints to the interface defined in EIA-679B Part B, the National Renewable Security Standard. This standard supports a variety of conditional access scrambling systems. Entitlement management messages (EMMs) for such scrambling systems are carried in the cable out of band channel as defined by SCTE DVS 167 and DVS 178. Other data transfer mechanisms such as the signaling methods of the DOCSIS version 1.1 cable modem standard may be supported in the Host device. A cable operator is able to upgrade security in response to a breach by replacing the POD modules, without requiring any change in the host device, or, if the POD module supports the option, by replacing the CA smart card.

The interface will support Emergency Alert messages transmitted over the out of band channel to the POD module and then delivered by the POD module over the interface to the host device using the format defined in SCTE DVS 208. It may also support Interactive Program Guide services, Impulse Pay-Per-View services, Video-on-Demand, and other messaging and interactive services. It supports both one way and two way cable systems, as well as host devices that incorporate DOCSIS modems or telephone modems. This standard defines the physical interface, signal timing, the link interface, and the application interface. It includes the extended channel specification, power management specifications, initialization procedures, and firmware upgrade methods.

DVS 301: OpenCable Copy Protection System Specification

This copy protection specification defines the means to protect high value content on the interface between the Point of Deployment (POD) Removable Security Module and the OpenCable Host device (Host). Content, which is delivered with copying permitted, e.g., free access off-air broadcast content, is not copy protected and the means described in this specification do not apply to it. Such content may be encrypted from the headend to POD but will be delivered in the clear on the POD Host Interface. Conversely only "copying permitted" content will be delivered in the clear (unencrypted) from the headend to POD and so will be output in the clear from the POD to Host with CCI=00.

The objective of copy protection is to secure protected content against unauthorized access throughout the entire delivery chain from source to display. Program providers have deployed means to secure content from the source to the cable headend, and cable systems have similarly deployed secure systems from the headend to home. Cable set-tops use copy protection technology to protect content on the analog and digital outputs to consumer displays.

With the introduction of the POD Module, cable security will terminate in the POD. A means is needed to prevent unauthorized access on the POD-Host interface. This document specifies such a means. Basically, the POD Module shall decrypt services under control of the headend and shall reencrypt content for the purpose of copy protection across the interface between the POD Module and Host device.

DVS 313: Digital Cable Network Interface Standard

This standard defines the characteristics and normative specifications for the network interface between a cable television plant and commercially available consumer equipment that is used to access multi-channel television programming. The interface is also compatible with existing set-top terminal equipment owned by cable operators and with terminal equipment developed via the OpenCable™ specification process. In this standard the Cable Network Interface is defined as the interface between the cable drop and the input terminals of the first device located on the subscriber's premises, regardless of whether that device is owned by the subscriber or the cable operator. A coaxial-based broadband access network is assumed. This may take the form of either an all-coax or hybrid fiber/ coax (HFC) network. The generic term "cable network" is used here to cover all cases.

Chapter

8

AES Standards

Jerry C. Whitaker, editor

8.1 Introduction

The following documents relating to audio technologies in general, and digital audio in particular, are available from the Audio Engineering Society. For more information, contact the AES at http://www.aes.org.

8.2 Informational Documents

AES-2id-1996 (r2001): Guidelines for the Use of the AES3 Interface

This document provides guidelines for the use of AES3, AES Recommended Practice for Digital Audio Engineering—Serial transmission format for two-channel linearly represented digital audio data, together with AES5, AES Recommended Practice For professional digital audio applications employing pulse-code modulation—Preferred sampling frequencies, AES11, AES Recommended Practice for Digital Audio Engineering—Synchronization of digital audio equipment in studio operations, and AES18, AES Recommended Practice for Digital Audio Engineering—Format for the user data channel of the AES digital audio interface.

AES-3id-2001 (Revision of AES-3id-1995): Transmission of AES3 Formatted Data by Unbalanced Coaxial Cable

This document contains information regarding cables, cable equalizers, and receiver circuits including adaptors to or from standard AES3 equipment and cabling where it is required to transmit AES3 formatted signals over long distances (up to 1000 m), or in a video installation using analog video distribution equipment. It is not intended to be an alternative electrical specification to AES3, which is based on balanced, shielded, twisted-pair cable transmission over distances of up to 100 m. The information is based on studies and laboratory experiments discussed in a series of technical reports that have been partly summarized and included.

AES-4id-2001: Characterization and Measurement of Surface Scattering Uniformity

This document provides guidelines for characterizing the uniformity of scattering produced by surfaces from measurements or predictions of scattered polar responses. In this context, the surface scattering is quantified in terms of a single diffusion coefficient. The diffusion coefficient is a measure of quality designed to be used by producers and users of surfaces that, either deliberately or accidentally, diffuse sound. It is also intended for use when needed by developers and users of geometric room acoustic models. The diffusion coefficient is not intended, however, to be blindly used as an input to current diffusion algorithms in geometric room acoustic models. The diffusion coefficient characterizes the sound reflected from a surface in terms of the uniformity of the scattered polar distribution. The information document details a free-field characterization method.

AES-5id-1997: Loudspeaker Modeling and Measurement—Frequency and Angular Resolution for Measuring, Presenting, and Predicting Loudspeaker Polar Data

This document provides guidelines for measuring, presenting and predicting polar data from a single acoustic source or from an array of acoustic sources. It describes and quantifies measurement resolution, presentation resolution, prediction techniques and measurement environments. The information presented here is based on objective measurements and does not take subjective or psychoacoustic criteria into account.

AES-6id-2000: Personal Computer Audio Quality Measurements

This document focuses on the measurement of audio quality specifications in a PC environment. Each specification listed has a definition and an example measurement technique. Also included is a detailed description of example test setups to measure each specification.

AES-10id-1995: Engineering Guidelines for the Multichannel-Audio Digital Interface (MADI) AES10

This document provides guidance for areas of application of the MADI standard (AES10) that might be unclear. It is not intended to replace AES10, but to supplement it in those areas that are not suitable for definition in a standards document.

8.3 Project Reports

AES-R1-1997: Specifications for Audio on High-Capacity Media

This document is the report of Task group SC-02-M, a task force within the Audio Engineering Society Standards Committee (AESSC), which studied the future of high-capacity audio media (such as DVD) for over one year with input from more than 80 persons and organizations.

AES-R3-2001: Compatibility for Patch Panels of Tip-Ring-Sleeve Connectors

This report covers the concentric connectors known as phone plugs and jacks that are widely used in the audio industry for the interconnection of sound system components as general-use consumer tip-ring-sleeve (TRS) types as well as for the interconnection of broadcast and other professional systems, such as professional TRS types. Because differing dimensions specified by various standards often result in mechanical and, as a consequence, electrical incompatibilities among the various TRS jacks and plugs, this report surveys the standards, connectors, and possible incompatibility problems.

8.4 Standards and Recommended Practices

AES2-1984 r1997: Specification of Loudspeaker Components used in Professional Audio and Sound Reinforcement

This document is a recommended practice for describing and specifying loudspeaker components used in professional audio and sound-reinforcement systems. These components include high-frequency drivers, high- and mid- frequency horns, low-frequency drivers, and low-frequency enclosures. For drivers, specifications are given for describing frequency response, impedance, distortion, and power handling. For horns and enclosures, specifications are given for describing directional characteristics and additional pertinent performance data. For all components, specifications as are given for describing necessary physical and mechanical characteristics, such as hardware, mounting data, size, and weight. Appendices supporting the text give guidelines for making proper free-field measurements, sizing of baffles for low-frequency measurements, a method for producing the specified noise signal used in power testing, and a summary of required information.

AES3-1992 (r1997): Serial Transmission Format for Two-Channel Linearly Represented Digital Audio Data

The format provides for the serial digital transmission of two channels of periodically sampled and uniformly quantized audio signals on a single shielded twisted wire pair. The transmission rate is such that samples of audio data, one from each channel, are transmitted in time division multiplex in one sample period. Provision is made for the transmission of both user and interface related data as well as of timing related data, which may be used for editing and other purposes. It is expected that the format will be used to convey audio data that have been sampled at any of the sampling frequencies recognized by the AES5, Recommended Practice for Professional Digital Audio Applications Employing Pulse-Code Modulation—Preferred Sampling Frequencies.

AES5-1998: Preferred Sampling Frequencies for Applications Employing Pulse-Code Modulation

A sampling frequency of 48 kHz is recommended for the origination, processing, and interchange of audio programs employing pulse-code modulation. Recognition is also given to the use of a 44.1-kHz sampling frequency related to certain consumer digital applications, the use of a 32-kHz sampling frequency for transmission-related applications, and the use of a 96-kHz sampling frequency for applications requiring a higher bandwidth or more relaxed anti-alias filtering.

AES6-1982 r1997: Method for Measurement of Weighted Peak Flutter of Sound Recording and Reproducing Equipment

Weighted peak flutter is measured using a 3150-Hz tone transmitted through the equipment. The tone is frequency demodulated, frequency-response weighted, peak-to-peak detected, time-response weighted, and read on a meter as the zero-to-peak (one-half of peak-to-peak) values. Results are reported as "weighted peak of the recorder (or reproducer, or recording/reproducing system): +/– __ percent." A toleranced graph and table give the frequency-response weighting (approximately a 6-dB-per-octave drop above and below 4 Hz, with an additional drop below 0.5 Hz.) A toleranced table gives the time-response weighting (the reference is the peak-to-peak amplitude of a 4-Hz sine wave; the test signal is a series of unidirectional pulses of the same peak-to-peak amplitude, spaced 1 second apart; a pulse length of 60 ms gives 90-percent response; the fall between 100-ms pulses is a 40-percent reading.)

Good engineering practices are given for the meter design. The rationale for this standards is given in an appendix. This standard, originally published as IEEE Std-193, has technical requirements identical to standards IEC Pub. 386, CCIR 409-2, and DIN 45 507. An ANSI version is available as S4.3-1982.

AES7-2000: Method of Measuring Recorded Fluxivity of Magnetic Sound Records at Medium Wavelengths

This standard specifies a method of measuring the recorded flux per unit track width, called *fluxivity*, of a magnetically recorded sinusoidal test signal at medium wavelengths for all magnetic sound record formats, by using a high-efficiency magnetic reproducing head. It also specifies the equipment needed to implement this method. An ANSI version is available as S4.6-1982.

AES10-1991 (r1997): Serial Multichannel Audio Digital Interface (MADI)

This standard describes the data organization for a multi-channel-audio digital interface. It includes a bit level description, features in common with the AES3-1985 two channel format, and the data rates required for its utilization. The specification provides for the serial digital transmission of 56 channels of linearly represented digital audio data at a common sampling frequency within the range 32 kHz to 48 kHz (+/– 12.5 percent) having a resolution of up to 24 bits per channel.

The format makes possible the transmission and reception of the complete 28-bit channel word (excluding preamble) as specified in the document AES3-1985, providing for the validity, user, channel status, and parity information allowable under that Standard. The transmission format is of the asynchronous simplex type and is specified for a single 75 ohm coaxial cable point-to-point interconnection, although the use of fibre-optic medium is also possible. An ANSI version is available as S4.43-1991.

AES11-1997: Synchronization of Digital Audio Equipment in Studio Operations

This standard provides a systematic approach to the synchronization of digital audio signals. Recommendations are made concerning the accuracy of sample clocks as embodied in the interface signal and the use of this format as a convenient synchronization reference where signals must be rendered co-timed for digital processing. Synchronism is defined, and limits are given which take account of relevant timing uncertainties encountered in an audio studio.

AES14-1992 (r1998): Application of Connectors, Part 1—XLR-Type Polarity and Gender

This standard specifies a common scheme for wiring the connectors used in audio systems, particularly to avoid the inversion of absolute polarity among the items in the analog signal chain. An ANSI version is available as S4.48-1992.

AES15-1991 (r1997): Sound-Reinforcement Systems Communications Interface (PA-422)

This standard specifies the electrical characteristics of a balanced-voltage circuit for the interchange of serial binary signals for the control of sound-reinforcement systems. It provides for interchange among data terminal equipment (DTE), that is, computers and microprocessors, and data circuit terminating equipment (DCE). PA-422 is a mnemonic, signifying professional audio implementation of Electronics Industries Association EIA-422-A. Device control language is provided in an annex. An ANSI version is available as S4.49-1991.

AES17-1998 (revision of AES17-1991): Measurement of Digital Audio Equipment

This standard provides methods for specifying and verifying the performance of digital audio equipment. Many tests are substantially identical to those used when testing analog equipment. However, because of the unique requirements of digital audio equipment and the effects of its imperfections, additional tests are necessary.

AES18-1996 (revision of AES18-1992): Format for the User Data Channel of the AES Digital Audio Interface

This standard describes a method of formatting the user data channels provided within the digital audio serial interface format (AES3). The transmission format is an adaptation of the packet-based high-level data link control (HDLC) communications protocol and provides for the transmission of ancillary data that may or may not be time related to the audio signal. The data rate is constant within a range of +/- 12.5 percent of a sampling frequency of 48 kHz. The standard also provides a data priority and management strategy to ensure that adequate capacity is available for downstream data insertion.

AES19-1992 r1998: Measurement of the Lowest Resonance Frequency of Loudspeaker Cones

This standard test method is intended to determine the frequency of lowest resonance of a loudspeaker cone. Such information is used for engineering design and for quality control. The method has been developed to improve correlation of measurement between cone manufacturers and loudspeaker manufacturers. An ANSI version is available as S4.30-1992.

AES20-1996: Subjective Evaluation of Loudspeakers

This standard is a set of recommendations for subjective evaluation of high-performance loudspeaker systems. It is believed that, for certain audio components including loudspeakers, subjective evaluation is a necessary adjunct to objective measurements. The strong influence of listening conditions, program material and of individual evaluators is recognized. This document seeks, therefore, to assist in avoiding testing errors rather than to attempt to establish a correct procedure.

AES22-1997: Storage of Polyester-Base Magnetic Tape

This standard provides recommendations concerning the storage conditions, storage facilities, enclosures, and inspection for recorded polyester-base magnetic tapes in roll form. It covers analog and digital tape and includes tape made for audio, video, instrumentation, and computer use.

AES24-1-1999, (Revision of AES24-1-1995): Application Protocol for Controlling and Monitoring Audio Devices via Digital Data Networks—Part 1: Principles, Formats, and Basic Procedures

This document describes the architecture of AES24, the name assigned to an extensible application protocol for controlling and monitoring audio devices via local area networks, and, when possible in the future, devices designed for other media.

AES26-1995: Conservation of the Polarity of Audio Signals

This document standardizes the polarity of the signals at the various interface points between different items of equipment, in particular from the acoustical, electrical, and the magnetic aspects. Each item of equipment complies separately with the polarity requirements for the input and output signals.

AES27-1996: Managing Recorded Audio Materials Intended for Examination

This document specifies recommended practices for safekeeping, conveyance, inspection, description, and labeling of audio recordings offered as evidence in criminal investigations, in criminal or civil proceedings, or in other forensic applications. It does not cover analysis of magnetic tapes or other recording media for the purposes of authenticity determination, talker identification, copyright violation, enhancement of oral conversations or other signals, or otherwise characterizing signals recorded on such tapes.

AES28-1997: Method for Estimating Life Expectancy of Compact Discs (CD-ROM), Based on Effects of Temperature and Relative Humidity

This standard specifies test methods for estimating the storage life expectancy (LE) of information stored on compact discs (CD-ROM). Only the effects of temperature and relative humidity are considered. Block error rate (BLER) is the measured response and the end-of-life criterion. An Eyring model is developed from accelerated test results. Data are normalized to 25 degree C and 50 percent relative humidity, and the LE, percent compliance, and confidence intervals at these conditions are calculated.

AES31-3-1999: Audio-File Transfer and Exchange—Part 3: Simple Project Interchange

This standard provides a convention for expressing edit data in text form in a manner that enables simple and accurate computer parsing while retaining human readability. It also describes a method for expressing time-code information in character notation.

AES33-1999: Database of Multiple-Program Connection Configuration

This document specifies the purpose and procedures for the maintenance of data on an AESSC database of connectors for multiple-program applications. Purchasers of this standard obtain full access to the database using their names and order numbers for identification.

AES35-2000: Method for Estimating Life Expectancy of Magneto-Optical (MO) Disks, Based on Effects of Temperature and Relative Humidity

This standard specifies test methods for estimating the life expectancy of information stored on magneto-optical (M-O) disks. Only the effects of temperature and relative humidity on the media are considered.

AES38-2000: Life Expectancy of Information Stored in Recordable Compact Disc Systems—Method for Estimating, Based on Effects of Temperature and Relative Humidity

This standard specifies test methods for estimating the life expectancy of information stored in recordable compact disc systems. Only the effects of temperature and relative humidity on the media are considered.

AES41-2000: Recoding Data Set for Audio Bit-Rate Reduction

This document describes a format for the data to be transmitted to identify a number of popular audio bit-rate reduction techniques. Provision is also made for the transmission of additional ancillary information. This document also describes a mechanism whereby the data derived from the coded signal can be transported with the decoded signal so that the data may be recovered and used to aid a subsequent re-encoding process.

AES42-2001: Digital Interface for Microphones

This standard describes an extension of the existing digital audio interface AES3 to provide a digital interface for microphones.

AES43-2000: Criteria for the Authentication of Analog Audio Tape Recordings

The purpose of this standard is to formulate a standard scientific procedure for the authentication of audio tape recordings intended to be offered as evidence or otherwise utilized in civil, criminal, or other fact finding proceedings.

AES45-2001: Connectors for Loudspeaker-Level Patch Panels

This standard complements IEC 60268-12 including amendments 1 and 2, extending the standardization of application of AES45-series connectors to their use for loudspeaker-level patch panels.

Chapter

9

References in ATSC Standards and Recommended Practices

Jerry C. Whitaker, editor

9.1 Introduction

This chapter represents a compilation of the normative and informative references used in ATSC Standards and Recommended Practices approved as of the date of publication. For the purposes of this document, those references specifically listed as "normative" are categorized as "normative." All other references are categorized as "informative." These terms are defined as follows:

- **normative reference:** A reference used to provide material that is critical to the practice of a standard. It represents a document, without which the standard could not be implemented. For example, in a standard that requires video MPEG compression, reference to ISO 13818-2 is critical, because without it the compression scheme is not defined. Thus, the normative references become an integral part of the standard. They can be thought of as indirect pointers to material that could otherwise literally have been put inline in the text of the standard.
- **informative reference**. A reference that provides helpful, often critical, information related to a standard, but is not required to practice it. In other words, the standard is still complete and implementable without all of its informative references. For example, in a standard that requires video compression, it may be helpful to read ISO 13818-1 (Systems), but it is not required if you otherwise understood 13818-2. So, it would be an informative

9.1a A/52: Digital Audio Compression Standard (AC-3)

This document specifies coded representation of audio information and the decoding process, as well as information on the encoding process. The coded representation specified is suitable for use in digital audio transmission and storage applications, and may convey from 1 to 5 full bandwidth audio channels, along with a low frequency enhancement channel. A wide range of encoded bit-rates is supported by this specification. Typical applications of digital audio compression are in satellite or terrestrial audio broadcasting, delivery of audio over metallic or optical cables, or storage of audio on magnetic, optical, semiconductor, or other storage media.

Normative References

There are none.

Informative References

Ehmer, R H.: "Masking of Tones vs. Noise Bands," *J. Acoust. Soc. Am.*, vol. 31, pp 1253–1256, September 1959.

Ehmer, R. H.: "Masking Patterns of Tones," *J. Acoust. Soc. Am.*, vol. 31, pp. 1115–1120, August 1959.

Moore, B. C. J., and B. R. Glasberg: "Formulae Describing Frequency Selectivity as a Function of Frequency and Level, and Their Use in Calculating Excitation Patterns," *Hearing Research*, vol. 28, pp. 209–225, 1987.

Todd, C. et. al.: "AC-3: Flexible Perceptual Coding for Audio Transmission and Storage", AES 96th Convention, Preprint 3796, Audio Engineering Society, New York, N.Y., February 1994.

Zwicker, E.: "Subdivision of the Audible Frequency Range into Critical Bands (Frequenzgruppen)," *J. Acoust. Soc. of Am.*, vol. 33, pg. 248, February 1961.

9.1b A/53B: ATSC Digital Television Standard (Revision B)

The Digital Television Standard describes the system characteristics of the advanced television (ATV) system. The document and its normative annexes provide detailed specification of the parameters of the system including the video encoder input scanning formats and the pre-processing and compression parameters of the video encoder, the audio encoder input signal format and the pre-processing and compression parameters of the audio encoder, the service multiplex and transport layer characteristics and normative specifications, and the VSB RF/transmission subsystem. The system is modular in concept and the specifications for each of the modules are provided in the appropriate annex.

Normative References

AES 3-1992 (ANSI S4.40-1992): "AES Recommended Practice for Digital Audio Engineering—Serial Transmission Format for Two-Channel Linearly Represented Digital Audio Data," Audio Engineering Society, New York, N.Y., 1992.

ANSI S1.4-1983: Specification for Sound Level Meters.

ATSC Standard A/52 (1995): *Digital Audio Compression* (AC-3), Advanced Television Systems Committee, Washington, D.C., 1995.

ATSC Standard A/65 (1997): *Program and System Information Protocol for Terrestrial Broadcast and Cable*, Advanced Television Systems Committee, Washington, D.C., 1997.

IEC 651 (1979): Sound Level Meters.

IEC 804 (1985): Amendment 1 (1989) Integrating/Averaging Sound Level Meters.

ISO/IEC 13818-1: 1996/Amd. 1: 1997 (E) Amendment 1.

ISO/IEC 13818-1: 1996/Amd. 2: 1997 (E) Amendment 2.

ISO/IEC 13818-1: 1996/Amd. 3: 1997 (E) Amendment 3.

ISO/IEC 13818-1: 1996/Amd. 4: 1997 (E) Amendment 4.

ISO/IEC 13818-1: 1996/Cor. 1: 1997 (E) Technical Corrigendum 1.

ISO/IEC 13818-2: 1996/Cor. 1: 1997 (E) MPEG-2 Video Technical Corrigendum 1.

ISO/IEC 13818-2: 1996/Cor. 2: 1997 (E) MPEG-2 Video Technical Corrigendum 2.

ISO/IEC CD 13818-4: MPEG Committee Draft (1994), MPEG-2 Compliance.

ISO/IEC IS 13818-1: International Standard (1996), MPEG-2 Systems.

ISO/IEC IS 13818-2: International Standard (1996), MPEG-2 Video.

Informative References

47 CFR Part 15: FCC Rules.

ATSC Document A/54 (1995): *Guide to the Use of the ATSC Digital Television Standard,* Advanced Television Systems Committee, Washington, D.C., 1995.

EIA IS-23: *EIA Interim Standard for RF Interface Specification for Television Receiving Devices and Cable Television Systems.*

EIA IS-105: *EIA Interim Standard for a Decoder Interface Specification for Television Receiving Devices and Cable Television Decoders.*

EIA IS-132: *EIA Interim Standard for Channelization of Cable Television.*

ITU-R BT.601-4 (1994): Encoding parameters of digital television for studios.

SMPTE 274M (1995): *Standard for Television, 1920 × 1080 Scanning and Interface*, Society of Motion Picture and Television Engineers, White Plains, N.Y., 1995.

SMPTE 296M (1997): *Standard for Television, 1280 × 720 Scanning, Analog and Digital Representation, and Analog Interface*, Society of Motion Picture and Television Engineers, White Plains, N.Y., 1997.

9.1c A/54: Guide to the Use of the ATSC Digital Television Standard

This guide provides an overview and tutorial of the system characteristics of the advanced television (ATV) system defined by ATSC Standard A/53, ATSC Digital Television.

Normative References

There are none.

Informative References

There are none.

9.1d A/57: Program/Episode/Version Identification

The Program/Episode/Version Identification (Program Identifier) standard provides a means of uniquely defining a program, episode, version, and source within the MPEG-2 syntax. The standard provides for a program identifier data packet that may be inserted into the transport stream at random intervals.

Normative References

There are none.

Informative References

ATSC Standard A/53 (1995): *Digital Television Standard*, Advanced Television Systems Committee, Washington, D.C., 1996.

ATSC Standard A/55 (1996): *Program Guide for Digital Television*, Advanced Television Systems Committee, Washington, D.C., 1996.

ISO/IEC IS 13818-1: International Standard (1994), MPEG-2 Systems.

9.1e A/58: Harmonization with DVB SI in the use of the ATSC Digital Television Standard

This document specifies a recommended practice for use of the ATSC Digital Television Standard to ensure interoperability internationally at the transport level with the European Digital Video Broadcast (DVB) project as standardized by the European Telecommunications Standards Institute (ETSI). Guidelines for use of the Digital Television Standard are outlined which ensure avoidance of conflict with DVB transport in the areas of packet identifier (PID) usage and assignment of user private values for descriptor tags and table identifiers.

Normative Reference

There are none.

Informative Reference

ATSC Standard A/53 (1995): *ATSC Digital Television Standard*, Advanced Television Systems Committee, Washington, D.C., 1994.

ETSI ETS 300 468: "Digital Broadcasting Systems for Television, Sound and Data Services; Specification for Service Information (SI) in Digital Video Broadcasting (DVB) Systems."

9.1f A/63: Standard for Coding 25/50 Hz Video

This document describes the characteristics for the video subsystem of a digital television system operating at 25 Hz and 50 Hz frame rates.

Normative References

ISO/IEC IS 11172-2: International Standard, MPEG-1 Video.

ISO/IEC IS 13818-1: International Standard, MPEG-2 Systems.

ISO/IEC IS 13818-2: International Standard, MPEG-2 Video.

Informative References

ITU-R BT.601-4 (1994): *Encoding Parameters of Digital Television for Studios.*

SMPTE 274M (1995): *Standard for Television, 1920 × 1080 Scanning and Interface*, Society of Motion Picture and Television Engineers, White Plains, N.Y., 1995.

SMPTE 296M (1995): *Standard for Television, 1280 × 720 Scanning and Interface*, Society of Motion Picture and Television Engineers, White Plains, N.Y., 1994.

9.1g A/64: Transmission Measurement and Compliance For Digital Television

This document describes methods for testing, monitoring, and measurement of the transmission subsystem intended for use in the digital television (DTV) system, including specifications for maximum out-of-band emissions, parameters affecting the quality of the inband signal, symbol error tolerance, phase noise and jitter, power, power measure, frequency offset, and stability. In addition, it describes the condition of the RF symbol stream upon loss of MPEG packets. (The ATSC approved a revision to this document on May 30, 2000, that includes the revised FCC DTV emission mask.

Normative References

ATSC Standard A/52: *Digital Audio Compression* (AC-3), Advanced Television Systems Committee, Washington, D.C., 1995.

ATSC Standard A/53: *ATSC Digital Television Standard*, Advanced Television Systems Committee, Washington, D.C, 1996.

FCC Rules and Regulations, Parts 73 and 76, Federal Communications Commission, Washington, D.C.

IEEE Standard 100-1992: *The New IEEE Standard Dictionary of Electrical and Electronic Terms*, IEEE, New York, N.Y., 1992.

ITU-R Recommendation BT.1206: *Spectrum Shaping Limits for Digital Terrestrial Television Broadcasting.*

Informative References

There are none.

9.1h A/65A: Program and System Information Protocol for Terrestrial Broadcast and Cable

The Program and System Information Protocol Standard provides a methodology for transporting digital television system information and electronic program guide data. The standard includes an amendment that provides new functionality known as Directed Channel Change (DCC). This new feature will allow broadcasters to tailor programming or advertising based upon parameters defined by the viewer such as: postal, zip or location code, program identifier, demographic category, and content subject category. Potential applications include customized programming services, commercials based upon demographics, and localized weather and traffic reports.

Normative References

ATSC Standard A/52 (1995): *Digital Audio Compression* (AC-3), Advanced Television Systems Committee, Washington, D.C., 1995.

ATSC Standard A/53 (1995): *ATSC Digital Television Standard*, Advanced Television Systems Committee, Washington, D.C., 1995.

ATSC Standard A/57 (1996): *Program/Episode/Version Identification*, Advanced Television Systems Committee, Washington, D.C., 1996.

EIA-708A: *Specification for Advanced Television Closed Captioning* (ATVCC), Electronic Industries Association.

EIA-752: *Specification for Transport of Transmission Signal Identifier* (TSID) *Using Extended Data Service*, Electronic Industries Association.

EIA-766: *Specification for U.S. Region Rating Table* (RRT) *and Content Advisory Descriptor for Transport of Content Advisory Information Using ATSC A/65 Program and System Information Protocol* (PSIP), Electronic Industries Association.

Federal Information Processing Standard, FIPS Pub 6-4: *Counties and Equivalent Entities of the U.S.*, Its Possessions, and Associated Areas—90 Aug 31, U.S. Government Printing Office, Washington, D.C..

ISO CD 639.2: *Code for the Representation of Names of Languages: alpha-3 code*, Committee Draft, dated December 1994.

ISO/IEC 10646-1:1993: Information Technology—Universal Multiple-Octet Coded Character Set (UCS)—Part 1: *Architecture and Basic Multilingual Plane.*

ISO/IEC 8859: Information Processing—8-bit Single-Octet Coded Character Sets, Parts 1 through 10.

ITU-T Rec. H.222.0 | ISO/IEC 13818-1:1996, Information Technology—Generic Coding of Moving Pictures and Associated Audio—Part 1: *Systems.*

ITU-T Rec. H.262 | ISO/IEC 13818-2:1996, Information Technology—Generic Coding of Moving Pictures and Associated Audio—Part 2: *Video.*

U.S. Code of Federal Regulations (CFR) Title 47, 47CFR11: *Emergency Alert System* (EAS), U.S. Government Printing Office, Washington, D.C.

Informative References

ATSC Standard A/55 (1996): *Program Guide for Digital Television*, Advanced Television Systems Committee, Washington, D.C., 1996.

ATSC Standard A/56 (1996): *System Information for Digital Television*, Advanced Television Systems Committee, Washington, D.C., 1996.

Digital Video Transmission Standard for Cable Television, SCTE DVS-031, rev. 2, 29 May 1997.

EIA 608A: *Recommended Practice for Line 21 Data Service*, Electronic Industries Association.

ISO 639: *Code for the Representation of Names of Languages*, 1988.

Record of Test Results for Digital HDTV Grand Alliance System, September 8, 1995, Advanced Television Test Center, Alexandria, VA.

9.1i A/68: Use of ATSC A/65A PSIP Standard in Taiwan

This standard specifies the use of character sets, rating regions, and major channel numbers for the implementation of ATSC A/65A *Program and System Information Protocol* (PSIP) in Taiwan.

Normative References

ATSC Standard A/65A (2000): *Program and System Information Protocol for Terrestrial Broadcast and Cable*, Advanced Television Systems Committee, Washington, D.C., 2000.

Institute for Information Industry Technical Report C-26: *The Code Mapping Table of Chinese Character Glyphs for Computer Applications* (The Big 5 Code Table), May 1984.

ISO/IEC 10646-1:2000, Information Technology—Universal Multiple-Octet Coded Character Set (UCS)—Part 1: *Architecture and Basic Multilingual Plane.*

The Unicode Standard, Version 3.0, The Unicode Consortium, Addison-Wesley Pub., April 2000.

Informative References

There are none.

9.1j A/70: Conditional Access System for Terrestrial Broadcast

This document defines a standard for the Conditional Access system for ATSC terrestrial broadcasting to enable broadcasters to fully utilize the capabilities of digital broadcasting. This standard is based, whenever possible, on existing open standards and defines the building blocks necessary to ensure interoperability. The ATSC CA module is replaceable; to ensure that ATSC hosts are protected against obsolescence as security is upgraded. This standard applies to all CA vendors that supply CA service on behalf of an ATSC service provider. An overview of the CA standard is given in Annex C. (This document includes an Amendment that the ATSC approved on May 30, 2000.)

Normative References

ATSC A/66 (23 July 1999): *Technical Corrigendum No. 1 to ATSC Standard A/65, Program and System Information Protocol for Terrestrial Broadcast and Cable*, Advanced Television Systems Committee, Washington, D.C., 1999.

ATSC Standard A/53 (1995): *ATSC Digital Television Standard*, Advanced Television Systems Committee, Washington, D.C., 1995.

ATSC Standard A/65 (1997): *Program and System Information Protocol for Terrestrial Broadcast and Cable* (PSIP), Advanced Television Systems Committee, Washington, D.C., 1997.

CENELEC EN 50221: *Common Interface Specification for Conditional Access and other Digital Video Broadcasting Decoder Applications*, February 1997. (Certain sections are normative).

DES CBC Packet Encryption, SCTE DVS 042, Society of Cable and Television Engineers, 25 October 1996.

EIA 679-B (1999): *National Renewable Security Standard* (NRSS), Electronic Industry Association.

FIPS PUB 46-2 (1993): *Specification for the Data Encryption Standard*, National Institute of Standards and Technology, Gaithersburg, MD.

FIPS PUB 81 (1980): *DES Modes of Operation*, National Institute of Standards and Technology, Gaithersburg, MD.

ITU-T Rec. H.222.0 | ISO/IEC 13818-1:1996, Information Technology—Generic Coding of Moving Pictures and Associated Audio—Part 1: *Systems*.

NBS PUB 500-20 (1980): *Validating the Correctness of Hardware Implementations of the NBS DES*, National Bureau of Standards.

Informative References

ETR 289 ed.1 (1996-10), Digital Video Broadcasting (DVB): *Support for Use of Scrambling and Conditional Access* (CA) *within Digital Broadcast Systems*, European Telecommunications Standards Institute.

FIPS PUB 74 (1981): *Guidelines for Implementing and Using NBS DES*, National Institute of Standards and Technology, Gaithersburg, MD.

TS 101 197-1 V1.1.1 (1997-06): *Technical Specification of DVB Simulcrypt*, 1 June 1997, European Telecommunications Standards Institute.

9.1k A/75: ATSC Recommended Practice for Developing DTV Field Test Plans

This document presents the objectives of, and general methodology for conducting field tests of over-the-air terrestrial digital television (DTV) systems. The scope of the work includes reception, demodulation, and recovery of the transmitted data. The scope of the work herein is not concerned with the decoded data or analog signals except when these signals are used as a means to determine that the data has been correctly recovered.

Normative References

47 CFR 73.686–Coverage Measurements

47 CFR 73.622–DTV Allocations

47 CFR 73.625–DTV Coverage of Principal Community and Antenna Systems

Informative References

ATSC Field Test Vehicle Design Information, Gary Sgrignoli, Zenith Corporation, December 20, 1999.

ATV Test Procedures Manual, ACATS/SSWP-2/Document SSWP2-0601, December 15, 1993.

ATV Test Procedures Manual: Field Test of the Grand Alliance Prototype, ACATS/SSWP-2/ Document SSWP2-1389, June 28, 1995.

ETR290: Measurement Guidelines for DVB Systems, European Broadcasting Union, May, 1997.

General Description of Field Tests, ABERT/SET, July 2, 2000.

General Field Test Plan for Digital Television Propagation, Model HDTV Station Project, Washington, D.C., July 22, 1999.

Indoor Field Test Plan for Digital Television Propagation, Model HDTV Station Project, Washington, D.C., July 22, 1999.

9.11 A/80: Modulation & Coding Requirements For Digital TV Applications Over Satellite

This document defines a standard for modulation and coding of data delivered over satellite for digital television contribution and distribution applications. The data can be a collection of program material including video, audio, data, multimedia, or other material. It includes the ability to handle multiplexed bit streams in accordance with the MPEG-2 systems layer, but it is not limited to this format and makes provision for arbitrary types of data as well. QPSK, 8PSK and 16 QAM modulation modes are included, as well as a range of forward error correction techniques.

Normative References

ETS 300 421 (v1.1.2, 1997-08): Digital Video Broadcasting (DVB); *Framing Structure, Channel Coding and Modulation for 11/12 GHz Satellite Services*.

Informative References

IESS-308 (Rev 9, 30 Nov 1998): *Intelsat Earth Station Standards*, Performance Characteristics For Intermediate Data Rate Digital Carriers Using Convolutional Encoding/Viterbi Encoding and QPSK Modulation.

IESS-310 (Rev 1, 30 Nov 1998): *Intelsat Earth Station Standards*, Performance Characteristics For Intermediate Data Rate Digital Carriers Using Rate 2/3 TCM / 8PSK and Reed-Solomon Outer Coding.

EN 301 210 (v1.1.1, 1999-03): Digital Video Broadcasting (DVB); *Framing Structure, Channel Coding and Modulation for Digital Satellite News Gathering (DSNG) and Other Contribution Applications by Satellite*.

9.1m A/90:ATSC Data Broadcast Standard

The ATSC Data Broadcast Standard defines protocols for data transmission compatible with digital multiplex bit streams constructed in accordance with ISO/IEC 13818-1 (MPEG-2 systems). The standard supports data services that are both TV program related and non-program related. Applications may include enhanced television, webcasting, and streaming video services. Data broadcasting receivers may include PCs, televisions, set-top boxes, or other devices. The standard provides mechanisms for download of data, delivery of datagrams, and streaming data.

Normative References

Amendment 1:1997 to ISO/IEC 13818-1:1996—*Registration procedure for "copyright identifier"*.

Amendment 2:1997 to ISO/IEC 13818-1:1996—*Registration procedure for "format identifier."*

Amendment 2:2000 to ISO/IEC 13818-6—*Additions to support Synchronized Download Services, Opportunistic Data Services and Resource Announcement in Broadcast and Interactive Services.*

ATSC Standard A/53 (1995): *ATSC Digital Television Standard*, Advanced Television Systems Committee, Washington, D.C., 1995.

ATSC Standard A/65A (2000): *Program and System Information Protocol for Terrestrial Broadcast and Cable*, Advanced Television Systems Committee, Washington, D.C., 2000.

IEEE 802-1990: *IEEE Standards for Local and Metropolitan Area Networks—Overview and Architecture*, IEEE, New York, N.Y., 1990.

IETF RFC 1112: *Host Extensions for IP Multicasting.*

IETF RFC 2396: *Uniform Resource Identifiers* (URI), Generic Syntax, 1998.

ISO/IEC 13818-1 | ITU-T Rec. H.222.0:1996, Information Technology—Generic Coding of Moving Pictures and Associated Audio—Part 1: *Systems.*

ISO/IEC 13818-2 | ITU-T Rec. H262, 1995: Informaton Technology—Generic Coding of Moving Pictures and Associated Studio—Part 2: *Video.*

ISO/IEC 13818-6, 1998: *MPEG-2 Digital Storage Media Command and Control*, Chapters 2, 4, 5, 6, 7, 9 and 11.

ISO/IEC 8802-2, 1998: Information Technology—Telecommunications and Information Exchange between Systems—Local and Metropolitan Area Networks—Specific Requirements—Part 2: *Logical Link Control.*

ISO/IEC DIS 16500-7, 1998: Information Technology—Generic Digital Audio-Visual Systems—Part 7: *Basic Security Tools.*

ISO/IEC/TR3 8802-1:1997: Information Technology—Telecommunications and Information Exchange between Systems—Local and Metropolitan Area Networks—Specific Requirements—Part 1: *Overview of Local Area Network Standards.*

SCTE DVS 051: *Methods for Asynchronous Data Services Transport.*

SCTE DVS 132: *Standard Method for Isochronous Data Service Transport*, Chapter 5.

SMPTE 325M (1999): *Standard for Television—Opportunistic Data Broadcast Flow Control*, Society of Motion Picture and Television Engineers, White Plains, N.Y., 1999.

Technical Corrigendum 1:1999 to ISO/IEC 13818-1:1996.

Informative References

ATSC T3-512: ATSC Data Broadcast Standard Implementation Guidelines (draft).

ETSI EN 301 192 V1.2.1: Digital Video Broadcasting (DVB); *DVB Specification for Data Broadcasting* (1999-6).

9.1n A/91 Implementation Guidelines for the ATSC Data Broadcast Standard

This document provides a set of guidelines for the use and implementation of the ATSC Data Broadcast Standard as described in ATSC Standard, A/90 (2000), "Data Broadcast Standard." As such, they facilitate the efficient and reliable implementation of data broadcast services.

Normative References

There are none.

Informative References

ATSC Standard A/90 (2000): *ATSC Data Broadcast Standard*, Advanced Television Systems Committee, Washington, D.C., 2000.

ISO/IEC 13818-1: Information Technology—Generic Coding of Moving Pictures and Associated Audio Information, Part 1: *Systems*—International Standard.

ISO/IEC 13818-2: Information Technology—Generic Coding of Moving Pictures and Associated Audio Information, Part 2: *Video*—International Standard.

ISO/IEC 13818-6: Information Technology—Generic Coding of Moving Pictures and Associated Audio Information, Part 6: *Extension for Digital Storage Media Command and Control* (DSM-CC)—International Standard.

IETF RFC 791: "Internet Protocol", J. Postel, 01.09.1981.

ATSC Standard A/65A (2000): *Program and System Information Protocol (PSIP) for Terrestrial Broadcast and Cable*, Advanced Television Systems Committee, Washington, D.C., 2000.

IETF RFC 1112: *Host Extensions for IP Multicasting*.

ISO/IEC 8802-2 (ANSI/IEEE Standard 802.2): Information Technology—Telecommunications and Information Exchange Between Systems, Local and Metropolitan Area Networks, Specific Requirements, Part 2: *Logical Link Control*.

IETF RFC1042: *A Standard for the Transmission of IP Datagrams over IEEE 802 Networks*," J. Postel and J. Reynolds, February 1988.

9.2 Master Listings of References

The following sections represent a compilation of the documents listed previously in alphabetical order.

9.2a Normative References (alphabetical order)

47 CFR 73.622—*DTV Allocations*

47 CFR 73.625—*DTV Coverage of Principal Community and Antenna Systems*

47 CFR 73.686—*Coverage Measurements*

AES 3-1992 (ANSI S4.40-1992): "AES Recommended Practice for Digital Audio Engineering—Serial Transmission Format for Two-Channel Linearly Represented Digital Audio Data," Audio Engineering Society, New York, N.Y., 1992.

Amendment 1:1997 to ISO/IEC 13818-1:1996—*Registration procedure for "copyright identifier."*

Amendment 2:1997 to ISO/IEC 13818-1:1996—*Registration procedure for "format identifier."*

Amendment 2:2000 to ISO/IEC 13818-6—*Additions to support Synchronized Download Services, Opportunistic Data Services and Resource Announcement in Broadcast and Interactive Services.*

ANSI S1.4-1983: *Specification for Sound Level Meters.*

ATSC A/66 (23 July 1999): *Technical Corrigendum No. 1 to ATSC Standard A/65, Program and System Information Protocol for Terrestrial Broadcast and Cable*, Advanced Television Systems Committee, Washington, D.C., 1999.

ATSC Standard A/52A (2001): *Digital Audio Compression* (AC-3), Advanced Television Systems Committee, Washington, D.C., 1995.

ATSC Standard A/53B (2001): *ATSC Digital Television Standard*, Advanced Television Systems Committee, Washington, D.C., 1995.

ATSC Standard A/57 (1996): *Program/Episode/Version Identification*, Advanced Television Systems Committee, Washington, D.C., 1996.

ATSC Standard A/65 (1997): *Program and System Information Protocol for Terrestrial Broadcast and Cable*, Advanced Television Systems Committee, Washington, D.C., 1997.

ATSC Standard A/65A (2000): *Program and System Information Protocol for Terrestrial Broadcast and Cable*, Advanced Television Systems Committee, Washington, D.C., 2000.

CENELEC EN 50221: *Common Interface Specification for Conditional Access and other Digital Video Broadcasting Decoder Applications*, February 1997.

DES CBC Packet Encryption, SCTE DVS 042, Society of Cable Telecommunications Engineers, 25 October 1996.

EIA 679-B (1999): *National Renewable Security Standard* (NRSS), Electronic Industries Association.

EIA-708A: *Specification for Advanced Television Closed Captioning* (ATVCC), Electronic Industries Association.

EIA-752: *Specification for Transport of Transmission Signal Identifier* (TSID) *Using Extended Data Service*, Electronic Industries Association.

EIA-766: *Specification for U.S. Region Rating Table* (RRT) *and Content Advisory Descriptor for Transport of Content Advisory Information Using ATSC A/65 Program and System Information Protocol* (PSIP), Electronic Industries Association.

ETS 300 421 (v1.1.2, 1997-08): Digital Video Broadcasting (DVB); *Framing Structure, Channel Coding and Modulation for 11/12 GHz Satellite Services.*

FCC Rules and Regulations, Parts 73 and 76, Federal Communications Commission, Washington, D.C.

Federal Information Processing Standard, FIPS Pub 6-4: *Counties and Equivalent Entities of the U.S., Its Possessions, and Associated Areas—90 Aug 31*, U.S. Government Printing Office, Washington, D.C.

FIPS PUB 46-2 (1993): *Specification for the Data Encryption Standard*, National Institute of Standards and Technology, Gaithersburg, MD.

FIPS PUB 81 (1980): *DES Modes of Operation*, National Institute of Standards and Technology, Gaithersburg, MD.

IEC 651 (1979): *Sound Level Meters.*

IEC 804 (1985): Amendment 1 (1989) *Integrating/Averaging Sound Level Meters.*

IEEE 802-1990: *IEEE Standards for Local and Metropolitan Area Networks: Overview and Architecture*, IEEE, New York, N.Y., 1990.

IEEE Standard 100-1992: *The New IEEE Standard Dictionary of Electrical and Electronic Terms*, IEEE, New York, N.Y., 1992.

IETF RFC 1112: *Host Extensions for IP Multicasting.*

IETF RFC 2396: *Uniform Resource Identifiers* (URI), Generic Syntax, 1998.

ISO/IEC 13818-1 | ITU-T Rec. H.222.0:1996, Information Technology—Generic Coding of Moving Pictures and Associated Audio—Part 1: *Systems.*

ISO/IEC 13818-1: 1996/Amd. 1: 1997 (E) Amendment 1.

ISO/IEC 13818-1: 1996/Amd. 2: 1997 (E) Amendment 2.

ISO/IEC 13818-1: 1996/Amd. 3: 1997 (E) Amendment 3.

ISO/IEC 13818-1: 1996/Amd. 4: 1997 (E) Amendment 4.

ISO/IEC 13818-1: 1996/Cor. 1: 1997 (E) Technical Corrigendum 1.

ISO/IEC 13818-2 | ITU-T Rec. H262, 1995: Information Technology—Generic Coding of Moving Pictures and Associated Studio—Part 2: *Video.*

ISO/IEC 13818-2: 1996/Cor. 1: 1997 (E) MPEG-2 Video Technical Corrigendum 1.

ISO/IEC 13818-2: 1996/Cor. 2: 1997 (E) MPEG-2 Video Technical Corrigendum 2.

ISO/IEC 13818-6, 1998: *MPEG-2 Digital Storage Media Command and Control*, Chapters 2, 4, 5, 6, 7, 9 and 11.

ISO/IEC 8802-2, 1998: Information Technology—Telecommunications and Information Exchange between Systems—Local and Metropolitan Area Networks—Specific Requirements—Part 2: *Logical Link Control.*

ISO/IEC 8859: *Information Processing—8-bit Single-Octet Coded Character Sets*, Parts 1 through 10.

ISO/IEC CD 13818-4: MPEG Committee Draft (1994), *MPEG-2 Compliance.*

ISO/IEC DIS 16500-7, 1998: Information Technology—Generic Digital Audio-Visual Systems—Part 7: *Basic Security Tools.*

ISO/IEC IS 11172-2: International Standard, *MPEG-1 Video.*

ISO/IEC IS 13818-1: International Standard (1996), *MPEG-2 Systems.*

ISO/IEC IS 13818-2: International Standard (1996), *MPEG-2 Video.*

ISO/IEC/TR3 8802-1:1997: Information Technology—Telecommunications and Information Exchange between Systems—Local and Metropolitan Area Networks—Specific Requirements—Part 1: *Overview of Local Area Network Standards.*

ITU-R Recommendation BT.1206: *Spectrum Shaping Limits for Digital Terrestrial Television Broadcasting.*

ITU-T Rec. H.222.0 | ISO/IEC 13818-1:1996, Information Technology—Generic Coding of Moving Pictures and Associated Audio—Part 1: *Systems.*

ITU-T Rec. H.262 | ISO/IEC 13818-2:1996, Information Technology—Generic Coding of Moving Pictures and Associated Audio—Part 2: *Video.*

NBS PUB 500-20 (1980): *Validating the Correctness of Hardware Implementations of the NBS DES*, National Bureau of Standards.

SCTE DVS 051: *Methods for Asynchronous Data Services Transport.*

SCTE DVS 132: *Standard Method for Isochronous Data Service Transport*, Chapter 5.

SMPTE 325M (1999): *Standard for Television—Opportunistic Data Broadcast Flow Control*, Society of Motion Picture and Television Engineers, White Plains, N.Y., 1999.

Technical Corrigendum 1:1999 to ISO/IEC 13818-1:1996.

The Unicode Standard, Version 3.0, The Unicode Consortium, Addison-Wesley Pub., April 2000.

U.S. Code of Federal Regulations (CFR) Title 47, 47CFR11: *Emergency Alert System* (EAS), U.S. Government Printing Office, Washington, D.C.

Listing of Organizations Referenced

Advanced Television Systems Committee, Washington, D.C.

American National Standards Institute.

Audio Engineering Society, New York, N.Y.

CENELEC.

Electronic Industries Association.

Federal Communications Commission, Washington, D.C.

Institute of Electrical and Electronics Engineers, Piscataway, N.J.

International Electrotechnical Commission.

International Standards Organization.

International Telecommunications Union, Geneva.

Internet Engineering Task Force, Reston, VA

Society of Cable Telecommunications Engineers.

Society of Motion Picture and Television Engineers, White Plains, N.Y.

U. S. Government.

Unicode Consortium.

9.2b Informative References (alphabetical order)

47 CFR Part 15: FCC Rules.

ATSC Standard A/53B (2001): *ATSC Digital Television Standard*, Advanced Television Systems Committee, Washington, D.C., 1994.

ATSC Standard A/54 (1995): *Guide to the Use of the ATSC Digital Television Standard*, Advanced Television Systems Committee, Washington, D.C., 1995.

ATSC Standard A/55 (1996): *Program Guide for Digital Television*, Advanced Television Systems Committee, Washington, D.C., 1996.

ATSC Standard A/56 (1996): *System Information for Digital Television*, Advanced Television Systems Committee, Washington, D.C., 1996.

ATSC Standard A/65A (2000): *Program and System Information Protocol* (PSIP) *for Terrestrial Broadcast and Cable*, Advanced Television Systems Committee, Washington, D.C., 2000.

ATSC Standard A/90 (2000): *ATSC Data Broadcast Standard*, Advanced Television Systems Committee, Washington, D.C., 2000.

ATSC T3-512: *ATSC Data Broadcast Standard Implementation Guidelines* (draft).

Digital Video Transmission Standard for Cable Television, SCTE DVS-031, rev. 2, 29 May 1997.

Ehmer, R H.: "Masking of Tones vs. Noise Bands," *J. Acoust. Soc. Am.*, vol. 31, pp 1253–1256, September 1959.

Ehmer, R. H.: "Masking Patterns of Tones," *J. Acoust. Soc. Am.*, vol. 31, pp. 1115–1120, August 1959.

EIA 608A:*Recommended Practice for Line 21 Data Service*, Electronic Industries Association.

EIA 608B: *Recommended Practice for Line 21 Data Service*, Electronic Industries Association.

EIA IS-105: *EIA Interim Standard for a Decoder Interface Specification for Television Receiving Devices and Cable Television Decoders.*

EIA IS-132: *EIA Interim Standard for Channelization of Cable Television.*

EIA IS-23: *EIA Interim Standard for RF Interface Specification for Television Receiving Devices and Cable Television Systems.*

EN 301 210 (v1.1.1, 1999-03): Digital Video Broadcasting (DVB); *Framing Structure, Channel Coding and Modulation for Digital Satellite News Gathering (DSNG) and Other Contribution Applications by Satellite.*

ETR 289 ed.1 (1996-10), Digital Video Broadcasting (DVB): *Support for Use of Scrambling and Conditional Access* (CA) *within Digital Broadcast Systems*, European Telecommunications Standards Institute.

ETSI EN 301 192 V1.2.1: Digital Video Broadcasting (DVB); *DVB Specification for Data Broadcasting* (1999-6).

ETSI ETS 300 468: "Digital Broadcasting Systems for Television, Sound and Data Services; Specification for Service Information (SI) in Digital Video Broadcasting (DVB) Systems."

FIPS PUB 74 (1981): *Guidelines for Implementing and Using NBS DES*, National Institute of Standards and Technology, Gaithersburg, MD.

IESS-308 (Rev 9, 30 Nov 1998): *Intelsat Earth Station Standards, Performance Characteristics For Intermediate Data Rate Digital Carriers Using Convolutional Encoding/Viterbi Encoding and QPSK Modulation.*

IESS-310 (Rev 1, 30 Nov 1998): *Intelsat Earth Station Standards, Performance Characteristics For Intermediate Data Rate Digital Carriers Using Rate 2/3 TCM / 8PSK and Reed-Solomon Outer Coding.*

IETF RFC 1112: *Host Extensions for IP Multicasting.*

IETF RFC 791: "Internet Protocol", J. Postel, 01.09.1981.

IETF RFC1042: "A Standard for the Transmission of IP Datagrams over IEEE 802 Networks," J. Postel and J. Reynolds, February 1988.

ISO/IEC 13818-1: Information Technology—Generic Coding of Moving Pictures and Associated Audio Information, Part 1: *Systems*—International Standard.

ISO/IEC 13818-2: Information Technology—Generic Coding of Moving Pictures and Associated Audio Information, Part 2: *Video*—International Standard.

ISO/IEC 13818-6: Information Technology—Generic Coding of Moving Pictures and Associated Audio Information, Part 6: *Extension for Digital Storage Media Command and Control* (DSM-CC)—International Standard.

ISO/IEC 8802-2 (ANSI/IEEE Standard 802.2): Information Technology—Telecommunications and Information Exchange Between Systems, Local and Metropolitan Area Networks, Specific Requirements, Part 2: *Logical Link Control.*

ISO/IEC IS 13818-1: International Standard (1994), *MPEG-2 Systems.*

ITU-R BT.601-4 (1994): *Encoding Parameters of Digital Television for Studios.*

Moore, B. C. J., and B. R. Glasberg: "Formulae Describing Frequency Selectivity as a Function of Frequency and Level, and Their Use in Calculating Excitation Patterns," *Hearing Research*, vol. 28, pp. 209–225, 1987.

Record of Test Results for Digital HDTV Grand Alliance System, September 8, 1995, Advanced Television Test Center, Alexandria, VA.

SMPTE 274M (1995): *Standard for Television, 1920 × 1080 Scanning and Interface*, Society of Motion Picture and Television Engineers, White Plains, N.Y., 1995.

SMPTE 296M (1995): *Standard for Television, 1280 × 720 Scanning and Interface*, Society of Motion Picture and Television Engineers, White Plains, N.Y., 1994.

SMPTE 296M (1997): *Standard for Television, 1280 × 720 Scanning, Analog and Digital Representation, and Analog Interface*, Society of Motion Picture and Television Engineers, White Plains, N.Y., 1997.

Todd, C. et. al.: "AC-3: Flexible Perceptual Coding for Audio Transmission and Storage", AES 96th Convention, Preprint 3796, Audio Engineering Society, New York, N.Y., February 1994.

TS 101 197-1 V1.1.1 (1997-06): *Technical Specification of DVB Simulcrypt*, 1 June 1997, European Telecommunications Standards Institute.

Zwicker, E.: "Subdivision of the Audible Frequency Range into Critical Bands (Frequenzgruppen)," *J. Acoust. Soc. of Am.*, vol. 33, pg. 248, February 1961.

Listing of Organizations Referenced

Advanced Television Systems Committee, Washington, D.C.

Advanced Television Test Center, Alexandria, VA.

Audio Engineering Society, New York, N.Y.

Electronics Industries Association.

European Telecommunications Standards Institute.

Hearing Research

Intelsat.

International Standards Organization.

International Telecommunications Union, Geneva.

Internet Engineering Task Force, Reston, VA

Journal of the Acoustical Society of America.

Society of Cable Telecommunications Engineers.

Society of Motion Picture and Television Engineers, White Plains, N.Y.

U.S. Government.

Chapter

10

The Electromagnetic Spectrum

John Norgard

10.1 Introduction[1]

The electromagnetic (EM) spectrum consists of all forms of EM radiation—EM waves (radiant energy) propagating through space, from DC to light to gamma rays. The EM spectrum can be arranged in order of frequency and/or wavelength into a number of regions, usually wide in extent, within which the EM waves have some specified common characteristics, such as characteristics relating to the production or detection of the radiation. A common example is the spectrum of the radiant energy in white light, as dispersed by a prism, to produce a "rainbow" of its constituent colors. Specific frequency ranges are often called *bands*; several contiguous frequency bands are usually called *spectrums*; and sub-frequency ranges within a band are sometimes called *segments*.

The EM spectrum can be displayed as a function of frequency (or wavelength). In air, frequency and wavelength are inversely proportional, $f = c/\lambda$ (where $c \approx 3 \times 10^8$ m/s, the speed of light in a vacuum). The MKS unit of frequency is the Hertz and the MKS unit of wavelength is the meter. Frequency is also measured in the following sub-units:

- Kilohertz, 1 kHz = 10^3 Hz
- Megahertz, 1 MHz = 10^6 Hz
- Gigahertz, 1 GHz = 10^9 Hz
- Terahertz, 1 THz = 10^{12} Hz
- Petahertz, 1 PHz = 10^{15} Hz
- Exahertz, 1 EHz = 10^{18} Hz

Or for very high frequencies, *electron volts*, 1 ev ~ 2.41×10^{14} Hz

Wavelength is also measured in the following sub-units:

- Centimeters, 1 cm = 10^{-2} m

1. This chapter is based on: Norgard, John: "The Electromagnetic Spectrum," in *Standard Handbook of Video and Television Engineering*, 3rd ed., Jerry C. Whitaker (ed.), McGraw-Hill, New York, N.Y., 2000. Used with permission. All rights reserved.

- Millimeters, 1 mm = 10^{-3} m
- Micrometers, 1 μm = 10^{-6} m (microns)
- Nanometers, 1 nm = 10^{-9} m
- Ångstroms, 1 Å = 10^{-10} m
- Picometers, 1 pm = 10^{-12} m
- Femtometers, 1 fm = 10^{-15} m
- Attometers, 1 am = 10^{-18} m

10.2 Spectral Sub-Regions

For convenience, the overall EM spectrum can be divided into three main sub-regions:

- *Optical spectrum*
- *DC to light spectrum*
- *Light to gamma ray spectrum*

These main sub-regions of the EM spectrum are next discussed. Note that the boundaries between some of the spectral regions are somewhat arbitrary. Certain spectral bands have no sharp edges and merge into each other, while other spectral segments overlap each other slightly.

10.2a Optical Spectrum

The optical spectrum is the "middle" frequency/wavelength region of the EM spectrum. It is defined here as the visible and near-visible regions of the EM spectrum and includes:

- The *infrared (IR)* band, circa 300 μm–0.7 μm (circa 1 THz–429 THz)
- The *visible light* band, 0.7 μm–0.4 μm (429 THz–750 THz)
- The *ultraviolet (UV)* band, 0.4 μm–circa 10 nm (750 THz–circa 30 PHz), approximately 100 ev

These regions of the EM spectrum are usually described in terms of their wavelengths.

Atomic and molecular radiation produce radiant light energy. Molecular radiation and radiation from hot bodies produce EM waves in the IR band. Atomic radiation (outer shell electrons) and radiation from arcs and sparks produce EM waves in the UV band.

Visible Light Band

In the "middle" of the optical spectrum is the visible light band, extending approximately from 0.4 μm (violet) up to 0.7 μm (red), i.e. from 750 THz (violet) down to 429 THz (red). EM radiation in this region of the EM spectrum, when entering the eye, gives rise to visual sensations (colors), according to the spectral response of the eye, which responds only to radiant energy in the visible light band extending from the extreme long wavelength edge of red to the extreme short wavelength edge of violet. (The spectral response of the eye is sometimes quoted as extending from 0.38 μm (violet) up to 0.75 or 0.78 μm (red); i.e., from 789 THz down to 400 or 385

THz.)This visible light band is further subdivided into the various colors of the rainbow, in decreasing wavelength/increasing frequency:

- Red, a primary color, peak intensity at 700.0 nm (429 THz)
- Orange
- Yellow
- Green, a primary color, peak intensity at 546.1 nm (549 THz)
- Cyan
- Blue, a primary color, peak intensity at 435.8 nm (688 THz)
- Indigo
- Violet

IR Band

The IR band is the region of the EM spectrum lying immediately below the visible light band. The IR band consists of EM radiation with wavelengths extending between the longest visible red (circa 0.7 μm) and the shortest microwaves (300 μm–1 mm), i.e., from circa 429 THz down to 1 THz–300 GHz. The IR band is further subdivided into the "near" (shortwave), "intermediate" (midwave), and "far" (longwave) IR segments as follows [2]:

- *Near* IR segment, 0.7 μm up to 3 μm (429 THz down to 100 THz)
- *Intermediate* IR segment, 3 μm up to 7 μm (100 THz down to 42.9 THz)
- *Far* IR segment, 7 μm up to 300 μm (42.9 THz down to 1 THz)
- Sub-millimeter band, 100 μm up to 1 mm (3 THz down to 300 GHz). Note that the sub-millimeter region of wavelengths is sometimes included in the very far region of the IR band.

EM radiation is produced by oscillating and rotating molecules and atoms. Therefore, all objects at temperatures above absolute zero emit EM radiation by virtue of their thermal motion (warmth) alone. Objects near room temperature emit most of their radiation in the IR band. However, even relatively cool objects emit some IR radiation; hot objects, such as incandescent filaments, emit strong IR radiation.

IR radiation is sometimes incorrectly called "radiant heat" because warm bodies emit IR radiation and bodies that absorb IR radiation are warmed. However, IR radiation is not itself "heat". This radiant energy is called "black body" radiation. Such waves are emitted by all material objects. For example, the background cosmic radiation (2.7K) emits microwaves; room temperature objects (293K) emit IR rays; the Sun (6000K) emits yellow light; the Solar Corona (1 million K) emits X rays.

IR astronomy uses the 1 μm to 1 mm part of the IR band to study celestial objects by their IR emissions. IR detectors are used in night vision systems, intruder alarm systems, weather fore-

---

2. Some reference texts use 2.5 mm (120 THz) as the breakpoint between the near and the intermediate IR bands and 10 mm (30 THz) as the breakpoint between the intermediate and the far IR bands. Also, 15 mm (20 Thz) is sometimes considered as the long wavelength end of the far IR band.

casting, and missile guidance systems. IR photography uses multilayered color film, with an IR sensitive emulsion in the wavelengths between 700–900 nm, for medical and forensic applications, and for aerial surveying.

UV Band

The UV band is the region of the EM spectrum lying immediately above the visible light band. The UV band consists of EM radiation with wavelengths extending between the shortest visible violet (circa 0.4 μm) and the longest X rays (circa 10 nm), i.e., from 750 THz—approximately 3 ev—up to circa 30 PHz—approximately 100 ev.[3]

The UV band is further subdivided into the "near" and the "far" UV segments as follows:

- *Near*" UV segment, circa 0.4 μm down to 100 nm (circa 750 THz up to 3 PHz, approximately 3 ev up to 10 ev)
- *Far* UV segment, 100 nm down to circa 10 nm, (3 PHz up to circa 30 PHz, approximately 10 ev up to 100 ev)

The far UV band is also referred to as the *vacuum UV band*, since air is opaque to all UV radiation in this region.

UV radiation is produced by electron transitions in atoms and molecules, as in a mercury discharge lamp. Radiation in the UV range is easily detected and can cause florescence in some substances, and can produce photographic and ionizing effects. In UV astronomy, the emissions of celestial bodies in the wavelength band between 50–320 nm are detected and analyzed to study the heavens. The hottest stars emit most of their radiation in the UV band.

10.2b DC to Light

Below the IR band are the lower frequency (longer wavelength) regions of the EM spectrum, subdivided generally into the following spectral bands (by frequency/wavelength):

- *Microwave* band, 300 GHz down to 300 MHz (1 mm up to 1 m). Some reference works define the lower edge of the microwave spectrum at 1 GHz.
- *Radio frequency (RF)* band, 300 MHz down to 10 kHz (1 m up to 30 Km)
- *Power (PF)/telephony* band, 10 kHz down to dc (30 Km up to ∞)

These regions of the EM spectrum are usually described in terms of their frequencies.

Radiations whose wavelengths are of the order of millimeters & centimeters are called *microwaves*, and those still longer are called radio frequency (RF) waves (or *Hertzian waves*).

Radiation from electronic devices produces EM waves in both the microwave and RF bands. Power frequency energy is generated by rotating machinery. Direct current (dc) is produced by batteries or rectified alternating current (ac).

Microwave Band

The microwave band is the region of wavelengths lying between the far IR/sub-millimeter region and the conventional RF region. The boundaries of the microwave band have not been definitely

3. Some references use 4, 5, or 6 nm as the upper edge of the UV band.

fixed, but it is commonly regarded as the region of the EM spectrum extending from about 1 mm up to 1 m in wavelengths, i.e. from 300 GHz down to 300 MHz. The microwave band is further sub-divided into the following segments:

- *Millimeter* waves, 300 GHz down to 30 GHz (1 mm up to 1 cm); the EHF band. (Some references consider the top edge of the millimeter region to stop at 100 GHz.)
- *Centimeter* waves, 30 GHz down to 3 GHz (1 cm up to 10 cm); the SHF band.

The microwave band usually includes the UHF band from 3 GHz down to 300 MHz (from 10 cm up to 1 m).

Microwaves are used in radar, in communication links spanning moderate distances, as radio carrier waves in television broadcasting, for mechanical heating, and cooking in microwave ovens.

Radio Frequency (RF) Band

The RF range of the EM spectrum is the wavelength band suitable for utilization in radio communications extending from 10 kHz up to 300 MHz (from 30 Km down to 1 m). (Some references consider the RF band as extending from 10 kHz to 300 GHz, with the microwave band as a subset of the RF band from 300 MHz to 300 GHz.)

Some of the radio waves in this band serve as the carriers of low-frequency audio signals; other radio waves are modulated by video and digital information. The *amplitude modulated* (AM) broadcasting band uses waves with frequencies between 550–1640 kHz; the *frequency modulated* (FM) broadcasting band uses waves with frequencies between 88–108 MHz.

In the U.S., the Federal Communications Commission (FCC) is responsible for assigning a range of frequencies to specific services. The International Telecommunications Union (ITU) coordinates frequency band allocation and cooperation on a worldwide basis.

Radio astronomy uses radio telescopes to receive and study radio waves naturally emitted by objects in space. Radio waves are emitted from hot gases (*thermal radiation*), from charged particles spiraling in magnetic fields (*synchrotron radiation*), and from excited atoms and molecules in space (*spectral lines*), such as the 21 cm line emitted by hydrogen gas.

Power Frequency (PF)/Telephone Band

The PF range of the EM spectrum is the wavelength band suitable for generating, transmitting, and consuming low frequency power, extending from 10 kHz down to dc (zero frequency), i.e., from 30Km up in wavelength. In the US, most power is generated at 60 Hz (some military and computer applications use 400 Hz); in other countries, including Europe, power is generated at 50 Hz.

Frequency Band Designations

The combined microwave, RF (Hertzian Waves), and power/telephone spectra are subdivided into the specific bands given in Table 10.1, which lists the international radio frequency band designations and the numerical designations. Note that the band designated (12) has no commonly used name or abbreviation.

The radar band often is considered to extend from the middle of the HF (7) band to the end of the EHF (11) band. The current US Tri-Service radar band designations are listed in Table 10.2.

Table 10.1 Frequency Band Designations

Description	Band Designation	Frequency	Wavelength
Extremely Low Frequency	ELF (1) Band	3 Hz up to 30 Hz	100 Mm down to 10 Mm
Super Low Frequency	SLF (2) Band	30 Hz up to 300 Hz	10 Mm down to 1 Mm
Ultra Low Frequency	ULF (3) Band	300 Hz up to 3 kHz	1 Mm down to 100 Km
Very Low Frequency	VLF (4) Band	3 kHz up to 30 kHz	100 Km down to 10 Km
Low Frequency	LF (5) Band	30 kHz up to 300 kHz	10 Km down to 1 Km
Medium Frequency	MF (6) Band	300 kHz up to 3 MHz	1 Km down to 100 m
High Frequency	HF (7) Band	3 MHz up to 30 MHz	100 m down to 10 m
Very High Frequency	VHF (8) Band	30 MHz up to 300 MHz	10 m down to 1 m
Ultra High Frequency	UHF (9) Band	300 MHz up to 3 GHz	1 m down to 10 cm
Super High Frequency	SHF (10) Band	3 GHz up to 30 GHz	10 cm down to 1 cm
Extremely High Frequency	EHF (11) Band	30 GHz up to 300 GHz	1 cm down to 1 mm
—	(12) Band	300 GHz up to 3 THz	1 mm down to 100 μ

Table 10.2 Radar Band Designations

Band	Frequency	Wavelength
A Band	0 Hz up to 250 MHz	∞ down to 1.2 m
B Band	250 MHz up to 500 MHz	1.2 m down to 60 cm
C Band	500 MHz up to 1 GHz	60 cm down to 30 cm
D Band	1 GHz up to 2 GHz	30 cm down to 15 cm
E Band	2 GHz up to 3 GHz	15 cm down to 10 cm
F Band	3 GHz up to 4 GHz	10 cm down to 7.5 cm
G Band	4 GHz up to 6 GHz	7.5 cm down to 5 cm
H Band	6 GHz up to 8 GHz	5 cm down to 3.75 cm
I Band	8 GHz up to 10 GHz	3.75 cm down to 3 cm
J Band	10 GHz up to 20 GHz	3 cm down to 1.5 cm
K Band	20 GHz up to 40 GHz	1.5 cm down to 7.5 mm
L Band	40 GHz up to 60 GHz	7.5 mm down to 5 mm)
M Band	60 GHz up to 100 GHz	5 mm down to 3 mm
N Band	100 GHz up to 200 GHz	3 mm down to 1.5 mm
O Band	200 GHz up to 300 GHz	1.5 mm down to 1 mm

An alternate and more detailed sub-division of the UHF (9), SHF (10), and EHF (11) bands is given in Table 10.3.

Several other frequency bands of interest (not exclusive) are listed in Tables 10.4–10.6.

Table 10.3 Detail of UHF, SHF, and EHF Band Designations

Band	Frequency	Wavelength
L Band	1.12 GHz up to 1.7 GHz	26.8 cm down to 17.6 cm
LS Band	1.7 GHz up to 2.6 GHz	17.6 cm down to 11.5 cm
S Band	2.6 GHz up to 3.95 GHz	11.5 cm down to 7.59 cm
C(G) Band	3.95 GHz up to 5.85 GHz	7.59 cm down to 5.13 cm
XN(J, XC) Band	5.85 GHz up to 8.2 GHz	5.13 cm down to 3.66 cm
XB(H, BL) Band	7.05 GHz up to 10 GHz	4.26 cm down to 3 cm
X Band	8.2 GHz up to 12.4 GHz	3.66 cm down to 2.42 cm
Ku(P) Band	12.4 GHz up to 18 GHz	2.42 cm down to 1.67 cm
K Band	18 GHz up to 26.5 GHz	1.67 cm down to 1.13 cm
V(R, Ka) Band	26.5 GHz up to 40 GHz	1.13 cm down to 7.5 mm
Q(V) Band	33 GHz up to 50 GHz	9.09 mm down to 6 mm
M(W) Band	50 GHz up to 75 GHz	6 mm down to 4 mm
E(Y) Band	60 GHz up to 90 GHz	5 mm down to 3.33 mm
F(N) Band	90 GHz up to 140 GHz	3.33 mm down to 2.14 mm
G(A)	140 GHz p to 220 GHz	2.14 mm down to 1.36 mm
R Band	220 GHz up to 325 GHz	1.36 mm down to 0.923 mm

Table 10.4 Low Frequency Bands of Interest

Band	Frequency
Sub-sonic band	0 Hz–10 Hz
Audio band	10 Hz–10 kHz
Ultra-sonic band	10 kHz and up

Table 10.5 Applications of Interest in the RF Band

Band	Frequency
Longwave broadcasting band	150–290 kHz
AM broadcasting band	550–1640 kHz (1.640 MHz), 107 channels, 10 kHz separation
International broadcasting band	3–30 MHz
Shortwave broadcasting band	5.95–26.1 MHz (8 bands)
VHF TV (Channels 2 - 4)	54–72 MHz
VHF TV (Channels 5 - 6)	76–88 MHz
FM broadcasting band	88–108 MHz
VHF TV (Channels 7 - 13)	174–216 MHz
UHF TV (Channels 14 - 69)	512–806 MHz

10.2c Light to Gamma Rays

Above the UV spectrum are the higher frequency (shorter wavelength) regions of the EM spectrum, subdivided generally into the following spectral bands (by frequency/wavelength):

- *X ray* band, approximately 10 ev up to 1 Mev (circa 10 nm down to circa 1 pm), circa 3 PHz up to circa 300 EHz
- *Gamma ray* band, approximately 1 Kev up to ∞ (circa 300 pm down to 0 m), circa 1 EHz up to ∞

These regions of the EM spectrum are usually described in terms of their photon energies in electron volts. Note that the bottom of the gamma ray band overlaps the top of the X ray band.

It should be pointed out that *cosmic "rays"* (from astronomical sources) are not EM waves (rays) and, therefore, are not part of the EM spectrum. Cosmic "rays" are high energy charged particles (electrons, protons, and ions) of extraterrestrial origin moving through space, which may have energies as high as 10^{20} ev. Cosmic "rays" have been traced to cataclysmic astrophysical/cosmological events, such as exploding stars and black holes. Cosmic "rays" are emitted by supernova remnants, pulsars, quasars, and radio galaxies. Comic "rays" that collide with molecules in the Earth's upper atmosphere produce secondary cosmic "rays" and gamma rays of high energy that also contribute to natural background radiation. These gamma rays are sometimes called "cosmic" or *secondary* gamma rays. Cosmic rays are a useful source of high-energy particles for certain scientific experiments.

Radiation from atomic inner shell excitations produces EM waves in the X ray band. Radiation from naturally radioactive nuclei produces EM waves in the gamma ray band.

X Ray Band

The X ray band is further sub-divided into the following segments:

- *Soft* X rays, approximately 10 ev up to 10 Kev (circa 10 nm down to 100 pm), circa 3 PHz up to 3 EHz
- *Hard* X rays, approximately 10 Kev up to 1Mev (100 pm down to circa 1 pm), 3 EHz up to circa 300 EHz

Because the physical nature of these rays was first unknown, this radiation was called X rays. The more powerful X rays are called hard X rays and are of high frequencies and, therefore, are more energetic; less powerful X rays are called soft X rays and have lower energies.

X rays are produced by transitions of electrons in the inner levels of excited atoms or by rapid deceleration of charged particles (*Brehmsstrahlung* or breaking radiation). An important source of X rays is *synchrotron radiation*. X rays can also be produced when high energy electrons from a heated filament cathode strike the surface of a target anode (usually tungsten) between which a high alternating voltage (approximately 100 kV) is applied.

X rays are a highly penetrating form of EM radiation and applications of X rays are based on their short wavelengths and their ability to easily pass through matter. X rays are very useful in crystallography for determining crystalline structure and in medicine for photographing the body. Because different parts of the body absorb X rays to a different extent, X rays passing through the body provide a visual image of its interior structure when striking a photographic plate. X rays are dangerous and can destroy living tissue. They can also cause severe skin burns. X rays are useful in the diagnosis and non-destructive testing of products for defects.

Table 10.6 Applications of Interest in the Microwave Band (up to 40 GHz):

Application	Frequency
Aero Navigation	0.96–1.215 GHz
GPS Down Link	1.2276 GHz
Military COM/Radar	1.35–1.40 GHz
Miscellaneous COM/Radar	1.40–1.71 GHz
L-Band Telemetry	1.435–1.535 GHz
GPS Down Link	1.57 GHz
Military COM (Troposcatter/Telemetry)	1.71–1.85 GHz
Commercial COM & Private LOS	1.85–2.20 GHz
Microwave Ovens	2.45 GHz
Commercial COM/Radar	2.45–2.69 GHz
Instructional TV	2.50–2.69 GHz
Military Radar (Airport Surveillance)	2.70–2.90 GHz
Maritime Navigation Radar	2.90–3.10 GHz
Miscellaneous Radars	2.90–3.70 GHz
Commercial C-Band SAT COM Down Link	3.70–4.20 GHz
Radar Altimeter	4.20–4.40 GHz
Military COM (Troposcatter)	4.40–4.99 GHz
Commercial Microwave Landing System	5.00–5.25 GHz
Miscellaneous Radars	5.25–5.925 GHz
C-Band Weather Radar	5.35–5.47 GHz
Commercial C-Band SAT COM Up Link	5.925–6.425 GHz
Commercial COM	6.425–7.125 GHz
Mobile TV Links	6.875–7.125 GHz
Military LOS COM	7.125–7.25 GHz
Military SAT COM Down Link	7.25–7.75 GHz
Military LOS COM	7.75–7.9 GHz
Military SAT COM Up Link	7.90–8.40 GHz
Miscellaneous Radars	8.50–10.55 GHz
Precision Approach Radar	9.00–9.20 GHz
X-Band Weather Radar (& Maritime Navigation Radar)	9.30–9.50 GHz
Police Radar	10.525 GHz
Commercial Mobile COM (LOS & ENG)	10.55–10.68 GHz
Common Carrier LOS COM	10.70–11.70 GHz
Commercial COM	10.70–13.25 GHz
Commercial Ku-Band SAT COM Down Link	11.70–12.20 GHz
DBS Down Link & Private LOS COM	12.20–12.70 GHz
ENG & LOS COM	12.75–13.25 GHz
Miscellaneous Radars & SAT COM	13.25–14.00 GHz
Commercial Ku-Band SAT COM Up Link	14.00–14.50 GHz
Military COM (LOS, Mobile, &Tactical)	14.50–15.35 GHz
Aero Navigation	15.40–15.70 GHz
Miscellaneous Radars	15.70–17.70 GHz
DBS Up Link	17.30–17.80 GHz
Common Carrier LOS COM	17.70–19.70 GHz
Commercial COM (SAT COM & LOS)	17.70–20.20 GHz
Private LOS COM	18.36–19.04 GHz
Military SAT COM	20.20–21.20 GHz
Miscellaneous COM	21.20–24.00 GHz
Police Radar	24.15 GHz
Navigation Radar	24.25–25.25 GHz
Military COM	25.25–27.50 GHz
Commercial COM	27.50–30.00 GHz
Military SAT COM	30.00–31.00 GHz
Commercial COM	31.00–31.20 GHz

Gamma Ray Band

The gamma ray band is sub-divided into the following segments:

- *Primary* gamma rays, approximately 1 Kev up to 1 Mev (circa 300 pm down to 300 fm), circa 1 EHz up to 1000 EHz
- *Secondary* gamma rays, approximately 1 Mev up to ∞ (300 fm down to 0 m), 1000 EHz up to ∞

Secondary gamma rays are created from collisions of high energy cosmic rays with particles in the Earth's upper atmosphere.

The primary gamma rays are further sub-divided into the following segments:

- *Soft* gamma rays, approximately 1 Kev up to circa 300 Kev (circa 300 pm down to circa 3 pm), circa 1 EHz up to circa 100 EHz
- *Hard* gamma rays, approximately 300 Kev up to 1 Mev (circa 3 pm down to 300 fm), circa 100 EHz up to 1000 EHz

Gamma rays are essentially very energetic X rays. The distinction between the two is based on their origin. X rays are emitted during atomic processes involving energetic electrons; gamma rays are emitted by excited nuclei or other processes involving sub-atomic particles.

Gamma rays are emitted by the nucleus of radioactive material during the process of natural radioactive decay as a result of transitions from high energy excited states to low energy states in atomic nuclei. Cobalt 90 is a common gamma ray source (with a half-life of 5.26 years). Gamma rays are also produced by the interaction of high energy electrons with matter. "Cosmic" gamma rays cannot penetrate the Earth's atmosphere.

Applications of gamma rays are found both in medicine and in industry. In medicine, gamma rays are used for cancer treatment, diagnoses, and prevention. Gamma ray emitting radioisotopes are used as tracers. In industry, gamma rays are used in the inspection of castings, seams, and welds.

10.3 Bibliography

Collocott, T. C., A. B. Dobson, and W. R. Chambers (eds.): *Dictionary of Science & Technology.*

Handbook of Physics, McGraw-Hill, New York, N.Y., 1958.

Judd, D. B., and G. Wyszecki: *Color in Business, Science and Industry,* 3rd ed., John Wiley and Sons, New York, N.Y.

Kaufman, Ed: *IES Illumination Handbook*, Illumination Engineering Society.

Lapedes, D. N. (ed.): *The McGraw-Hill Encyclopedia of Science & Technology*, 2nd ed., McGraw-Hill, New York, N.Y.

Norgard, John: "Electromagnetic Spectrum," *NAB Engineering Handbook*, 9th ed., Jerry C. Whitaker (ed.), National Association of Broadcasters, Washington, D.C., 1999.

Norgard, John: "Electromagnetic Spectrum," *The Electronics Handbook*, Jerry C. Whitaker (ed.), CRC Press, Boca Raton, Fla., 1996.

Stemson, A: *Photometry and Radiometry for Engineers*, John Wiley and Sons, New York, N.Y.

The Cambridge Encyclopedia, Cambridge University Press, 1990.

The Columbia Encyclopedia, Columbia University Press, 1993.

Webster's New World Encyclopedia, Prentice Hall, 1992.

Wyszecki, G., and W. S. Stiles: *Color Science, Concepts and Methods, Quantitative Data and Formulae*, 2nd ed., John Wiley and Sons, New York, N.Y.

Chapter

11

Frequency Assignment and Allocations

Jerry C. Whitaker, editor

11.1 Introduction

The Communications Act of 1934, as amended, provides for the regulation of interstate and foreign commerce in communication by wire or radio in the U.S.[1] This Act is printed in Title 47 of the U.S. Code, beginning with Section 151. The primary treaties and other international agreements in force relating to radiocommunication and to which the U.S. is a party are as follows:

- The International Telecommunication Convention, signed at Nairobi on November 6, 1982. The U.S. deposited its instrument of ratification on January 7, 1986.
- The Radio Regulations annexed to the International Telecommunication Convention, signed at Geneva on December 6, 1979 and entered into force with respect to the U.S. on January 1, 1982.
- The United States-Canada Agreement relating to the Coordination and Use of Radio Frequencies above 30 MHz, effected by an exchange of notes at Ottawa on October 24, 1962. A revision to the Technical Annex to the Agreement, made in October 1964 at Washington, was effected by an exchange of notes signed by the U.S. on June 16, 1965, and by Canada on June 24, 1965. The revision entered into force on June 24, 1965. A revision to this Agreement to add Arrangement E (Arrangement between the Department of Communications of Canada and the National Telecommunications and Information Administration and the Federal Communications Commission of the U.S. concerning the use of the 406.1 to 430 MHz band in Canada-U.S. border areas) was effected by an exchange of notes signed by the U.S. on February 26, 1982, and Canada on April 7, 1982.

---

1. This chapter is based on: *NTIA Manual of Regulations and Procedures for Federal Radio Frequency Management*, January 2000 Edition with May/September 2000 Revisions, National Telecommunications and Information Administration, U.S. Department of Commerce, Washington, D.C., 2000.

11.1a The International Telecommunication Union

The International Telecommunication Union is the international body responsible for international frequency allocations, worldwide telecommunications standards, and telecommunication development activities. At this writing, 185 countries were members of the ITU. The broad functions of the ITU are the regulation, coordination, and development of international telecommunications. The U.S. is an active member of the ITU and its work is considered critical to the interest of the United States.

The ITU is the oldest of the intergovernmental organizations that have become specialized agencies within the United Nations. The ITU was born with the spread of one of the great inventions of the 19th century, the telegraph, which crossed national frontiers to link major cities in Europe. International action was essential to establish an international telegraph network. It was necessary to reach agreement on the technical systems to be used, on uniform methods of handling messages, and on the collection of charges. A procedure of international accounting had to be set up.

First came bilateral understanding between bordering countries, then international agreement between regional groups of countries, ending in an inter-European association. Extra-European countries were progressively drawn in, and a truly international organization came into being. In 1865 the International Telegraph Union was created in Paris by the first International Telegraph Convention. The member countries agreed to a set of basic telegraph service regulations. These were modified later as a result of practical operating experience. At Vienna, in 1868, a permanent international bureau was created and established in Berne.

The international telephone service came much later and its progress was much slower. It was not until 1927, when radio provided the means to carry the human voice across the ocean from continent to continent, that this service became world-wide; nevertheless, in 1885, in Berlin, the first provisions concerning the international telephone service were drawn up.

When, at the end of the 19th century, wireless (radiotelegraphy) became practicable, it was seen at once to be an invaluable complement of telegraphy by wire and cable, since radio alone could provide telecommunication between land and ships at sea. The first International Radiotelegraph Convention was signed in Berlin in 1906 by twenty-nine countries. Nearly two decades later, in 1924 and 1925, at Conferences in Paris, the International Telephone Consultative Committee (CCIF) and the International Telegraph Consultative Committee (CCIT) were established. This was followed by the 1927 International Radiotelegraph Conference in Washington, D.C. in 1927, which was attended by 80 countries. It was a historical milestone in the development of radio because it was at this Conference that the Table of Frequency Allocations was first devised and the International Radio Consultative Committee (CCIR) was formed.

In 1932, two Plenipotentiary Conferences were held in Madrid: a Telegraph and Telephone Conference and a Radiotelegraph Conference. On that occasion, the two existing Conventions were amalgamated in a single International Telecommunication Convention, and the countries that signed and acceded to it renamed the Union the International Telecommunication Union (ITU) to indicate its broader scope. Four sets of Regulations were annexed to the Convention: telegraph, telephone, radio, and additional radio regulations.

A Plenipotentiary Conference met in Atlantic City, N.J., in 1947 to revise the Madrid Convention. It introduced important changes in the organization of the Union. The International Frequency Registration Board (IFRB) and the Administrative Council were created. Also, the ITU became the specialized agency within the United Nations in the sphere of telecommunications, and its headquarters was transferred from Berne to Geneva.

The Union remained essentially unchanged until 1992, when an Additional Plenipotentiary Conference in Geneva extensively restructured the ITU. The Nice Constitution and Convention of 1989, which had not been ratified, was used as the general model for the 1992 Conference. The CCIR, IFRB, and World Administrative Radio Conference (WARC) functions were incorporated into the Radiocommunication Sector (ITU-R); the CCITT and Telecommunication Conference functions were incorporated into the Telecommunication Standardization Sector (ITU-T); development activities were incorporated into the Telecommunication Development Sector (ITU-D); and the Secretariats were combined into one General Secretariat.

Purposes of the Union

The purposes of the Union are as follows:

- To promote the development and efficient operation of telecommunication facilities, in order to improve the efficiency of telecommunication services, their usefulness, and their general availability to the public
- Promote and offer technical assistance to developing countries in the field of telecommunications, to promote the mobilization of the human and financial resources needed to develop telecommunications, and to promote the extension of the benefits of new telecommunications technologies to people everywhere
- Promote, at the international level, the adoption of a broader approach to the issues of telecommunications in the global information economy and society

While the principal facilities of the ITU are in Geneva adjacent to the grounds of the United Nations, the Union also has a number of regional and sub-regional offices.

Structure of the Union

The ITU Constitution states that the Union shall comprise:

- The *Plenipotentiary Conference*, which is the supreme authority of the Union
- The *Council*, which acts on behalf of the Plenipotentiary Conference
- World conferences on international telecommunications
- The *Radiocommunication Sector*, including world and regional radiocommunication conferences, radiocommunication assemblies, and the Radio Regulations Board
- The *Telecommunication Standardization Sector*, including world telecommunication standardization conferences
- The *Telecommunication Development Sector*, including world and regional telecommunication development conferences
- The *General Secretariat*

11.1b The Federal Communications Commission

Congress, through adoption of the Communications Act of 1934, created the Federal Communications Commission (FCC) as an independent regulatory agency. Section I of the Act specifies that the FCC was created, "For the purpose of regulation of interstate and foreign commerce in communication by wire and radio so as to make available, so far as possible, to all the people of the United States a rapid, efficient, nationwide, and worldwide wire and radio communication service with adequate facilities at reasonable charges, for the purpose of the national defense, for the purpose of promoting the safety of life and property through the use of wire and radio communication, and for the purpose of securing a more effective execution of this policy by centralizing authority heretofore granted by law to several agencies and by granting additional authority with respect to interstate and foreign commerce in wire and radio communication."

The FCC is directed by five Commissioners appointed by the President, by and with the advice and consent of the Senate, for staggered five-year terms. No more than three can be members of the same political party. The President designates one Commissioner as Chairman. The Commissioners make their decisions collectively by formal vote although authority to act on routine matters is normally delegated to the staff.

The staff of the FCC performs day-to-day functions of the agency, including license and application processing, drafting of rulemaking items, enforcing rules and regulations, and formulating policy.

The Commission reorganized itself in 1995 to establish two new bureaus—Wireless Telecommunications and International—to reflect the changes in the industries it regulates. The staff is divided along functional lines into six operating bureaus and 10 remote offices.

11.2 National Table of Frequency Allocations

The *National Table of Frequency Allocations* is comprised of the U.S. Government Table of Frequency Allocations and the FCC Table of Frequency Allocations. The National Table indicates the normal national frequency allocation planning and the degree of conformity with the ITU table. When required in the national interest and consistent with national rights, as well as obligations undertaken by the U.S. to other countries that may be affected, additional uses of frequencies in any band may be authorized to meet service needs other than those provided for in the National Table.

Specific exceptions to the National Table of Frequency Allocations are as follows:

- A government frequency assignment may be authorized in a non-government band, as an exception, provided: a) the assignment is coordinated with the FCC, and b) no harmful interference will be caused to the service rendered by non-government stations, present or future.
- A non-government frequency assignment may be authorized in a government band, as an exception, provided: a) the assignment is coordinated with the IRAC, and b) no harmful interference will be caused to the service rendered by government stations, present or future.

In the case of bands shared by government and non-government services, frequency assignments therein are subject to coordination between the IRAC and the FCC, and no priority is recognized unless the terms of such priority are specifically defined in the National Table of Frequency Allocations or unless they are subject to mutually agreed arrangements in specific cases.

11.2a U.S. Government Table of Frequency Allocations

The U.S. Government Table of Frequency Allocations is used as a guide in the assignment of radio frequencies to government radio stations in the United States and Possessions. Exceptions to the table may be made by the IRAC after careful consideration to avoid harmful interference and to ensure compliance with the ITU radio regulations.

For the use of frequencies by government radio stations outside the U.S., government agencies are guided insofar as practicable by the ITU Table of Frequency Allocations and, where applicable, by the authority of the host government. Maximum practicable effort should be made to avoid the possibility of harmful interference to other authorized U.S. operations. If harmful interference is considered likely, it is incumbent upon the agency conducting the operation to coordinate with other U.S. users.

Application of the U.S. Government Table is subject to the recognition that:

- Below 25000 kHz the table is only applicable in the assignment of frequencies after September 5, 1961.
- Under Article 38 of the International Telecommunication Convention, administrations "retain their entire freedom with regard to military radio installations of their army, naval and air forces."
- Under No. 342 of the ITU Radio Regulations, administrations may assign frequencies in derogation of the ITU Table of Frequency Allocations "on the express condition that harmful interference shall not be caused to services carried on by stations operating in accordance with the provisions of the Convention and of these Regulations."

Some frequency assignments below 25000 kHz that were made before September 5, 1961, are not in conformity with the government table. Because of the exception mentioned previously, the status of these assignments can be determined only on a case-by-case basis. With this exception, the rules pertaining to the relative status between radio services are as follows:

- Primary and permitted services have equal rights, except that, in the preparation of frequency plans, the primary service, as compared with the permitted service, has prior choice of frequencies.
- Secondary services are on a non-interference basis to the primary and permitted services. Stations of a secondary service: (*a*) must not cause harmful interference to stations of primary or permitted services to which frequencies are already assigned or to which frequencies may be assigned at a later date; (*b*) cannot claim protection from harmful interference from stations of a primary or permitted service to which frequencies are already assigned or may be assigned at a later date; (*c*) can claim protection, however, from harmful interference from stations of the same or other secondary service(s) to which frequencies may be assigned at a later date.

Important definitions for terms used in the table include the following:

- *Additional allocation*, where a band is indicated in a footnote of the table as "also allocated" to a service in an area smaller than a region, or in a particular country. For example, an allocation that is added in this area or in this country to the service or services which are indicated in the table.

- *Alternative allocation*, where a band is indicated in a footnote of the table as "allocated" to one or more services in an area smaller than a region, or in a particular country. For example, an allocation that replaces, in this area or in this country, the allocation indicated in the table.
- *Different category of service*, where the allocation category (primary, permitted, or secondary) of the service in the table is changed. For example, the table reflects the allocation as Fixed, Mobile, and RADIOLOCATION; the category of these services are changed by the footnote to FIXED, MOBILE, and Radiolocation.
- An *allocation* or a footnote to the government table denoting relative status between radio services automatically applies to each assignment in the band to which the footnote or allocation pertains, unless at the time of a particular frequency assignment action a different provision is decided upon for the assignment concerned.
- A *priority note* reflecting the same provisions as an allocation or an applicable footnote to the government table is redundant and is not applied to frequency assignments.

An assignment that is in conformity with the service allocation (as amplified by pertinent footnotes) for the band in which it is contained takes precedence over assignments therein that are not in conformity unless, at the time of the frequency assignment action, a different provision is decided upon.

Where in the table a band is indicated as allocated to more than one service, such services are listed in the following order:

- *Primary services*, the names of which are printed in all capital letters (example: FIXED)
- *Permitted services*, the names of which are printed in "capitals between oblique strokes" (example: /RADIOLOCATION/)
- *Secondary services*, the names of which are printed in "normal characters" (example: Mobile)

Other details of the table include the following:

- The columns to the right of the double line show the national provisions; those to the left show the provisions of the ITU Table of Frequency Allocations.
- Column 1 indicates the national band limits.
- Column 2 indicates the government allocation, including all "US" and "G" footnotes considered to be applicable to the government nationally. Where the allocated service is followed by a function in parentheses, e.g., SPACE (space-to-Earth), the allocation is limited to the function shown.
- Column 3 indicates the non-government allocation including all "US" footnotes, and certain "NG" footnotes as contained in Part 2 of the FCC Rules and Regulations. Where the allocated service is followed by a function in parentheses, e.g., SPACE (space-to-Earth), the allocation is limited to the function shown. These data have been included in the Government Table for information purposes only.
- Column 4 contains such remarks as serve to amplify the government and non-government allocations or point up understanding between the FCC and IRAC/NTIA in respect thereof.
- The international footnotes shown in the columns to the left of the double line are applicable only in the relationships between the U.S. and other countries. An international footnote is applicable to the U.S. Table of Allocations if the number also appears in Columns 2 and 3 of

the U.S. table. The international footnote is then applicable to both government and non-government use.

The texts of footnotes in the table are listed in numerical order at the end of the table, in sections headed Government Footnotes, U.S. Footnotes, International Footnotes, and NG Footnotes. Because of space limitations, the footnotes are not included in this chapter. The complete set of footnotes is available from the National Telecommunications and Information Administration, Washington, D.C. (www.ntia.doc.gov).

The U.S. Government Table of Frequency Allocations is contained on the following pages.

0-130 kHz (VLF/LF)					
International Table			United States Table		FCC Rule Part(s)
Region 1	Region 2	Region 3	Federal Government	Non-Federal Government	
Below 9 (Not Allocated) S5.53 S5.54			Below 9 (Not Allocated) S5.53 S5.54		
9-14 RADIONAVIGATION			9-14 RADIONAVIGATION US18 US294		
14-19.95 FIXED MARITIME MOBILE S5.57 S5.55 S5.56			14-19.95 FIXED MARITIME MOBILE S5.57 US294	14-19.95 Fixed US294	International Fixed (23)
19.95-20.05 STANDARD FREQUENCY AND TIME SIGNAL (20 kHz)			19.95-20.05 STANDARD FREQUENCY AND TIME SIGNAL (20 kHz) US294		
20.05-70 FIXED MARITIME MOBILE S5.57			20.05-59 FIXED MARITIME MOBILE S5.57 US294	20.05-59 FIXED US294	International Fixed (23)
			59-61 STANDARD FREQUENCY AND TIME SIGNAL (60 kHz) US294		
S5.56 S5.58			61-70 FIXED MARITIME MOBILE S5.57 US294	61-70 FIXED US294	International Fixed (23)
70-72 RADIONAVIGATION S5.60 S5.56	70-90 FIXED MARITIME MOBILE S5.57 MARITIME RADIO- NAVIGATION S5.60 Radiolocation	70-72 RADIONAVIGATION S5.60 Fixed Maritime mobile S5.57 S5.59	70-90 FIXED MARITIME MOBILE S5.57 Radiolocation	70-90 FIXED Radiolocation	International Fixed (23) Private Land Mobile (90)
72-84 FIXED MARITIME MOBILE S5.57 RADIONAVIGATION S5.60 S5.56		72-84 FIXED MARITIME MOBILE S5.57 RADIONAVIGATION S5.60			

84-86 RADIONAVIGATION S5.60		84-86 RADIONAVIGATION S5.60 Fixed Maritime mobile S5.57 S5.59			
86-90 FIXED MARITIME MOBILE S5.57 RADIONAVIGATION S5.56	S5.61	86-90 FIXED MARITIME MOBILE S5.57 RADIONAVIGATION S5.60	S5.60 US294	S5.60 US294	
90-110 RADIONAVIGATION S5.62 Fixed S5.64			90-110 RADIONAVIGATION S5.62 US18 US104 US294		Private Land Mobile (90)
110-112 FIXED MARITIME MOBILE RADIONAVIGATION S5.64	110-130 FIXED MARITIME MOBILE MARITIME RADIO- NAVIGATION S5.60 Radiolocation	110-112 FIXED MARITIME MOBILE RADIONAVIGATION S5.60 S5.64	110-130 FIXED MARITIME MOBILE Radiolocation		International Fixed (23) Maritime (80) Private Land Mobile (90)
112-115 RADIONAVIGATION S5.60		112-117.6 RADIONAVIGATION S5.60 Fixed Maritime mobile S5.64 S5.65			
115-117.6 RADIONAVIGATION S5.60 Fixed Maritime mobile S5.64 S5.66					
117.6-126 FIXED MARITIME MOBILE RADIONAVIGATION S5.60 S5.64		117.6-126 FIXED MARITIME MOBILE RADIONAVIGATION S5.60 S5.64			
126-129 RADIONAVIGATION S5.60		126-129 RADIONAVIGATION S5.60 Fixed Maritime mobile S5.64 S5.65			
See next page for 129-130	S5.61 S5.64	See next page for 129-130	S5.60 S5.64 US294		

130-505 kHz (LF/MF)					
International Table			United States Table		Remarks
Region 1	Region 2	Region 3	Federal Government	Non-Federal Government	
129-130 FIXED MARITIME MOBILE RADIONAVIGATION S5.60 S5.64	See previous page for 110-130 kHz	129-130 FIXED MARITIME MOBILE RADIONAVIGATION S5.60 S5.64	See previous page for 110-130 kHz		See previous page for 110-130 kHz
130-148.5 FIXED MARITIME MOBILE S5.64 S5.67	130-160 FIXED MARITIME MOBILE S5.64	130-160 FIXED MARITIME MOBILE RADIONAVIGATION S5.64	130-160 FIXED MARITIME MOBILE S5.64 US294		International Fixed (23) Maritime (80)
148.5-255 BROADCASTING S5.68 S5.69 S5.70	160-190 FIXED	160-190 FIXED Aeronautical radionavigation	160-190 FIXED MARITIME MOBILE 459 US294	160-190 FIXED 459 US294	International Fixed (23)
	190-200 AERONAUTICAL RADIONAVIGATION		190-200 AERONAUTICAL RADIONAVIGATION US18 US226 US294		Aviation (87)
	200-275 AERONAUTICAL RADIONAVIGATION Aeronautical mobile	200-285 AERONAUTICAL RADIONAVIGATION Aeronautical mobile	200-275 AERONAUTICAL RADIONAVIGATION Aeronautical mobile US18 US294		
255-283.5 BROADCASTING AERONAUTICAL RADIONAVIGATION S5.70 S5.71	275-285 AERONAUTICAL RADIONAVIGATION Aeronautical mobile Maritime radionavigation (radiobeacons)		275-285 AERONAUTICAL RADIONAVIGATION Aeronautical mobile Maritime radionavigation (radiobeacons) US18 US294		
283.5-315 AERONAUTICAL RADIONAVIGATION MARITIME RADIONAVIGATION (radiobeacons) S5.73 S5.72 S5.74	285-315 AERONAUTICAL RADIONAVIGATION MARITIME RADIONAVIGATION (radiobeacons) S5.73		285-325 MARITIME RADIONAVIGATION (radiobeacons) S5.73 Aeronautical radionavigation (radiobeacons)		

315-325 AERONAUTICAL RADIONAVIGATION Maritime radionavigation (radiobeacons) S5.73 S5.72 S5.75	315-325 MARITIME RADIONAVIGATION (radiobeacons) S5.73 Aeronautical radionavigation	315-325 AERONAUTICAL RADIONAVIGATION MARITIME RADIONAVIGATION (radiobeacons) S5.73	US18 US294 G121		
325-405 AERONAUTICAL RADIONAVIGATION S5.72	325-335 AERONAUTICAL RADIONAVIGATION Aeronautical mobile Maritime radionavigation (radiobeacons)	325-405 AERONAUTICAL RADIONAVIGATION Aeronautical mobile	325-335 AERONAUTICAL RADIONAVIGATION (radiobeacons) Aeronautical mobile Maritime radionavigation (radiobeacons) US18 US294		
	335-405 AERONAUTICAL RADIONAVIGATION Aeronautical mobile		335-405 AERONAUTICAL RADIONAVIGATION (radiobeacons) Aeronautical mobile US18 US294		
405-415 RADIONAVIGATION S5.76 S5.72	405-415 RADIONAVIGATION S5.76 Aeronautical mobile		405-415 RADIONAVIGATION S5.76 Aeronautical mobile US18 US294		Maritime (80) Aviation (87)
415-435 MARITIME MOBILE S5.79 AERONAUTICAL RADIONAVIGATION S5.72	415-495 MARITIME MOBILE S5.79 S5.79A Aeronautical radionavigation S5.80 S5.77 S5.78 S5.81 S5.82		415-435 MARITIME MOBILE S5.79 AERONAUTICAL RADIONAVIGATION S5.80 US294		
435-495 MARITIME MOBILE S5.79 S5.79A Aeronautical radionavigation S5.72 S5.81 S5.82			435-495 MARITIME MOBILE S5.79 Aeronautical radionavigation 471 472A US231 US294	435-495 MARITIME MOBILE S5.79 471 472A US231 US294	Maritime (80)
495-505 MOBILE (distress and calling) S5.83			495-505 MOBILE (distress and calling) 472		

505-2107 kHz (MF)					
International Table			United States Table		Remarks
Region 1	Region 2	Region 3	Federal Government	Non-Federal Government	
505-526.5 MARITIME MOBILE S5.79 S5.79A S5.84 AERONAUTICAL RADIONAVIGATION S5.72 S5.81	505-510 MARITIME MOBILE S5.79 S5.81	505-526.5 MARITIME MOBILE S5.79 S5.79A S5.84 AERONAUTICAL RADIONAVIGATION Aeronautical mobile Land mobile S5.81	505-510 MARITIME MOBILE S5.79 471		Maritime (80)
	510-525 MOBILE S5.79A S5.84 AERONAUTICAL RADIONAVIGATION		510-525 MARITIME MOBILE (ships only) 474 AERONAUTICAL RADIONAVIGATION (radiobeacons) US14 US18 US225		Maritime (80) Aviation (87)
526.5-1606.5 BROADCASTING S5.87 S5.87A	525-535 BROADCASTING S5.86 AERONAUTICAL RADIONAVIGATION	526.5-535 BROADCASTING Mobile S5.88	525-535 AERONAUTICAL RADIONAVIGATION (radiobeacons) MOBILE US221 US18 US239		Aviation (87) Private Land Mobile (90)
	535-1605 BROADCASTING	535-1606.5 BROADCASTING	535-1605	535-1605 BROADCASTING US321	Radio Broadcasting (AM) (73) Auxiliary Broadcasting (74) Alaska Fixed (80) Private Land Mobile (90) 1610 kHz Travelers Information Service
	1605-1625 BROADCASTING S5.89 S5.90		1605-1615 MOBILE US221 US238 G127	1605-1705 BROADCASTING 480 US238 US299 US321 NG128	
1606.5-1625 FIXED MARITIME MOBILE S5.90 LAND MOBILE S5.92		1606.5-1800 FIXED MOBILE RADIOLOCATION RADIONAVIGATION			
			1615-1625 US238 US299		
1625-1635 RADIOLOCATION S5.93	1625-1705 FIXED MOBILE BROADCASTING S5.89 Radiolocation S5.90		1625-1705 Radiolocation US238 US299		
1635-1800 FIXED MARITIME MOBILE S5.90 LAND MOBILE					

S5.92 S5.96	1705-1800 FIXED MOBILE RADIOLOCATION AERONAUTICAL RADIONAVIGATION	S5.91	1705-1800 FIXED MOBILE RADIOLOCATION US240		International Fixed (23) Maritime (80) Private Land Mobile (90)
1800-1810 RADIOLOCATION S5.93	1800-1850 AMATEUR	1800-2000 AMATEUR FIXED MOBILE except aeronautical mobile RADIONAVIGATION Radiolocation	1800-1900	1800-1900 AMATEUR	Amateur (97)
1810-1850 AMATEUR S5.98 S5.99 S5.100 S5.101					
1850-2000 FIXED MOBILE except aeronautical mobile S5.92 S5.96 S5.103	1850-2000 AMATEUR FIXED MOBILE except aeronautical mobile RADIOLOCATION RADIONAVIGATION S5.102	S5.97	1900-2000 RADIOLOCATION US290		Private Land Mobile (90) Amateur (97)
2000-2025 FIXED MOBILE except aeronautical mobile (R) S5.92 S5.103	2000-2065 FIXED MOBILE		2000-2065 FIXED MOBILE US340	2000-2065 MARITIME MOBILE NG19 US340	Maritime (80)
2025-2045 FIXED MOBILE except aeronautical mobile (R) Meteorological aids S5.104 S5.92 S5.103					
2045-2160 FIXED MARITIME MOBILE LAND MOBILE S5.92	2065-2107 MARITIME MOBILE S5.105 S5.106		2065-2107 MARITIME MOBILE S5.105 US296 US340		
	See next page for 2107-2170 kHz		See next page for 2107-2170 kHz		See next page for 2107-2170 kHz

2107-3230 kHz (MF/HF)					
International Table			United States Table		Remarks
Region 1	Region 2	Region 3	Federal Government	Non-Federal Government	
See previous page for 2045-2160 kHz	2107-2170 FIXED MOBILE		2107-2170 FIXED MOBILE US340	2107-2170 FIXED LAND MOBILE MARITIME MOBILE NG19 US340	International Fixed (23) Maritime (80) Aviation (87) Private Land Mobile (90)
2160-2170 RADIOLOCATION S5.93 S5.107					
2170-2173.5 MARITIME MOBILE			2170-2173.5 MARITIME MOBILE (telephony) US340	2170-2173.5 MARITIME MOBILE US340	Maritime (80)
2173.5-2190.5 MOBILE (distress and calling) S5.108 S5.109 S5.110 S5.111			2173.5-2190.5 MOBILE (distress and calling) S5.108 S5.109 S5.110 S5.111 US279 US340		Maritime (80) Aviation (87)
2190.5-2194 MARITIME MOBILE			2190.5-2194 MARITIME MOBILE (telephony) US340	2190.5-2194 MARITIME MOBILE US340	Maritime (80)
2194-2300 FIXED MOBILE except aeronautical mobile (R) S5.92 S5.103 S5.112	2194-2300 FIXED MOBILE S5.112		2194-2495 FIXED MOBILE US340	2194-2495 FIXED LAND MOBILE MARITIME MOBILE NG19 US340	International Fixed (23) Maritime (80) Aviation (87) Private Land Mobile (90)
2300-2498 FIXED MOBILE except aeronautical mobile (R) BROADCASTING S5.113 S5.103	2300-2495 FIXED MOBILE BROADCASTING S5.113				
	2495-2501 STANDARD FREQUENCY AND TIME SIGNAL (2500 kHz)		2495-2501 STANDARD FREQUENCY AND TIME SIGNAL (2500 kHz) US340		
2498-2501 STANDARD FREQUENCY AND TIME SIGNAL (2500 kHz)					

<table>
<tr><td colspan="2">2501-2502
STANDARD FREQUENCY AND TIME SIGNAL
Space research</td><td>2501-2502
STANDARD FREQUENCY
AND TIME SIGNAL

US340 G106</td><td>2501-2502
STANDARD FREQUENCY
AND TIME SIGNAL

US340</td><td></td></tr>
<tr><td rowspan="2">2502-2625
FIXED
MOBILE except aeronautical mobile (R)

S5.92 S5.103 S5.114</td><td>2502-2505
STANDARD FREQUENCY AND TIME SIGNAL</td><td colspan="2">2502-2505
STANDARD FREQUENCY AND TIME SIGNAL

US340</td><td></td></tr>
<tr><td rowspan="3">2505-2850
FIXED
MOBILE</td><td rowspan="3">2505-2850
FIXED
MOBILE

US285 US340</td><td rowspan="3">2505-2850
FIXED
LAND MOBILE
MARITIME MOBILE

US285 US340</td><td rowspan="3">International Fixed (23)
Maritime (80)
Aviation (87)
Private Land Mobile (90)</td></tr>
<tr><td>2625-2650
MARITIME MOBILE
MARITIME RADIONAVIGATION

S5.92</td></tr>
<tr><td>2650-2850
FIXED
MOBILE except aeronautical mobile (R)

S5.92 S5.103</td></tr>
<tr><td colspan="2">2850-3025
AERONAUTICAL MOBILE (R)

S5.111 S5.115</td><td colspan="2">2850-3025
AERONAUTICAL MOBILE (R)

S5.111 S5.115 US283 US340</td><td>Aviation (87)</td></tr>
<tr><td colspan="2">3025-3155
AERONAUTICAL MOBILE (OR)</td><td colspan="2">3025-3155
AERONAUTICAL MOBILE (OR)

US340</td><td></td></tr>
<tr><td colspan="2">3155-3200
FIXED
MOBILE except aeronautical mobile (R)

S5.116 S5.117</td><td colspan="2" rowspan="2">3155-3230
FIXED
MOBILE except aeronautical mobile (R)

US340</td><td rowspan="2">International Fixed (23)
Maritime (80)
Aviation (87)
Private Land Mobile (90)</td></tr>
<tr><td colspan="2">3200-3230
FIXED
MOBILE except aeronautical mobile (R)
BROADCASTING S5.113

S5.116</td></tr>
</table>

<table>
<tr><th colspan="6">3230-5060 kHz (HF)</th></tr>
<tr><th colspan="3">International Table</th><th colspan="2">United States Table</th><th rowspan="2">Remarks</th></tr>
<tr><th>Region 1</th><th>Region 2</th><th>Region 3</th><th>Federal Government</th><th>Non-Federal Government</th></tr>
<tr><td colspan="3">3230-3400
FIXED
MOBILE except aeronautical mobile
BROADCASTING S5.113

S5.116 S5.118</td><td colspan="2">3230-3400
FIXED
MOBILE except aeronautical mobile
Radiolocation

US340</td><td>International Fixed (23)
Maritime (80)
Aviation (87)
Private Land Mobile (90)</td></tr>
<tr><td colspan="3">3400-3500
AERONAUTICAL MOBILE (R)</td><td colspan="2">3400-3500
AERONAUTICAL MOBILE (R)

US283 US340</td><td>Aviation (87)</td></tr>
<tr><td rowspan="2">3500-3800
AMATEUR S5.120
FIXED
MOBILE except aeronautical mobile

S5.92</td><td>3500-3750
AMATEUR S5.120

S5.119</td><td rowspan="3">3500-3900
AMATEUR S5.120
FIXED
MOBILE</td><td rowspan="5">3500-4000

US340</td><td rowspan="5">3500-4000
AMATEUR S5.120

US340</td><td rowspan="5">Amateur (97)</td></tr>
<tr><td rowspan="4">3750-4000
AMATEUR S5.120
FIXED
MOBILE except aeronautical mobile (R)

S5.122 S5.124 S5.125</td></tr>
<tr><td>3800-3900
FIXED
AERONAUTICAL MOBILE (OR)
LAND MOBILE</td></tr>
<tr><td>3900-3950
AERONAUTICAL MOBILE (OR)

S5.123</td><td>3900-3950
AERONAUTICAL MOBILE
BROADCASTING</td></tr>
<tr><td>3950-4000
FIXED
BROADCASTING</td><td>3950-4000
FIXED
BROADCASTING

S5.126</td></tr>
<tr><td colspan="3">4000-4063
FIXED
MARITIME MOBILE S5.127

S5.126</td><td colspan="2">4000-4063
MARITIME MOBILE

US236 US340</td><td rowspan="2">International Fixed (23)
Maritime (80)</td></tr>
<tr><td colspan="3">4063-4438
MARITIME MOBILE S5.79A S5.109 S5.110 S5.130 S5.131 S5.132

S5.128 S5.129</td><td colspan="2">4063-4438
MARITIME MOBILE S5.109 S5.110 S5.130 S5.132

US82 US296 US340</td></tr>
</table>

4438-4650 FIXED MOBILE except aeronautical mobile (R)		4438-4650 FIXED MOBILE except aeronautical mobile	4438-4650 FIXED MOBILE except aeronautical mobile (R) US340		International Fixed (23) Maritime (80) Aviation (87) Private Land Mobile (90)
4650-4700 AERONAUTICAL MOBILE (R)			4650-4700 AERONAUTICAL MOBILE (R) US282 US283 US340		Aviation (87)
4700-4750 AERONAUTICAL MOBILE (OR)			4700-4750 AERONAUTICAL MOBILE (OR) US340		
4750-4850 FIXED AERONAUTICAL MOBILE (OR) LAND MOBILE BROADCASTING S5.113	4750-4850 FIXED MOBILE except aeronautical mobile (R) BROADCASTING S5.113	4750-4850 FIXED BROADCASTING S5.113 Land mobile	4750-4850 FIXED MOBILE except aeronautical mobile (R) US340		International Fixed (23) Maritime (80) Aviation (87)
4850-4995 FIXED LAND MOBILE BROADCASTING S5.113			4850-4995 FIXED MOBILE US340	4850-4995 FIXED US340	
4995-5003 STANDARD FREQUENCY AND TIME SIGNAL (5000 kHz)			4995-5003 STANDARD FREQUENCY AND TIME SIGNAL (5000 kHz) US340		
5003-5005 STANDARD FREQUENCY AND TIME SIGNAL Space research			5003-5005 STANDARD FREQUENCY AND TIME SIGNAL US340 G106	5003-5005 STANDARD FREQUENCY AND TIME SIGNAL US340	
5005-5060 FIXED BROADCASTING S5.113			5005-5060 FIXED US340		International Fixed (23) Maritime (80) Aviation (87) Private Land Mobile (90)

5060-9040 kHz (HF)					
International Table			United States Table		Remarks
Region 1	Region 2	Region 3	Federal Government	Non-Federal Government	
5060-5250 FIXED Mobile except aeronautical mobile S5.133			5060-5450 FIXED Mobile except aeronautical mobile US212 US340		International Fixed (23) Maritime (80) Aviation (87) Private Land Mobile (90)
5250-5450 FIXED MOBILE except aeronautical mobile					
5450-5480 FIXED AERONAUTICAL MOBILE (OR) LAND MOBILE	5450-5480 AERONAUTICAL MOBILE (R)	5450-5480 FIXED AERONAUTICAL MOBILE (OR) LAND MOBILE	5450-5480 AERONAUTICAL MOBILE (R) US283 US340		Aviation (87)
5480-5680 AERONAUTICAL MOBILE (R) S5.111 S5.115			5480-5680 AERONAUTICAL MOBILE (R) S5.111 S5.115 US283 US340		
5680-5730 AERONAUTICAL MOBILE (OR) S5.111 S5.115			5680-5730 AERONAUTICAL MOBILE (OR) S5.111 S5.115 US340		
5730-5900 FIXED LAND MOBILE	5730-5900 FIXED MOBILE except aeronautical mobile (R)	5730-5900 FIXED Mobile except aeronautical mobile (R)	5730-5950 FIXED MOBILE except aeronautical mobile (R) US340		International Fixed (23) Maritime (80) Aviation (87)
5900-5950 BROADCASTING S5.134 S5.136					
5950-6200 BROADCASTING			5950-6200 BROADCASTING US340		Radio Broadcast (HF) (73)
6200-6525 MARITIME MOBILE S5.109 S5.110 S5.130 S5.132 S5.137			6200-6525 MARITIME MOBILE S5.109 S5.110 S5.130 S5.132 US82 US296 US340		Maritime (80)
6525-6685 AERONAUTICAL MOBILE (R)			6525-6685 AERONAUTICAL MOBILE (R) US283 US340		Aviation (87)

6685-6765 AERONAUTICAL MOBILE (OR)			6685-6765 AERONAUTICAL MOBILE (OR) US340		
6765-7000 FIXED Land mobile S5.139 S5.138			6765-7000 FIXED Mobile S5.138 US340		ISM Equipment (18) International Fixed (23) Aviation (87)
7000-7100 AMATEUR S5.120 AMATEUR-SATELLITE S5.140 S5.141			7000-7100 US340	7000-7100 AMATEUR S5.120 AMATEUR-SATELLITE US340	Amateur (97)
7100-7300 BROADCASTING	7100-7300 AMATEUR S5.120 S5.142	7100-7300 BROADCASTING	7100-7300 US340	7100-7300 AMATEUR S5.120 S5.142 US340	
7300-7350 BROADCASTING S5.134 S5.143			7300-8100 FIXED Mobile US340		International Fixed (23) Maritime (80) Aviation (87) Private Land Mobile (90)
7350-8100 FIXED Land mobile S5.144					
8100-8195 FIXED MARITIME MOBILE			8100-8195 MARITIME MOBILE US236 US340		Maritime (80)
8195-8815 MARITIME MOBILE S5.109 S5.110 S5.132 S5.145 S5.111			8195-8815 MARITIME MOBILE S5.109 S5.110 S5.132 S5.145 S5.111 US82 US296 US340		
8815-8965 AERONAUTICAL MOBILE (R)			8815-8965 AERONAUTICAL MOBILE (R) US340		Aviation (87)
8965-9040 AERONAUTICAL MOBILE (OR)			8965-9040 AERONAUTICAL MOBILE (OR) US340		

9040-13410 kHz (HF)					
International Table			United States Table		Remarks
Region 1	Region 2	Region 3	Federal Government	Non-Federal Government	
9040-9400 FIXED			9040-9500 FIXED US340		International Fixed (23) Maritime (80) Aviation (87)
9400-9500 BROADCASTING S5.134 S5.146					
9500-9900 BROADCASTING S5.147			9500-9900 BROADCASTING S5.147 S5.148 US235 US340		International Fixed (23) Radio Broadcast (HF) (73)
9900-9995 FIXED			9900-9995 FIXED US340		International Fixed (23) Aviation (87)
9995-10003 STANDARD FREQUENCY AND TIME SIGNAL (10000 kHz) S5.111			9995-10003 STANDARD FREQUENCY AND TIME SIGNAL (10000 kHz) S5.111 US340		
10003-10005 STANDARD FREQUENCY AND TIME SIGNAL Space research S5.111			10003-10005 STANDARD FREQUENCY AND TIME SIGNAL S5.111 US340 G106	10003-10005 STANDARD FREQUENCY AND TIME SIGNAL S5.111 US340	
10005-10100 AERONAUTICAL MOBILE (R) S5.111			10005-10100 AERONAUTICAL MOBILE (R) S5.111 US283 US340		Aviation (87)
10100-10150 FIXED Amateur S5.120			10100-10150 US247 US340	10100-10150 AMATEUR S5.120 US247 US340	Amateur (97)
10150-11175 FIXED Mobile except aeronautical mobile (R)			10150-11175 FIXED Mobile except aeronautical mobile (R) US340		International Fixed (23) Aviation (87)
11175-11275 AERONAUTICAL MOBILE (OR)			11175-11275 AERONAUTICAL MOBILE (OR) US340		

<table>
<tr><td>11275-11400
AERONAUTICAL MOBILE (R)</td><td colspan="2">11275-11400
AERONAUTICAL MOBILE (R)

US283 US340</td><td>Aviation (87)</td></tr>
<tr><td>11400-11600
FIXED</td><td colspan="2" rowspan="2">11400-11650
FIXED

US340</td><td rowspan="2">International Fixed (23)
Aviation (87)</td></tr>
<tr><td>11600-11650
BROADCASTING S5.134

S5.146</td></tr>
<tr><td>11650-12050
BROADCASTING

S5.147</td><td colspan="2">11650-12050
BROADCASTING

US235 US340</td><td>International Fixed (23)
Radio Broadcast (HF) (73)</td></tr>
<tr><td>12050-12100
BROADCASTING S5.134

S5.146</td><td colspan="2" rowspan="2">12050-12230
FIXED

US340</td><td rowspan="2">International Fixed (23)
Aviation (87)</td></tr>
<tr><td>12100-12230
FIXED</td></tr>
<tr><td>12230-13200
MARITIME MOBILE S5.109 S5.110 S5.132 S5.145</td><td colspan="2">12230-13200
MARITIME MOBILE S5.109 S5.110 S5.132 S5.145

US82 US296 US340</td><td>International Fixed (23)
Maritime (80)</td></tr>
<tr><td>13200-13260
AERONAUTICAL MOBILE (OR)</td><td colspan="2">13200-13260
AERONAUTICAL MOBILE (OR)

US340</td><td></td></tr>
<tr><td>13260-13360
AERONAUTICAL MOBILE (R)</td><td colspan="2">13260-13360
AERONAUTICAL MOBILE (R)

US283 US340</td><td>Aviation (87)</td></tr>
<tr><td>13360-13410
FIXED
RADIO ASTRONOMY

S5.149</td><td>13360-13410
RADIO ASTRONOMY

S5.149 G115</td><td>13360-13410
RADIO ASTRONOMY

S5.149</td><td></td></tr>
</table>

13410-17900 kHz (HF)					
International Table			United States Table		Remarks
Region 1	Region 2	Region 3	Federal Government	Non-Federal Government	
13410-13570 FIXED Mobile except aeronautical mobile (R) S5.150			13410-13570 FIXED Mobile except aeronautical mobile (R) S5.150 US340	13410-13570 FIXED S5.150 US340	ISM Equipment (18) International Fixed (23) Aviation (87)
13570-13600 BROADCASTING S5.134 S5.151			13570-13600 FIXED Mobile except aeronautical mobile (R) US340	13570-13600 FIXED US340	International Fixed (23) Aviation (87)
13600-13800 BROADCASTING			13600-13800 BROADCASTING S5.148 US340		International Fixed (23) Radio Broadcast (HF) (73)
13800-13870 BROADCASTING S5.134 S5.151			13800-14000 FIXED Mobile except aeronautical mobile (R) US340	13800-14000 FIXED US340	International Fixed (23) Aviation (87)
13870-14000 FIXED Mobile except aeronautical mobile (R)					
14000-14250 AMATEUR S5.120 AMATEUR-SATELLITE			14000-14350 US340	14000-14250 AMATEUR S5.120 AMATEUR-SATELLITE US340	Amateur (97)
14250-14350 AMATEUR S5.120 S5.152				14250-14350 AMATEUR S5.120 US340	
14350-14990 FIXED Mobile except aeronautical mobile (R)			14350-14990 FIXED Mobile except aeronautical mobile (R) US340	14350-14990 FIXED US340	International Fixed (23) Aviation (87)
14990-15005 STANDARD FREQUENCY AND TIME SIGNAL (15000 kHz) S5.111			14990-15005 STANDARD FREQUENCY AND TIME SIGNAL (15000 kHz) S5.111 US340		

15005-15010 STANDARD FREQUENCY AND TIME SIGNAL Space research	15005-15010 STANDARD FREQUENCY AND TIME SIGNAL US340 G106	15005-15010 STANDARD FREQUENCY AND TIME SIGNAL US340	
15010-15100 AERONAUTICAL MOBILE (OR)	15010-15100 AERONAUTICAL MOBILE (OR) US340		
15100-15600 BROADCASTING	15100-15600 BROADCASTING S5.148 US340		International Fixed (23) Radio Broadcast (HF) (73)
15600-15800 BROADCASTING S5.134 S5.146	15600-16360 FIXED US340		International Fixed (23) Aviation (87)
15800-16360 FIXED S5.153			
16360-17410 MARITIME MOBILE S5.109 S5.110 S5.132 S5.145	16360-17410 MARITIME MOBILE S5.109 S5.110 S5.132 S5.145 US82 US296 US340		Maritime (80)
17410-17480 FIXED	17410-17550 FIXED US340		International Fixed (23) Aviation (87)
17480-17550 BROADCASTING S5.134 S5.146			
17550-17900 BROADCASTING	17550-17900 BROADCASTING S5.148 US340		International Fixed (23) Radio Broadcast (HF) (73)

17900-22855 kHz (HF)					
International Table			United States Table		Remarks
Region 1	Region 2	Region 3	Federal Government	Non-Federal Government	
17900-17970 AERONAUTICAL MOBILE (R)			17900-17970 AERONAUTICAL MOBILE (R) US283 US340		Aviation (87)
17970-18030 AERONAUTICAL MOBILE (OR)			17970-18030 AERONAUTICAL MOBILE (OR) US340		
18030-18052 FIXED			18030-18068 FIXED US340		International Fixed (23) Maritime (80)
18052-18068 FIXED Space research					
18068-18168 AMATEUR S5.120 AMATEUR-SATELLITE S5.154			18068-18168 US340	18068-18168 AMATEUR S5.120 AMATEUR-SATELLITE US340	International Fixed (23) Amateur (97)
18168-18780 FIXED Mobile except aeronautical mobile			18168-18780 FIXED Mobile US340		International Fixed (23) Maritime (80) Aviation (87)
18780-18900 MARITIME MOBILE			18780-18900 MARITIME MOBILE US82 US296 US340		International Fixed (23) Maritime (80)
18900-19020 BROADCASTING S5.134 S5.146			18900-19680 FIXED US340		International Fixed (23) Aviation (87)
19020-19680 FIXED					
19680-19800 MARITIME MOBILE S5.132			19680-19800 MARITIME MOBILE S5.132 US340		Maritime (80)
19800-19990 FIXED			19800-19990 FIXED US340		International Fixed (23) Aviation (87)

19990-19995 STANDARD FREQUENCY AND TIME SIGNAL Space research S5.111	19990-19995 STANDARD FREQUENCY AND TIME SIGNAL Space research G106 S5.111 US340	19990-19995 STANDARD FREQUENCY AND TIME SIGNAL Space research S5.111 US340	
19995-20010 STANDARD FREQUENCY AND TIME SIGNAL (20000 kHz) S5.111	19995-20010 STANDARD FREQUENCY AND TIME SIGNAL (20000 kHz) S5.111 US340 G106	19995-20010 STANDARD FREQUENCY AND TIME SIGNAL (20000 kHz) S5.111 US340	
20010-21000 FIXED Mobile	20010-21000 FIXED Mobile US340	20010-21000 FIXED US340	
21000-21450 AMATEUR S5.120 AMATEUR-SATELLITE	21000-21450 US340	21000-21450 AMATEUR S5.120 AMATEUR-SATELLITE US340	Amateur (97)
21450-21850 BROADCASTING	21450-21850 BROADCASTING S5.148 US340		International Fixed (23) Radio Broadcast (HF) (73)
21850-21870 FIXED S5.155A S5.155	21850-21924 FIXED US340		International Fixed (23) Aviation (87)
21870-21924 FIXED S5.155B			
21924-22000 AERONAUTICAL MOBILE (R)	21924-22000 AERONAUTICAL MOBILE (R) US340		Aviation (87)
22000-22855 MARITIME MOBILE S5.132 S5.156	22000-22855 MARITIME MOBILE S5.132 US82 US296 US340		International Fixed (23) Maritime (80)

22855-26175 kHz (HF)					
International Table			United States Table		Remarks
Region 1	Region 2	Region 3	Federal Government	Non-Federal Government	
22855-23000 FIXED S5.156			22855-23000 FIXED US340		International Fixed (23) Aviation (87)
23000-23200 FIXED Mobile except aeronautical mobile (R) S5.156			23000-23200 FIXED Mobile except aeronautical mobile (R) US340	23000-23200 FIXED US340	
23200-23350 FIXED S5.156A AERONAUTICAL MOBILE (OR)			23200-23350 AERONAUTICAL MOBILE (OR) US340		
23350-24000 FIXED MOBILE except aeronautical mobile S5.157			23350-24890 FIXED MOBILE except aeronautical mobile US340	23350-24890 FIXED US340	International Fixed (23) Aviation (87)
24000-24890 FIXED LAND MOBILE					
24890-24990 AMATEUR S5.120 AMATEUR-SATELLITE			24890-24990 US340	24890-24990 AMATEUR S5.120 AMATEUR-SATELLITE US340	Amateur (97)
24990-25005 STANDARD FREQUENCY AND TIME SIGNAL (25000 kHz)			24990-25005 STANDARD FREQUENCY AND TIME SIGNAL (25000 kHz) US340		
25005-25010 STANDARD FREQUENCY AND TIME SIGNAL Space research			25005-25010 STANDARD FREQUENCY AND TIME SIGNAL US340 G106	25005-25010 STANDARD FREQUENCY AND TIME SIGNAL US340	
25010-25070 FIXED MOBILE except aeronautical mobile			25010-25070 US340	25010-25070 LAND MOBILE US340 NG112	Private Land Mobile (90)

25070-25210 MARITIME MOBILE	25070-25210 MARITIME MOBILE US82 US281 US296 US340	25070-25210 MARITIME MOBILE US82 US281 US296 US340 NG112	Maritime (80) Private Land Mobile (90)
25210-25550 FIXED MOBILE except aeronautical mobile	25210-25330 US340	25210-25330 LAND MOBILE US340	Private Land Mobile (90)
	25330-25550 FIXED MOBILE except aeronautical mobile US340	25330-25550 US340	
25550-25670 RADIO ASTRONOMY S5.149	25550-25670 RADIO ASTRONOMY US74 S5.149		
25670-26100 BROADCASTING	25670-26100 BROADCASTING US25 US340		Radio Broadcast (HF) (73) Remote Pickup (74D)
26100-26175 MARITIME MOBILE S5.132	26100-26175 MARITIME MOBILE S5.132 US340		Auxiliary Broadcasting (74) Maritime (80)

25175-33 MHz (HF/VHF)					
International Table			United States Table		Remarks
Region 1	Region 2	Region 3	Federal Government	Non-Federal Government	
26175-27500 FIXED MOBILE except aeronautical mobile S5.150			26175-26480 US340	26175-26480 LAND MOBILE US340	Auxiliary Broadcasting (74)
			26480-26950 FIXED MOBILE except aeronautical mobile US10 US340	26480-26950 US10 US340	
			26950-27410 S5.150 US340	26950-26960 FIXED S5.150 US340	ISM Equipment (18) International Fixed (23)
				26960-27230 MOBILE except aeronautical mobile S5.150 US340	ISM Equipment (18) Personal Radio (95)
				27230-27410 FIXED MOBILE except aeronautical mobile S5.150 US340	ISM Equipment (18) Private Land Mobile (90) Personal Radio (95)
27500-28000 METEOROLOGICAL AIDS FIXED MOBILE			27410-27540 US340	27410-27540 FIXED LAND MOBILE US340	Private Land Mobile (90)
			27540-28000 FIXED MOBILE US298 US340	27540-28000 US298 US340	

28-29.7 AMATEUR AMATEUR-SATELLITE	28-29.89 US340	28-29.7 AMATEUR AMATEUR-SATELLITE US340	Amateur (97)
29.7-30.005 FIXED MOBILE		29.7-29.8 LAND MOBILE US340	Private Land Mobile (90)
		29.8-29.89 FIXED US340	International Fixed (23) Aviation (87)
	29.89-29.91 FIXED MOBILE US340	29.89-29.91 US340	
	29.91-30 US340	29.91-30 FIXED US340	International Fixed (23) Aviation (87)
30.005-30.01 SPACE OPERATION (satellite identification) FIXED MOBILE SPACE RESEARCH	30-30.56 FIXED MOBILE	30-30.56	
30.01-37.5 FIXED MOBILE	30.56-32	30.56-32 FIXED LAND MOBILE NG124	Private Land Mobile (90)
	32-33 FIXED MOBILE	32-33	

33-50 MHz (VHF)					
International Table			United States Table		Remarks
Region 1	Region 2	Region 3	Federal Government	Non-Federal Government	
See previous page for 30.01-37.5 MHz			33-34	33-34 FIXED LAND MOBILE NG124	Private Land Mobile (90)
			34-35 FIXED MOBILE	34-35	
			35-36	35-36 FIXED LAND MOBILE	Public Mobile (22) Private Land Mobile (90)
			36-37 FIXED MOBILE US220	36-37 US220	
			37-37.5	37-37.5 LAND MOBILE NG124	Private Land Mobile (90)
37.5-38.25 FIXED MOBILE Radio astronomy S5.149			37.5-38 Radio astronomy S5.149	37.5-38 LAND MOBILE Radio astronomy S5.149 NG59 NG124	
			38-38.25 FIXED MOBILE RADIO ASTRONOMY S5.149 US81	38-38.25 RADIO ASTRONOMY S5.149 US81	
38.25-39.986 FIXED MOBILE			38.25-39 FIXED MOBILE	38.25-39	
			39-40	39-40 LAND MOBILE NG124	Private Land Mobile (90)
39.986-40.02 FIXED MOBILE Space research			40-42 FIXED MOBILE	40-40.98	ISM Equipment (18)

40.02-40.98 FIXED MOBILE S5.150				S5.150 US210	
40.98-41.015 FIXED MOBILE Space research S5.160 S5.161				40.98-42	
41.015-44 FIXED MOBILE S5.160 S5.161			S5.150 US210 US220	US220	
			42-46.6	42-43.69 FIXED LAND MOBILE NG124 NG141	Public Mobile (22) Private Land Mobile (90)
				43.69-46.6 LAND MOBILE NG124 NG141	Private Land Mobile (90)
44-47 FIXED MOBILE S5.162 S5.162A			46.6-47 FIXED MOBILE	46.6-47	
47-68 BROADCASTING	47-50 FIXED MOBILE	47-50 FIXED MOBILE BROADCASTING	47-49.6	47-49.6 LAND MOBILE NG124	Private Land Mobile (90)
			49.6-50 FIXED MOBILE	49.6-50	
S5.162A S5.163 S5.164 S5.165 S5.169 S5.171	See next page for 50-68 MHz		See next page for 50-73 MHz	See next page for 50-72 MHz	See next page for 50-72 MHz

50-123.5875 (VHF)					
International Table			United States Table		Remarks
Region 1	Region 2	Region 3	Federal Government	Non-Federal Government	
See previous page for 47-68 MHz	50-54 AMATEUR S5.166 S5.167 S5.168 S5.170		50-73	50-54 AMATEUR	Amateur (97)
	54-68 BROADCASTING Fixed Mobile S5.172	54-68 FIXED MOBILE BROADCASTING		54-72 BROADCASTING NG128 NG149	Broadcast Radio (TV) (73) Auxiliary Broadcasting (74)
68-74.8 FIXED MOBILE except aeronautical mobile S5.149 S5.174 S5.175 S5.177 S5.179	68-72 BROADCASTING Fixed Mobile S5.173	68-74.8 FIXED MOBILE S5.149 S5.176 S5.179			
	72-73 FIXED MOBILE			72-73 FIXED MOBILE NG3 NG49 NG56	Public Mobile (22) Private Land Mobile (90) Personal Radio (95)
	73-74.6 RADIO ASTRONOMY S5.178		73-74.6 RADIO ASTRONOMY US74		
	74.6-74.8 FIXED MOBILE		74.6-74.8 FIXED MOBILE US273		Private Land Mobile (90)
74.8-75.2 AERONAUTICAL RADIONAVIGATION S5.180 S5.181			74.8-75.2 AERONAUTICAL RADIONAVIGATION S5.180		Aviation (87)
75.2-87.5 FIXED MOBILE except aeronautical mobile	75.2-75.4 FIXED MOBILE S5.179		75.2-75.4 FIXED MOBILE US273		Private Land Mobile (90)

S5.175 S5.179 S5.184 S5.187	75.4-76 FIXED MOBILE	75.4-87 FIXED MOBILE S5.149 S5.182 S5.183 S5.188	75.4-88	75.4-76 FIXED MOBILE NG3 NG49 NG56	Public Mobile (22) Private Land Mobile (90) Personal Radio (95)
	76-88 BROADCASTING Fixed Mobile S5.185			76-88 BROADCASTING NG128 NG129 NG149	Broadcast Radio (TV) (73) Auxiliary Broadcasting (74)
87.5-100 BROADCASTING S5.190	88-100 BROADCASTING	87-100 FIXED MOBILE BROADCASTING	88-108 US93	88-108 BROADCASTING US93 NG2 NG128 NG129	Broadcast Radio (FM) (73) Auxiliary Broadcasting (74)
100-108 BROADCASTING S5.192 S5.194					
108-117.975 AERONAUTICAL RADIONAVIGATION S5.197			108-117.975 AERONAUTICAL RADIONAVIGATION US93 G126		Note: Footnote G126 states that DGPS stations may be authorized on a primary basis in this band, but the FCC has not yet addressed this footnote.
117.975-137 AERONAUTICAL MOBILE (R) S5.111 S5.198 S5.199 S5.200 S5.201 S5.202 S5.203 S5.203A S5.203B			117.975-121.9375 AERONAUTICAL MOBILE (R) S5.111 S5.199 S5.200 591 US26 US28		Aviation (87)
			121.9375-123.0875 591 US30 US31 US33 US80 US102 US213	121.9375-123.0875 AERONAUTICAL MOBILE 591 US30 US31 US33 US80 US102 US213	
			123.0875-123.5875 AERONAUTICAL MOBILE S5.200 591 US32 US33 US112		
			See next page for 123.5875-137 MHz		See next page for 123.5875-137 MHz

<table>
<tr><th colspan="6">123.5875-148 MHz (VHF)</th></tr>
<tr><th colspan="3">International Table</th><th colspan="2">United States Table</th><th rowspan="2">Remarks</th></tr>
<tr><th>Region 1</th><th>Region 2</th><th>Region 3</th><th>Federal Government</th><th>Non-Federal Government</th></tr>
<tr><td colspan="3" rowspan="4">See previous page for 117.975-137 MHz</td><td colspan="2">123.5875-128.8125
AERONAUTICAL MOBILE (R)
591 US26</td><td rowspan="3">Aviation (87)</td></tr>
<tr><td>128.8125-132.0125
591</td><td>128.8125-132.0125
AERONAUTICAL MOBILE (R)
591</td></tr>
<tr><td colspan="2">132.0125-136.00
AERONAUTICAL MOBILE (R)
591 US26</td></tr>
<tr><td>136-137
591 US244</td><td>136-137
AERONAUTICAL MOBILE (R)
591 US244</td><td>Satellite Communications (25)
Aviation (87)</td></tr>
<tr><td colspan="3">137-137.025
SPACE OPERATION (space-to-Earth)
METEOROLOGICAL-SATELLITE (space-to-Earth)
MOBILE-SATELLITE (space-to-Earth) S5.208A S5.209
SPACE RESEARCH (space-to-Earth)
Fixed
Mobile except aeronautical mobile (R)
S5.204 S5.205 S5.206 S5.207 S5.208</td><td colspan="2">137-137.025
SPACE OPERATION (space-to-Earth)
METEOROLOGICAL-SATELLITE (space-to-Earth)
MOBILE-SATELLITE (space-to-Earth) 599B US318 US319 US320
SPACE RESEARCH (space-to-Earth)
599A</td><td rowspan="3">Satellite Communications (25)</td></tr>
<tr><td colspan="3">137.025-137.175
SPACE OPERATION (space-to-Earth)
METEOROLOGICAL-SATELLITE (space-to-Earth)
SPACE RESEARCH (space-to-Earth)
Fixed
Mobile-satellite (space-to-Earth) S5.208A S5.209
Mobile except aeronautical mobile (R)
S5.204 S5.205 S5.206 S5.207 S5.208</td><td colspan="2">137.025-137.175
SPACE OPERATION (space-to-Earth)
METEOROLOGICAL-SATELLITE (space-to-Earth)
SPACE RESEARCH (space-to-earth)
Mobile-satellite (space-to-Earth) 599B US318 US319 US320
599A</td></tr>
<tr><td colspan="3">137.175-137.825
SPACE OPERATION (space-to-Earth)
METEOROLOGICAL-SATELLITE (space-to-Earth)
MOBILE-SATELLITE (space-to-Earth) S5.208A S5.209
SPACE RESEARCH (space-to-Earth)
Fixed
Mobile except aeronautical mobile (R)
S5.204 S5.205 S5.206 S5.207 S5.208</td><td colspan="2">137.175-137.825
SPACE OPERATION (space-to-Earth)
METEOROLOGICAL-SATELLITE (space-to-Earth)
MOBILE-SATELLITE (space-to-Earth) 599B US318 US319 US320
SPACE RESEARCH (space-to-Earth)
599A</td></tr>
</table>

137.825-138 SPACE OPERATION (space-to-Earth) METEOROLOGICAL-SATELLITE (space-to-Earth) SPACE RESEARCH (space-to-Earth) Fixed Mobile-satellite (space-to-Earth) S5.208A S5.209 Mobile except aeronautical mobile (R) S5.204 S5.205 S5.206 S5.207 S5.208			137.825-138 SPACE OPERATION (space-to-Earth) METEOROLOGICAL-SATELLITE (space-to-Earth) SPACE RESEARCH (space-to-Earth) Mobile-satellite (space-to-Earth) 599B US318 US319 US320 599A		
138-143.6 AERONAUTICAL MOBILE (OR) S5.210 S5.211 S5.212 S5.214	138-143.6 FIXED MOBILE RADIOLOCATION Space research (space-to-Earth)	138-143.6 FIXED MOBILE Space research (space-to-Earth) S5.207 S5.213	138-144 FIXED MOBILE	138-144	
143.6-143.65 AERONAUTICAL MOBILE (OR) SPACE RESEARCH (space-to-Earth) S5.211 S5.212 S5.214	143.6-143.65 FIXED MOBILE RADIOLOCATION SPACE RESEARCH (space-to-Earth)	143.6-143.65 FIXED MOBILE SPACE RESEARCH (space-to-Earth) S5.207 S5.213			
143.65-144 AERONAUTICAL MOBILE (OR) S5.210 S5.211 S5.212 S5.214	143.65-144 FIXED MOBILE RADIOLOCATION Space research (space-to-Earth)	143.65-144 FIXED MOBILE Space research (space-to-Earth) S5.207 S5.213	US10 G30	US10	
144-146 AMATEUR S5.120 AMATEUR-SATELLITE S5.216			144-148	144-146 AMATEUR 510 AMATEUR-SATELLITE	Amateur (97)
146-148 FIXED MOBILE except aeronautical mobile (R)	146-148 AMATEUR S5.217	146-148 AMATEUR FIXED MOBILE S5.217		146-148 AMATEUR	

<table>
<tr><th colspan="6">148-162.0125 MHz (VHF)</th></tr>
<tr><th colspan="3">International Table</th><th colspan="2">United States Table</th><th rowspan="2">Remarks</th></tr>
<tr><th>Region 1</th><th>Region 2</th><th>Region 3</th><th>Federal Government</th><th>Non-Federal Government</th></tr>
<tr><td>148-149.9
FIXED
MOBILE except aeronautical mobile (R)
MOBILE-SATELLITE (Earth-to-space) S5.209

S5.218 S5.219 S5.221</td><td colspan="2">148-149.9
FIXED
MOBILE
MOBILE-SATELLITE (Earth-to-space) S5.209

S5.218 S5.219 S5.221</td><td>148-149.9
FIXED
MOBILE
MOBILE-SATELLITE (Earth-to-space) 599B US319 US320 US323 US325

S5.218 608A US10 G30</td><td>148-149.9
MOBILE-SATELLITE (Earth-to-space) 599B US319 US320 US323 US325

S5.218 608A US10</td><td rowspan="2">Satellite Communications (25)</td></tr>
<tr><td colspan="3">149.9-150.05
MOBILE-SATELLITE (Earth-to-space) S5.209 S5.224A
RADIONAVIGATION-SATELLITE S5.224B

S5.220 S5.222 S5.223</td><td colspan="2">149.9-150.05
MOBILE-SATELLITE (Earth-to-space) 599B US319 US322
RADIONAVIGATION-SATELLITE

S5.223 608B</td></tr>
<tr><td rowspan="3">150.05-153
FIXED
MOBILE except aeronautical mobile
RADIO ASTRONOMY

S5.149</td><td colspan="2" rowspan="6">150.05-156.7625
FIXED
MOBILE

S5.225 S5.226 S5.227</td><td>150.05-150.8
FIXED
MOBILE

US216 G30</td><td>150.05-150.8

US216</td><td></td></tr>
<tr><td>150.8-152.855

US216</td><td>150.8-152.855
FIXED
LAND MOBILE

US216 NG4 NG51 NG112 NG124</td><td>Public Mobile (22)
Private Land Mobile (90)</td></tr>
<tr><td rowspan="2">152.855-154</td><td rowspan="2">152.855-154
LAND MOBILE

NG4 NG124</td><td rowspan="2">Auxiliary Broadcasting (74)
Private Land Mobile (90)</td></tr>
<tr><td>153-154
FIXED
MOBILE except aeronautical mobile (R)
Meteorological aids</td></tr>
<tr><td rowspan="2">154-156.7625
FIXED
MOBILE except aeronautical mobile (R)

S5.226 S5.227</td><td>154-156.2475

S5.226</td><td>154-156.2475
FIXED
LAND MOBILE

S5.226 NG112 NG117 NG124 NG148</td><td>Maritime (80)
Private Land Mobile (90)
Personal Radio (95)</td></tr>
<tr><td>156.2475-157.0375</td><td>156.2475-157.0375
MARITIME MOBILE</td><td></td></tr>
</table>

156.7625-156.8375 MARITIME MOBILE (distress and calling) S5.111 S5.226		(continued) S5.226 S5.227 US77 US106 US107 US266	(continued) S5.226 S5.227 US77 US106 US107 US266 NG117	
156.8375-174 FIXED MOBILE except aeronautical mobile S5.226 S5.229	156.8375-174 FIXED MOBILE S5.226 S5.230 S5.231 S5.232	157.0375-157.1875 MARITIME MOBILE S5.226 US214 US266 G109	157.0375-157.1875 S5.226 US214 US266	Private Land Mobile (90)
		157.1875-157.45 S5.226 US223 US266	157.1875-157.45 LAND MOBILE MARITIME MOBILE S5.226 US223 US266 NG111	Maritime (80) Private Land Mobile (90)
		157.45-161.575 S5.226 US266	157.45-161.575 FIXED LAND MOBILE S5.226 US266 NG6 NG28 NG70 NG111 NG112 NG124 NG148 NG155	Public Mobile (22) Maritime (80) Private Land Mobile (90)
		161.575-161.625 S5.226 US77	161.575-161.625 MARITIME MOBILE S5.226 US77 NG6 NG17	Public Mobile (22) Maritime (80)
		161.625-161.775 S5.226	161.625-161.775 LAND MOBILE S5.226 NG6	Public Mobile (22) Auxiliary Broadcasting (74)
		161.775-162.0125 S5.226 US266	161.775-162.0125 LAND MOBILE MARITIME MOBILE S5.226 US266 NG6	Public Mobile (22) Maritime (80) Private Land Mobile (90)
		See next page for 162.0125-174 MHz		See next page for 162.0125-174 MHz

162.0125-322 MHz (VHF/UHF)					
International Table			United States Table		Remarks
Region 1	Region 2	Region 3	Federal Government	Non-Federal Government	
See previous page for 156.8375-174 MHz			162.0125-173.2 FIXED MOBILE S5.226 US8 US11 US13 US216 US223 US300 US312 G5	162.0125-173.2 S5.226 US8 US11 US13 US216 US223 US300 US312	Auxiliary Broadcasting (74) Private Land Mobile (90)
			173.2-173.4	173.2-173.4 FIXED Land mobile	Private Land Mobile (90)
			173.4-174 FIXED MOBILE G5	173.4-174	
174-223 BROADCASTING	174-216 BROADCASTING Fixed Mobile S5.234	174-223 FIXED MOBILE BROADCASTING	174-216	174-216 BROADCASTING NG115 NG128 NG149	Broadcast Radio (TV) (73) Auxiliary Broadcasting (74)
	216-220 FIXED MARITIME MOBILE Radiolocation S5.241 S5.242		216-220 MARITIME MOBILE Fixed Radiolocation S5.241 G2 Aeronautical mobile Land mobile US210 US229 US274 US317	216-220 MARITIME MOBILE Fixed Aeronautical mobile Land mobile US210 US229 US274 US317 NG152	Maritime (80) Private Land Mobile (90) Personal Radio (95) Amateur (97) Note: 216-220 MHz will become a mixed-use band in January 2002
	220-225 AMATEUR FIXED MOBILE Radiolocation S5.241		220-222 FIXED LAND MOBILE Radiolocation S5.241 G2 US335	220-222 FIXED LAND MOBILE US335	Private Land Mobile (90)
S5.235 S5.237 S5.243		S5.233 S5.238 S5.240 S5.245	222-225 Radiolocation S5.241 G2	222-225 AMATEUR	Amateur (97)

223-230 BROADCASTING Fixed Mobile S5.243 S5.246 S5.247		223-230 FIXED MOBILE BROADCASTING AERONAUTICAL RADIONAVIGATION Radiolocation S5.250			
	225-235 FIXED MOBILE		225-235 FIXED MOBILE G27	225-235	
230-235 FIXED MOBILE S5.247 S5.251 S5.252		230-235 FIXED MOBILE AERONAUTICAL RADIONAVIGATION S5.250			
235-267 FIXED MOBILE S5.111 S5.199 S5.252 S5.254 S5.256			235-267 FIXED MOBILE S5.111 S5.199 S5.256 G27 G100	235-267 S5.111 S5.199 S5.256	
267-272 FIXED MOBILE Space operation (space-to-Earth) S5.254 S5.257			267-322 FIXED MOBILE G27 G100	267-322	
272-273 SPACE OPERATION (space-to-Earth) FIXED MOBILE S5.254					
273-312 FIXED MOBILE S5.254					
312-315 FIXED MOBILE Mobile-satellite (Earth-to-space) S5.254 S5.255					
See next page for 315-322 MHz					

322-410 MHz (UHF)					
International Table			United States Table		Remarks
Region 1	Region 2	Region 3	Federal Government	Non-Federal Government	
315-322 FIXED MOBILE S5.254			See previous page for 267-322 MHz		See previous page for 267-322 MHz
322-328.6 FIXED MOBILE RADIO ASTRONOMY S5.149			322-328.6 FIXED MOBILE S5.149 G27	322-328.6 S5.149	
328.6-335.4 AERONAUTICAL RADIONAVIGATION S5.258 S5.259			328.6-335.4 AERONAUTICAL RADIONAVIGATION S5.258		
335.4-387 FIXED MOBILE S5.254			335.4-399.9 FIXED MOBILE G27 G100	335.4-399.9	
387-390 FIXED MOBILE Mobile-satellite (space-to-Earth) S5.208A S5.254 S5.255					
390-399.9 FIXED MOBILE S5.254					
399.9-400.05 MOBILE-SATELLITE (Earth-to-space) S5.209 A5.224A RADIONAVIGATION-SATELLITE S5.222 S5.224B S5.260 S5.220			399.9-400.05 MOBILE-SATELLITE (Earth-to-space) US319 US322 RADIONAVIGATION-SATELLITE S5.260		
400.05-400.15 STANDARD FREQUENCY AND TIME SIGNAL-SATELLITE (400.1 MHz) S5.261 S5.262			400.05-400.15 STANDARD FREQUENCY AND TIME SIGNAL-SATELLITE (400.1 MHz) S5.261		

400.15-401 METEOROLOGICAL AIDS METEOROLOGICAL-SATELLITE (space-to-Earth) MOBILE-SATELLITE (space-to-Earth) S5.208A S5.209 SPACE RESEARCH (space-to-Earth) S5.263 Space operation (space-to-Earth) S5.262 S5.264	400.15-401 METEOROLOGICAL AIDS (radiosonde) METEOROLOGICAL-SATELLITE (space-to-Earth) MOBILE-SATELLITE (space-to-Earth) 599B US319 US320 US324 SPACE RESEARCH (space-to-Earth) S5.263 Space operation (space-to-Earth) 647B US70	400.15-401 METEOROLOGICAL AIDS (radiosonde) MOBILE-SATELLITE (space-to-Earth) 599B US319 US320 US324 SPACE RESEARCH (space-to-Earth) S5.263 Space operation (space-to-Earth) 647B US70	Satellite Communications (25)
401-402 METEOROLOGICAL AIDS SPACE OPERATION (space-to-Earth) EARTH EXPLORATION-SATELLITE (Earth-to-space) METEOROLOGICAL-SATELLITE (Earth-to-space) Fixed Mobile except aeronautical mobile	401-402 METEOROLOGICAL AIDS (radiosonde) SPACE OPERATION (space-to-Earth) Earth exploration-satellite (Earth-to-space) Meteorological-satellite (Earth-to-space) US70		
402-403 METEOROLOGICAL AIDS EARTH EXPLORATION-SATELLITE (Earth-to-space) METEOROLOGICAL-SATELLITE (Earth-to-space) Fixed Mobile except aeronautical mobile	402-403 METEOROLOGICAL AIDS (radiosonde) US70 Earth exploration-satellite (Earth-to-space) Meteorological-satellite (Earth-to-space) US345		Personal Radio (95)
403-406 METEOROLOGICAL AIDS Fixed Mobile except aeronautical mobile	403-406 METEOROLOGICAL AIDS (radiosonde) US70 US345 G6	403-406 METEOROLOGICAL AIDS (radiosonde) US70 US345	
406-406.1 MOBILE-SATELLITE (Earth-to-space) S5.266 S5.267	406-406.1 MOBILE-SATELLITE (Earth-to-space) S5.266 S5.267		
406.1-410 FIXED MOBILE except aeronautical mobile RADIO ASTRONOMY S5.149	406.1-410 FIXED MOBILE RADIO ASTRONOMY US74 US13 US117 G5 G6	406.1-410 RADIO ASTRONOMY US74 US13 US117	

<table>
<tr><th colspan="6">410-470 MHz (UHF)</th></tr>
<tr><th colspan="3">International Table</th><th colspan="2">United States Table</th><th rowspan="2">Remarks</th></tr>
<tr><th>Region 1</th><th>Region 2</th><th>Region 3</th><th>Federal Government</th><th>Non-Federal Government</th></tr>
<tr><td colspan="3">410-420
FIXED
MOBILE except aeronautical mobile
SPACE RESEARCH (space-to-space) S5.268</td><td>410-420
FIXED
MOBILE
SPACE RESEARCH
(space-to-space) S5.268
US13 G5</td><td>410-420
US13</td><td>Note: In this manual a primary allocation has been added for SPACE RESEARCH (space-to-space). At this time, the FCC has not adopted this change.</td></tr>
<tr><td colspan="3">420-430
FIXED
MOBILE except aeronautical mobile
Radiolocation
S5.269 S5.270 S5.271</td><td rowspan="3">420-450
RADIOLOCATION G2
S5.286 US7 US87 US217 US228 US230 G8</td><td rowspan="3">420-450
Amateur
S5.282 S5.286 US7 US87 US217 US228 US230 NG135</td><td rowspan="3">Private Land Mobile (90)
Amateur (97)</td></tr>
<tr><td>430-440
AMATEUR
RADIOLOCATION
S5.138 S5.271 S5.272 S5.273 S5.274 S5.275 S5.276 S5.277 S5.280 S5.281 S5.282 S5.283</td><td colspan="2">430-440
RADIOLOCATION
Amateur
S5.271 S5.276 S5.277 S5.278 S5.279 S5.281 S5.282</td></tr>
<tr><td colspan="3">440-450
FIXED
MOBILE except aeronautical mobile
Radiolocation
S5.269 S5.270 S5.271 S5.284 S5.285 S5.286</td></tr>
<tr><td colspan="3" rowspan="2">450-455
FIXED
MOBILE
S5.209 S5.271 S5.286 S5.286A S5.286B S5.286C S5.286D S5.286E</td><td>450-454
S5.286 US87</td><td>450-454
LAND MOBILE
S5.286 US87 NG112 NG124</td><td>Auxiliary Broadcasting (74)
Private Land Mobile (90)</td></tr>
<tr><td rowspan="2">454-456</td><td>454-455
FIXED
LAND MOBILE
NG12 NG112 NG148</td><td>Public Mobile (22)
Maritime (80)</td></tr>
<tr><td>455-456
FIXED
MOBILE
S5.209 S5.271 S5.286A S5.286B S5.286C S5.286E</td><td>455-456
FIXED
MOBILE
MOBILE-SATELLITE
(Earth-to-space) S5.286A S5.286B S5.286C
S5.209</td><td>455-456
FIXED
MOBILE
S5.209 S5.271 S5.286A S5.286B S5.286C S5.286E</td><td>455-456
LAND MOBILE</td><td>Auxiliary Broadcasting (74)</td></tr>
</table>

456-459 FIXED MOBILE S5.271 S5.287 S5.288			456-460 S5.288 669	456-460 FIXED LAND MOBILE S5.288 669 NG112 NG124 NG148	Public Mobile (22) Maritime (80) Private Land Mobile (90)
459-460 FIXED MOBILE S5.209 S5.271 S5.286A S5.286B S5.286C S5.286E	459-460 FIXED MOBILE MOBILE-SATELLITE (Earth-to-space) S5.286A S5.286B S5.286C S5.209	459-460 FIXED MOBILE S5.209 S5.271 S5.286A S5.286B S5.286C S5.286E			
460-470 FIXED MOBILE Meteorological-satellite (space-to-Earth) S5.287 S5.288 S5.289 S5.290			460-470 Meteorological-satellite (space-to-Earth) S5.288 S5.289 669 US201 US209 US216	460-462.5375 FIXED LAND MOBILE S5.289 US201 US209 NG124	Private Land Mobile (90)
				462.5375-462.7375 LAND MOBILE S5.289 US201	Personal Radio (95)
				462.7375-467.5375 FIXED LAND MOBILE S5.289 669 US201 US209 US216 NG124	Private Land Mobile (90)
				467.5375-467.7375 LAND MOBILE S5.289 669 US201	Personal Radio (95)
				467.7375-470 FIXED LAND MOBILE S5.288 S5.289 US201 US216 NG124	Private Land Mobile (90)

470-849 MHz (UHF)					
International Table			United States Table		Remarks
Region 1	Region 2	Region 3	Federal Government	Non-Federal Government	
470-790 BROADCASTING	470-512 BROADCASTING Fixed Mobile S5.292 S5.293	470-585 FIXED MOBILE BROADCASTING S5.291 S5.298	470-608	470-512 FIXED BROADCASTING LAND MOBILE NG66 NG114 NG127 NG128 NG149	Public Mobile (22) Broadcast Radio (TV) (73) Auxiliary Broadcasting (74) Private Land Mobile (90)
	512-608 BROADCASTING S5.297	585-610 FIXED MOBILE BROADCASTING RADIONAVIGATION S5.149 S5.305 S5.306 S5.307		512-608 BROADCASTING NG128 NG149	Broadcast Radio (TV) (73) Auxiliary Broadcasting (74)
	608-614 RADIO ASTRONOMY Mobile-satellite except aeronautical mobile-satellite (Earth-to-space)	610-890 FIXED MOBILE BROADCASTING	608-614 LAND MOBILE US350 RADIO ASTRONOMY US74 US246		Personal (95)
	614-806 BROADCASTING Fixed Mobile		614-890	614-698 BROADCASTING NG128 NG149	Broadcast Radio (TV) (73) Auxiliary Broadcast. (74)
				698-746 BROADCASTING NG128 NG149	Broadcast Radio (TV) (73) Auxiliary Broadcast. (74) Note: Band to be reallocated and auctioned by Sept. 30, 2002.
				746-764 FIXED MOBILE BROADCASTING NG128 NG159	Wireless Communications (27) Broadcast Radio (TV) (73) Auxiliary Broadcasting (74) Private Land Mobile (90)

				764-776 FIXED MOBILE NG128 NG158 NG159	Auxiliary Broadcasting (74) Private Land Mobile (90)
S5.149 S5.291A S5.294 S5.296 S5.300 S5.302 S5.304 S5.306 S5.311 S5.312 790-862 FIXED BROADCASTING				776-794 FIXED MOBILE BROADCASTING NG128 NG159	Wireless Communications (27) Broadcast Radio (TV) (73) Auxiliary Broadcast. (74) Private Land Mobile (90)
	S5.293 S5.309 S5.311			794-806 FIXED MOBILE NG128 NG158 NG159	Auxiliary Broadcasting (74) Private Land Mobile (90)
	806-890 FIXED MOBILE BROADCASTING			806-821 FIXED LAND MOBILE NG30 NG31 NG43 NG63	Public Mobile (22) Private Land Mobile (90)
				821-824 LAND MOBILE NG30 NG43 NG63	Private Land Mobile (90)
S5.312 S5.314 S5.315				824-849 FIXED LAND MOBILE NG30 NG43 NG63 NG151	Public Mobile (22)
S5.316 S5.319 S5.321 See next page for 862-890 MHz	S5.317 S5.318	S5.149 S5.305 S5.306 S5.307 S5.311 S5.320		See next page for 849-894 MHz	See next page for 866-896 MHz

<table>
<tr><th colspan="6">849-941 MHz (UHF)</th></tr>
<tr><th colspan="3">International Table</th><th colspan="2">United States Table</th><th rowspan="2">Remarks</th></tr>
<tr><th>Region 1</th><th>Region 2</th><th>Region 3</th><th>Federal Government</th><th>Non-Federal Government</th></tr>
<tr><td rowspan="3">See previous pages for
470-862 MHz</td><td rowspan="5">See previous pages for
614-890 MHz</td><td rowspan="5">See previous pages for
585-890 MHz</td><td rowspan="5">See previous pages for
614-890 MHz</td><td>See previous pages for
614-849 MHz</td><td>See previous pages for
614-849 MHz</td></tr>
<tr><td>849-851
AERONAUTICAL MOBILE

NG30 NG63</td><td>Public Mobile (22)</td></tr>
<tr><td>851-866
FIXED
LAND MOBILE

NG30 NG31 NG63</td><td>Public Mobile (22)
Private Land Mobile (90)</td></tr>
<tr><td rowspan="2">862-890
FIXED
MOBILE except aeronautical mobile
BROADCASTING S5.322

S5.319 S5.323</td><td>866-869
LAND MOBILE

NG30 NG63</td><td>Private Land Mobile (90)</td></tr>
<tr><td>869-894
FIXED
LAND MOBILE

US116 US268 NG30 NG63 NG151</td><td rowspan="2">Public Mobile (22)</td></tr>
<tr><td rowspan="3">890-942
FIXED
MOBILE except aeronautical mobile
BROADCASTING S5.322
Radiolocation</td><td rowspan="3">890-902
FIXED
MOBILE except aeronautical mobile
Radiolocation

S5.318 S5.325</td><td rowspan="3">890-942
FIXED
MOBILE
BROADCASTING
Radiolocation</td><td rowspan="3">890-902

US116 US268 G2</td><td>894-896
AERONAUTICAL MOBILE

US116 US268</td></tr>
<tr><td>896-901
FIXED
LAND MOBILE

US116 US268</td><td>Private Land Mobile (90)</td></tr>
<tr><td>901-902
FIXED
MOBILE

US116 US268</td><td>Personal Communications (24)</td></tr>
</table>

	902-928 FIXED Amateur Mobile except aeronautical mobile Radiolocation S5.150 S5.325 S5.326		902-928 RADIOLOCATION G59 S5.150 US215 US218 US267 US275 G11	902-928 S5.150 US215 US218 US267 US275	ISM Equipment (18) Private Land Mobile (90) Amateur (97)
	928-942 FIXED MOBILE except aeronautical mobile Radiolocation		928-932	928-929 FIXED US116 US215 US268 NG120	Public Mobile (22) Private Land Mobile (90) Fixed Microwave (101)
				929-930 FIXED LAND MOBILE US116 US215 US268	Private Land Mobile (90)
				930-931 FIXED MOBILE US116 US215 US268	Personal Communications (24)
			US116 US215 US268 G2	931-932 FIXED LAND MOBILE US116 US215 US268	Public Mobile (22)
			932-935 FIXED US215 US268 G2	932-935 FIXED US215 US268 NG120	Public Mobile (22) Fixed Microwave (101)
			935-940 US116 US215 US268 G2	935-940 FIXED LAND MOBILE US116 US215 US268	Private Land Mobile (90)
			940-941 US116 US268 G2	940-941 FIXED MOBILE US116 US268	Personal Communications (24)
S5.323	S5.325	S5.327	See next page for 941-944 MHz		See next page for 941-944 MHz

941-1429 MHz (UHF)					
International Table			United States Table		Remarks
Region 1	Region 2	Region 3	Federal Government	Non-Federal Government	
See previous page for 890-942 MHz	See previous page for 928-942 MHz	See previous page for 890-942 MHz	941-944 FIXED US268 US301 US302 G2	941-944 FIXED US268 US301 US302 NG120	Public Mobile (22) Fixed Microwave (101)
942-960 FIXED MOBILE except aeronautical mobile BROADCASTING S5.322 S5.323	942-960 FIXED MOBILE	942-960 FIXED MOBILE BROADCASTING S5.320	944-960	944-960 FIXED NG120	Public Mobile (22) International Fixed (23) Auxiliary Broadcast. (74) Fixed Microwave (101)
960-1215 AERONAUTICAL RADIONAVIGATION S5.328			960-1215 AERONAUTICAL RADIONAVIGATION S5.328 US224		Aviation (87)
1215-1240 EARTH EXPLORATION-SATELLITE (active) RADIOLOCATION RADIONAVIGATION-SATELLITE (space-to-Earth) S5.329 SPACE RESEARCH (active) S5.330 S5.331 S5.332			1215-1240 RADIOLOCATION S5.333 G56 RADIONAVIGATION-SATELLITE (space-to- Earth)	1215-1240 S5.333	
1240-1260 EARTH EXPLORATION-SATELLITE (active) RADIOLOCATION RADIONAVIGATION-SATELLITE (space-to-Earth) S5.329 SPACE RESEARCH (active) Amateur S5.330 S5.331 S5.332 S5.334 S5.335			1240-1300 RADIOLOCATION S5.333 G56 S5.334	1240-1300 Amateur S5.282 S5.333 S5.334	Amateur (97)
1260-1300 EARTH EXPLORATION-SATELLITE (active) RADIOLOCATION SPACE RESEARCH (active) Amateur S5.282 S5.330 S5.331 S5.332 S5.334 S5.335					

1300-1350 AERONAUTICAL RADIONAVIGATION S5.337 Radiolocation S5.149		1300-1350 AERONAUTICAL RADIO-NAVIGATION S5.337 Radiolocation G2 S5.149	1300-1350 AERONAUTICAL RADIO-NAVIGATION S5.337 S5.149	Aviation (87)
1350-1400 FIXED MOBILE RADIOLOCATION S5.149 S5.338 S5.339	1350-1400 RADIOLOCATION S5.149 S5.334 S5.339	1350-1390 FIXED MOBILE RADIOLOCATION G2 S5.149 S5.334 S5.339 US311 G27 G114	1350-1390 S5.149 S5.334 S5.339	
		1390-1395 RADIOLOCATION G2 Fixed Mobile S5.149 S5.339 US311 US351 G27 G114	1390-1395 S5.149 S5.339 US351	Note: 1390-1395 MHz became non-Federal Government exclusive spectrum in January 1999
		1395-1400 LAND MOBILE US350 S5.149 S5.339 US311 US351	1395-1400 LAND MOBILE US350 S5.149 S5.339 US311 US351	Personal (95)
1400-1427 EARTH EXPLORATION-SATELLITE (passive) RADIO ASTRONOMY SPACE RESEARCH (passive) S5.340 S5.341		1400-1427 EARTH EXPLORATION-SATELLITE (passive) RADIO ASTRONOMY US74 SPACE RESEARCH (passive) S5.341 US246		
1427-1429 SPACE OPERATION (Earth-to-space) FIXED MOBILE except aeronautical mobile S5.341		1427-1429 SPACE OPERATION (Earth-to-space) FIXED MOBILE except aeronautical mobile S5.341 G30	1427-1429 SPACE OPERATION (Earth-to-space) Fixed (telemetry) Land mobile (telemetry and telecommand) S5.341	Satellite Communications (25) Private Land Mobile (90) Note: 1427-1429 MHz became non-Federal Government exclusive spectrum in January 1999

<table>
<tr><th colspan="6">1429-1610 MHz (UHF)</th></tr>
<tr><th colspan="3">International Table</th><th colspan="2">United States Table</th><th rowspan="2">Remarks</th></tr>
<tr><th>Region 1</th><th>Region 2</th><th>Region 3</th><th>Federal Government</th><th>Non-Federal Government</th></tr>
<tr><td rowspan="2">1429-1452
FIXED
MOBILE except aeronautical mobile

S5.341 S5.342</td><td colspan="2" rowspan="2">1429-1452
FIXED
MOBILE S5.343

S5.341</td><td>1429-1432
LAND MOBILE US350

S5.341 US352</td><td>1429-1432
LAND MOBILE US350
Fixed (telemetry)
Land mobile (telemetry and telecommand)

S5.341 US352</td><td>Private Land Mobile (90)
Personal (95)</td></tr>
<tr><td>1432-1435
FIXED
MOBILE

S5.341 G30</td><td>1432-1435
Fixed (telemetry)
Land mobile (telemetry and telecommand)

S5.341</td><td>Private Land Mobile (90)
Note: 1432-1435 MHz became mixed-use spectrum In January 1999.</td></tr>
<tr><td>1452-1492
FIXED
MOBILE except aeronautical mobile
BROADCASTING S5.345 S5.347
BROADCASTING-SATELLITE S5.345 S5.347

S5.341 S5.342</td><td colspan="2">1452-1492
FIXED
MOBILE S5.343
BROADCASTING S5.345 S5.347
BROADCASTING-SATELLITE S5.345 S5.347

S5.341 S5.344</td><td colspan="2" rowspan="2">1435-1525
MOBILE (aeronautical telemetry)

S5.341 US78</td><td rowspan="2">Aviation (87)</td></tr>
<tr><td>1492-1525
FIXED
MOBILE except aeronautical mobile

S5.341 S5.342</td><td>1492-1525
FIXED
MOBILE S5.343
MOBILE-SATELLITE (space-to-Earth) S5.348A

S5.341 S5.344 S5.348</td><td>1492-1525
FIXED
MOBILE

S5.341 S5.348A</td></tr>
<tr><td>1525-1530
SPACE OPERATION (space-to-Earth)
FIXED
MOBILE-SATELLITE (space-to-Earth)
Earth exploration-satellite
Mobile except aeronautical mobile S5.349

S5.341 S5.342 S5.350 S5.351 S5.352A S5.354</td><td>1525-1530
SPACE OPERATION (space-to-Earth)
MOBILE-SATELLITE (space-to-Earth)
Earth exploration-satellite
Fixed
Mobile S5.343

S5.341 S5.351 S5.354</td><td>1525-1530
SPACE OPERATION (space-to-Earth)
FIXED
MOBILE-SATELLITE (space-to-Earth)
Earth exploration-satellite
Mobile S5.349

S5.341 S5.351 S5.352A S5.354</td><td colspan="2">1525-1530
MOBILE-SATELLITE (space-to-Earth)
Mobile (aeronautical telemetry)

S5.341 S5.351 US78</td><td>Satellite Communications (25)
Aviation (87)</td></tr>
</table>

<table>
<tr><td>1530-1535
SPACE OPERATION
(space-to-Earth)
MOBILE-SATELLITE
(space-to-Earth) S5.353A
Earth exploration-satellite
Fixed
Mobile except aeronautical
mobile

S5.341 S5.342 S5.351 S5.354</td><td>1530-1535
SPACE OPERATION (space-to-Earth)
MOBILE-SATELLITE (space-to-Earth) S5.353A
Earth exploration-satellite
Fixed
Mobile S5.343

S5.341 S5.351 S5.354</td><td>1530-1535
MOBILE-SATELLITE (space-to-Earth)
MARITIME MOBILE-SATELLITE (space-to-Earth)
Mobile (aeronautical telemetry)

S5.341 S5.351 US78 US315</td><td></td></tr>
<tr><td colspan="2" rowspan="5">1535-1559
MOBILE-SATELLITE (space-to-Earth)

S5.341 S5.351 S5.353A S5.354 S5.355 S5.356 S5.357 S5.357A S5.359 S5.362A</td><td>1535-1544
MOBILE-SATELLITE (space-to-Earth)
MARITIME MOBILE-SATELLITE (space-to-Earth)

S5.341 S5.351 US315</td><td rowspan="2">Satellite
Communications (25)
Maritime (80)</td></tr>
<tr><td>1544-1545
MOBILE-SATELLITE (space-to-Earth)

S5.341 S5.356</td></tr>
<tr><td>1545-1549.5
AERONAUTICAL MOBILE-SATELLITE (R)
(space-to-Earth)
Mobile-satellite (space-to-Earth)

S5.341 S5.351 US308 US309</td><td rowspan="3">Aviation (87)</td></tr>
<tr><td>1549.5-1558.5
AERONAUTICAL MOBILE-SATELLITE (R)
(space-to-Earth)
MOBILE-SATELLITE (space-to-Earth)

S5.341 S5.351 US308 US309</td></tr>
<tr><td>1558.5-1559
AERONAUTICAL MOBILE-SATELLITE (R)
(space-to-Earth)

S5.341 S5.351 US308 US309</td></tr>
<tr><td colspan="2">1559-1610
AERONAUTICAL RADIONAVIGATION
RADIONAVIGATION-SATELLITE (space-to-Earth)

S5.341 S5.355 S5.359 S5.363</td><td>1559-1610
AERONAUTICAL RADIONAVIGATION
RADIONAVIGATION-SATELLITE (space-to-Earth)

S5.341 US208 US260 G126</td><td>Note: Footnote G126 states that DGPS stations may be authorized on a primary basis in this band, but the FCC has not yet addressed this footnote.</td></tr>
</table>

1610-1670 MHz (UHF)					
International Table			United States Table		Remarks
Region 1	Region 2	Region 3	Federal Government	Non-Federal Government	
1610-1610.6 MOBILE-SATELLITE (Earth-to-space) AERONAUTICAL RADIONAVIGATION S5.341 S5.355 S5.359 S5.363 S5.364 S5.366 S5.367 S5.368 S5.369 S5.371 S5.372	1610-1610.6 MOBILE-SATELLITE (Earth-to-space) AERONAUTICAL RADIONAVIGATION RADIODETERMINATION-SATELLITE (Earth-to-space) S5.341 S5.364 S5.366 S5.367 S5.368 S5.370 S5.372	1610-1610.6 MOBILE-SATELLITE (Earth-to-space) AERONAUTICAL RADIONAVIGATION Radiodetermination-Satellite (Earth-to-space) S5.341 S5.355 S5.359 S5.364 S5.366 S5.367 S5.368 S5.369 S5.372	1610-1610.6 MOBILE-SATELLITE (Earth-to-space) US319 AERONAUTICAL RADIONAVIGATION US260 RADIODETERMINATION-SATELLITE(Earth-to-space) S5.341 S5.364 S5.366 S5.367 S5.368 S5.372 US208		Satellite Communications (25) Aviation (87)
1610.6-1613.8 MOBILE-SATELLITE (Earth-to-space) RADIO ASTRONOMY AERONAUTICAL RADIONAVIGATION S5.149 S5.341 S5.355 S5.359 S5.363 S5.364 S5.366 S5.367 S5.368 S5.369 S5.371 S5.372	1610.6-1613.8 MOBILE-SATELLITE (Earth-to-space) RADIO ASTRONOMY AERONAUTICAL RADIONAVIGATION RADIODETERMINATION-SATELLITE (Earth-to-space) S5.149 S5.341 S5.364 S5.366 S5.367 S5.368 S5.370 S5.372	1610.6-1613.8 MOBILE-SATELLITE (Earth-to-space) RADIO ASTRONOMY AERONAUTICAL RADIONAVIGATION Radiodetermination-satellite (Earth-to-space) S5.149 S5.341 S5.355 S5.359 S5.364 S5.366 S5.367 S5.368 S5.369 S5.372	1610.6-1613.8 MOBILE-SATELLITE (Earth-to-space) US319 RADIO ASTRONOMY AERONAUTICAL RADIONAVIGATION US260 RADIODETERMINATION-SATELLITE (Earth-to-space) S5.341 S5.364 S5.366 S5.367 S5.368 S5.372 US208		
1613.8-1626.5 MOBILE-SATELLITE (Earth-to-space) AERONAUTICAL RADIONAVIGATION Mobile-satellite (space-to-Earth) S5.341 S5.355 S5.359 S5.363 S5.364 S5.365 S5.366 S5.367 S5.368 S5.369 S5.371 S5.372	1613.8-1626.5 MOBILE-SATELLITE (Earth-to-space) AERONAUTICAL RADIONAVIGATION RADIODETERMINATION-SATELLITE (Earth-to-space) Mobile-satellite (space-to-Earth) S5.341 S5.364 S5.365 S5.366 S5.367 S5.368 S5.370 S5.372	1613.8-1626.5 MOBILE-SATELLITE (Earth-to-space) AERONAUTICAL RADIONAVIGATION Mobile-satellite (space-to-Earth) Radiodetermination-satellite (Earth-to-space) S5.341 S5.355 S5.359 S5.364 S5.365 S5.366 S5.367 S5.368 S5.369 S5.372	1613.8-1626.5 MOBILE-SATELLITE (Earth-to-space) US319 AERONAUTICAL RADIONAVIGATION US260 RADIODETERMINATION-SATELLITE (Earth-to-space) Mobile-satellite (space-to-Earth) S5.341 S5.364 S5.365 S5.366 S5.367 S5.368 S5.372 US208		
1626.5-1660 MOBILE-SATELLITE (Earth-to-space)			1626.5-1645.5 MOBILE-SATELLITE (Earth-to-space) MARITIME MOBILE-SATELLITE (Earth-to-space) S5.341 S5.351 US315		Satellite Communications (25) Maritime (80)

	1645.5-1646.5 MOBILE-SATELLITE (Earth-to-space) S5.341 S5.375	
	1646.5-1651 AERONAUTICAL MOBILE-SATELLITE (R) (Earth-to-space) Mobile-satellite (Earth-to-space) S5.341 S5.351 US308 US309	Aviation (87)
S5.341 S5.351 S5.353A S5.354 S5.355 S5.357A S5.359 S5.362A S5.374 S5.375 S5.376	1651-1660 MOBILE-SATELLITE (Earth-to-space) AERONAUTICAL MOBILE-SATELLITE (R) (Earth-to-space) S5.341 S5.351 US308 US309	
1660-1660.5 MOBILE-SATELLITE (Earth-to-space) RADIO ASTRONOMY S5.149 S5.341 S5.351 S5.354 S5.362A S5.376A	1660-1660.5 AERONAUTICAL MOBILE-SATELLITE (R) (Earth-to-space) RADIO ASTRONOMY S5.149 S5.341 S5.351 US308 US309	
1660.5-1668.4 RADIO ASTRONOMY SPACE RESEARCH (passive) Fixed Mobile except aeronautical mobile S5.149 S5.341 S5.379 S5.379A	1660.5-1668.4 RADIO ASTRONOMY US74 SPACE RESEARCH (passive) S5.341 US246	
1668.4-1670 METEOROLOGICAL AIDS FIXED MOBILE except aeronautical mobile RADIO ASTRONOMY S5.149 S5.341	1668.4-1670 METEOROLOGICAL AIDS (radiosonde) RADIO ASTRONOMY US74 S5.149 S5.341 US99	

1670-2110 MHz (UHF)					
International Table			United States Table		Remarks
Region 1	Region 2	Region 3	Federal Government	Non-Federal Government	
1670-1675 METEOROLOGICAL AIDS FIXED METEOROLOGICAL-SATELLITE (space-to-Earth) MOBILE S5.380 S5.341			1670-1675 METEOROLOGICAL AIDS (radiosonde) METEOROLOGICAL-SATELLITE (space-to-Earth) S5.341 US211		Note: 1670-1675 MHz became mixed-use spectrum in January 1999
1675-1690 METEOROLOGICAL AIDS FIXED METEOROLOGICAL-SATELLITE (space-to-Earth) MOBILE except aeronautical mobile S5.341	1675-1690 METEOROLOGICAL AIDS FIXED METEOROLOGICAL-SATELLITE (space-to-Earth) MOBILE except aeronautical mobile MOBILE-SATELLITE (Earth-to-space) S5.341 S5.377	1675-1690 METEOROLOGICAL AIDS FIXED METEOROLOGICAL-SATELLITE (space-to-Earth) MOBILE except aeronautical mobile S5.341	1675-1700 METEOROLOGICAL AIDS (radiosonde) METEOROLOGICAL-SATELLITE (space-to-Earth) S5.289 S5.341 US211		
1690-1700 METEOROLOGICAL AIDS METEOROLOGICAL-SATELLITE (space-to-Earth) Fixed Mobile except aeronautical mobile S5.289 S5.341 S5.382	1690-1700 METEOROLOGICAL AIDS METEOROLOGICAL-SATELLITE (space-to-Earth) MOBILE-SATELLITE (Earth-to-space) S5.289 S5.341 S5.377 S5.381	1690-1700 METEOROLOGICAL AIDS METEOROLOGICAL-SATELLITE (space-to-Earth) S5.289 S5.341 S5.381			
1700-1710 FIXED METEOROLOGICAL-SATELLITE (space- to-Earth) MOBILE except aeronautical mobile S5.289 S5.341	1700-1710 FIXED METEOROLOGICAL-SATELLITE (space-to-Earth) MOBILE except aeronautical mobile MOBILE-SATELLITE (Earth-to-space) S5.289 S5.341 S5.377	1700-1710 FIXED METEOROLOGICAL-SATELLITE (space-to-Earth) MOBILE except aeronautical mobile S5.289 S5.341 S5.384	1700-1710 FIXED G118 METEOROLOGICAL-SATELLITE (space-to-Earth) S5.289 S5.341	1700-1710 METEOROLOGICAL-SATELLITE (space-to-Earth) Fixed S5.289 S5.341	
1710-1930 FIXED MOBILE S5.380			1710-1755 FIXED MOBILE S5.341 US256	1710-1755 S5.341 US256	Note: Proceeds from the auction of the 1710-1755 MHz mixed-use band are to be deposited not later than September 30, 2002.

S5.149 S5.341 S5.385 S5.386 S5.387 S5.388			1755-1850 FIXED MOBILE G42	1755-1850	
1930-1970 FIXED MOBILE S5.388	1930-1970 FIXED MOBILE Mobile-satellite (Earth-to-space) S5.388	1930-1970 FIXED MOBILE S5.388	1850-1990	1850-1990 FIXED MOBILE	RF Devices (15) Personal Communications (24) Fixed Microwave (101)
1970-1980 FIXED MOBILE S5.388					
1980-2010 FIXED MOBILE MOBILE-SATELLITE (Earth-to-space) S5.388 S5.389A S5.389B S5.389F			1990-2025	1990-2025 MOBILE-SATELLITE (Earth-to-space) NG156	Satellite Communications (25)
2010-2025 FIXED MOBILE S5.388	2010-2025 FIXED MOBILE MOBILE-SATELLITE (Earth-to-space) S5.388 S5.389C S5.389D S5.389E S5.390	2010-2025 FIXED MOBILE S5.388			
2025–2110 SPACE OPERATION (Earth-to-space) (space-to-space) EARTH EXPLORATION-SATELLITE (Earth-to-space) (space-to-space) FIXED MOBILE S5.391 SPACE RESEARCH (Earth-to-space) (space-to-space) S5.392			2025-2110 SPACE OPERATION (Earth-to-space) (space-to-space) EARTH EXPLORATION-SATELLITE (Earth-to-space) (space-to-space) SPACE RESEARCH (Earth-to-space) (space-to-space) S5.391 S5.392 US90 US222 US346 US347	2025-2110 FIXED NG23 NG118 MOBILE S5.391 S5.392 US90 US222 US346 US347	TV Auxiliary Broadcasting (74F) Cable TV Relay (78) Local TV Transmission (101J)

2110-2345 MHz (UHF)					
International Table			United States Table		Remarks
Region 1	Region 2	Region 3	Federal Government	Non-Federal Government	
2110-2120 FIXED MOBILE SPACE RESEARCH (deep space) (Earth-to-space) S5.388			2110-2120 US252	2110-2150 FIXED NG23 MOBILE US252 NG153	Public Mobile (22) Fixed Microwave (101) Note: 2110-2150 MHz must be auctioned by September 30, 2002.
2120-2160 FIXED MOBILE S5.388	2120-2160 FIXED MOBILE Mobile-satellite (space-to-Earth) S5.388	2120-2160 FIXED MOBILE S5.388	2120-2200	2150-2160 FIXED NG23	Domestic Public Fixed (21) Fixed Microwave (101)
2160-2170 FIXED MOBILE S5.388 S5.392A	2160-2170 FIXED MOBILE MOBILE-SATELLITE (space-to-Earth) S5.388 S5.389C S5.389D S5.389E S5.390	2160-2170 FIXED MOBILE S5.388		2160-2165 FIXED NG23 MOBILE NG153	Domestic Public Fixed (21) Public Mobile (22) Fixed Microwave (101)
2170-2200 FIXED MOBILE MOBILE-SATELLITE (space-to-Earth) S5.388 S5.389A S5.389F S5.392A				2165-2200 MOBILE-SATELLITE (space-to-Earth) NG23 NG168	Satellite Communications (25)
2200-2290 SPACE OPERATION (space-to-Earth) (space-to-space) EARTH EXPLORATION-SATELLITE (space-to-Earth) (space-to-space) FIXED MOBILE S5.391 SPACE RESEARCH (space-to-Earth) (space-to-space)			2200-2290 SPACE OPERATION (space-to-Earth) (space-to-space) EARTH EXPLORATION-SATELLITE (space-to-Earth) (space-to-space) FIXED (line-of-sight only)	2200-2290	

S5.392		MOBILE (line-of-sight only including aeronautical telemetry, but excluding flight testing of manned aircraft) SPACE RESEARCH (space-to-Earth) (space-to-space) S5.392 US303	US303	
2290-2300 FIXED MOBILE except aeronautical mobile SPACE RESEARCH (deep space) (space-to-Earth)		2290-2300 FIXED MOBILE except aeronautical mobile SPACE RESEARCH (deep space) (space-to-Earth)	2290-2300 SPACE RESEARCH (deep space) (space-to-Earth)	
2300-2450 FIXED MOBILE Amateur Radiolocation S5.150 S5.282 S5.395	2300-2450 FIXED MOBILE RADIOLOCATION Amateur S5.150 S5.282 S5.393 S5.394 S5.396	2300-2305 G123	2300-2305 Amateur	Amateur (97) Note: 2300-2305 MHz became non-Federal Government exclusive spectrum in August 1995
		2305-2310 US338 G123	2305-2310 FIXED MOBILE except aeronautical mobile RADIOLOCATION Amateur US338	Wireless Communications (27) Amateur (97)
		2310-2360 Fixed Mobile US339 Radiolocation G2 S5.396 US327 US328 G120	2310-2320 FIXED MOBILE US339 RADIOLOCATION BROADCASTING-SATELLITE US327 S5.396 US338	Wireless Communications (27)
			2320-2345 BROADCASTING-SATELLITE US327 Mobile US276 US328 S5.396	
		See next page	See next page for 2345-2450 MHz	See next page for 2345-2450 MHz

2345-2655 MHz (UHF)					
International Table			United States Table		Remarks
Region 1	Region 2	Region 3	Federal Government	Non-Federal Government	
See previous page for 2300-2450 MHz			See previous page for 2310-2360 MHz	2345-2360 FIXED MOBILE US339 RADIOLOCATION BROADCASTING-SATELLITE US327 S5.396	Wireless Communications (27)
			2360-2385 MOBILE US276 RADIOLOCATION G2 Fixed G120	2360-2385 MOBILE US276	
			2385-2390 MOBILE US276 RADIOLOCATION G2 Fixed G120	2385-2390 MOBILE US276	Note: 2385-2390 MHz will become non-Federal Government exclusive spectrum in January 2005
			2390-2400 G122	2390-2400 AMATEUR	RF Devices (15) Amateur (97)
			2400-2402 S5.150 G123	2400-2402 Amateur S5.150 S5.282	ISM Equipment (18) Amateur (97)
			2402-2417 S5.150 G122	2402-2417 AMATEUR S5.150 S5.282	RF Devices (15) ISM Equipment (18) Amateur (97)
			2417-2450 Radiolocation G2 S5.150 G124	2417-2450 Amateur S5.150 S5.282	ISM Equipment (18) Amateur (97)
2450-2483.5 FIXED MOBILE Radiolocation S5.150 S5.397	2450-2483.5 FIXED MOBILE RADIOLOCATION S5.150 S5.394		2450-2483.5 S5.150 US41	2450-2483.5 FIXED MOBILE Radiolocation S5.150 US41	ISM Equipment (18) Private Land Mobile (90) Fixed Microwave (101)

2483.5-2500 FIXED MOBILE MOBILE-SATELLITE (space-to-Earth) Radiolocation S5.150 S5.371 S5.397 S5.398 S5.399 S5.400 S5.402	2483.5-2500 FIXED MOBILE MOBILE-SATELLITE (space-to-Earth) RADIOLOCATION RADIODETERMINATION-SATELLITE (space-to-Earth) S5.398 S5.150 S5.402	2483.5-2500 FIXED MOBILE MOBILE-SATELLITE (space-to-Earth) RADIOLOCATION Radiodetermination-satellite (space-to-Earth) S5.398 S5.150 S5.400 S5.402	2483.5-2500 MOBILE-SATELLITE (space-to-Earth) US319 RADIODETERMINATION-SATELLITE (space-to-Earth) S5.398 S5.150 753F US41	2483.5-2500 MOBILE-SATELLITE (space-to-Earth) US319 RADIODETERMINATION-SATELLITE (space-to-Earth) S5.398 S5.150 753F US41 NG147	ISM Equipment (18) Satellite Communications (25) Private Land Mobile (90) Fixed Microwave (101)
2500-2520 FIXED S5.409 S5.410 S5.411 MOBILE except aeronautical mobile MOBILE-SATELLITE (space-to-Earth) S5.403 S5.405 S5.407 S5.408 S5.412 S5.414	2500-2520 FIXED S5.409 S5.411 FIXED-SATELLITE (space-to-Earth) S5.415 MOBILE except aeronautical mobile MOBILE-SATELLITE (space-to-Earth) S5.403 S5.404 S5.407 S5.414 S5.415A		2500-2655 S5.339 US205 US269	2500-2655 FIXED S5.409 S5.411 US205 FIXED-SATELLITE (space-to-Earth) NG102 BROADCASTING-SATELLITE NG101 S5.339 US269	Domestic Public Fixed (21) Auxiliary Broadcasting (74)
2520-2655 FIXED S5.409 S5.410 S5.411 MOBILE except aeronautical mobile BROADCASTING-SATELLITE S5.413 S5.416 S5.339 S5.403 S5.405 S5.408 S5.412 S5.417 S5.418	2520-2655 FIXED S5.409 S5.411 FIXED-SATELLITE (space-to-Earth) S5.415 MOBILE except aeronautical mobile BROADCASTING-SATELLITE S5.413 S5.416 S5.339 S5.403	2520-2535 FIXED S5.409 S5.411 FIXED-SATELLITE (space-to-Earth) S5.415 MOBILE except aeronautical mobile BROADCASTING-SATELLITE S5.413 S5.416 S5.403 S5.415A			
		2535-2655 FIXED S5.409 S5.411 MOBILE except aeronautical mobile BROADCASTING-SATELLITE S5.413 S5.416 S5.339 S5.418			

2655-3700 MHz (UHF/SHF)					
International Table			United States Table		Remarks
Region 1	Region 2	Region 3	Federal Government	Non-Federal Government	
2655-2670 FIXED S5.409 S5.410 S5.411 MOBILE except aeronautical mobile BROADCASTING-SATELLITE S5.413 S5.416 Earth exploration-satellite (passive) Radio astronomy Space research (passive) S5.149 S5.412 S5.417 S5.420	2655-2670 FIXED S5.409 S5.411 FIXED-SATELLITE (Earth-to-space) (space-to-Earth) S5.415 MOBILE except aeronautical mobile BROADCASTING-SATELLITE S5.413 S5.416 Earth exploration-satellite (passive) Radio astronomy Space research (passive) S5.149 S5.420	2655-2670 FIXED S5.409 S5.411 FIXED-SATELLITE (Earth-to-space) S5.415 MOBILE except aeronautical mobile BROADCASTING-SATELLITE S5.413 S5.416 Earth exploration-satellite (passive) Radio astronomy Space research (passive) S5.149 S5.420	2655-2690 Earth exploration-satellite (passive) Radio astronomy Space research (passive) US205 US269	2655-2690 FIXED US205 NG47 FIXED-SATELLITE (Earth-to-space) NG102 BROADCASTING-SATELLITE NG101 Earth exploration-satellite (passive) Radio astronomy Space research (passive) US269	
2670-2690 FIXED S5.409 S5.410 S5.411 MOBILE except aeronautical mobile MOBILE-SATELLITE (Earth-to-space) Earth exploration-satellite (passive) Radio astronomy Space research (passive) S5.149 S5.419 S5.420	2670-2690 FIXED S5.409 S5.411 FIXED-SATELLITE (Earth-to-space) (space-to-Earth) S5.415 MOBILE except aeronautical mobile MOBILE-SATELLITE (Earth-to-space) Earth exploration-satellite (passive) Radio astronomy Space research (passive) S5.149 S5.419 S5.420	2670-2690 FIXED S5.409 S5.411 FIXED-SATELLITE (Earth-to-space) S5.415 MOBILE except aeronautical mobile MOBILE-SATELLITE (Earth-to-space) Earth exploration-satellite (passive) Radio astronomy Space research (passive) S5.149 S5.419 S5.420 S5.420A			
2690-2700 EARTH EXPLORATION-SATELLITE (passive) RADIO ASTRONOMY SPACE RESEARCH (passive) S5.340 S5.421 S5.422			2690-2700 EARTH EXPLORATION-SATELLITE (passive) RADIO ASTRONOMY US74 SPACE RESEARCH (passive) US246		
2700-2900 AERONAUTICAL RADIONAVIGATION S5.337 Radiolocation S5.423 S5.424			2700-2900 AERONAUTICAL RADIO-NAVIGATION S5.337 METEOROLOGICAL AIDS Radiolocation G2 S5.423 US18 G15	2700-2900 S5.423 US18	

2900-3100 RADIONAVIGATION S5.426 Radiolocation S5.425 S5.427			2900-3100 MARITIME RADIONAVIGATION Radiolocation G56 S5.427 US44 US316	2900-3100 MARITIME RADIONAVIGATION Radiolocation S5.5427 US44 US316	Maritime (80)
3100-3300 RADIOLOCATION Earth exploration-satellite (active) Space research (active) S5.149 S5.428			3100-3300 RADIOLOCATION S5.333 US110 G59 S5.149	3100-3300 Radiolocation S5.333 US110 S5.149	
3300-3400 RADIOLOCATION S5.149 S5.429 S5.430	3300-3400 RADIOLOCATION Amateur Fixed Mobile S5.149 S5.430	3300-3400 RADIOLOCATION Amateur S5.149 S5.429	3300-3500 RADIOLOCATION US108 G31 S5.149	3300-3500 Amateur Radiolocation US108 S5.149 S5.282	Amateur (97)
3400-3600 FIXED FIXED-SATELLITE (space-to-Earth) Mobile Radiolocation S5.431	3400-3500 FIXED FIXED-SATELLITE (space-to-Earth) Amateur Mobile Radiolocation S5.433 S5.282 S5.432				
	3500-3700 FIXED FIXED-SATELLITE (space-to-Earth) MOBILE except aeronautical mobile Radiolocation S5.433 S5.435		3500-3650 RADIOLOCATION US110 G59 AERONAUTICAL RADIONAVIGATION (ground-based) G110 US245	3500-3600 Radiolocation US110	
3600-4200 FIXED FIXED-SATELLITE (space-to-Earth) Mobile				3600-3650 FIXED-SATELLITE (space-to-Earth) US245 Radiolocation US110	
			3650-3700 RADIOLOCATION US110 G59 AERONAUTICAL RADIONAVIGATION (ground-based) G110 US245	3650-3700 FIXED-SATELLITE (space-to-Earth) US245 Radiolocation US110	Note: 3650-3700 MHz became mixed-use spectrum in January 1999
	See next page for 3700-4200 MHz		See next page for 3700-4200 MHz		

3700-5650 MHz (SHF)					
International Table			United States Table		Remarks
Region 1	Region 2	Region 3	Federal Government	Non-Federal Government	
See previous page for 3600-4200 MHz	3700-4200 FIXED FIXED-SATELLITE (space-to-Earth) MOBILE except aeronautical mobile		3700-4200	3700-4200 FIXED NG41 FIXED-SATELLITE (space-to-Earth)	International Fixed (23) Satellite Communications (25) Fixed Microwave (101)
4200-4400 AERONAUTICAL RADIONAVIGATION S5.438 S5.437 S5.439 S5.440			4200-4400 AERONAUTICAL RADIONAVIGATION S5.440 US261		Aviation (87)
4400-4500 FIXED MOBILE			4400-4500 FIXED MOBILE	4400-4500	
4500-4800 FIXED FIXED-SATELLITE (space-to-Earth) S5.441 MOBILE			4500-4800 FIXED MOBILE US245	4500-4800 FIXED-SATELLITE (space-to-Earth) 792A US245	
4800-4990 FIXED MOBILE S5.442 Radio astronomy S5.149 S5.339 S5.443			4800-4940 FIXED MOBILE S5.149 US203	4800-4940 S5.149 US203	
			4940-4990 FIXED MOBILE S5.149 S5.339 US257	4940-4990 S5.149 S5.339 US257	Note: 4940-4990 MHz became non-Federal Government exclusive spectrum in March 1999
4990-5000 FIXED MOBILE except aeronautical mobile RADIO ASTRONOMY Space research (passive) S5.149			4990-5000 RADIO ASTRONOMY US74 Space research (passive) US246		
5000-5150 AERONAUTICAL RADIONAVIGATION S5.367 S5.444 S5.444A			5000-5250 AERONAUTICAL RADIONAVIGATION US260		Satellite Communications (25) Aviation (87)

5150-5250 AERONAUTICAL RADIONAVIGATION FIXED-SATELLITE (Earth-to-space) S5.447A S5.446 S5.447 S5.447B S5.447C	S5.446 733 796 797 US211 US307 G126		Note: Footnote G126 states that DGPS stations may be authorized on a primary basis in this band, but the FCC has not yet addressed this footnote.
5250-5255 EARTH EXPLORATION-SATELLITE (active) RADIOLOCATION SPACE RESEARCH S5.447D S5.448 S5.448A	5250-5350 RADIOLOCATION S5.333 US110 G59	5250-5350 Radiolocation S5.333 US110	
5255-5350 EARTH EXPLORATION-SATELLITE (active) RADIOLOCATION SPACE RESEARCH (active) S5.448 S5.448A			
5350-5460 EARTH EXPLORATION-SATELLITE (active) S5.448B AERONAUTICAL RADIONAVIGATION S5.449 Radiolocation	5350-5460 AERONAUTICAL RADIO- NAVIGATION S5.449 RADIOLOCATION G56 US48	5350-5460AERONAUTICAL RADIO- NAVIGATION S5.449 Radiolocation US48	Aviation (87)
5460-5470 RADIONAVIGATION S5.449 Radiolocation	5460-5470 RADIONAVIGATION S5.449 Radiolocation G56 US49 US65	5460-5470 RADIONAVIGATION S5.449 Radiolocation US49 US65	
5470-5650 MARITIME RADIONAVIGATION Radiolocation	5470-5600 MARITIME RADIONAVIGATION Radiolocation G56 US50 US65	5470-5600 MARITIME RADIONAVIGATION Radiolocation US50 US65	Maritime (80)
S5.450 S5.451 S5.452	5600-5650 MARITIME RADIONAVIGATION METEOROLOGICAL AIDS Radiolocation US51 G56 S5.452 US65	5600-5650 MARITIME RADIONAVIGATION METEOROLOGICAL AIDS Radiolocation US51 S5.452 US65	

5650-7250 MHz (SHF)					
International Table			United States Table		Remarks
Region 1	Region 2	Region 3	Federal Government	Non-Federal Government	
5650-5725 RADIOLOCATION Amateur Space research (deep space) S5.282 S5.451 S5.453 S5.454 S5.455			5650-5925 RADIOLOCATION G2	5650-5830 Amateur S5.150 S5.282	ISM Equipment (18) Amateur (97)
5725-5830 FIXED-SATELLITE (Earth-to-space) RADIOLOCATION Amateur S5.150 S5.451 S5.453 S5.455 S5.456	5725-5830 RADIOLOCATION Amateur S5.150 S5.453 S5.455				
5830-5850 FIXED-SATELLITE (Earth-to-space) RADIOLOCATION Amateur Amateur-satellite (space-to-Earth) S5.150 S5.451 S5.453 S5.455 S5.456	5830-5850 RADIOLOCATION Amateur Amateur-satellite (space-to-Earth) S5.150 S5.453 S5.455			5830-5850 Amateur Amateur-satellite (space-to-Earth) S5.150	
5850-5925 FIXED FIXED-SATELLITE (Earth-to-space) MOBILE S5.150	5850-5925 FIXED FIXED-SATELLITE (Earth-to-space) MOBILE Amateur Radiolocation S5.150	5850-5925 FIXED FIXED-SATELLITE (Earth-to-space) MOBILE Radiolocation S5.150	S5.150 US245	5850-5925 FIXED-SATELLITE (Earth-to-space) US245 MOBILE NG160 Amateur S5.150	ISM Equipment (18) Private Land Mobile (90) Amateur (97)
5925-6700 FIXED FIXED-SATELLITE (Earth-to-space) MOBILE			5925-6425	5925-6425 FIXED NG41 FIXED-SATELLITE (Earth-to-space)	International Fixed (23) Satellite Communications (25) Fixed Microwave (101)
			6425-6525 S5.440 S5.458	6425-6525 FIXED-SATELLITE (Earth-to-space) MOBILE S5.440 S5.458	Auxiliary Broadcasting (74) Cable TV Relay (78) Fixed Microwave (101)

S5.149 S5.440 S5.458	6525-6875 S5.458	6525-6875 FIXED FIXED-SATELLITE (Earth-to-space) 792A S5.458	Satellite Communications (25) Fixed Microwave (101)
6700-7075 FIXED FIXED-SATELLITE (Earth-to-space) (space-to-Earth) S5.441 MOBILE S5.458 S5.458A S5.458B S5.458C	6875-7125 S5.458	6875-7075 FIXED FIXED-SATELLITE (Earth-to-space) 792A MOBILE S5.458 NG118	Auxiliary Broadcasting (74) Cable TV Relay (78)
7075-7250 FIXED MOBILE S5.458 S5.459 S5.460		7075-7125 FIXED MOBILE S5.458 NG118	
	7125-7190 FIXED S5.458 US252 G116	7125-7190 S5.458 US252	
	7190-7235 FIXED SPACE RESEARCH (Earth-to-space) S5.458	7190-7250 S5.458	
	7235-7250 FIXED S5.458		

7250-8215 MHz (SHF)					
International Table			United States Table		Remarks
Region 1	Region 2	Region 3	Federal Government	Non-Federal Government	
7250-7300 FIXED FIXED-SATELLITE (space-to-Earth) MOBILE S5.461			7250-7300 FIXED-SATELLITE (space-to-Earth) MOBILE-SATELLITE (space-to-Earth) Fixed G117	7250-8025	
7300-7450 FIXED FIXED-SATELLITE (space-to-Earth) MOBILE except aeronautical mobile S5.461			7300-7450 FIXED FIXED-SATELLITE (space-to-Earth) Mobile-satellite (space-to-Earth) G117		
7450-7550 FIXED FIXED-SATELLITE (space-to-Earth) METEOROLOGICAL-SATELLITE (space-to-Earth) MOBILE except aeronautical mobile S5.461A			7450-7550 FIXED FIXED-SATELLITE (space-to-Earth) METEOROLOGICAL-SAT- ELLITE (space-to-Earth) Mobile-satellite (space-to-Earth) G104 G117		
7550-7750 FIXED FIXED-SATELLITE (space-to-Earth) MOBILE except aeronautical mobile			7550-7750 FIXED FIXED-SATELLITE (space-to-Earth) Mobile-satellite (space-to-Earth) G117		
7750-7850 FIXED METEOROLOGICAL-SATELLITE (space-to-Earth) S5.461B MOBILE except aeronautical mobile			7750-7900 FIXED		
7850-7900 FIXED MOBILE except aeronautical mobile					

7900-8025 FIXED FIXED-SATELLITE (Earth-to-space) MOBILE S5.461	7900-8025 FIXED-SATELLITE (Earth-to-space) MOBILE-SATELLITE (Earth-to-space) Fixed G117		
8025-8175 EARTH EXPLORATION-SATELLITE (space-to-Earth) FIXED FIXED-SATELLITE (Earth-to-space) MOBILE S5.463 S5.462A	8025-8175 EARTH EXPLORATION-SATELLITE (space-to-Earth) FIXED FIXED-SATELLITE (Earth-to-space) Mobile-satellite (Earth-to-space) (no airborne transmissions) US258 G117	8025-8175 US258	
8175-8215 EARTH EXPLORATION-SATELLITE (space-to-Earth) FIXED FIXED-SATELLITE (Earth-to-space) METEOROLOGICAL-SATELLITE (Earth-to-space) MOBILE S5.463 S5.462A	8175-8215 EARTH EXPLORATION-SATELLITE (space-to- Earth) FIXED FIXED-SATELLITE (Earth-to-space) METEOROLOGICAL-SATELLITE (Earth-to-space) Mobile-satellite (Earth-to-space) (no airborne transmissions) US258 G104 G117	8175-8215	

8215-10000 MHz (SHF)					
International Table			United States Table		Remarks
Region 1	Region 2	Region 3	Federal Government	Non-Federal Government	
8215-8400 EARTH EXPLORATION-SATELLITE (space-to-Earth) FIXED FIXED-SATELLITE (Earth-to-space) MOBILE S5.463 S5.462A			8215-8400 EARTH EXPLORATION-SATELLITE (space-to-Earth) FIXED FIXED-SATELLITE (Earth-to-space) Mobile-satellite (Earth-to-space) (no airborne transmissions) US258 G117	8215-8400 US258	
8400-8500 FIXED MOBILE except aeronautical mobile SPACE RESEARCH (space-to-Earth) S5.465 S5.466 S5.467			8400-8450 FIXED SPACE RESEARCH (space-to-Earth) (deep space only)	8400-88450	
			8450-8500 FIXED SPACE RESEARCH (space-to-Earth)	8450-8500 SPACE RESEARCH (space-to-Earth)	
8500-8550 RADIOLOCATION S5.468 S5.469			8500-9000 RADIOLOCATION S5.333 US110 G59	8500-9000 Radiolocation S5.333 US110	
8550-8650 EARTH EXPLORATION-SATELLITE (active) RADIOLOCATION SPACE RESEARCH (active) S5.468 S5.469 S5.469A					
8650-8750 RADIOLOCATION S5.468 S5.469					
8750-8850 RADIOLOCATION AERONAUTICAL RADIONAVIGATION S5.470 S5.471					

8850-9000 RADIOLOCATION MARITIME RADIONAVIGATION S5.472 S5.473	US53	US53	
9000-9200 AERONAUTICAL RADIONAVIGATION S5.337 Radiolocation S5.471	9000-9200 AERONAUTICAL RADIO- NAVIGATION S5.337 Radiolocation G2 US48 US54 G19	9000-9200 AERONAUTICAL RADIO- NAVIGATION S5.337 Radiolocation US48 US54	Aviation (87)
9200-9300 RADIOLOCATION MARITIME RADIONAVIGATION S5.472 S5.473 S5.474	9200-9300 MARITIME RADIO- NAVIGATION S5.472 Radiolocation US110 G59 S5.474	9200-9300 MARITIME RADIO- NAVIGATION S5.472 Radiolocation US110 S5.474	
9300-9500 RADIONAVIGATION S5.476 Radiolocation S5.427 S5.474 S5.475	9300-9500 RADIONAVIGATION S5.476 US66 Radiolocation US51 G56 Meteorological aids S5.427 S5.474 US67 US71	9300-9500 RADIONAVIGATION S5.476 US66 Radiolocation US51 Meteorological aids S5.427 S5.474 US67 US71	
9500-9800 EARTH EXPLORATION-SATELLITE (active) RADIOLOCATION RADIONAVIGATION SPACE RESEARCH (active) S5.476A	9500-10000 RADIOLOCATION S5.333 US110 S5.479	9500-10000 Radiolocation S5.333 US110 S5.479	
9800-10000 RADIOLOCATION Fixed S5.477 S5.478 S5.479			

10-12.7 GHz (SHF)					
International Table			United States Table		Remarks
Region 1	Region 2	Region 3	Federal Government	Non-Federal Government	
10-10.45 FIXED MOBILE RADIOLOCATION Amateur S5.479	10-10.45 RADIOLOCATION Amateur S5.479 S5.480	10-10.45 FIXED MOBILE RADIOLOCATION Amateur S5.479	10-10.45 RADIOLOCATION S5.479 US58 US108 G32	10-10.45 Radiolocation Amateur S5.479 US58 US108 NG42	Private Land Mobile (90) Amateur (97)
10.45-10.5 RADIOLOCATION Amateur Amateur-satellite S5.481			10.45-10.5 RADIOLOCATION US58 US108 G32	10.45-10.5 Radiolocation Amateur Amateur-satellite US58 US108 NG42 NG134	
10.5-10.55 FIXED MOBILE Radiolocation	10.5-10.55 FIXED MOBILE RADIOLOCATION		10.5-10.55 RADIOLOCATION US59		Private Land Mobile (90)
10.55-10.6 FIXED MOBILE except aeronautical mobile Radiolocation			10.55-10.6	10.55-10.6 FIXED	Fixed Microwave (101)
10.6-10.68 EARTH EXPLORATION-SATELLITE (passive) FIXED MOBILE except aeronautical mobile RADIO ASTRONOMY SPACE RESEARCH (passive) Radiolocation S5.149 S5.482			10.6-10.68 EARTH EXPLORATION-SATELLITE (passive) SPACE RESEARCH (passive) US265 US277	10.6-10.68 EARTH EXPLORATION-SATELLITE (passive) FIXED SPACE RESEARCH (passive) US265 US277	
10.68-10.7 EARTH EXPLORATION-SATELLITE (passive) RADIO ASTRONOMY SPACE RESEARCH (passive) S5.340 S5.483			10.68-10.7 EARTH EXPLORATION-SATELLITE (passive) RADIO ASTRONOMY US74 SPACE RESEARCH (passive) US246		
10.7-11.7 FIXED FIXED-SATELLITE (space-to-Earth) S5.441 S5.484A (Earth-to-space) S5.484 MOBILE except aeronautical mobile	10.7-11.7 FIXED FIXED-SATELLITE (space-to-Earth) S5.441 S5.484A MOBILE except aeronautical mobile		10.7-11.7 US211	10.7-11.7 FIXED NG41 FIXED-SATELLITE (space-to-Earth) S5.441 US211 NG104	International Fixed (23) Satellite Communications (25) Fixed Microwave (101)

11.7-12.5 FIXED MOBILE except aeronautical mobile BROADCASTING BROADCASTING-SATELLITE S5.487 S5.487A S5.492	11.7-12.1 FIXED S5.486 FIXED-SATELLITE (space-to-Earth) S5.484A Mobile except aeronautical mobile S5.485 S5.488	11.7-12.2 FIXED MOBILE except aeronautical mobile BROADCASTING BROADCASTING-SATELLITE S5.487 S5.487A S5.492	11.7-12.1 S5.486	11.7-12.2 FIXED-SATELLITE (space-to-Earth) NG143 NG145 Mobile except aeronautical mobile S5.486 S5.488	Satellite Communications (25) Fixed Microwave (101)
	12.1-12.2 FIXED-SATELLITE (space-to-Earth) S5.484A S5.485 S5.488 S5.489		12.1-12.2		
	12.2-12.7 FIXED MOBILE except aeronautical mobile BROADCASTING BROADCASTING-SATELLITE S5.487A S5.488 S5.490 S5.492	12.2-12.5 FIXED MOBILE except aeronautical mobile BROADCASTING S5.484A S5.487 S5.491	12.2-12.7 S5.490	12.2-12.7 FIXED BROADCASTING-SATELLITE S5.488 S5.490	International Fixed (23) Direct Broadcast Satellite (100) Fixed Microwave (101)
12.5-12.75 FIXED-SATELLITE (space-to-Earth) S5.484A (Earth-to-space) S5.494 S5.495 S5.496	See next page for 12.7-12.75 GHz	12.5-12.75 FIXED FIXED-SATELLITE (space-to-Earth) S5.484A MOBILE except aeronautical mobile BROADCASTING-SATELLITE S5.493	See next page for 12.7-12.75 GHz		See next page for 12.7-12.75 GHz

12.7-14.5 GHz (SHF)					
International Table			United States Table		Remarks
Region 1	Region 2	Region 3	Federal Government	Non-Federal Government	
See previous page for 12.5-12.75 GHz	12.7-12.75 FIXED FIXED-SATELLITE (Earth-to-space) MOBILE except aeronautical mobile	See previous page for 12.5-12.75 GHz	12.7-12.75	12.7-12.75 FIXED FIXED-SATELLITE (Earth-to-space) MOBILE NG53 NG118	Auxiliary Broadcasting (74) Cable TV Relay (78) Fixed Microwave (101)
12.75-13.25 FIXED FIXED-SATELLITE (Earth-to-space) S5.441 MOBILE Space research (deep space) (space-to-Earth)			12.75-13.25 US251	12.75-13.25 FIXED FIXED-SATELLITE (Earth-to-space) S5.441 NG104 MOBILE US251 NG53 NG118	
13.25-13.4 EARTH EXPLORATION-SATELLITE (active) AERONAUTICAL RADIONAVIGATION S5.497 SPACE RESEARCH (active) S5.498A S5.499			13.25-13.4 AERONAUTICAL RADIONAVIGATION S5.497 Space research (Earth-to-space)		Aviation (87)
13.4-13.75 EARTH EXPLORATION-SATELLITE (active) RADIOLOCATION SPACE RESEARCH S5.501A Standard frequency and time signal-satellite (Earth-to-space) S5.499 S5.500 S5.501 S5.501B			13.4-13.75 RADIOLOCATION S5.333 US110 G59 Space research Standard frequency and time signal-satellite (Earth-to-space)	13.4-13.75 Radiolocation S5.333 US110 Space research Standard frequency and time signal-satellite (Earth-to-space)	Private Land Mobile (90)
13.75-14 FIXED-SATELLITE (Earth-to-space) S5.484A RADIOLOCATION Standard frequency and time signal-satellite (Earth-to-space) Space research S5.499 S5.500 S5.501 S5.502 S5.503 S5.503A			13.75-14 RADIOLOCATION US110 G59 Standard frequency and time signal-satellite (Earth-to-space) Space research US337 S5.502 S5.503 S5.503A	13.75-14 FIXED-SATELLITE (Earth-to-space) US337 Radiolocation US110 Standard frequency and time signal-satellite (Earth-to-space) Space research S5.502 S5.503 S5.503A	Satellite Communications (25) Private Land Mobile (90)

<table>
<tr><td colspan="3" rowspan="2">14-14.25
FIXED-SATELLITE (Earth-to-space) S5.484A S5.506
RADIONAVIGATION S5.504
Mobile-satellite (Earth-to-space) except aeronautical mobile-satellite
Space research

S5.505</td><td>14-14.2
RADIONAVIGATION US292
Space research</td><td>14-14.2
FIXED-SATELLITE (Earth-to-space)
RADIONAVIGATION US292
Land mobile-satellite (Earth-to-space)
Space research</td><td>Satellite Communications (25)
Maritime (80)
Aviation (87)</td></tr>
<tr><td rowspan="3">14.2-14.4</td><td rowspan="3">14.2-14.4
FIXED-SATELLITE (Earth-to-space)
Land mobile-satellite (Earth-to-space)
Mobile except aeronautical mobile</td><td rowspan="3">Satellite Communications (25)
Fixed Microwave (101)</td></tr>
<tr><td colspan="3">14.25-14.3
FIXED-SATELLITE (Earth-to-space) S5.484A S5.506
RADIONAVIGATION S5.504
Mobile-satellite (Earth-to-space) except aeronautical mobile-satellite
Space research

S5.505 S5.508 S5.509</td></tr>
<tr><td>14.3-14.4
FIXED
FIXED-SATELLITE (Earth-to- space) S5.484A S5.506
MOBILE except aeronautical mobile
Mobile-satellite (Earth-to-space) except aeronautical mobile-satellite
Radionavigation-satellite</td><td>14.3-14.4
FIXED-SATELLITE (Earth-to-space) S5.484A S5.506
Mobile-satellite (Earth-to-space) except aeronautical mobile-satellite
Radionavigation-satellite</td><td>14.3-14.4
FIXED
FIXED-SATELLITE (Earth-to-space) S5.484A S5.506
MOBILE except aeronautical mobile
Mobile-satellite (Earth-to-space) except aeronautical mobile-satellite
Radionavigation-satellite</td></tr>
<tr><td colspan="3">14.4-14.47
FIXED
FIXED-SATELLITE (Earth-to-space) S5.484A S5.506
MOBILE except aeronautical mobile
Mobile-satellite (Earth-to-space) except aeronautical mobile-satellite
Space research (space-to-Earth)</td><td>14.4-14.47
Fixed
Mobile</td><td>14.4-14.47
FIXED-SATELLITE (Earth-to-space)
Land mobile-satellite (Earth-to-space)</td><td rowspan="2">Satellite Communications (25)</td></tr>
<tr><td colspan="3">14.47-14.5
FIXED
FIXED-SATELLITE (Earth-to-space) S5.484A S5.506
MOBILE except aeronautical mobile
Mobile-satellite (Earth-to-space) except aeronautical mobile-satellite
Radio astronomy

S5.149</td><td>14.47-14.5
Fixed
Mobile

S5.149 US203</td><td>14.47-14.5
FIXED-SATELLITE (Earth-to-space)
Land mobile-satellite (Earth-to-space)

S5.149 US203</td></tr>
</table>

14.5-18.3 GHz (SHF)					
International Table			United States Table		Remarks
Region 1	Region 2	Region 3	Federal Government	Non-Federal Government	
14.5-14.8 FIXED FIXED-SATELLITE (Earth-to-space) S5.510 MOBILE Space research			14.5-14.7145 FIXED Mobile Space research	14.5-15.1365	
14.8-15.35 FIXED MOBILE Space research S5.339			14.7145-15.1365 MOBILE Fixed Space research US310	14.7145-15.1365 US310	
			15.1365-15.35 FIXED Mobile Space research S5.339 US211	15.1365-15.35 S5.339 US211	
15.35-15.4 EARTH EXPLORATION-SATELLITE (passive) RADIO ASTRONOMY SPACE RESEARCH (passive) S5.340 S5.511			15.35-15.4 EARTH EXPLORATION-SATELLITE (passive) RADIO ASTRONOMY US74 SPACE RESEARCH (passive) US246		
15.4-15.43 AERONAUTICAL RADIONAVIGATION S5.511D			15.4-15.7 AERONAUTICAL RADIONAVIGATION US260 733 797 US211		Aviation (87)
15.43-15.63 FIXED SATELLITE (space-to-Earth) (Earth-to-space) S5.511A AERONAUTICAL RADIONAVIGATION S5.511C					
15.63-15.7 AERONAUTICAL RADIONAVIGATION S5.511D					
15.7-16.6 RADIOLOCATION S5.512 S5.513			15.7-16.6 RADIOLOCATION US110 G59	15.7-17.2 Radiolocation US110	Private Land Mobile (90)

16.6-17.1 RADIOLOCATION Space research (deep space) (Earth-to-space) S5.512 S5.513			16.6-17.1 RADIOLOCATION US110 G59 Space research (deep space) (Earth-to-space)		
17.1-17.2 RADIOLOCATION S5.512 S5.513			17.1-17.2 RADIOLOCATION US110 G59		
17.2-17.3 EARTH EXPLORATION-SATELLITE (active) RADIOLOCATION SPACE RESEARCH (active) S5.512 S5.513 S5.513A			17.2-17.3 RADIOLOCATION US110 G59 Earth exploration-satellite (active) Space research (active)	17.2-17.3 Radiolocation US110 Earth exploration-satellite (active) Space research (active)	
17.3-17.7 FIXED-SATELLITE (Earth-to-space) S5.516 Radiolocation S5.514	17.3-17.7 FIXED-SATELLITE (Earth-to-space) S5.516 BROADCASTING- SATELLITE Radiolocation S5.514 S5.515 S5.517	17.3-17.7 FIXED-SATELLITE (Earth-to-space) S5.516 Radiolocation S5.514	17.3-17.7 Radiolocation US259 G59	17.3-17.7 FIXED-SATELLITE (Earth-to-space) US271 BROADCASTING- SATELLITE NG163 US259	Satellite Communications (25) Direct Broadcast Satellite (100)
17.7-18.1 FIXED FIXED-SATELLITE (space-to-Earth) S5.484A (Earth-to-space) S5.516 MOBILE	17.7-17.8 FIXED FIXED-SATELLITE (space-to-Earth) (Earth-to-space) S5.516 BROADCASTING- SATELLITE Mobile S5.518 S5.515 S5.517	17.7-18.1 FIXED FIXED-SATELLITE (space-to-Earth) S5.484A (Earth-to-space) S5.516 MOBILE	17.7-17.8	17.7-17.8 FIXED FIXED-SATELLITE (Earth-to-space) US271 NG144	Satellite Communications (25) Auxiliary Broadcasting (74) Cable TV Relay (78) Fixed Microwave (101)
	17.8-18.1 FIXED FIXED-SATELLITE (space-to-Earth) S5.484A (Earth-to-space) S5.516 MOBILE		17.8-18.3 FIXED-SATELLITE (space-to-Earth) G117 S5.519 US334	17.8-18.3 FIXED S5.519 US334 NG144	Auxiliary Broadcasting (74) Cable TV Relay (78) Fixed Microwave (101)
18.1-18.4 FIXED FIXED-SATELLITE (space-to-Earth) S5.484A (Earth-to-space) S5.520 MOBILE S5.519 S5.521			See next page for 18.3-18.6 GHz	See next page for 18.3-18.58 GHz	See next page for 18.3-18.58 GHz

<table>
<tr><th colspan="6">18.3-22.5 GHz (SHF)</th></tr>
<tr><th colspan="3">International Table</th><th colspan="2">United States Table</th><th rowspan="2">Remarks</th></tr>
<tr><th>Region 1</th><th>Region 2</th><th>Region 3</th><th>Federal Government</th><th>Non-Federal Government</th></tr>
<tr><td colspan="3">See previous page for 18.1-18.4 GHz</td><td rowspan="2">18.3-18.6
FIXED-SATELLITE
(space-to-Earth) G117

US334</td><td>18.3-18.58
FIXED
FIXED-SATELLITE
(space- to-Earth) NG164

US334 NG144</td><td>Satellite Communications (25)
Auxiliary Broadcast. (74)
Cable TV Relay (78)
Fixed Microwave (101)</td></tr>
<tr><td colspan="3">18.4-18.6
FIXED
FIXED-SATELLITE (space-to-Earth) S5.484A
MOBILE</td><td>18.58-18.6
FIXED-SATELLITE
(space- to-Earth) NG164

US334 NG144</td><td rowspan="3">Satellite Communications (25)</td></tr>
<tr><td>18.6-18.8
FIXED
FIXED-SATELLITE
(space-to-Earth) S5.523
MOBILE except
aeronautical mobile
Earth exploration-satellite
(passive)
Space research (passive)

S5.522</td><td>18.6-18.8
EARTH EXPLORATION-
SATELLITE (passive)
FIXED
FIXED-SATELLITE
(space-to-Earth) S5.523
MOBILE except
aeronautical mobile
SPACE RESEARCH
(passive)

S5.222</td><td>18.6-18.8
FIXED
FIXED-SATELLITE
(space-to-Earth) S5.523
MOBILE except
aeronautical mobile
Space research (passive)

S5.522</td><td>18.6-18.8
EARTH EXPLORATION-
SATELLITE (passive)
FIXED-SATELLITE
(space-to-Earth) US255
G117
SPACE RESEARCH
(passive)

US254 US334</td><td>18.6-18.8
EARTH EXPLORATION-
SATELLITE (passive)
FIXED-SATELLITE
(space-to-Earth) US255
NG164
SPACE RESEARCH
(passive)

US254 US334 NG144</td></tr>
<tr><td colspan="3">18.8-19.3
FIXED
FIXED-SATELLITE (space-to-Earth) S5.523A
MOBILE</td><td rowspan="3">18.8-20.2
FIXED-SATELLITE
(space-to-Earth) G117</td><td>18.8-19.3
FIXED-SATELLITE
(space-to-Earth) NG165

US334 NG144</td></tr>
<tr><td colspan="3">19.3-19.7
FIXED
FIXED-SATELLITE (space-to-Earth) (Earth-space) S5.523B S5.523C S5.523D S5.523E
MOBILE</td><td>19.3-19.7
FIXED
FIXED-SATELLITE
(space- to-Earth) NG166

US334 NG144</td><td>Satellite Communications (25)
Auxiliary Broadcast (74)
Cable TV Relay (78)
Fixed Microwave (101)</td></tr>
<tr><td>19.7-20.1
FIXED-SATELLITE
(space-to-Earth) S5.484A
Mobile-satellite
(space-to-Earth)

S5.524</td><td>19.7-20.1
FIXED-SATELLITE
(space-to-Earth) S5.484A
MOBILE-SATELLITE
(space-to-Earth)

S5.524 S5.525 S5.526 S5.527
S5.528 S5.529</td><td>19.7-20.1
FIXED-SATELLITE
(space-to-Earth) S5.484A
Mobile-satellite
(space-to-Earth)

S5.524</td><td>19.7-20.1
FIXED-SATELLITE
(space-to-Earth)
MOBILE-SATELLITE
(space-to-Earth)

S5.525 S5.526 S5.527
S5.528 S5.529 US334</td><td>Satellite Communications (25)</td></tr>
</table>

20.1-20.2 FIXED-SATELLITE (space-to-Earth) S5.484A MOBILE-SATELLITE (space-to-Earth) S5.524 S5.525 S5.526 S5.527 S5.528			US334	20.1-20.2 FIXED-SATELLITE (space-to-Earth) MOBILE-SATELLITE (space-to-Earth) S5.525 S5.526 S5.527 S5.528 US334	
20.2-21.2 FIXED-SATELLITE (space-to-Earth) MOBILE-SATELLITE (space-to-Earth) Standard frequency and time signal-satellite (space-to-Earth) S5.524			20.2-21.2 FIXED-SATELLITE (space-to-Earth) MOBILE-SATELLITE (space-to-Earth) Standard frequency and time signal-satellite (space-to-Earth) G117	20.2-21.2 Standard frequency and time signal-satellite (space-to-Earth)	
21.2-21.4 EARTH EXPLORATION-SATELLITE (passive) FIXED MOBILE SPACE RESEARCH (passive)			21.2-21.4 EARTH EXPLORATION-SATELLITE (passive) FIXED MOBILE SPACE RESEARCH (passive) US263		Fixed Microwave (101)
21.4-22 FIXED MOBILE BROADCASTING- SATELLITE S5.530	21.4-22 FIXED MOBILE	21.4-22 FIXED MOBILE BROADCASTING- SATELLITE S5.530 S5.531	21.4-22 FIXED MOBILE		
22-22.21 FIXED MOBILE except aeronautical mobile S5.149			22-22.21 FIXED MOBILE except aeronautical mobile S5.149		
22.21-22.5 EARTH EXPLORATION-SATELLITE (passive) FIXED MOBILE except aeronautical mobile RADIO ASTRONOMY SPACE RESEARCH (passive) S5.149 S5.532			22.21-22.5 EARTH EXPLORATION-SATELLITE (passive) FIXED MOBILE except aeronautical mobile RADIO ASTRONOMY SPACE RESEARCH (passive) S5.149 US263		

22.5-27.5 GHz (SHF)					
International Table			United States Table		Remarks
Region 1	Region 2	Region 3	Federal Government	Non-Federal Government	
22.5-22.55 FIXED MOBILE			22.5-22.55 FIXED MOBILE US211		Fixed Microwave (101)
22.55-23.55 FIXED INTER-SATELLITE MOBILE S5.149			22.55-23.55 FIXED INTER-SATELLITE MOBILE S5.149 US278		Satellite Communications (25) Fixed Microwave (101)
23.55-23.6 FIXED MOBILE			23.55-23.6 FIXED MOBILE		Fixed Microwave (101)
23.6-24 EARTH EXPLORATION-SATELLITE (passive) RADIO ASTRONOMY SPACE RESEARCH (passive) S5.340			23.6-24 EARTH EXPLORATION-SATELLITE (passive) RADIO ASTRONOMY US74 SPACE RESEARCH (passive) US246		
24-24.05 AMATEUR AMATEUR-SATELLITE S5.150			24-24.05 S5.150 US211	24-24.05 AMATEUR AMATEUR-SATELLITE S5.150 US211	ISM Equipment (18) Amateur (97)
24.05-24.25 RADIOLOCATION Amateur Earth exploration-satellite (active) S5.150			24.05-24.25 RADIOLOCATION US110 G59 Earth exploration-satellite (active) S5.150	24.05-24.25 Radiolocation US110 Amateur Earth exploration-satellite (active) S5.150	ISM Equipment (18) Private Land Mobile (90) Amateur (97)
24.25-24.45 FIXED	24.25-24.45 RADIONAVIGATION	24.25-24.45 RADIONAVIGATION FIXED MOBILE	24.25-24.45	24.25-24.45 FIXED	Fixed Microwave (101)
24.45-24.75 FIXED INTER-SATELLITE	24.45-24.65 INTER-SATELLITE RADIONAVIGATION S5.533	24.45-24.65 FIXED INTER-SATELLITE MOBILE RADIONAVIGATION S5.533	24.45-24.65 INTER-SATELLITE RADIONAVIGATION S5.533		Satellite Communications (25)

	24.65-24.75 INTER-SATELLITE RADIOLOCATION-SAT-ELLITE (Earth-to-space)	24.65-24.75 FIXED INTER-SATELLITE MOBILE S5.533 S5.534	24.65-24.75 INTER-SATELLITE RADIOLOCATION-SATELLITE (Earth-to-space)		
24.75-25.25 FIXED	24.75-25.25 FIXED-SATELLITE (Earth-to-space) S5.535	24.75-25.25 FIXED FIXED-SATELLITE (Earth-to-space) S5.535 MOBILE S5.534	24.75-25.05 RADIONAVIGATION	24.75-25.05 FIXED-SATELLITE (Earth-to-space) NG167 RADIONAVIGATION	Satellite Communications (25) Aviation (87)
			25.05-25.25	25.05-25.25 FIXED-SATELLITE (Earth-to-space) NG167 FIXED	Satellite Communications (25) Fixed Microwave (101)
25.25-25.5 FIXED INTER-SATELLITE S5.536 MOBILE Standard frequency and time signal-satellite (Earth-to-space)			25.25-25.5 FIXED INTER-SATELLITE S5.536 MOBILE Standard frequency and time signal-satellite (Earth-to-space)	25.25-27 Standard frequency and time signal-satellite (Earth-to-space) Earth exploration-satellite (space-to-space)	Note: In this manual a primary allocation has been added for the inter-satellite service in the bands 25.25-25.5, 25.5.-27 and 27-27.5 GHz. The footnote S5.536 changes the direction indicator for the earth-exploration satellite service allocation in the 25.5-27 GHz band from space-to-space to space-to-Earth. At this time, the FCC has not adopted these changes.
25.5-27 EARTH EXPLORATION-SATELLITE (space-to-Earth) S5.536A S5.536B FIXED INTER-SATELLITE S5.536 MOBILE Standard frequency and time signal-satellite (Earth-to-space)			25.5-27 FIXED INTER-SATELLITE S5.536 MOBILE Earth exploration-satellite (space-to-Earth) Standard frequency and time signal-satellite (Earth-to- space)		
27-27.5 FIXED INTER-SATELLITE S5.536 MOBILE	27-27.5 FIXED FIXED-SATELLITE (Earth-to-space) INTER-SATELLITE S5.536 S5.537 MOBILE		27-27.5 FIXED INTER-SATELLITE S5.536 MOBILE	27-27.5 Earth exploration-satellite (space-to-space)	

<table>
<tr><th colspan="6">27.5-32 GHz (SHF/EHF)</th></tr>
<tr><th colspan="3">International Table</th><th colspan="2">United States Table</th><th rowspan="2">Remarks</th></tr>
<tr><th>Region 1</th><th>Region 2</th><th>Region 3</th><th>Federal Government</th><th>Non-Federal Government</th></tr>
<tr><td colspan="3">27.5-28.5
FIXED
FIXED-SATELLITE (Earth-to-space) S5.484A S5.539
MOBILE

S5.538 S5.540</td><td rowspan="5">27.5-30</td><td rowspan="3">27.5-29.5
FIXED
FIXED-SATELLITE
(Earth-to-space)
MOBILE</td><td rowspan="3">Satellite
Communications (25)
Fixed Microwave (101)</td></tr>
<tr><td colspan="3">28.5-29.1
FIXED
FIXED-SATELLITE (Earth-to-space) S5.484A S5.523A S5.539
MOBILE
Earth exploration-satellite (Earth-to-space) S5.541

S5.540</td></tr>
<tr><td colspan="3">29.1-29.5
FIXED
FIXED-SATELLITE (Earth-to-space) S5.523C S5.523E S5.535A S5.539 S5.541A
MOBILE
Earth exploration-satellite (Earth-to-space) S5.541

S5.540</td></tr>
<tr><td>29.5-29.9
FIXED-SATELLITE (Earth-to-
space) S5.484A S5.539
Earth exploration-satellite
(Earth-to-space) S5.541
Mobile-satellite
(Earth-to-space)

S5.540 S5.542</td><td>29.5-29.9
FIXED-SATELLITE (Earth-to-space) S5.484A S5.539
MOBILE-SATELLITE
(Earth-to-space)
Earth exploration-satellite
(Earth-to-space) S5.541

S5.525 S5.526 S5.527 S5.529
S5.540 S5.542</td><td>29.5-29.9
FIXED-SATELLITE (Earth-to-space) S5.484A S5.539
Earth exploration-satellite
(Earth-to-space) S5.541
Mobile-satellite
(Earth-to-space)

S5.540 S5.542</td><td>29.5-29.9
FIXED-SATELLITE
(Earth-to-space)
MOBILE-SATELLITE
(Earth-to-space)

S5.525 S5.526 S5.527
S5.529</td><td rowspan="2">Satellite
Communications (25)</td></tr>
<tr><td colspan="3">29.9-30
FIXED-SATELLITE (Earth-to-space) S5.484A S5.539
MOBILE-SATELLITE (Earth-to-space)
Earth exploration-satellite (Earth-to-space) S5.541 S5.543

S5.525 S5.526 S5.527 S5.538 S5.540 S5.542</td><td>29.9-30
FIXED-SATELLITE
(Earth-to-space)
MOBILE-SATELLITE
(Earth-to-space)

S5.525 S5.526 S5.527
S5.543</td></tr>
</table>

<table>
<tr>
<td colspan="3">30-31
FIXED-SATELLITE (Earth-to-space)
MOBILE-SATELLITE (Earth-to-space)
Standard frequency and time signal-satellite (space-to-Earth)

S5.542</td>
<td>30-31
FIXED-SATELLITE
(Earth-to-space)
MOBILE-SATELLITE
(Earth-to-space)
Standard frequency and time signal-satellite
(space-to- Earth)

G117</td>
<td>30-31
Standard frequency and time signal-satellite
(space-to- Earth)</td>
<td></td>
</tr>
<tr>
<td colspan="3">31-31.3
FIXED
MOBILE
Standard frequency and time signal-satellite (space-to-Earth)
Space research S5.544 S5.545

S5.149</td>
<td>31-31.3
Standard frequency and time signal-satellite (space-to-Earth)

S5.149 US211</td>
<td>31-31.3
FIXED
MOBILE
Standard frequency and time signal-satellite (space-to-Earth)

S5.149 US211</td>
<td>Fixed Microwave (101)</td>
</tr>
<tr>
<td colspan="3">31.3-31.5
EARTH EXPLORATION-SATELLITE (passive)
RADIO ASTRONOMY
SPACE RESEARCH (passive)

S5.340</td>
<td colspan="2" rowspan="2">31.3-31.8
EARTH EXPLORATION-SATELLITE (passive)
RADIO ASTRONOMY US74
SPACE RESEARCH (passive)

US246</td>
<td rowspan="2"></td>
</tr>
<tr>
<td>31.5-31.8
EARTH EXPLORATION-SATELLITE (passive)
RADIO ASTRONOMY
SPACE RESEARCH (passive)
Fixed
Mobile except aeronautical mobile

S5.149 S5.546</td>
<td>31.5-31.8
EARTH EXPLORATION-SATELLITE (passive)
RADIO ASTRONOMY
SPACE RESEARCH (passive)

S5.340</td>
<td>31.5-31.8
EARTH EXPLORATION-SATELLITE (passive)
RADIO ASTRONOMY
SPACE RESEARCH (passive)
Fixed
Mobile except aeronautical mobile

S5.149</td>
</tr>
<tr>
<td colspan="3">31.8-32
FIXED S5.547A
RADIONAVIGATION
SPACE RESEARCH (deep space) (space-to-Earth)

S5.547 S5.547B S5.548</td>
<td>31.8-32
RADIONAVIGATION US69
SPACE RESEARCH (deep space)(space-to-Earth) US262

S5.548 US211</td>
<td>31.8-32
SPACE RESEARCH (deep space)(space-to-Earth)
US262

S5.548 US211</td>
<td></td>
</tr>
</table>

32-40 GHz (EHF)					
International Table			United States Table		Remarks
Region 1	Region 2	Region 3	Federal Government	Non-Federal Government	
32-32.3 FIXED S5.547A INTER-SATELLITE RADIONAVIGATION SPACE RESEARCH (deep space) (space-to-Earth) S5.547 S5.547C S5.548			32-32.3 INTER-SATELLITE US278 RADIONAVIGATION US69 SPACE RESEARCH (deep space)(space-to-Earth) US262 S5.548	32-32.3 INTER-SATELLITE US278 SPACE RESEARCH (deep space)(space-to-Earth) US262 S5.548	
32.3-33 FIXED S5.547A INTER-SATELLITE RADIONAVIGATION S5.547 S5.547D S5.548			32.3-33 INTER-SATELLITE US278 RADIONAVIGATION US69 S5.548		
33-33.4 FIXED S5.547A RADIONAVIGATION S5.547 S5.547E			33-33.4 RADIONAVIGATION US69		
33.4-34.2 RADIOLOCATION S5.549			33.4-36 RADIOLOCATION US110 G34	33.4-36 Radiolocation US110	Private Land Mobile (90)
34.2-34.7 RADIOLOCATION SPACE RESEARCH (deep space) (Earth-to-space) S5.549					
34.7-35.2 RADIOLOCATION Space research S5.550 S5.549					
35.2-35.5 METEOROLOGICAL AIDS RADIOLOCATION S5.549					

35.5-36 METEOROLOGICAL AIDS EARTH EXPLORATION-SATELLITE (active) RADIOLOCATION SPACE RESEARCH (active) S5.549 S5.551A	S5.551 US252	S5.551 US252	
36-37 EARTH EXPLORATION-SATELLITE (passive) FIXED MOBILE SPACE RESEARCH (passive) S5.149	36-37 EARTH EXPLORATION-SATELLITE (passive) FIXED MOBILE SPACE RESEARCH (passive) US263 US342		
37-37.5 FIXED MOBILE SPACE RESEARCH (space-to-Earth)	37-38 FIXED MOBILE SPACE RESEARCH (space-to-Earth)	37-37.6 FIXED MOBILE	
37.5-38 FIXED FIXED-SATELLITE (space-to-Earth) MOBILE SPACE RESEARCH (space-to-Earth) Earth exploration-satellite (space-to-Earth)		37.6-38.6 FIXED FIXED-SATELLITE (space-to-Earth) MOBILE	Satellite Communications (25)
38-39.5 FIXED FIXED-SATELLITE (space-to-Earth) MOBILE Earth exploration-satellite (space-to-Earth)	38-38.6 FIXED MOBILE		
	38.6-39.5 US291	38.6-39.5 FIXED FIXED-SATELLITE (space-to-Earth) MOBILE US291	Auxiliary Broadcasting (74) Fixed Microwave (101)
39.5-40 FIXED FIXED-SATELLITE (space-to-Earth) MOBILE MOBILE-SATELLITE (space-to-Earth) Earth exploration-satellite (space-to-Earth)	39.5-40 FIXED-SATELLITE (space-to-Earth) MOBILE-SATELLITE (space-to-Earth) US291 G117	39.5-40 FIXED FIXED-SATELLITE (space-to-Earth) MOBILE MOBILE-SATELLITE (space-to-Earth) US291	

40-50.2 GHz (EHF)					
International Table			United States Table		Remarks
Region 1	Region 2	Region 3	Federal Government	Non-Federal Government	
40-40.5 EARTH EXPLORATION-SATELLITE (Earth-to-space) FIXED FIXED-SATELLITE (space-to-Earth) MOBILE MOBILE-SATELLITE (space-to-Earth) SPACE RESEARCH (Earth-to-space) Earth exploration-satellite (space-to-Earth)			40-40.5 EARTH EXPLORATION-SATELLITE (Earth-to- space) FIXED-SATELLITE (space-to-Earth) MOBILE-SATELLITE (space-to-Earth) SPACE RESEARCH (Earth-to-space) Earth exploration-satellite (space-to-Earth) G117	40-40.5 FIXED-SATELLITE (space-to-Earth) MOBILE-SATELLITE (space-to-Earth)	Satellite Communications (25)
40.5-42.5 FIXED BROADCASTING BROADCASTING-SATELLITE Mobile S5.551B S5.551D	40.5-42.5 FIXED FIXED-SATELLITE (space-to-Earth) S5.551B S5.551E BROADCASTING BROADCASTING-SATELLITE Mobile S5.551C S5.551F		40.5-42.5 US211	40.5-41 FIXED-SATELLITE (space-to-Earth) BROADCASTING BROADCASTING-SATELLITE Mobile Fixed US211	
				41-42.5 FIXED BROADCASTING BROADCASTING-SATELLITE MOBILE US211	
42.5-43.5 FIXED FIXED-SATELLITE (Earth-to-space) S5.552 MOBILE except aeronautical mobile RADIO ASTRONOMY S5.149			42.5-43.5 FIXED FIXED-SATELLITE (Earth- to-space) MOBILE except aeronautical mobile RADIO ASTRONOMY US342	42.5-43.5 RADIO ASTRONOMY US342	

43.5-47 MOBILE S5.553 MOBILE-SATELLITE RADIONAVIGATION RADIONAVIGATION-SATELLITE S5.554	43.5-45.5 FIXED-SATELLITE (Earth-to-space) MOBILE-SATELLITE (Earth-to-space) G117	43.5-45.5	
	45.5-46.9 MOBILE MOBILE-SATELLITE (Earth-to-space) RADIONAVIGATION-SATELLITE S5.554		RF Devices (15)
	46.9-47 MOBILE MOBILE-SATELLITE (Earth-to-space) RADIONAVIGATION-SATELLITE S5.554	46.9-47 MOBILE MOBILE-SATELLITE (Earth-to-space) RADIONAVIGATION-SATELLITE FIXED S5.554	
47-47.2 AMATEUR AMATEUR-SATELLITE	47-48.2	47-47.2 AMATEUR AMATEUR-SATELLITE	Amateur (97)
47.2-50.2 FIXED FIXED-SATELLITE (Earth-to-space) S5.552 MOBILE S5.149 S5.340 S5.552A S5.555		47.2-48.2 FIXED FIXED-SATELLITE (Earth-to-space) US297 MOBILE US264 S5.555	
	48.2-50.2 FIXED FIXED-SATELLITE (Earth-to-space) US297 MOBILE US264 S5.555 US342		Satellite Communications (25)

50.2-65 GHz (EHF)					
International Table			United States Table		Remarks
Region 1	Region 2	Region 3	Federal Government	Non-Federal Government	
50.2-50.4 EARTH EXPLORATION-SATELLITE (passive) SPACE RESEARCH (passive) S5.340 S5.555A			50.2-50.4 EARTH EXPLORATION-SATELLITE (passive) FIXED MOBILE SPACE RESEARCH (passive) US263		
50.4-51.4 FIXED FIXED-SATELLITE (Earth-to-space) MOBILE Mobile-satellite (Earth-to-space)			50.4-51.4 FIXED FIXED-SATELLITE (Earth-to-space) MOBILE MOBILE-SATELLITE (Earth-to-space) G117	50.4-51.4 FIXED FIXED-SATELLITE (Earth-to-space) MOBILE MOBILE-SATELLITE (Earth-to-space)	
51.4-52.6 FIXED MOBILE S5.547 S5.556			51.4-54.25 EARTH EXPLORATION-SATELLITE (passive) SPACE RESEARCH (passive) RADIO ASTRONOMY US246		
52.6-54.25 EARTH EXPLORATION-SATELLITE (passive) SPACE RESEARCH (passive) S5.340 S5.556					
54.25-55.78 EARTH EXPLORATION-SATELLITE (passive) INTER-SATELLITE S5.556A SPACE RESEARCH (passive) S5.556B			54.25-58.2 EARTH EXPLORATION-SATELLITE (passive) FIXED INTER-SATELLITE MOBILE 909 SPACE RESEARCH (passive)		
55.78-56.9 EARTH EXPLORATION-SATELLITE (passive) FIXED INTER-SATELLITE S5.556A MOBILE S5.558 SPACE RESEARCH (passive) S5.547 S5.557					

56.9-57 EARTH EXPLORATION-SATELLITE (passive) FIXED INTER-SATELLITE S5.558A MOBILE S5.558 SPACE RESEARCH (passive) S5.547 S5.557	US263	
57-58.2 EARTH EXPLORATION-SATELLITE (passive) FIXED INTER-SATELLITE S5.556A MOBILE S5.558 SPACE RESEARCH (passive) S5.547 S5.557		
58.2-59 EARTH EXPLORATION-SATELLITE (passive) FIXED MOBILE SPACE RESEARCH (passive) S5.547 S5.556	58.2-59 EARTH EXPLORATION-SATELLITE (passive) SPACE RESEARCH (passive) RADIO ASTRONOMY US246	
59-59.3 EARTH EXPLORATION-SATELLITE (passive) FIXED INTER-SATELLITE S5.556A MOBILE S5.558 RADIOLOCATION S5.559 SPACE RESEARCH (passive)	59-64 FIXED INTER-SATELLITE MOBILE 909 RADIOLOCATION S5.138	RF Devices (15) ISM Equipment (18)
59.3-64 FIXED INTER-SATELLITE MOBILE S5.558 RADIOLOCATION S5.559 S5.138		
64-65 FIXED INTER-SATELLITE MOBILE except aeronautical mobile S5.547 S5.556	64-65 EARTH EXPLORATION-SATELLITE (passive) RADIO ASTRONOMY SPACE RESEARCH (passive) US246	

65-95 GHz (EHF)					
International Table			United States Table		Remarks
Region 1	Region 2	Region 3	Federal Government	Non-Federal Government	
65-66 EARTH EXPLORATION-SATELLITE FIXED INTER-SATELLITE MOBILE except aeronautical mobile SPACE RESEARCH S5.547			65-66 EARTH EXPLORATION-SATELLITE SPACE RESEARCH Fixed Mobile		
66-71 INTER-SATELLITE MOBILE S5.553 S5.558 MOBILE-SATELLITE RADIONAVIGATION RADIONAVIGATION-SATELLITE S5.554			66-71 MOBILE S5.553 MOBILE-SATELLITE RADIONAVIGATION RADIONAVIGATION-SATELLITE S5.554		
71-74 FIXED FIXED-SATELLITE (Earth-to-space) MOBILE MOBILE-SATELLITE (Earth-to-space) S5.149 S5.556			71-74 FIXED FIXED-SATELLITE (Earth-to-space) MOBILE MOBILE-SATELLITE (Earth-to-space) US270		
74-75.5 FIXED FIXED-SATELLITE (Earth-to-space) MOBILE Space research (space-to-Earth)			74-75.5 FIXED FIXED-SATELLITE (Earth-to-space) US297 MOBILE		
75.5-76 AMATEUR AMATEUR-SATELLITE Space research (space-to-Earth)			75.5-76	75.5-76 AMATEUR AMATEUR-SATELLITE	Amateur (97)
76-81 RADIOLOCATION Amateur Amateur-satellite Space research (space-to-Earth)			76-81 RADIOLOCATION	76-77 RADIOLOCATION Amateur	RF Devices (15)
				77-77.5 RADIOLOCATION Amateur Amateur-satellite	Amateur (97)

S5.560	S5.560	77.5-78 RADIOLOCATION AMATEUR AMATEUR-SATELLITE 78-81 RADIOLOCATION Amateur Amateur-satellite S5.560	
81-84 FIXED FIXED-SATELLITE (space-to-Earth) MOBILE MOBILE-SATELLITE (space-to-Earth) Space research (space-to-Earth)	81-84 FIXED FIXED-SATELLITE (space-to-Earth) MOBILE MOBILE-SATELLITE (space-to-Earth)		
84-86 FIXED MOBILE BROADCASTING BROADCASTING-SATELLITE S5.561	84-86 FIXED MOBILE S5.561 US211	84-86 FIXED MOBILE BROADCASTING BROADCASTING- SATELLITE S5.561 US211	
86-92 EARTH EXPLORATION-SATELLITE (passive) RADIO ASTRONOMY SPACE RESEARCH (passive) S5.340	86-92 EARTH EXPLORATION-SATELLITE (passive) RADIO ASTRONOMY US74 SPACE RESEARCH (passive) US246		
92-94 FIXED FIXED-SATELLITE (Earth-to-space) MOBILE RADIOLOCATION S5.149 S5.556 94-94.1 EARTH EXPLORATION-SATELLITE (active) RADIOLOCATION SPACE RESEARCH (active) S5.562 See next page for 94.1-95 GHz	92-95 FIXED FIXED-SATELLITE (Earth-to-space) MOBILE RADIOLOCATION S5.149		

95-150 GHz (EHF)					
International Table			United States Table		Remarks
Region 1	Region 2	Region 3	Federal Government	Non-Federal Government	
94.1-95 FIXED FIXED-SATELLITE (Earth-to-space) MOBILE RADIOLOCATION			See previous page for 92-95 GHz		See previous page for 92-95 GHz
95-100 MOBILE S5.553 MOBILE-SATELLITE RADIONAVIGATION RADIONAVIGATION-SATELLITE Radiolocation S5.149 S5.554 S5.555			95-100 MOBILE S5.553 MOBILE-SATELLITE RADIONAVIGATION RADIONAVIGATION-SATELLITE Radiolocation S5.149 S5.554		
100-102 EARTH EXPLORATION-SATELLITE (passive) FIXED MOBILE SPACE RESEARCH (passive) S5.341			100-102 EARTH EXPLORATION-SATELLITE (passive) SPACE RESEARCH (passive) S5.341 US246		
102-105 FIXED FIXED-SATELLITE (space-to-Earth) MOBILE S5.341			102-105 FIXED FIXED-SATELLITE (space-to-Earth) S5.341 US211		
105-116 EARTH EXPLORATION-SATELLITE (passive) RADIO ASTRONOMY SPACE RESEARCH (passive) S5.340 S5.341			105-116 EARTH EXPLORATION-SATELLITE (passive) RADIO ASTRONOMY US74 SPACE RESEARCH (passive) S5.341 US246		
116-119.98 EARTH EXPLORATION-SATELLITE (passive) FIXED INTER-SATELLITE MOBILE S5.558 SPACE RESEARCH (passive) S5.341			116-119.98 EARTH EXPLORATION-SATELLITE (passive) FIXED INTER-SATELLITE MOBILE S5.558 SPACE RESEARCH (passive) S5.341 US211 US263		

<table>
<tr><td>119.98-120.02
EARTH EXPLORATION-SATELLITE (passive)
FIXED
INTER-SATELLITE
MOBILE S5.558
SPACE RESEARCH (passive)
Amateur

S5.341</td><td colspan="2">119.98-120.02
EARTH EXPLORATION-SATELLITE (passive)
FIXED
INTER-SATELLITE
MOBILE S5.558
SPACE RESEARCH (passive)
Amateur

S5.341 US211 US263</td><td></td></tr>
<tr><td>120.02-126
EARTH EXPLORATION-SATELLITE (passive)
FIXED
INTER-SATELLITE
MOBILE S5.558
SPACE RESEARCH (passive)

S5.138</td><td colspan="2">120.02-126
EARTH EXPLORATION-SATELLITE (passive)
FIXED
INTER-SATELLITE
MOBILE S5.558
SPACE RESEARCH (passive)

S5.138 US211 US263</td><td>ISM Equipment (18)</td></tr>
<tr><td>126-134
FIXED
INTER-SATELLITE
MOBILE S5.558
RADIOLOCATION S5.559</td><td colspan="2">126-134
FIXED
INTER-SATELLITE
MOBILE 909
RADIOLOCATION S5.559</td><td></td></tr>
<tr><td>134-142
MOBILE S5.553
MOBILE-SATELLITE
RADIONAVIGATION
RADIONAVIGATION-SATELLITE
Radiolocation

S5.149 S5.340 S5.554 S5.555</td><td colspan="2">134-142
MOBILE S5.553
MOBILE-SATELLITE
RADIONAVIGATION
RADIONAVIGATION-SATELLITE
Radiolocation

S5.149 S5.554 S5.555 917</td><td></td></tr>
<tr><td>142-144
AMATEUR
AMATEUR-SATELLITE</td><td>142-144</td><td>142-144
AMATEUR
AMATEUR-SATELLITE</td><td rowspan="2">Amateur (97)</td></tr>
<tr><td>144-149
RADIOLOCATION
Amateur
Amateur-satellite

S5.149 S5.555</td><td>144-149
RADIOLOCATION

S5.149 S5.555</td><td>144-149
RADIOLOCATION
Amateur
Amateur-satellite

S5.149 S5.555</td></tr>
<tr><td>149-150
FIXED
FIXED-SATELLITE (space-to-Earth)
MOBILE</td><td colspan="2">149-150
FIXED
FIXED-SATELLITE (space-to-Earth)
MOBILE</td><td></td></tr>
</table>

150-202 GHz (EHF)					
International Table			United States Table		Remarks
Region 1	Region 2	Region 3	Federal Government	Non-Federal Government	
150-151 EARTH EXPLORATION-SATELLITE (passive) FIXED FIXED-SATELLITE (space-to-Earth) MOBILE SPACE RESEARCH (passive) S5.149 S5.385			150-151 EARTH EXPLORATION-SATELLITE (passive) FIXED FIXED-SATELLITE (space-to-Earth) MOBILE SPACE RESEARCH (passive) S5.149 S5.385 US263		
151-156 FIXED FIXED-SATELLITE (space-to-Earth) MOBILE			151-164 FIXED FIXED-SATELLITE (space-to-Earth) US211		
156-158 EARTH EXPLORATION-SATELLITE (passive) FIXED FIXED-SATELLITE (space-to-Earth) MOBILE					
158-164 FIXED FIXED-SATELLITE (space-to-Earth) MOBILE					
164-168 EARTH EXPLORATION-SATELLITE (passive) RADIO ASTRONOMY SPACE RESEARCH (passive)			164-168 EARTH EXPLORATION-SATELLITE (passive) RADIO ASTRONOMY SPACE RESEARCH (passive) US246		
168-170 FIXED MOBILE			168-170 FIXED MOBILE		
170-174.5 FIXED INTER-SATELLITE MOBILE S5.558 S5.149 S5.385			170-174.5 FIXED INTER-SATELLITE MOBILE 909 S5.149 S5.385		

174.5-176.5 EARTH EXPLORATION-SATELLITE (passive) FIXED INTER-SATELLITE MOBILE S5.558 SPACE RESEARCH (passive) S5.149 S5.385	174.5-176.5 EARTH EXPLORATION-SATELLITE (passive) FIXED INTER-SATELLITE MOBILE 909 SPACE RESEARCH (passive) S5.149 S5.385 US263	
176.5-182 FIXED INTER-SATELLITE MOBILE S5.558 S5.149 S5.385	176.5-182 FIXED INTER-SATELLITE MOBILE 909 S5.149 S5.385 US211	
182-185 EARTH EXPLORATION-SATELLITE (passive) RADIO ASTRONOMY SPACE RESEARCH (passive) S5.340 S5.563	182-185 EARTH EXPLORATION-SATELLITE (passive) RADIO ASTRONOMY SPACE RESEARCH (passive) US246	
185-190 FIXED INTER-SATELLITE MOBILE S5.558 S5.149 S5.385	185-190 FIXED INTER-SATELLITE MOBILE 909 S5.149 S5.385 US211	
190-200 MOBILE S5.553 MOBILE-SATELLITE RADIONAVIGATION RADIONAVIGATION-SATELLITE S5.341 S5.554	190-200 MOBILE S5.553 MOBILE-SATELLITE RADIONAVIGATION RADIONAVIGATION-SATELLITE S5.341 S5.554	
200-202 EARTH EXPLORATION-SATELLITE (passive) FIXED MOBILE SPACE RESEARCH (passive) S5.341	200-202 EARTH EXPLORATION-SATELLITE (passive) FIXED MOBILE SPACE RESEARCH (passive) S5.341 US263	

202-400 GHz (EHF)					
International Table			United States Table		Remarks
Region 1	Region 2	Region 3	Federal Government	Non-Federal Government	
202-217 FIXED FIXED-SATELLITE (Earth-to-space) MOBILE S5.341			202-217 FIXED FIXED-SATELLITE (Earth-to-space) MOBILE S5.341		
217-231 EARTH EXPLORATION-SATELLITE (passive) RADIO ASTRONOMY SPACE RESEARCH (passive) S5.340 S5.341			217-231 EARTH EXPLORATION-SATELLITE (passive) RADIO ASTRONOMY US74 SPACE RESEARCH (passive) S5.341 US246		
231-235 FIXED FIXED-SATELLITE (space-to-Earth) MOBILE Radiolocation			231-235 FIXED FIXED-SATELLITE (space-to-Earth) MOBILE Radiolocation US211		
235-238 EARTH EXPLORATION-SATELLITE (passive) FIXED FIXED-SATELLITE (space-to-Earth) MOBILE SPACE RESEARCH (passive)			235-238 EARTH EXPLORATION-SATELLITE (passive) FIXED FIXED-SATELLITE (space-to-Earth) MOBILE SPACE RESEARCH (passive) US263		
238-241 FIXED FIXED-SATELLITE (space-to-Earth) MOBILE Radiolocation			238-241 FIXED FIXED-SATELLITE (space-to-Earth) MOBILE Radiolocation		
241-248 RADIOLOCATION Amateur Amateur-satellite S5.138			241-248 RADIOLOCATION S5.138	241-248 RADIOLOCATION Amateur Amateur-satellite S5.138	ISM Equipment (18) Amateur (97)
248-250 AMATEUR AMATEUR-SATELLITE			248-250	248-250 AMATEUR AMATEUR-SATELLITE	Amateur (97)

250-252 EARTH EXPLORATION-SATELLITE (passive) SPACE RESEARCH (passive) S5.149 S5.555	250-252 EARTH EXPLORATION-SATELLITE (passive) SPACE RESEARCH (passive) S5.149 S5.555	
252-265 MOBILE S5.553 MOBILE-SATELLITE RADIONAVIGATION RADIONAVIGATION-SATELLITE S5.149 S5.385 S5.554 S5.555 S5.564	252-265 MOBILE S5.553 MOBILE-SATELLITE RADIONAVIGATION RADIONAVIGATION-SATELLITE S5.149 S5.385 S5.554 S5.555 US211	
265-275 FIXED FIXED-SATELLITE (Earth-to-space) MOBILE RADIO ASTRONOMY S5.149	265-275 FIXED FIXED-SATELLITE (Earth-to-space) MOBILE RADIO ASTRONOMY S5.149	
275-400 (Not Allocated) S5.565	275-300 FIXED MOBILE S5.565	
275-400 (Not Allocated) S5.565	300-400 (Not allocated) S5.565	Amateur (97)

Chapter

12

Dictionary of Electronics Terms

Jerry C. Whitaker, editor

12.1 Terms Relating to Digital Television

The following definitions apply to digital television in general, and the ATSC DTV system in particular [1–6]:

16 VSB Vestigial sideband modulation with 16 discrete amplitude levels.

8 VSB Vestigial sideband modulation with 8 discrete amplitude levels.

access unit A coded representation of a presentation unit. In the case of audio, an access unit is the coded representation of an audio frame. In the case of video, an access unit includes all the coded data for a picture, and any *stuffing* that follows it, up to but not including the start of the next access unit.

anchor frame A video frame that is used for prediction. *I*-frames and *P*-frames are generally used as anchor frames, but *B*-frames are never anchor frames.

asynchronous transfer mode (ATM) A digital signal protocol for efficient transport of both constant-rate and variable-rate information in broadband digital networks. The ATM digital stream consists of fixed-length packets called *cells*, each containing 53 8-bit bytes (a 5-byte header and a 48-byte information payload).

bidirectional pictures (B-pictures or **B-frames)** Pictures that use both future and past pictures as a reference. This technique is termed *bidirectional prediction*. *B*-pictures provide the most compression. *B*-pictures do not propagate coding errors as they are never used as a reference.

bit rate The rate at which the compressed bit stream is delivered from the channel to the input of a decoder.

block An 8-by-8 array of DCT coefficients representing luminance or chrominance information.

byte-aligned A bit stream operational condition. A bit in a coded bit stream is byte-aligned if its position is a multiple of 8-bits from the first bit in the stream.

channel A medium that stores or transports a digital television stream.

coded representation A data element as represented in its encoded form.

compression The reduction in the number of bits used to represent an item of data.

constant bit rate The operating condition where the bit rate is constant from start to finish of the compressed bit stream.

conventional definition television (CDTV) This term is used to signify the *analog* NTSC television system as defined in ITU-R Rec. 470. (*See also standard definition television* and ITU-R Rec. 1125.)

CRC Cyclic redundancy check, an algorithm used to verify the correctness of data.

decoded stream The decoded reconstruction of a compressed bit stream.

decoder An embodiment of a decoding process.

decoding (process) The process defined in the ATSC digital television standard that reads an input coded bit stream and outputs decoded pictures or audio samples.

decoding time-stamp (DTS) A field that may be present in a PES packet header which indicates the time that an access unit is decoded in the system target decoder.

***D*-frame** A frame coded according to an MPEG-1 mode that uses dc coefficients only.

DHTML (dynamic HTML) A term used by some vendors to describe the combination of HTML, style sheets, and scripts that enable the animation of web pages.

DOM (document object model) A platform- and language-neutral interface that allows programs and scripts to dynamically access and update the content, structure, and style of documents. The document can be further processed and the results of that processing can be incorporated back into the presented page.

digital storage media (DSM) A digital storage or transmission device or system.

discrete cosine transform (DCT) A mathematical transform that can be perfectly undone and which is useful in image compression.

editing A process by which one or more compressed bit streams are manipulated to produce a new compressed bit stream. Conforming edited bit streams are understood to meet the requirements defined in the ATSC digital television standard.

elementary stream (ES) A generic term for one of the coded video, coded audio, or other coded bit streams. One elementary stream is carried in a sequence of PES packets.

elementary stream clock reference (ESCR) A time stamp in the PES stream from which decoders of PES streams may derive timing.

encoder An embodiment of an encoding process.

encoding (process) A process that reads a stream of input pictures or audio samples and produces a valid coded bit stream as defined in the ATSC digital television standard.

entitlement control message (ECM) Private conditional access information that specifies control words and possibly other stream-specific, scrambling, and/or control parameters.

entitlement management message (EMM) Private conditional access information that specifies the authorization level or the services of specific decoders. They may be addressed to single decoders or groups of decoders.

entropy coding The process of variable-length lossless coding of the digital representation of a signal to reduce redundancy.

entry point A point in a coded bit stream after which a decoder can become properly initialized and commence syntactically correct decoding. The first transmitted picture after an entry point is either an *I*-picture or a *P*-picture. If the first transmitted picture is not an *I*-picture, the decoder may produce one or more pictures during acquisition.

event A collection of elementary streams with a common time base, an associated start time, and an associated end time.

field For an interlaced video signal, a *field* is the assembly of alternate lines of a frame. Therefore, an interlaced frame is composed of two fields, a top field and a bottom field.

frame Lines of spatial information of a video signal. For progressive video, these lines contain samples starting from one time instant and continuing through successive lines to the bottom of the frame. For interlaced video, a frame consists of two fields, a top field and a bottom field. One of these fields will commence one field later than the other.

group of pictures (GOP) One or more pictures in sequence.

high-definition television (HDTV) An imaging system with a resolution of approximately twice that of conventional television in both the horizontal (H) and vertical (V) dimensions, and a picture aspect ratio (H × V) of 16:9. ITU-R Rec. 1125 further defines "HDTV quality" as the delivery of a television picture that is subjectively identical with the interlaced HDTV studio standard.

high level A range of allowed picture parameters defined by the MPEG-2 video coding specification that corresponds to high-definition television.

HTML (hypertext markup language) A collection of tags typically used in the development of Web pages.

HTTP (hypertext transfer protocol) A set of instructions for communication between a server and a World Wide Web client.

Huffman coding A type of source coding that uses codes of different lengths to represent symbols which have unequal likelihood of occurrence.

intra-coded pictures (*I*-pictures or *I*-frames) Pictures that are coded using information present only in the picture itself and not depending on information from other pictures. *I*-pictures provide a mechanism for random access into the compressed video data. *I*-pictures employ transform coding of the pixel blocks and provide only moderate compression.

layer One of the levels in the data hierarchy of the DTV video and system specifications.

level A range of allowed picture parameters and combinations of picture parameters.

macroblock In the advanced television system, a macroblock consists of four blocks of luminance and one each C_r and C_b block.

main level A range of allowed picture parameters defined by the MPEG-2 video coding specification, with maximum resolution equivalent to ITU-R Rec. 601.

main profile A subset of the syntax of the MPEG-2 video coding specification that is supported over a large range of applications.

MIME (multipart/signed, multipart/encrypted content-types) A protocol for allowing e-mail messages to contain various types of media (text, audio, video, images, etc.).

motion vector A pair of numbers that represent the vertical and horizontal displacement of a region of a reference picture for prediction purposes.

MPEG Standards developed by the ISO/IEC JTC1/SC29 WG11, *Moving Picture Experts Group.* MPEG may also refer to the Group itself.

MPEG-1 ISO/IEC standards 11172-1 (Systems), 11172-2 (Video), 11172-3 (Audio), 11172-4 (Compliance Testing), and 11172-5 (Technical Report).

MPEG-2 ISO/IEC standards 13818-1 (Systems), 13818-2 (Video), 13818-3 (Audio), and 13818-4 (Compliance).

pack A header followed by zero or more packets; a layer in the ATSC DTV system coding syntax.

packet A header followed by a number of contiguous bytes from an elementary data stream; a layer in the ATSC DTV system coding syntax.

packet data Contiguous bytes of data from an elementary data stream present in the packet.

packet identifier (PID) A unique integer value used to associate elementary streams of a program in a single or multi-program transport stream.

padding A method to adjust the average length of an audio frame in time to the duration of the corresponding PCM samples by continuously adding a slot to the audio frame.

payload The bytes that follow the header byte in a packet. The transport stream packet header and adaptation fields are not payload.

PES packet The data structure used to carry elementary stream data. It consists of a packet header followed by PES packet payload.

PES stream A stream of PES packets, all of whose payloads consist of data from a single elementary stream, and all of which have the same stream identification.

picture Source, coded, or reconstructed image data. A source or reconstructed picture consists of three rectangular matrices representing the luminance and two chrominance signals.

pixel "Picture element" or "pel." A pixel is a digital sample of the color intensity values of a picture at a single point.

predicted pictures (*P*-pictures or *P*-frames) Pictures that are coded with respect to the nearest *previous I* or *P*-picture. This technique is termed *forward prediction. P*-pictures provide more compression than *I*-pictures and serve as a reference for future *P*-pictures or *B*-pictures. *P*-pictures can propagate coding errors when *P*-pictures (or *B*-pictures) are predicted from prior *P*-pictures where the prediction is flawed.

presentation time-stamp (PTS) A field that may be present in a PES packet header that indicates the time that a presentation unit is presented in the system target decoder.

presentation unit (PU) A decoded audio access unit or a decoded picture.

profile A defined subset of the syntax specified in the MPEG-2 video coding specification.

program A collection of program elements. Program elements may be elementary streams. Program elements need not have any defined time base; those that do have a common time base and are intended for synchronized presentation.

program clock reference (PCR) A time stamp in the transport stream from which decoder timing is derived.

program element A generic term for one of the elementary streams or other data streams that may be included in the program.

program specific information (PSI) Normative data that is necessary for the demultiplexing of transport streams and the successful regeneration of programs.

quantizer A processing step that intentionally reduces the precision of DCT coefficients

random access The process of beginning to read and decode the coded bit stream at an arbitrary point.

scrambling The alteration of the characteristics of a video, audio, or coded data stream in order to prevent unauthorized reception of the information in a clear form.

slice A series of consecutive macroblocks.

source stream A single, non-multiplexed stream of samples before compression coding.

splicing The concatenation performed on the system level or two different elementary streams. It is understood that the resulting stream must conform totally to the ATSC digital television standard.

standard definition television (SDTV) This term is used to signify a *digital* television system in which the quality is approximately equivalent to that of NTSC. This equivalent quality may be achieved from pictures sourced at the 4:2:2 level of ITU-R Rec. 601 and subjected to processing as part of bit rate compression. The results should be such that when judged across a representative sample of program material, subjective equivalence with NTSC is achieved. Also called standard digital television.

start codes 32-bit codes embedded in the coded bit stream that are unique. They are used for several purposes, including identifying some of the layers in the coding syntax.

STD input buffer A first-in, first-out buffer at the input of a system target decoder (STD) for storage of compressed data from elementary streams before decoding.

still picture A video sequence containing exactly one coded picture that is intra-coded. This picture has an associated PTS and the presentation time of succeeding pictures, if any, is later than that of the still picture by at least two picture periods.

system clock reference (SCR) A time stamp in the program stream from which decoder timing is derived.

system header A data structure that carries information summarizing the system characteristics of the ATSC digital television standard multiplexed bit stream.

system target decoder (STD) A hypothetical reference model of a decoding process used to describe the semantics of the ATSC digital television standard multiplexed bit stream.

time-stamp A term that indicates the time of a specific action such as the arrival of a byte or the presentation of a presentation unit.

transport stream packet header The leading fields in a transport stream packet.

UHTTP (unidirectional hypertext transfer protocol) A is a simple, robust, one-way resource transfer protocol that is designed to efficiently deliver resource data in a one-way broadcast-only environment. This resource transfer protocol is appropriate for IP multicast over the television vertical blanking interval (IP-VBI), in an IP multicast carried in MPEG-2, or in other unidirectional transport systems.

UUID (universally unique identifier) An identifier that is unique across both space and time, with respect to the space of all UUIDs. Also known as GUID (globally unique identifier).

variable bit rate An operating mode where the bit rate varies with time during the decoding of a compressed bit stream.

video buffering verifier (VBV) A hypothetical decoder that is conceptually connected to the output of an encoder. Its purpose is to provide a constraint on the variability of the data rate that an encoder can produce.

video sequence An element represented by a sequence header, one or more groups of pictures, and an end of sequence code in the data stream.

12.2 General Electronics Terms

absolute delay The amount of time a signal is delayed. The delay may be expressed in time or number of pulse events.

absolute zero The lowest temperature theoretically possible, –273.16°C. *Absolute zero* is equal to zero degrees Kelvin.

absorption The transference of some or all of the energy contained in an electromagnetic wave to the substance or medium in which it is propagating or upon which it is incident.

absorption auroral The loss of energy in a radio wave passing through an area affected by solar auroral activity.

ac coupling A method of coupling one circuit to another through a capacitor or transformer so as to transmit the varying (ac) characteristics of the signal while blocking the static (dc) characteristics.

ac/dc coupling Coupling between circuits that accommodates the passing of both ac and dc signals (may also be referred to as simply dc coupling).

accelerated life test A special form of reliability testing performed by an equipment manufacturer. The unit under test is subjected to stresses that exceed those typically experienced in normal operation. The goal of an *accelerated life test* is to improve the reliability of products shipped by forcing latent failures in components to become evident before the unit leaves the factory.

accelerating electrode The electrode that causes electrons emitted from an electron gun to accelerate in their journey to the screen of a cathode ray tube.

accelerating voltage The voltage applied to an electrode that accelerates a beam of electrons or other charged particles.

acceptable reliability level The maximum number of failures allowed per thousand operating hours of a given component or system.

acceptance test The process of testing newly purchased equipment to ensure that it is fully compliant with contractual specifications.

access The point at which entry is gained to a circuit or facility.

acquisition time In a communication system, the amount of time required to attain synchronism.

active Any device or circuit that introduces gain or uses a source of energy other than that inherent in the signal to perform its function.

adapter A fitting or electrical connector that links equipment that cannot be connected directly.

adaptive A device able to adjust or react to a condition or application, as an *adaptive circuit.* This term usually refers to filter circuits.

adaptive system A general name for a system that is capable of reconfiguring itself to meet new requirements.

adder A device whose output represents the sum of its inputs.

adjacent channel interference Interference to communications caused by a transmitter operating on an adjacent radio channel. The sidebands of the transmitter mix with the carrier being received on the desired channel, resulting in noise.

admittance A measure of how well alternating current flows in a conductor. It is the reciprocal of *impedance* and is expressed in *siemens*. The real part of admittance is *conductance*; the imaginary part is *susceptance*.

AFC (automatic frequency control) A circuit that automatically keeps an oscillator on frequency by comparing the output of the oscillator with a standard frequency source or signal.

air core An inductor with no magnetic material in its core.

algorithm A prescribed finite set of well-defined rules or processes for the solution of a problem in a finite number of steps.

alignment The adjustment of circuit components so that an entire system meets minimum performance values. For example, the stages in a radio are aligned to ensure proper reception.

allocation The planned use of certain facilities and equipment to meet current, pending, and/or forecasted circuit- and carrier-system requirements.

alternating current (ac) A continuously variable current, rising to a maximum in one direction, falling to zero, then reversing direction and rising to a maximum in the other direction, then falling to zero and repeating the cycle. Alternating current usually follows a sinusoidal growth and decay curve. Note that the correct usage of the term *ac* is lower case.

alternator A generator that produces alternating current electric power.

ambient electromagnetic environment The radiated or conducted electromagnetic signals and noise at a specific location and time.

ambient level The magnitude of radiated or conducted electromagnetic signals and noise at a specific test location when equipment-under-test is not powered.

ambient temperature The temperature of the surrounding medium, typically air, that comes into contact with an apparatus. Ambient temperature may also refer simply to room temperature.

American National Standards Institute (ANSI) A nonprofit organization that coordinates voluntary standards activities in the U.S.

American Wire Gauge (AWG) The standard American method of classifying wire diameter.

ammeter An instrument that measures and records the amount of current in amperes flowing in a circuit.

amp (A) An abbreviation of the term *ampere*.

ampacity A measure of the current carrying capacity of a power cable. *Ampacity* is determined by the maximum continuous-performance temperature of the insulation, by the heat generated in the cable (as a result of conductor and insulation losses), and by the heat-dissipating properties of the cable and its environment.

ampere (amp) The standard unit of electric current.

ampere per meter The standard unit of magnetic field strength.

ampere-hour The energy that is consumed when a current of one ampere flows for a period of one hour.

ampere-turns The product of the number of turns of a coil and the current in amperes flowing through the coil.

amplification The process that results when the output of a circuit is an enlarged reproduction of the input signal. Amplifiers may be designed to provide amplification of voltage, current, or power, or a combination of these quantities.

amplification factor In a vacuum tube, the ratio of the change in plate voltage to the change in grid voltage that causes a corresponding change in plate current. Amplification factor is expressed by the Greek letter μ (*mu*).

amplifier (1—general) A device that receives an input signal and provides as an output a magnified replica of the input waveform. **(2—audio)** An amplifier designed to cover the normal audio frequency range (20 Hz to 20 kHz). **(3—balanced)** A circuit with two identical connected signal branches that operate in phase opposition, with input and output connections each balanced to ground. **(4—bridging)** An amplifying circuit featuring high input impedance to prevent loading of the source. **(5—broadband)** An amplifier capable of operating over a specified broad band of frequencies with acceptably small amplitude variations as a function of frequency. **(6—buffer)** An amplifier stage used to isolate a frequency-sensitive circuit from variations in the load presented by following stages. **(7—linear)** An amplifier in which the instantaneous output signal is a linear function of the corresponding input signal. **(8—magnetic)** An amplifier incorporating a control device dependent on magnetic saturation. A small dc signal applied to a control circuit triggers a large change in operating impedance and, hence, in the output of the circuit. **(9—microphone)** A circuit that amplifies the low level output from a microphone to make it sufficient to be used as an input signal to a power amplifier or another stage in a modulation circuit. Such a circuit is commonly known as a *preamplifier.* **(10—push-pull)** A balanced amplifier with two similar amplifying units connected in phase opposition in order to cancel undesired harmonics and minimize distortion. **(11—tuned radio frequency)** An amplifier tuned to a particular radio frequency or band so that only selected frequencies are amplified.

amplifier operating class (1—general) The operating point of an amplifying stage. The operating point, termed the operating *class*, determines the period during which current flows in the output. **(2—class A)** An amplifier in which output current flows during the whole of the input current cycle. **(3—class AB)** An amplifier in which the output current flows for more than half but less than the whole of the input cycle. **(4—class B)** An amplifier in which output current is cut off at zero input signal; a half-wave rectified output is produced. **(5—class C)** An amplifier in which output current flows for less than half the input cycle. **(6—class D)** An amplifier operating in a pulse-only mode.

amplitude The magnitude of a signal in voltage or current, frequently expressed in terms of *peak*, *peak-to-peak*, or *root-mean-square* (RMS). The actual amplitude of a quantity at a particular instant often varies in a sinusoidal manner.

amplitude distortion A distortion mechanism occurring in an amplifier or other device when the output amplitude is not a linear function of the input amplitude under specified conditions.

amplitude equalizer A corrective network that is designed to modify the amplitude characteristics of a circuit or system over a desired frequency range.

amplitude-versus-frequency distortion The distortion in a transmission system caused by the nonuniform attenuation or gain of the system with respect to frequency under specified conditions.

analog carrier system A carrier system whose signal amplitude, frequency, or phase is varied continuously as a function of a modulating input.

anode (1 — general) A positive pole or element. **(2—vacuum tube)** The outermost positive element in a vacuum tube, also called the *plate.* **(3—battery)** The positive element of a battery or cell.

anodize The formation of a thin film of oxide on a metallic surface, usually to produce an insulating layer.

antenna (1—general) A device used to transmit or receive a radio signal. An antenna is usually designed for a specified frequency range and serves to couple electromagnetic energy from a transmission line to and/or from the free space through which it travels. Directional antennas concentrate the energy in a particular horizontal or vertical direction. **(2—aperiodic)** An antenna that is not periodic or resonant at particular frequencies, and so can be used over a wide band of frequencies. **(3—artificial)** A device that behaves, so far as the transmitter is concerned, like a proper antenna, but does not radiate any power at radio frequencies. **(4—broadband)** An antenna that operates within specified performance limits over a wide band of frequencies, without requiring retuning for each individual frequency. **(5—Cassegrain)** A double reflecting antenna, often used for ground stations in satellite systems. **(6—coaxial)** A dipole antenna made by folding back on itself a quarter wavelength of the outer conductor of a coaxial line, leaving a quarter wavelength of the inner conductor exposed. **(7—corner)** An antenna within the angle formed by two plane-reflecting surfaces. **(8—dipole)** A center-fed antenna, one half-wavelength long. **(9—directional)** An antenna designed to receive or emit radiation more efficiently in a particular direction. **(10—dummy)** An artificial antenna, designed to accept power from the transmitter but not to radiate it. **(11—ferrite)** A common AM broadcast receive antenna that uses a small coil mounted on a short rod of ferrite material. **(12—flat top)** An antenna in which all the horizontal components are in the same horizontal plane. **(13—folded dipole)** A radiating device consisting of two ordinary half-wave dipoles joined at their outer ends and fed at the center of one of the dipoles. **(14—horn reflector)** A radiator in which the feed horn extends into a parabolic reflector, and the power is radiated through a window in the horn. **(15—isotropic)** A theoretical antenna in free space that transmits or receives with the same efficiency in all directions. **(16—log-periodic)** A broadband directional antenna incorporating an array of dipoles of different lengths, the length and spacing between dipoles increasing logarithmically away from the feeder element. **(17—long wire)** An antenna made up of one or more conductors in a straight line pointing in the required direction with a total length of several wavelengths at the operating frequency. **(18—loop)** An antenna consisting of one or more turns of wire in the same or parallel planes. **(19—nested rhombic)** An assembly of two rhombic antennas, one smaller than the other, so that the complete diamond-shaped antenna fits inside the area occupied by the larger unit. **(20—omnidirectional)** An antenna whose radiating or receiving properties are the same in all horizontal plane directions. **(21—periodic)** A resonant antenna designed for use at a particular frequency. **(22—quarter-wave)** A dipole antenna whose length is equal to one quarter of a wavelength at the operating frequency. **(23—rhombic)** A large diamond-shaped antenna, with sides of the diamond several wavelengths long. The rhombic antenna is fed at one of the corners, with directional efficiency in the direction of the diagonal. **(24—series fed)** A vertical antenna that is fed at its lower end. **(25—shunt fed)** A vertical antenna whose base is grounded, and is fed at a specified point above ground. The point at which the antenna is fed above ground determines the operating impedance. **(26—steerable)** An antenna so constructed that its major lobe may readily be changed in direction. **(27—top-loaded)** A vertical antenna capacitively loaded at its upper end, often by simple enlargement or the attachment of a disc or plate. **(28—turnstile)** An antenna with one or more tiers of horizontal dipoles, crossed at right angles to each other and with excitation of the dipoles in phase quadrature. **(29—whip)** An antenna constructed of a thin semiflexible metal rod or tube, fed at its base. **(30—Yagi)** A directional antenna constructed of a series of dipoles cut to specific lengths. *Director* elements are placed in front of the active dipole and *reflector* elements are placed behind the active element.

antenna array A group of several antennas coupled together to yield a required degree of directivity.

antenna beamwidth The angle between the *half-power* points (3 dB points) of the main lobe of the antenna pattern when referenced to the peak power point of the antenna pattern. *Antenna beamwidth* is measured in degrees and normally refers to the horizontal radiation pattern.

antenna directivity factor The ratio of the power flux density in the desired direction to the average value of power flux density at crests in the antenna directivity pattern in the interference section.

antenna factor A factor that, when applied to the voltage appearing at the terminals of measurement equipment, yields the electrical field strength at an antenna. The unit of antenna factor is volts per meter per measured volt.

antenna gain The ratio of the power required at the input of a theoretically perfect omnidirectional reference antenna to the power supplied to the input of the given antenna to produce the same field at the same distance. When not specified otherwise, the figure expressing the gain of an antenna refers to the gain in the direction of the radiation main lobe. In services using *scattering* modes of propagation, the full gain of an antenna may not be realizable in practice and the apparent gain may vary with time.

antenna gain-to-noise temperature For a satellite earth terminal receiving system, a figure of merit that equals G/T, where G is the gain in dB of the earth terminal antenna at the receive frequency, and T is the equivalent noise temperature of the receiving system in Kelvins.

antenna matching The process of adjusting an antenna matching circuit (or the antenna itself) so that the input impedance of the antenna is equal to the characteristic impedance of the transmission line.

antenna monitor A device used to measure the ratio and phase between the currents flowing in the towers of a directional AM broadcast station.

antenna noise temperature The temperature of a resistor having an available noise power per unit bandwidth equal to that at the antenna output at a specified frequency.

antenna pattern A diagram showing the efficiency of radiation in all directions from the antenna.

antenna power rating The maximum continuous-wave power that can be applied to an antenna without degrading its performance.

antenna preamplifier A small amplifier, usually mast-mounted, for amplifying weak signals to a level sufficient to compensate for down-lead losses.

apparent power The product of the root-mean-square values of the voltage and current in an alternating-current circuit without a correction for the phase difference between the voltage and current.

arc A sustained luminous discharge between two or more electrodes.

arithmetic mean The sum of the values of several quantities divided by the number of quantities, also referred to as the *average*.

armature winding The winding of an electrical machine, either a motor or generator, in which current is induced.

array (1—antenna) An assembly of several directional antennas so placed and interconnected that directivity may be enhanced. **(2—broadside)** An antenna array whose elements are all in the same plane, producing a major lobe perpendicular to the plane. **(3—colinear)** An antenna array whose elements are in the same line, either horizontal or vertical. **(4—end-fire)** An antenna array whose elements are in parallel rows, one behind the other, producing a major lobe perpendicular to the plane in which individual elements are placed. **(5—linear)** An antenna array whose elements are arranged end-to-end. **(6—stacked)** An antenna array whose elements are stacked, one above the other.

artificial line An assembly of resistors, inductors, and capacitors that simulates the electrical characteristics of a transmission line.

assembly A manufactured part made by combining several other parts or subassemblies.

assumed values A range of values, parameters, levels, and other elements assumed for a mathematical model, hypothetical circuit, or network, from which analysis, additional estimates, or calculations will be made. The range of values, while not measured, represents the best engineering judgment and is generally derived from values found or measured in real circuits or networks of the same generic type, and includes projected improvements.

atmosphere The gaseous envelope surrounding the earth, composed largely of oxygen, carbon dioxide, and water vapor. The atmosphere is divided into four primary layers: *troposphere*, *stratosphere*, *ionosphere*, and *exosphere*.

atmospheric noise Radio noise caused by natural atmospheric processes, such as lightning.

attack time The time interval in seconds required for a device to respond to a control stimulus.

attenuation The decrease in amplitude of an electrical signal traveling through a transmission medium caused by dielectric and conductor losses.

attenuation coefficient The rate of decrease in the amplitude of an electrical signal caused by attenuation. The *attenuation coefficient* can be expressed in decibels or nepers per unit length. It may also be referred to as the *attenuation constant*.

attenuation distortion The distortion caused by attenuation that varies over the frequency range of a signal.

attenuation-limited operation The condition prevailing when the received signal amplitude (rather than distortion) limits overall system performance.

attenuator A fixed or adjustable component that reduces the amplitude of an electrical signal without causing distortion.

atto A prefix meaning one *quintillionth*.

attraction The attractive force between two unlike magnetic poles (N/S) or electrically charged bodies (+/-).

attributes The characteristics of equipment that aid planning and circuit design.

automatic frequency control (AFC) A system designed to maintain the correct operating frequency of a receiver. Any drift in tuning results in the production of a control voltage, which is used to adjust the frequency of a local oscillator so as to minimize the tuning error.

automatic gain control (AGC) An electronic circuit that compares the level of an incoming signal with a previously defined standard and automatically amplifies or attenuates the signal so it arrives at its destination at the correct level.

autotransformer A transformer in which both the primary and secondary currents flow through one common part of the coil.

auxiliary power An alternate source of electric power, serving as a back-up for the primary utility company ac power.

availability A measure of the degree to which a system, subsystem, or equipment is operable and not in a stage of congestion or failure at any given point in time.

avalanche effect The effect obtained when the electric field across a barrier region is sufficiently strong for electrons to collide with *valence electrons*, thereby releasing more electrons and giving a cumulative multiplication effect in a semiconductor.

average life The mean value for a normal distribution of product or component lives, generally applied to mechanical failures resulting from "wear-out."

B

back emf A voltage induced in the reverse direction when current flows through an inductance. *Back emf* is also known as *counter-emf.*

back scattering A form of wave scattering in which at least one component of the scattered wave is deflected opposite to the direction of propagation of the incident wave.

background noise The total system noise in the absence of information transmission, independent of the presence or absence of a signal.

backscatter The deflection or reflection of radiant energy through angles greater than 90° with respect to the original angle of travel.

backscatter range The maximum distance from which backscattered radiant energy can be measured.

backup A circuit element or facility used to replace an element that has failed.

backup supply A redundant power supply that takes over if the primary power supply fails.

balance The process of equalizing the voltage, current, or other parameter between two or more circuits or systems.

balanced A circuit having two sides (conductors) carrying voltages that are symmetrical about a common reference point, typically ground.

balanced circuit A circuit whose two sides are electrically equal in all transmission respects.

balanced line A transmission line consisting of two conductors in the presence of ground capable of being operated in such a way that when the voltages of the two conductors at all transverse planes are equal in magnitude and opposite in polarity with respect to ground, the currents in the two conductors are equal in magnitude and opposite in direction.

balanced modulator A modulator that combines the information signal and the carrier so that the output contains the two sidebands without the carrier.

balanced three-wire system A power distribution system using three conductors, one of which is balanced to have a potential midway between the potentials of the other two.

balanced-to-ground The condition when the impedance to ground on one wire of a two-wire circuit is equal to the impedance to ground on the other wire.

balun (balanced/unbalanced) A device used to connect balanced circuits with unbalanced circuits.

band A range of frequencies between a specified upper and lower limit.

band elimination filter A filter having a single continuous attenuation band, with neither the upper nor lower cut-off frequencies being zero or infinite. A *band elimination filter* may also be referred to as a *band-stop*, *notch*, or *band reject* filter.

bandpass filter A filter having a single continuous transmission band with neither the upper nor the lower cut-off frequencies being zero or infinite. A bandpass filter permits only a specific band of frequencies to pass; frequencies above or below are attenuated.

bandwidth The range of signal frequencies that can be transmitted by a communications channel with a defined maximum loss or distortion. Bandwidth indicates the information-carrying capacity of a channel.

bandwidth expansion ratio The ratio of the necessary bandwidth to the baseband bandwidth.

bandwidth-limited operation The condition prevailing when the frequency spectrum or bandwidth, rather than the amplitude (or power) of the signal, is the limiting factor in communication capability. This condition is reached when the system distorts the shape of the waveform beyond tolerable limits.

bank A group of similar items connected together in a specified manner and used in conjunction with one another.

bare A wire conductor that is not enameled or enclosed in an insulating sheath.

baseband The band of frequencies occupied by a signal before it modulates a carrier wave to form a transmitted radio or line signal.

baseband channel A channel that carries a signal without modulation, in contrast to a *passband* channel.

baseband signal The original form of a signal, unchanged by modulation.

bath tub The shape of a typical graph of component failure rates: high during an initial period of operation, falling to an acceptable low level during the normal usage period, and then rising again as the components become time-expired.

battery A group of several cells connected together to furnish current by conversion of chemical, thermal, solar, or nuclear energy into electrical energy. A single cell is itself sometimes also called a battery.

bay A row or suite of racks on which transmission, switching, and/or processing equipment is mounted.

Bel A unit of power measurement, named in honor of Alexander Graham Bell. The commonly used unit is one tenth of a Bel, or a decibel (dB). One Bel is defined as a tenfold increase in power. If an amplifier increases the power of a signal by 10 times, the power gain of the amplifier is equal to 1 Bel or 10 *decibels* (dB). If power is increased by 100 times, the power gain is 2 Bels or 20 decibels.

bend A transition component between two elements of a transmission waveguide.

bending radius The smallest bend that may be put into a cable under a stated pulling force. The bending radius is typically expressed in inches.

bias A dc voltage difference applied between two elements of an active electronic device, such as a vacuum tube, transistor, or integrated circuit. Bias currents may or may not be drawn, depending on the device and circuit type.

bidirectional An operational qualification which implies that the transmission of information occurs in both directions.

bifilar winding A type of winding in which two insulated wires are placed side by side. In some components, bifilar winding is used to produce balanced circuits.

bipolar A signal that contains both positive-going and negative-going amplitude components. A bipolar signal may also contain a zero amplitude state.

bleeder A high resistance connected in parallel with one or more filter capacitors in a high voltage dc system. If the power supply load is disconnected, the capacitors discharge through the bleeder.

block diagram An overview diagram that uses geometric figures to represent the principal divisions or sections of a circuit, and lines and arrows to show the path of a signal, or to show program functionalities. It is not a *schematic*, which provides greater detail.

blocking capacitor A capacitor included in a circuit to stop the passage of direct current.

BNC An abbreviation for *bayonet Neill-Concelman*, a type of cable connector used extensively in RF applications (named for its inventor).

Boltzmann's constant 1.38×10^{-23} joules.

bridge A type of network circuit used to match different circuits to each other, ensuring minimum transmission impairment.

bridging The shunting or paralleling of one circuit with another.

broadband The quality of a communications link having essentially uniform response over a given range of frequencies. A communications link is said to be *broadband* if it offers no perceptible degradation to the signal being transported.

buffer A circuit or component that isolates one electrical circuit from another.

burn-in The operation of a device, sometimes under extreme conditions, to stabilize its characteristics and identify latent component failures before bringing the device into normal service.

bus A central conductor for the primary signal path. The term bus may also refer to a signal path to which a number of inputs may be connected for feed to one or more outputs.

busbar A main dc power bus.

bypass capacitor A capacitor that provides a signal path that effectively shunts or bypasses other components.

bypass relay A switch used to bypass the normal electrical route of a signal or current in the event of power, signal, or equipment failure.

C

cable An electrically and/or optically conductive interconnecting device.

cable loss Signal loss caused by passing a signal through a coaxial cable. Losses are the result of resistance, capacitance, and inductance in the cable.

cable splice The connection of two pieces of cable by joining them mechanically and closing the joint with a weather-tight case or sleeve.

cabling The wiring used to interconnect electronic equipment.

calibrate The process of checking, and adjusting if necessary, a test instrument against one known to be set correctly.

calibration The process of identifying and measuring errors in instruments and/or procedures.

capacitance The property of a device or component that enables it to store energy in an electrostatic field and to release it later. A capacitor consists of two conductors separated by an insulating material. When the conductors have a voltage difference between them, a charge will be stored in the electrostatic field between the conductors.

capacitor A device that stores electrical energy. A capacitor allows the apparent flow of alternating current, while blocking the flow of direct current. The degree to which the device permits ac current flow depends on the frequency of the signal and the size of the capacitor. Capacitors are used in filters, delay-line components, couplers, frequency selectors, timing elements, voltage transient suppression, and other applications.

carrier A single frequency wave that, prior to transmission, is modulated by another wave containing information. A carrier may be modulated by manipulating its amplitude and/or frequency in direct relation to one or more applied signals.

carrier frequency The frequency of an unmodulated oscillator or transmitter. Also, the average frequency of a transmitter when a signal is frequency modulated by a symmetrical signal.

cascade connection A tandem arrangement of two or more similar component devices or circuits, with the output of one connected to the input of the next.

cascaded An arrangement of two or more circuits in which the output of one circuit is connected to the input of the next circuit.

cathode ray tube (CRT) A vacuum tube device, usually glass, that is narrow at one end and widens at the other to create a surface onto which images can be projected. The narrow end contains the necessary circuits to generate and focus an electron beam on the luminescent screen at the other end. CRTs are used to display pictures in TV receivers, video monitors, oscilloscopes, computers, and other systems.

cell An elementary unit of communication, of power supply, or of equipment.

Celsius A temperature measurement scale, expressed in degrees C, in which water freezes at 0°C and boils at 100°C. To convert to degrees Fahrenheit, multiply by 0.555 and add 32. To convert to Kelvins add 273 (approximately).

center frequency In frequency modulation, the resting frequency or initial frequency of the carrier before modulation.

center tap A connection made at the electrical center of a coil.

channel The smallest subdivision of a circuit that provides a single type of communication service.

channel decoder A device that converts an incoming modulated signal on a given channel back into the source-encoded signal.

channel encoder A device that takes a given signal and converts it into a form suitable for transmission over the communications channel.

channel noise level The ratio of the channel noise at any point in a transmission system to some arbitrary amount of circuit noise chosen as a reference. This ratio is usually expressed in *decibels above reference noise*, abbreviated *dBrn*.

channel reliability The percent of time a channel is available for use in a specific direction during a specified period.

channelization The allocation of communication circuits to channels and the forming of these channels into groups for higher order multiplexing.

characteristic The property of a circuit or component.

characteristic impedance The impedance of a transmission line, as measured at the driving point, if the line were of infinite length. In such a line, there would be no standing waves. The *characteristic impedance* may also be referred to as the *surge impedance*.

charge The process of replenishing or replacing the electrical charge in a secondary cell or storage battery.

charger A device used to recharge a battery. Types of charging include: (1) *constant voltage charge*, (2) *equalizing charge*, and (3) *trickle charge*.

chassis ground A connection to the metal frame of an electronic system that holds the components in a place. The chassis ground connection serves as the ground return or electrical common for the system.

circuit Any closed path through which an electrical current can flow. In a *parallel circuit*, components are connected between common inputs and outputs such that all paths are parallel to each other. The same voltage appears across all paths. In a *series circuit*, the same current flows through all components.

circuit noise level The ratio of the circuit noise at some given point in a transmission system to an established reference, usually expressed in decibels above the reference.

circuit reliability The percentage of time a circuit is available to the user during a specified period of scheduled availability.

circular mil The measurement unit of the cross-sectional area of a circular conductor. A *circular mil* is the area of a circle whose diameter is one mil, or 0.001 inch.

clear channel A transmission path wherein the full bandwidth is available to the user, with no portions of the channel used for control, framing, or signaling. Can also refer to a classification of AM broadcast station.

clipper A limiting circuit which ensures that a specified output level is not exceeded by restricting the output waveform to a maximum peak amplitude.

clipping The distortion of a signal caused by removing a portion of the waveform through restriction of the amplitude of the signal by a circuit or device.

coax A short-hand expression for *coaxial cable*, which is used to transport high-frequency signals.

coaxial cable A transmission line consisting of an inner conductor surrounded first by an insulating material and then by an outer conductor, either solid or braided.The mechanical dimensions of the cable determine its *characteristic impedance*.

coherence The correlation between the phases of two or more waves.

coherent The condition characterized by a fixed phase relationship among points on an electromagnetic wave.

coherent pulse The condition in which a fixed phase relationship is maintained between consecutive pulses during pulse transmission.

cold joint A soldered connection that was inadequately heated, with the result that the wire is held in place by rosin flux, not solder. A cold joint is sometimes referred to as a *dry joint*.

comb filter An electrical filter circuit that passes a series of frequencies and rejects the frequencies in between, producing a frequency response similar to the teeth of a comb.

common A point that acts as a reference for circuits, often equal in potential to the local ground.

common mode Signals identical with respect to amplitude, frequency, and phase that are applied to both terminals of a cable and/or both the input and reference of an amplifier.

common return A return path that is common to two or more circuits, and returns currents to their source or to ground.

common return offset The dc common return potential difference of a line.

communications system A collection of individual communications networks, transmission systems, relay stations, tributary stations, and terminal equipment capable of interconnection and interoperation to form an integral whole. The individual components must serve a common purpose, be technically compatible, employ common procedures, respond to some form of control, and, in general, operate in unison.

commutation A successive switching process carried out by a commutator.

commutator A circular assembly of contacts, insulated one from another, each leading to a different portion of the circuit or machine.

compatibility The ability of diverse systems to exchange necessary information at appropriate levels of command directly and in usable form. Communications equipment items are compatible if signals can be exchanged between them without the addition of buffering or translation for the specific purpose of achieving workable interface connections, and if the equipment or systems being interconnected possess comparable performance characteristics, including the suppression of undesired radiation.

complex wave A waveform consisting of two or more sinewave components. At any instant of time, a complex wave is the algebraic sum of all its sinewave components.

compliance For mechanical systems, a property which is the reciprocal of stiffness.

component An assembly, or part thereof, that is essential to the operation of some larger circuit or system. A *component* is an immediate subdivision of the assembly to which it belongs.

COMSAT The *Communications Satellite Corporation*, an organization established by an act of Congress in 1962. COMSAT launches and operates the international satellites for the INTELSAT consortium of countries.

concentricity A measure of the deviation of the center conductor position relative to its ideal location in the exact center of the dielectric cross-section of a coaxial cable.

conditioning The adjustment of a channel in order to provide the appropriate transmission characteristics needed for data or other special services.

conditioning equipment The equipment used to match transmission levels and impedances, and to provide equalization between facilities.

conductance A measure of the capability of a material to conduct electricity. It is the reciprocal of *resistance* (ohm) and is expressed in *siemens*. (Formerly expressed as *mho*.)

conducted emission An electromagnetic energy propagated along a conductor.

conduction The transfer of energy through a medium, such as the conduction of electricity by a wire, or of heat by a metallic frame.

conduction band A partially filled or empty atomic energy band in which electrons are free to move easily, allowing the material to carry an electric current.

conductivity The conductance per unit length.

conductor Any material that is capable of carrying an electric current.

configuration A relative arrangement of parts.

connection A point at which a junction of two or more conductors is made.

connector A device mounted on the end of a wire or fiber optic cable that mates to a similar device on a specific piece of equipment or another cable.

constant-current source A source with infinitely high output impedance so that output current is independent of voltage, for a specified range of output voltages.

constant-voltage charge A method of charging a secondary cell or storage battery during which the terminal voltage is kept at a constant value.

constant-voltage source A source with low, ideally zero, internal impedance, so that voltage will remain constant, independent of current supplied.

contact The points that are brought together or separated to complete or break an electrical circuit.

contact bounce The rebound of a contact, which temporarily opens the circuit after its initial *make*.

contact form The configuration of a contact assembly on a relay. Many different configurations are possible from simple *single-make* contacts to complex arrangements involving *breaks* and *makes*.

contact noise A noise resulting from current flow through an electrical contact that has a rapidly varying resistance, as when the contacts are corroded or dirty.

contact resistance The resistance at the surface when two conductors make contact.

continuity A continuous path for the flow of current in an electrical circuit.

continuous wave An electromagnetic signal in which successive oscillations of the waves are identical.

control The supervision that an operator or device exercises over a circuit or system.

control grid The grid in an electron tube that controls the flow of current from the cathode to the anode.

convention A generally acceptable symbol, sign, or practice in a given industry.

Coordinated Universal Time (UTC) The time scale, maintained by the BIH (Bureau International de l'Heure) that forms the basis of a coordinated dissemination of standard frequencies and time signals.

copper loss The loss resulting from the heating effect of current.

corona A bluish luminous discharge resulting from ionization of the air near a conductor carrying a voltage gradient above a certain *critical level.*

corrective maintenance The necessary tests, measurements, and adjustments required to remove or correct a fault.

cosmic noise The random noise originating outside the earth's atmosphere.

coulomb The standard unit of electric quantity or charge. One *coulomb* is equal to the quantity of electricity transported in 1 second by a current of 1 ampere.

Coulomb's Law The attraction and repulsion of electric charges act on a line between them. The charges are inversely proportional to the square of the distance between them, and proportional to the product of their magnitudes. (Named for the French physicist Charles-Augustine de Coulomb, 1736–1806.)

counter-electromotive force The effective electromotive force within a system that opposes the passage of current in a specified direction.

couple The process of linking two circuits by inductance, so that energy is transferred from one circuit to another.

coupled mode The selection of either ac or dc coupling.

coupling The relationship between two components that enables the transfer of energy between them. Included are *direct coupling* through a direct electrical connection, such as a wire; *capacitive coupling* through the capacitance formed by two adjacent conductors; and *inductive coupling* in which energy is transferred through a magnetic field. Capacitive coupling is also called *electrostatic coupling*. Inductive coupling is often referred to as *electromagnetic coupling*.

coupling coefficient A measure of the electrical coupling that exists between two circuits. The *coupling coefficient* is equal to the ratio of the mutual impedance to the square root of the product of the self impedances of the coupled circuits.

cross coupling The coupling of a signal from one channel, circuit, or conductor to another, where it becomes an undesired signal.

crossover distortion A distortion that results in an amplifier when an irregularity is introduced into the signal as it crosses through a zero reference point. If an amplifier is properly designed and biased, the upper half cycle and lower half cycle of the signal coincide at the zero crossover reference.

crossover frequency The frequency at which output signals pass from one channel to the other in a *crossover network.* At the *crossover frequency* itself, the outputs to each side are equal.

crossover network A type of filter that divides an incoming signal into two or more outputs, with higher frequencies directed to one output, and lower frequencies to another.

crosstalk Undesired transmission of signals from one circuit into another circuit in the same system. Crosstalk is usually caused by unintentional capacitive (ac) coupling.

crosstalk coupling The ratio of the power in a disturbing circuit to the induced power in the disturbed circuit, observed at a particular point under specified conditions. Crosstalk coupling is typically expressed in dB.

crowbar A short-circuit or low resistance path placed across the input to a circuit, usually for protective purposes.

CRT (cathode ray tube) A vacuum tube device that produces light when energized by the electron beam generated inside the tube. A CRT includes an electron gun, deflection mechanism, and phosphor-covered faceplate.

crystal A solidified form of a substance that has atoms and molecules arranged in a symmetrical pattern.

crystal filter A filter that uses piezoelectric crystals to create resonant or antiresonant circuits.

crystal oscillator An oscillator using a piezoelectric crystal as the tuned circuit that controls the resonant frequency.

crystal-controlled oscillator An oscillator in which a piezoelectric-effect crystal is coupled to a tuned oscillator circuit in such a way that the crystal pulls the oscillator frequency to its own natural frequency and does not allow frequency drift.

current (1—general) A general term for the transfer of electricity, or the movement of electrons or *holes*. **(2—alternating)** An electric current that is constantly varying in amplitude and periodically reversing direction. **(3—average)** The arithmetic mean of the instantaneous values of current, averaged over one complete half cycle. **(4—charging)** The current that flows in to charge a capacitor when it is first connected to a source of electric potential. **(5—direct)** Electric current that flows in one direction only. **(6—eddy)** A wasteful current that flows in the core of a transformer and produces heat. *Eddy currents* are largely eliminated through the use of laminated cores. **(7—effective)** The ac current that will produce the same effective heat in a resistor as is produced by dc. If the ac is sinusoidal, the *effective current* value is 0.707 times the peak ac value. **(8—fault)** The current that flows between conductors or to ground during a fault condition. **(9—ground fault)** A fault current that flows to ground. **(10—ground return)** A current that returns through the earth. **(11—lagging)** A phenomenon observed in an inductive circuit where alternating current lags behind the voltage that produces it. **(12—leading)** A phenomenon observed in a capacitive circuit where alternating current leads the voltage that produces it. **(13—magnetizing)** The current in a transformer primary winding that is just sufficient to magnetize the core and offset iron losses. **(14—neutral)** The current that flows in the neutral conductor of an unbalanced polyphase power circuit. If correctly balanced, the neutral would carry no net current. **(15—peak)** The maximum value reached by a varying current during one cycle. **(16—pick-up)** The minimum current at which a relay just begins to operate. **(17—plate)** The anode current of an electron tube. **(18—residual)** The vector sum of the currents in the phase wires of an unbalanced polyphase power circuit. **(19—space)** The total current flowing through an electron tube.

current amplifier A low output impedance amplifier capable of providing high current output.

current probe A sensor, clamped around an electrical conductor, in which an induced current is developed from the magnetic field surrounding the conductor. For measurements, the current probe is connected to a suitable test instrument.

current transformer A transformer-type of instrument in which the primary carries the current to be measured and the secondary is in series with a low current ammeter. A current transformer is used to measure high values of alternating current.

current-carrying capacity A measure of the maximum current that can be carried continuously without damage to components or devices in a circuit.

cut-off frequency The frequency above or below which the output current in a circuit is reduced to a specified level.

cycle The interval of time or space required for a periodic signal to complete one period.

cycles per second The standard unit of frequency, expressed in Hertz (one cycle per second).

D

damped oscillation An oscillation exhibiting a progressive diminution of amplitude with time.

damping The dissipation and resultant reduction of any type of energy, such as electromagnetic waves.

dB (decibel) A measure of voltage, current, or power gain equal to 0.1 Bel. Decibels are given by the equations 20 log V_{out}/V_{in}, 20 log I_{out}/I_{in}, or 10 log P_{out}/P_{in}.

dBk A measure of power relative to 1 kilowatt. 0 dBk equals 1 kW.

dBm (decibels above 1 milliwatt) A logarithmic measure of power with respect to a reference power of one milliwatt.

dBmv A measure of voltage gain relative to 1 millivolt at 75 ohms.

dBr The power difference expressed in dB between any point and a reference point selected as the *zero relative transmission level* point. A power expressed in *dBr* does not specify the absolute power; it is a relative measurement only.

dBu A term that reflects comparison between a measured value of voltage and a reference value of 0.775 V, expressed under conditions in which the impedance at the point of measurement (and of the reference source) are not considered.

dbV A measure of voltage gain relative to 1 V.

dBW A measure of power relative to 1 watt. 0 dBW equals 1 W.

dc An abbreviation for *direct current*. Note that the preferred usage of the term *dc* is lower case.

dc amplifier A circuit capable of amplifying dc and slowly varying alternating current signals.

dc component The portion of a signal that consists of direct current. This term may also refer to the average value of a signal.

dc coupled A connection configured so that both the signal (ac component) and the constant voltage on which it is riding (dc component) are passed from one stage to the next.

dc coupling A method of coupling one circuit to another so as to transmit the static (dc) characteristics of the signal as well as the varying (ac) characteristics. Any dc offset present on the input signal is maintained and will be present in the output.

dc offset The amount that the dc component of a given signal has shifted from its correct level.

dc signal bounce Overshoot of the proper dc voltage level resulting from multiple ac couplings in a signal path.

de-energized A system from which sources of power have been disconnected.

deca A prefix meaning *ten*.

decay The reduction in amplitude of a signal on an exponential basis.

decay time The time required for a signal to fall to a certain fraction of its original value.

decibel (dB) One tenth of a Bel. The decibel is a logarithmic measure of the ratio between two powers.

decode The process of recovering information from a signal into which the information has been encoded.

decoder A device capable of deciphering encoded signals. A decoder interprets input instructions and initiates the appropriate control operations as a result.

decoupling The reduction or removal of undesired coupling between two circuits or stages.

deemphasis The reduction of the high-frequency components of a received signal to reverse the preemphasis that was placed on them to overcome attenuation and noise in the transmission process.

defect An error made during initial planning that is normally detected and corrected during the development phase. Note that a *fault* is an error that occurs in an in-service system.

deflection The control placed on electron direction and motion in CRTs and other vacuum tube devices by varying the strengths of electrostatic (electrical) or electromagnetic fields.

degradation In susceptibility testing, any undesirable change in the operational performance of a test specimen. This term does not necessarily mean malfunction or catastrophic failure.

degradation failure A failure that results from a gradual change in performance characteristics of a system or part with time.

delay The amount of time by which a signal is delayed or an event is retarded.

delay circuit A circuit designed to delay a signal passing through it by a specified amount.

delay distortion The distortion resulting from the difference in phase delays at two frequencies of interest.

delay equalizer A network that adjusts the velocity of propagation of the frequency components of a complex signal to counteract the delay distortion characteristics of a transmission channel.

delay line A transmission network that increases the propagation time of a signal traveling through it.

delta connection A common method of joining together a three-phase power supply, with each phase across a different pair of the three wires used.

delta-connected system A 3-phase power distribution system where a single-phase output can be derived from each of the adjacent pairs of an equilateral triangle formed by the service drop transformer secondary windings.

demodulator Any device that recovers the original signal after it has modulated a high-frequency carrier. The output from the unit may be in baseband composite form.

demultiplexer (demux) A device used to separate two or more signals that were previously combined by a compatible multiplexer and are transmitted over a single channel.

derating factor An operating safety margin provided for a component or system to ensure reliable performance. A *derating allowance* also is typically provided for operation under extreme environmental conditions, or under stringent reliability requirements.

desiccant A drying agent used for drying out cable splices or sensitive equipment.

design A layout of all the necessary equipment and facilities required to make a special circuit, piece of equipment, or system work.

design objective The desired electrical or mechanical performance characteristic for electronic circuits and equipment.

detection The rectification process that results in the modulating signal being separated from a modulated wave.

detectivity The reciprocal of *noise equivalent power.*

detector A device that converts one type of energy into another.

device A functional circuit, component, or network unit, such as a vacuum tube or transistor.

dewpoint The temperature at which moisture will condense out.

diagnosis The process of locating errors in software, or equipment faults in hardware.

diagnostic routine A software program designed to trace errors in software, locate hardware faults, or identify the cause of a breakdown.

dielectric An insulating material that separates the elements of various components, including capacitors and transmission lines. Dielectric materials include air, plastic, mica, ceramic, and Teflon. A dielectric material must be an insulator. (*Teflon* is a registered trademark of Du Pont.)

dielectric constant The ratio of the capacitance of a capacitor with a certain dielectric material to the capacitance with a vacuum as the dielectric. The *dielectric constant* is considered a measure of the capability of a dielectric material to store an electrostatic charge.

dielectric strength The potential gradient at which electrical breakdown occurs.

differential amplifier An input circuit that rejects voltages that are the same at both input terminals but amplifies any voltage difference between the inputs. Use of a differential amplifier causes any signal present on both terminals, such as common mode hum, to cancel itself.

differential dc The maximum dc voltage that can be applied between the differential inputs of an amplifier while maintaining linear operation.

differential gain The difference in output amplitude (expressed in percent or dB) of a small high frequency sinewave signal at two stated levels of a low frequency signal on which it is superimposed.

differential phase The difference in output phase of a small high frequency sinewave signal at two stated levels of a low frequency signal on which it is superimposed.

differential-mode interference An interference source that causes a change in potential of one side of a signal transmission path relative to the other side.

diffuse reflection The scattering effect that occurs when light, radio, or sound waves strike a rough surface.

diffusion The spreading or scattering of a wave, such as a radio wave.

diode A semiconductor or vacuum tube with two electrodes that passes electric current in one direction only. Diodes are used in rectifiers, gates, modulators, and detectors.

direct coupling A coupling method between stages that permits dc current to flow between the stages.

direct current An electrical signal in which the direction of current flow remains constant.

discharge The conversion of stored energy, as in a battery or capacitor, into an electric current.

discontinuity An abrupt nonuniform point of change in a transmission circuit that causes a disruption of normal operation.

discrete An individual circuit component.

discrete component A separately contained circuit element with its own external connections.

discriminator A device or circuit whose output amplitude and polarity vary according to how much the input signal varies from a standard or from another signal. A discriminator can be used to recover the modulating waveform in a frequency modulated signal.

dish An antenna system consisting of a parabolic shaped reflector with a signal feed element at the focal point. Dish antennas commonly are used for transmission and reception from microwave stations and communications satellites.

dispersion The wavelength dependence of a parameter.

display The representation of text and images on a cathode-ray tube, an array of light-emitting diodes, a liquid-crystal readout, or another similar device.

display device An output unit that provides a visual representation of data.

distortion The difference between the wave shape of an original signal and the signal after it has traversed a transmission circuit.

distortion-limited operation The condition prevailing when the shape of the signal, rather than the amplitude (or power), is the limiting factor in communication capability. This condition is reached when the system distorts the shape of the waveform beyond tolerable limits. For linear systems, *distortion-limited* operation is equivalent to *bandwidth-limited* operation.

disturbance The interference with normal conditions and communications by some external energy source.

disturbance current The unwanted current of any irregular phenomenon associated with transmission that tends to limit or interfere with the interchange of information.

disturbance power The unwanted power of any irregular phenomenon associated with transmission that tends to limit or interfere with the interchange of information.

disturbance voltage The unwanted voltage of any irregular phenomenon associated with transmission that tends to limit or interfere with the interchange of information.

diversity receiver A receiver using two antennas connected through circuitry that senses which antenna is receiving the stronger signal. Electronic gating permits the stronger source to be routed to the receiving system.

documentation A written description of a program. *Documentation* can be considered as any record that has permanence and can be read by humans or machines.

down-lead A lead-in wire from an antenna to a receiver.

downlink The portion of a communication link used for transmission of signals from a satellite or airborne platform to a surface terminal.

downstream A specified signal modification occurring after other given devices in a signal path.

downtime The time during which equipment is not capable of doing useful work because of malfunction. This does not include preventive maintenance time. In other words, *downtime* is measured from the occurrence of a malfunction to the correction of that malfunction.

drift A slow change in a nominally constant signal characteristic, such as frequency.

drift-space The area in a klystron tube in which electrons drift at their entering velocities and form electron *bunches*.

drive The input signal to a circuit, particularly to an amplifier.

driver An electronic circuit that supplies an isolated output to drive the input of another circuit.

drop-out value The value of current or voltage at which a relay will cease to be operated.

dropout The momentary loss of a signal.

dropping resistor A resistor designed to carry current that will make a required voltage available.

duplex separation The frequency spacing required in a communications system between the *forward* and *return* channels to maintain interference at an acceptably low level.

duplex signaling A configuration permitting signaling in both transmission directions simultaneously.

duty cycle The ratio of operating time to total elapsed time of a device that operates intermittently, expressed in percent.

dynamic A situation in which the operating parameters and/or requirements of a given system are continually changing.

dynamic range The maximum range or extremes in amplitude, from the lowest to the highest (noise floor to system clipping), that a system is capable of reproducing. The dynamic range is expressed in dB against a reference level.

dynamo A rotating machine, normally a dc generator.

dynamotor A rotating machine used to convert dc into ac.

E

earth A large conducting body with no electrical potential, also called *ground.*

earth capacitance The capacitance between a given circuit or component and a point at ground potential.

earth current A current that flows to earth/ground, especially one that follows from a fault in the system. *Earth current* may also refer to a current that flows in the earth, resulting from ionospheric disturbances, lightning, or faults on power lines.

earth fault A fault that occurs when a conductor is accidentally grounded/earthed, or when the resistance to earth of an insulator falls below a specified value.

earth ground A large conducting body that represents *zero level* in the scale of electrical potential. An *earth ground* is a connection made either accidentally or by design between a conductor and earth.

earth potential The potential taken to be the arbitrary zero in a scale of electric potential.

effective ground A connection to ground through a medium of sufficiently low impedance and adequate current-carrying capacity to prevent the buildup of voltages that might be hazardous to equipment or personnel.

effective resistance The increased resistance of a conductor to an alternating current resulting from the *skin effect*, relative to the direct-current resistance of the conductor. Higher frequencies tend to travel only on the outer skin of the conductor, whereas dc flows uniformly through the entire area.

efficiency The useful power output of an electrical device or circuit divided by the total power input, expressed in percent.

electric Any device or circuit that produces, operates on, transmits, or uses electricity.

electric charge An excess of either electrons or protons within a given space or material.

electric field strength The magnitude, measured in volts per meter, of the electric field in an electromagnetic wave.

electric flux The amount of electric charge, measured in coulombs, across a dielectric of specified area. *Electric flux* may also refer simply to electric lines of force.

electricity An energy force derived from the movement of negative and positive electric charges.

electrode An electrical terminal that emits, collects, or controls an electric current.

electrolysis A chemical change induced in a substance resulting from the passage of electric current through an electrolyte.

electrolyte A nonmetallic conductor of electricity in which current is carried by the physical movement of ions.

electromagnet An iron or steel core surrounded by a wire coil. The core becomes magnetized when current flows through the coil but loses its magnetism when the current flow is stopped.

electromagnetic compatibility The capability of electronic equipment or systems to operate in a specific electromagnetic environment, at designated levels of efficiency and within a defined margin of safety, without interfering with itself or other systems.

electromagnetic field The electric and magnetic fields associated with radio and light waves.

electromagnetic induction An electromotive force created with a conductor by the relative motion between the conductor and a nearby magnetic field.

electromagnetism The study of phenomena associated with varying magnetic fields, electromagnetic radiation, and moving electric charges.

electromotive force (EMF) An electrical potential, measured in volts, that can produce the movement of electrical charges.

electron A stable elementary particle with a negative charge that is mainly responsible for electrical conduction. Electrons move when under the influence of an electric field. This movement constitutes an *electric current.*

electron beam A stream of emitted electrons, usually in a vacuum.

electron gun A hot cathode that produces a finely focused stream of fast electrons, which are necessary for the operation of a vacuum tube, such as a cathode ray tube. The gun is made up of a hot cathode electron source, a control grid, accelerating anodes, and (usually) focusing electrodes.

electron lens A device used for focusing an electron beam in a cathode ray tube. Such focusing can be accomplished by either magnetic forces, in which external coils are used to create the proper magnetic field within the tube, or electrostatic forces, where metallic plates within the tube are charged electrically in such a way as to control the movement of electrons in the beam.

electron volt The energy acquired by an electron in passing through a potential difference of one volt in a vacuum.

electronic A description of devices (or systems) that are dependent on the flow of electrons in electron tubes, semiconductors, and other devices, and not solely on electron flow in ordinary wires, inductors, capacitors, and similar passive components.

Electronic Industries Association (EIA) A trade organization, based in Washington, DC, representing the manufacturers of electronic systems and parts, including communications systems. The association develops standards for electronic components and systems.

electronic switch A transistor, semiconductor diode, or a vacuum tube used as an on/off switch in an electrical circuit. Electronic switches can be controlled manually, by other circuits, or by computers.

electronics The field of science and engineering that deals with electron devices and their utilization.

electroplate The process of coating a given material with a deposit of metal by electrolytic action.

electrostatic The condition pertaining to electric charges that are at rest.

electrostatic field The space in which there is electric stress produced by static electric charges.

electrostatic induction The process of inducing static electric charges on a body by bringing it near other bodies that carry high electrostatic charges.

element A substance that consists of atoms of the same atomic number. Elements are the basic units in all chemical changes other than those in which *atomic changes*, such as fusion and fission, are involved.

EMI (electromagnetic interference) Undesirable electromagnetic waves that are radiated unintentionally from an electronic circuit or device into other circuits or devices, disrupting their operation.

emission (1—radiation) The radiation produced, or the production of radiation by a radio transmitting system. The emission is considered to be a *single emission* if the modulating signal and other characteristics are the same for every transmitter of the radio transmitting system and the spacing between antennas is not more than a few wavelengths. **(2—cathode)** The release of electrons from the cathode of a vacuum tube. **(3—parasitic)** A spurious radio frequency emission unintentionally generated at frequencies that are independent of the carrier frequency being amplified or modulated. **(4—secondary)** In an electron tube, emission of electrons by a plate or grid because of bombardment by *primary emission* electrons from the cathode of the tube. **(5—spurious)** An emission outside the radio frequency band authorized for a transmitter. **(6—thermonic)** An emission from a cathode resulting from high temperature.

emphasis The intentional alteration of the frequency-amplitude characteristics of a signal to reduce the adverse effects of noise in a communication system.

empirical A conclusion not based on pure theory, but on practical and experimental work.

emulation The use of one system to imitate the capabilities of another system.

enable To prepare a circuit for operation or to allow an item to function.

enabling signal A signal that permits the occurrence of a specified event.

encode The conversion of information from one form into another to obtain characteristics required by a transmission or storage system.

encoder A device that processes one or more input signals into a specified form for transmission and/or storage.

energized The condition when a circuit is switched on, or powered up.

energy spectral density A frequency-domain description of the energy in each of the frequency components of a pulse.

envelope The boundary of the family of curves obtained by varying a parameter of a wave.

envelope delay The difference in absolute delay between the fastest and slowest propagating frequencies within a specified bandwidth.

envelope delay distortion The maximum difference or deviation of the envelope-delay characteristic between any two specified frequencies.

envelope detection A demodulation process that senses the shape of the modulated RF envelope. A diode detector is one type of envelop detection device.

environmental An equipment specification category relating to temperature and humidity.

EQ (equalization) network A network connected to a circuit to correct or control its transmission frequency characteristics.

equalization (EQ) The reduction of frequency distortion and/or phase distortion of a circuit through the introduction of one or more networks to compensate for the difference in attenuation, time delay, or both, at the various frequencies in the transmission band.

equalize The process of inserting in a line a network with complementary transmission characteristics to those of the line, so that when the loss or delay in the line and that in the equalizer are combined, the overall loss or delay is approximately equal at all frequencies.

equalizer A network that corrects the transmission-frequency characteristics of a circuit to allow it to transmit selected frequencies in a uniform manner.

equatorial orbit The plane of a satellite orbit which coincides with that of the equator of the primary body.

equipment A general term for electrical apparatus and hardware, switching systems, and transmission components.

equipment failure The condition when a hardware fault stops the successful completion of a task.

equipment ground A protective ground consisting of a conducting path to ground of noncurrent carrying metal parts.

equivalent circuit A simplified network that emulates the characteristics of the real circuit it replaces. An equivalent circuit is typically used for mathematical analysis.

equivalent noise resistance A quantitative representation in resistance units of the spectral density of a noise voltage generator at a specified frequency.

error A collective term that includes all types of inconsistencies, transmission deviations, and control failures.

excitation The current that energizes field coils in a generator.

expandor A device with a nonlinear gain characteristic that acts to increase the gain more on larger input signals than it does on smaller input signals.

extremely high frequency (EHF) The band of microwave frequencies between the limits of 30 GHz and 300 GHz (wavelengths between 1 cm and 1 mm).

extremely low frequency The radio signals with operating frequencies below 300 Hz (wavelengths longer than 1000 km).

F

fail-safe operation A type of control architecture for a system that prevents improper functioning in the event of circuit or operator failure.

failure A detected cessation of ability to perform a specified function or functions within previously established limits. A *failure* is beyond adjustment by the operator by means of controls normally accessible during routine operation of the system. (This requires that measurable limits be established to define "satisfactory performance".)

failure effect The result of the malfunction or failure of a device or component.

failure in time (FIT) A unit value that indicates the reliability of a component or device. One failure in time corresponds to a failure rate of 10^{-9} per hour.

failure mode and effects analysis (FMEA) An iterative documented process performed to identify basic faults at the component level and determine their effects at higher levels of assembly.

failure rate The ratio of the number of actual failures to the number of times each item has been subjected to a set of specified stress conditions.

fall time The length of time during which a pulse decreases from 90 percent to 10 percent of its maximum amplitude.

farad The standard unit of capacitance equal to the value of a capacitor with a potential of one volt between its plates when the charge on one plate is one coulomb and there is an equal and opposite charge on the other plate. The farad is a large value and is more commonly expressed in *microfarads* or *picofarads*. The *farad* is named for the English chemist and physicist Michael Faraday (179–1867).

fast frequency shift keying (FFSK) A system of digital modulation where the digits are represented by different frequencies that are related to the baud rate, and where transitions occur at the zero crossings.

fatigue The reduction in strength of a metal caused by the formation of crystals resulting from repeated flexing of the part in question.

fault A condition that causes a device, a component, or an element to fail to perform in a required manner. Examples include a short-circuit, broken wire, or intermittent connection.

fault to ground A fault caused by the failure of insulation and the consequent establishment of a direct path to ground from a part of the circuit that should not normally be grounded.

fault tree analysis (FTA) An iterative documented process of a systematic nature performed to identify basic faults, determine their causes and effects, and establish their probabilities of occurrence.

feature A distinctive characteristic or part of a system or piece of equipment, usually visible to end users and designed for their convenience.

Federal Communications Commission (FCC) The federal agency empowered by law to regulate all interstate radio and wireline communications services originating in the United States, including radio, television, facsimile, telegraph, data transmission, and telephone systems. The agency was established by the Communications Act of 1934.

feedback The return of a portion of the output of a device to the input. *Positive feedback* adds to the input, *negative feedback* subtracts from the input.

feedback amplifier An amplifier with the components required to feed a portion of the output back into the input to alter the characteristics of the output signal.

feedline A transmission line, typically coaxial cable, that connects a high frequency energy source to its load.

femto A prefix meaning *one quadrillionth* (10^{-15}).

ferrite A ceramic material made of powdered and compressed ferric oxide, plus other oxides (mainly cobalt, nickel, zinc, yttrium-iron, and manganese). These materials have low eddy current losses at high frequencies.

ferromagnetic material A material with low relative permeability and high coercive force so that it is difficult to magnetize and demagnetize. Hard ferromagnetic materials retain magnetism well, and are commonly used in permanent magnets.

fidelity The degree to which a system, or a portion of a system, accurately reproduces at its output the essential characteristics of the signal impressed upon its input.

field strength The strength of an electric, magnetic, or electromagnetic field.

filament A wire that becomes hot when current is passed through it, used either to emit light (for a light bulb) or to heat a cathode to enable it to emit electrons (for an electron tube).

film resistor A type of resistor made by depositing a thin layer of resistive material on an insulating core.

filter A network that passes desired frequencies but greatly attenuates other frequencies.

filtered noise White noise that has been passed through a filter. The power spectral density of filtered white noise has the same shape as the transfer function of the filter.

fitting A coupling or other mechanical device that joins one component with another.

fixed A system or device that is not changeable or movable.

flashover An arc or spark between two conductors.

flashover voltage The voltage between conductors at which flashover just occurs.

flat face tube The design of CRT tube with almost a flat face, giving improved legibility of text and reduced reflection of ambient light.

flat level A signal that has an equal amplitude response for all frequencies within a stated range.

flat loss A circuit, device, or channel that attenuates all frequencies of interest by the same amount, also called *flat slope*.

flat noise A noise whose power per unit of frequency is essentially independent of frequency over a specified frequency range.

flat response The performance parameter of a system in which the output signal amplitude of the system is a faithful reproduction of the input amplitude over some range of specified input frequencies.

floating A circuit or device that is not connected to any source of potential or to ground.

fluorescence The characteristic of a material to produce light when excited by an external energy source. Minimal or no heat results from the process.

flux The electric or magnetic lines of force resulting from an applied energy source.

flywheel effect The characteristic of an oscillator that enables it to sustain oscillations after removal of the control stimulus. This characteristic may be desirable, as in the case of a phase-locked loop employed in a synchronous system, or undesirable, as in the case of a voltage-controlled oscillator.

focusing A method of making beams of radiation converge on a target, such as the face of a CRT.

Fourier analysis A mathematical process for transforming values between the frequency domain and the time domain. This term also refers to the decomposition of a time-domain signal into its frequency components.

Fourier transform An integral that performs an actual transformation between the frequency domain and the time domain in Fourier analysis.

frame A segment of an analog or digital signal that has a repetitive characteristic, in that corresponding elements of successive *frames* represent the same things.

free electron An electron that is not attached to an atom and is, thus, mobile when an electromotive force is applied.

free running An oscillator that is not controlled by an external synchronizing signal.

free-running oscillator An oscillator that is not synchronized with an external timing source.

frequency The number of complete cycles of a periodic waveform that occur within a given length of time. Frequency is usually specified in cycles per second (*Hertz*). Frequency is the reciprocal of wavelength. The higher the frequency, the shorter the wavelength. In general, the higher the frequency of a signal, the more capacity it has to carry information, the smaller an antenna is required, and the more susceptible the signal is to absorption by the atmosphere and by physical structures. At microwave frequencies, radio signals take on a *line-of-sight* characteristic and require highly directional and focused antennas to be used successfully.

frequency accuracy The degree of conformity of a given signal to the specified value of a frequency.

frequency allocation The designation of radio-frequency bands for use by specific radio services.

frequency content The band of frequencies or specific frequency components contained in a signal.

frequency converter A circuit or device used to change a signal of one frequency into another of a different frequency.

frequency coordination The process of analyzing frequencies in use in various bands of the spectrum to achieve reliable performance for current and new services.

frequency counter An instrument or test set used to measure the frequency of a radio signal or any other alternating waveform.

frequency departure An unintentional deviation from the nominal frequency value.

frequency difference The algebraic difference between two frequencies. The two frequencies can be of identical or different nominal values.

frequency displacement The end-to-end shift in frequency that may result from independent frequency translation errors in a circuit.

frequency distortion The distortion of a multifrequency signal caused by unequal attenuation or amplification at the different frequencies of the signal. This term may also be referred to as *amplitude distortion*.

frequency domain A representation of signals as a function of frequency, rather than of time.

frequency modulation (FM) The modulation of a carrier signal so that its instantaneous frequency is proportional to the instantaneous value of the modulating wave.

frequency multiplier A circuit that provides as an output an exact multiple of the input frequency.

frequency offset A frequency shift that occurs when a signal is sent over an analog transmission facility in which the modulating and demodulating frequencies are not identical. A channel with frequency offset does not preserve the waveform of a transmitted signal.

frequency response The measure of system linearity in reproducing signals across a specified bandwidth. Frequency response is expressed as a frequency range with a specified amplitude tolerance in dB.

frequency response characteristic The variation in the transmission performance (gain or loss) of a system with respect to variations in frequency.

frequency reuse A technique used to expand the capacity of a given set of frequencies or channels by separating the signals either geographically or through the use of different polarization techniques. Frequency reuse is a common element of the *frequency coordination* process.

frequency selectivity The ability of equipment to separate or differentiate between signals at different frequencies.

frequency shift The difference between the frequency of a signal applied at the input of a circuit and the frequency of that signal at the output.

frequency shift keying (FSK) A commonly-used method of digital modulation in which a one and a zero (the two possible states) are each transmitted as separate frequencies.

frequency stability A measure of the variations of the frequency of an oscillator from its mean frequency over a specified period of time.

frequency standard An oscillator with an output frequency sufficiently stable and accurate that it is used as a reference.

frequency-division multiple access (FDMA) The provision of multiple access to a transmission facility, such as an earth satellite, by assigning each transmitter its own frequency band.

frequency-division multiplexing (FDM) The process of transmitting multiple analog signals by an orderly assignment of frequency slots, that is, by dividing transmission bandwidth into several narrow bands, each of which carries a single communication and is sent simultaneously with others over a common transmission path.

full duplex A communications system capable of transmission simultaneously in two directions.

full-wave rectifier A circuit configuration in which both positive and negative half-cycles of the incoming ac signal are rectified to produce a unidirectional (dc) current through the load.

functional block diagram A diagram illustrating the definition of a device, system, or problem on a logical and functional basis.

functional unit An entity of hardware and/or software capable of accomplishing a given purpose.

fundamental frequency The lowest frequency component of a complex signal.

fuse A protective device used to limit current flow in a circuit to a specified level. The fuse consists of a metallic link that melts and opens the circuit at a specified current level.

fuse wire A fine-gauge wire made of an alloy that overheats and melts at the relatively low temperatures produced when the wire carries overload currents. When used in a fuse, the wire is called a fuse (or fusible) link.

G

gain An increase or decrease in the level of an electrical signal. Gain is measured in terms of decibels or number-of-times of magnification. Strictly speaking, *gain* refers to an increase in level. Negative numbers, however, are commonly used to denote a decrease in level.

gain-bandwidth The gain times the frequency of measurement when a device is biased for maximum obtainable gain.

gain/frequency characteristic The gain-versus-frequency characteristic of a channel over the bandwidth provided, also referred to as *frequency response*.

gain/frequency distortion A circuit defect in which a change in frequency causes a change in signal amplitude.

galvanic A device that produces direct current by chemical action.

gang The mechanical connection of two or more circuit devices so that they can all be adjusted simultaneously.

gang capacitor A variable capacitor with more than one set of moving plates linked together.

gang tuning The simultaneous tuning of several different circuits by turning a single shaft on which ganged capacitors are mounted.

ganged One or more devices that are mechanically coupled, normally through the use of a shared shaft.

gas breakdown The ionization of a gas between two electrodes caused by the application of a voltage that exceeds a threshold value. The ionized path has a low impedance. Certain types of circuit and line protectors rely on gas breakdown to divert hazardous currents away from protected equipment.

gas tube A protection device in which a sufficient voltage across two electrodes causes a gas to ionize, creating a low impedance path for the discharge of dangerous voltages.

gas-discharge tube A gas-filled tube designed to carry current during gas breakdown. The gas-discharge tube is commonly used as a protective device, preventing high voltages from damaging sensitive equipment.

gauge A measure of wire diameter. In measuring wire gauge, the lower the number, the thicker the wire.

Gaussian distribution A statistical distribution, also called the *normal* distribution. The graph of a Gaussian distribution is a bell-shaped curve.

Gaussian noise Noise in which the distribution of amplitude follows a Gaussian model, that is, the noise is random but distributed about a reference voltage of zero.

Gaussian pulse A pulse that has the same form as its own Fourier transform.

generator A machine that converts mechanical energy into electrical energy, or one form of electrical energy into another form.

geosynchronous The attribute of a satellite in which the relative position of the satellite as viewed from the surface of a given planet is stationary. For earth, the geosynchronous position is 22,300 miles above the planet.

getter A metal used in vaporized form to remove residual gas from inside an electron tube during manufacture.

giga A prefix meaning one billion.

gigahertz (GHz) A measure of frequency equal to one billion cycles per second. Signals operating above 1 gigahertz are commonly known as *microwaves*, and begin to take on the characteristics of visible light.

glitch A general term used to describe a wide variety of momentary signal discontinuities.

graceful degradation An equipment failure mode in which the system suffers reduced capability, but does not fail altogether.

graticule A fixed pattern of reference markings used with oscilloscope CRTs to simplify measurements. The graticule may be etched on a transparent plate covering the front of the

CRT or, for greater accuracy in readings, may be electrically generated within the CRT itself.

grid (1—general) A mesh electrode within an electron tube that controls the flow of electrons between the cathode and plate of the tube. **(2—bias)** The potential applied to a grid in an electron tube to control its center operating point. **(3—control)** The grid in an electron tube to which the input signal is usually applied. **(4—screen)** The grid in an electron tube, typically held at a steady potential, that screens the control grid from changes in anode potential. **(5—suppressor)** The grid in an electron tube near the anode (plate) that suppresses the emission of secondary electrons from the plate.

ground An electrical connection to earth or to a common conductor usually connected to earth.

ground clamp A clamp used to connect a ground wire to a ground rod or system.

ground loop An undesirable circulating ground current in a circuit grounded via multiple connections or at multiple points.

ground plane A conducting material at ground potential, physically close to other equipment, so that connections may be made readily to ground the equipment at the required points.

ground potential The point at zero electric potential.

ground return A conductor used as a path for one or more circuits back to the ground plane or central facility ground point.

ground rod A metal rod driven into the earth and connected into a mesh of interconnected rods so as to provide a low resistance link to ground.

ground window A single-point interface between the integrated ground plane of a building and an isolated ground plane.

ground wire A copper conductor used to extend a good low-resistance earth ground to protective devices in a facility.

grounded The connection of a piece of equipment to earth via a low resistance path.

grounding The act of connecting a device or circuit to ground or to a conductor that is grounded.

group delay A condition where the different frequency elements of a given signal suffer differing propagation delays through a circuit or a system. The delay at a lower frequency is different from the delay at a higher frequency, resulting in a time-related distortion of the signal at the receiving point.

group delay time The rate of change of the total phase shift of a waveform with angular frequency through a device or transmission facility.

group velocity The speed of a pulse on a transmission line.

guard band A narrow bandwidth between adjacent channels intended to reduce interference or crosstalk.

H

half-wave rectifier A circuit or device that changes only positive or negative half-cycle inputs of alternating current into direct current.

Hall effect The phenomenon by which a voltage develops between the edges of a current-carrying metal strip whose faces are perpendicular to an external magnetic field.

hard-wired Electrical devices connected through physical wiring.

harden The process of constructing military telecommunications facilities so as to protect them from damage by enemy action, especially *electromagnetic pulse* (EMP) radiation.

hardware Physical equipment, such as mechanical, magnetic, electrical, or electronic devices or components.

harmonic A periodic wave having a frequency that is an integral multiple of the fundamental frequency. For example, a wave with twice the frequency of the fundamental is called the *second harmonic*.

harmonic analyzer A test set capable of identifying the frequencies of the individual signals that make up a complex wave.

harmonic distortion The production of harmonics at the output of a circuit when a periodic wave is applied to its input. The level of the distortion is usually expressed as a percentage of the level of the input.

hazard A condition that could lead to danger for operating personnel.

headroom The difference, in decibels, between the typical operating signal level and a peak overload level.

heat loss The loss of useful electrical energy resulting from conversion into unwanted heat.

heat sink A device that conducts heat away from a heat-producing component so that it stays within a safe working temperature range.

heater In an electron tube, the filament that heats the cathode to enable it to emit electrons.

hecto A prefix meaning 100.

henry The standard unit of electrical inductance, equal to the self-inductance of a circuit or the mutual inductance of two circuits when there is an induced electromotive force of one volt and a current change of one ampere per second. The symbol for inductance is *H*, named for the American physicist Joseph Henry (1797–1878).

hertz (Hz) The unit of frequency that is equal to one cycle per second. Hertz is the reciprocal of the *period*, the interval after which the same portion of a periodic waveform recurs. Hertz was named for the German physicist Heinrich R. Hertz (1857–1894).

heterodyne The mixing of two signals in a nonlinear device in order to produce two additional signals at frequencies that are the sum and difference of the original frequencies.

heterodyne frequency The sum of, or the difference between, two frequencies, produced by combining the two signals together in a modulator or similar device.

heterodyne wavemeter A test set that uses the heterodyne principle to measure the frequencies of incoming signals.

high-frequency loss Loss of signal amplitude at higher frequencies through a given circuit or medium. For example, high frequency loss could be caused by passing a signal through a coaxial cable.

high Q An inductance or capacitance whose ratio of reactance to resistance is high.

high tension A high voltage circuit.

high-pass filter A network that passes signals of higher than a specified frequency but attenuates signals of all lower frequencies.

homochronous Signals whose corresponding significant instants have a constant but uncontrolled phase relationship with each other.

horn gap A lightning arrester utilizing a gap between two horns. When lightning causes a discharge between the horns, the heat produced lengthens the arc and breaks it.

horsepower The basic unit of mechanical power. One horsepower (hp) equals 550 foot-pounds per second or 746 watts.

hot A charged electrical circuit or device.

hot dip galvanized The process of galvanizing steel by dipping it into a bath of molten zinc.

hot standby System equipment that is fully powered but not in service. A *hot standby* can rapidly replace a primary system in the event of a failure.

hum Undesirable coupling of the 60 Hz power sine wave into other electrical signals and/or circuits.

HVAC An abbreviation for *heating, ventilation, and air conditioning* system.

hybrid system A communication system that accommodates both digital and analog signals.

hydrometer A testing device used to measure specific gravity, particularly the specific gravity of the dilute sulphuric acid in a lead-acid storage battery, to learn the state of charge of the battery.

hygrometer An instrument that measures the relative humidity of the atmosphere.

hygroscopic The ability of a substance to absorb moisture from the air.

hysteresis The property of an element evidenced by the dependence of the value of the output, for a given excursion of the input, upon the history of prior excursions and direction of the input. Originally, *hysteresis* was the name for magnetic phenomena only—the lagging of flux density behind the change in value of the magnetizing flux—but now, the term is also used to describe other inelastic behavior.

hysteresis loop The plot of magnetizing current against magnetic flux density (or of other similarly related pairs of parameters), which appears as a loop. The area within the loop is proportional to the power loss resulting from hysteresis.

hysteresis loss The loss in a magnetic core resulting from hysteresis.

I

I^2R loss The power lost as a result of the heating effect of current passing through resistance.

idling current The current drawn by a circuit, such as an amplifier, when no signal is present at its input.

image frequency A frequency on which a carrier signal, when heterodyned with the local oscillator in a superheterodyne receiver, will cause a sum or difference frequency that is the same as the intermediate frequency of the receiver. Thus, a signal on an *image frequency* will be demodulated along with the desired signal and will interfere with it.

impact ionization The ionization of an atom or molecule as a result of a high energy collision.

impedance The total passive opposition offered to the flow of an alternating current. *Impedance* consists of a combination of resistance, inductive reactance, and capacitive reactance. It is the vector sum of resistance and reactance ($R + jX$) or the vector of magnitude Z at an angle θ.

impedance characteristic A graph of the impedance of a circuit showing how it varies with frequency.

impedance irregularity A discontinuity in an impedance characteristic caused, for example, by the use of different coaxial cable types.

impedance matching The adjustment of the impedances of adjoining circuit components to a common value so as to minimize reflected energy from the junction and to maximize energy transfer across it. Incorrect adjustment results in an *impedance mismatch*.

impedance matching transformer A transformer used between two circuits of different impedances with a turns ratio that provides for maximum power transfer and minimum loss by reflection.

impulse A short high energy surge of electrical current in a circuit or on a line.

impulse current A current that rises rapidly to a peak then decays to zero without oscillating.

impulse excitation The production of an oscillatory current in a circuit by impressing a voltage for a relatively short period compared with the duration of the current produced.

impulse noise A noise signal consisting of random occurrences of energy spikes, having random amplitude and bandwidth.

impulse response The amplitude-versus-time output of a transmission facility or device in response to an impulse.

impulse voltage A unidirectional voltage that rises rapidly to a peak and then falls to zero, without any appreciable oscillation.

in-phase The property of alternating current signals of the same frequency that achieve their peak positive, peak negative, and zero amplitude values simultaneously.

incidence angle The angle between the perpendicular to a surface and the direction of arrival of a signal.

increment A small change in the value of a quantity.

induce To produce an electrical or magnetic effect in one conductor by changing the condition or position of another conductor.

induced current The current that flows in a conductor because a voltage has been induced across two points in, or connected to, the conductor.

induced voltage A voltage developed in a conductor when the conductor passes through magnetic lines of force.

inductance The property of an inductor that opposes any change in a current that flows through it. The standard unit of inductance is the *Henry.*

induction The electrical and magnetic interaction process by which a changing current in one circuit produces a voltage change not only in its own circuit (*self inductance*) but also in other circuits to which it is linked magnetically.

inductive A circuit element exhibiting inductive reactance.

inductive kick A voltage surge produced when a current flowing through an inductance is interrupted.

inductive load A load that possesses a net inductive reactance.

inductive reactance The reactance of a circuit resulting from the presence of inductance and the phenomenon of induction.

inductor A coil of wire, usually wound on a core of high permeability, that provides high inductance without necessarily exhibiting high resistance.

inert An inactive unit, or a unit that has no power requirements.

infinite line A transmission line that appears to be of infinite length. There are no reflections back from the far end because it is terminated in its characteristic impedance.

infra low frequency (ILF) The frequency band from 300 Hz to 3000 Hz.

inhibit A control signal that prevents a device or circuit from operating.

injection The application of a signal to an electronic device.

input The waveform fed into a circuit, or the terminals that receive the input waveform.

insertion gain The gain resulting from the insertion of a transducer in a transmission system, expressed as the ratio of the power delivered to that part of the system following the transducer to the power delivered to that same part before insertion. If more than one component is involved in the input or output, the particular component used must be specified. This ratio is usually expressed in decibels. If the resulting number is negative, an *insertion loss* is indicated.

insertion loss The signal loss within a circuit, usually expressed in decibels as the ratio of input power to output power.

insertion loss-vs.-frequency characteristic The amplitude transfer characteristic of a system or component as a function of frequency. The amplitude response may be stated as actual gain, loss, amplification, or attenuation, or as a ratio of any one of these quantities at a particular frequency, with respect to that at a specified reference frequency.

inspection lot A collection of units of product from which a sample is drawn and inspected to determine conformance with acceptability criteria.

instantaneous value The value of a varying waveform at a given instant of time. The value can be in volts, amperes, or phase angle.

Institute of Electrical and Electronics Engineers (IEEE) The organization of electrical and electronics scientists and engineers formed in 1963 by the merger of the Institute of Radio Engineers (IRE) and the American Institute of Electrical Engineers (AIEE).

instrument multiplier A measuring device that enables a high voltage to be measured using a meter with only a low voltage range.

instrument rating The range within which an instrument has been designed to operate without damage.

insulate The process of separating one conducting body from another conductor.

insulation The material that surrounds and insulates an electrical wire from other wires or circuits. *Insulation* may also refer to any material that does not ionize easily and thus presents a large impedance to the flow of electrical current.

insulator A material or device used to separate one conducting body from another.

intelligence signal A signal containing information.

intensity The strength of a given signal under specified conditions.

interconnect cable A short distance cable intended for use between equipment (generally less than 3 m in length).

interface A device or circuit used to interconnect two pieces of electronic equipment.

interface device A unit that joins two interconnecting systems.

interference emission An emission that results in an electrical signal being propagated into and interfering with the proper operation of electrical or electronic equipment.

interlock A protection device or system designed to remove all dangerous voltages from a machine or piece of equipment when access doors or panels are opened or removed.

intermediate frequency A frequency that results from combining a signal of interest with a signal generated within a radio receiver. In superheterodyne receivers, all incoming signals are converted to a single intermediate frequency for which the amplifiers and filters of the receiver have been optimized.

intermittent A noncontinuous recurring event, often used to denote a problem that is difficult to find because of its unpredictable nature.

intermodulation The production, in a nonlinear transducer element, of frequencies corresponding to the sums and differences of the fundamentals and harmonics of two or more frequencies that are transmitted through the transducer.

intermodulation distortion (IMD) The distortion that results from the mixing of two input signals in a nonlinear system. The resulting output contains new frequencies that represent the sum and difference of the input signals and the sums and differences of their harmonics. IMD is also called *intermodulation noise*.

intermodulation noise In a transmission path or device, the noise signal that is contingent upon modulation and demodulation, resulting from nonlinear characteristics in the path or device.

internal resistance The actual resistance of a source of electric power. The total electromotive force produced by a power source is not available for external use; some of the energy is used in driving current through the source itself.

International Standards Organization (ISO) An international body concerned with worldwide standardization for a broad range of industrial products, including telecommunications equipment. Members are represented by national standards organizations, such as ANSI (American National Standards Institute) in the United States. ISO was established in 1947 as a specialized agency of the United Nations.

International Telecommunications Union (ITU) A specialized agency of the United Nations established to maintain and extend international cooperation for the maintenance, development, and efficient use of telecommunications. The union does this through standards and recommended regulations, and through technical and telecommunications studies.

International Telecommunications Satellite Consortium (Intelsat) A nonprofit cooperative of member nations that owns and operates a satellite system for international and, in many instances, domestic communications.

interoperability The condition achieved among communications and electronics systems or equipment when information or services can be exchanged directly between them or their users, or both.

interpolate The process of estimating unknown values based on a knowledge of comparable data that falls on both sides of the point in question.

interrupting capacity The rating of a circuit breaker or fuse that specifies the maximum current the device is designed to interrupt at its rated voltage.

interval The points or numbers lying between two specified endpoints.

inverse voltage The effective value of voltage across a rectifying device, which conducts a current in one direction during one half cycle of the alternating input, during the half cycle when current is not flowing.

inversion The change in the polarity of a pulse, such as from positive to negative.

inverter A circuit or device that converts a direct current into an alternating current.

ionizing radiation The form of electromagnetic radiation that can turn an atom into an ion by knocking one or more of its electrons loose. Examples of ionizing radiation include X rays, gamma rays, and cosmic rays

***IR* drop** A drop in voltage because of the flow of current (I) through a resistance (R), also called *resistance drop.*

***IR* loss** The conversion of electrical power to heat caused by the flow of electrical current through a resistance.

isochronous A signal in which the time interval separating any two significant instants is theoretically equal to a specified unit interval or to an integral multiple of the unit interval.

isolated ground A ground circuit that is isolated from all equipment framework and any other grounds, except for a single-point external connection.

isolated ground plane A set of connected frames that are grounded through a single connection to a ground reference point. That point and all parts of the frames are insulated from any other ground system in a building.

isolated pulse A pulse uninfluenced by other pulses in the same signal.

isophasing amplifier A timing device that corrects for small timing errors.

isotropic A quantity exhibiting the same properties in all planes and directions.

J

jack A receptacle or connector that makes electrical contact with the mating contacts of a plug. In combination, the plug and jack provide a ready means for making connections in electrical circuits.

jacket An insulating layer of material surrounding a wire in a cable.

jitter Small, rapid variations in a waveform resulting from fluctuations in a supply voltage or other causes.

joule The standard unit of work that is equal to the work done by one newton of force when the point at which the force is applied is displaced a distance of one meter in the direction of the force. The *joule* is named for the English physicist James Prescott Joule (1818-1889).

Julian date A chronological date in which days of the year are numbered in sequence. For example, the first day is 001, the second is 002, and the last is 365 (or 366 in a leap year).

K

Kelvin (K) The standard unit of thermodynamic temperature. Zero degrees Kelvin represents *absolute zero*. Water freezes at 273 K and water boils at 373 K under standard pressure conditions.

kilo A prefix meaning one thousand.

kilohertz (kHz) A unit of measure of frequency equal to 1,000 Hz.

kilovar A unit equal to one thousand volt-amperes.

kilovolt (kV) A unit of measure of electrical voltage equal to 1,000 V.

kilowatt A unit equal to one thousand watts.

Kirchoff's Law At any point in a circuit, there is as much current flowing into the point as there is flowing away from it.

klystron (1—general) A family of electron tubes that function as microwave amplifiers and oscillators. Simplest in form are two-cavity klystrons in which an electron beam passes through a cavity that is excited by a microwave input, producing a velocity-modulated beam which passes through a second cavity a precise distance away that is coupled to a tuned circuit, thereby producing an amplified output of the original input signal frequency. If part of the output is fed back to the input, an oscillator can be the result. **(2—multi-cavity)** An amplifier device for UHF and microwave signals based on velocity modulation of an electron beam. The beam is directed through an input cavity, where the input RF signal polarity initializes a *bunching effect* on electrons in the beam. The bunching effect excites subsequent cavities, which increase the bunching through an energy flywheel concept. Finally, the beam passes to an output cavity that couples the amplified signal to the load (antenna system). The beam falls onto a collector element that forms the return path for the current and dissipates the heat resulting from electron beam bombardment. **(3—reflex)** A klystron with only one cavity. The action is the same as in a two-cavity klystron but the beam is reflected back into the cavity in which it was first excited, after being sent out to a reflector. The one cavity, therefore, acts both as the original exciter (or buncher) and as the collector from which the output is taken.

knee In a response curve, the region of maximum curvature.

ku band Radio frequencies in the range of 15.35 GHz to 17.25 GHz, typically used for satellite telecommunications.

L

ladder network A type of filter with components alternately across the line and in the line.

lag The difference in phase between a current and the voltage that produced it, expressed in electrical degrees.

lagging current A current that lags behind the alternating electromotive force that produced it. A circuit that produces a *lagging current* is one containing inductance alone, or whose effective impedance is inductive.

lagging load A load whose combined inductive reactance exceeds its capacitive reactance. When an alternating voltage is applied, the current lags behind the voltage.

laminate A material consisting of layers of the same or different materials bonded together and built up to the required thickness.

latitude An angular measurement of a point on the earth above or below the equator. The equator represents 0°, the north pole +90°, and the south pole –90°.

layout A proposed or actual arrangement or allocation of equipment.

LC circuit An electrical circuit with both inductance (*L*) and capacitance (*C*) that is resonant at a particular frequency.

LC ratio The ratio of inductance to capacitance in a given circuit.

lead An electrical wire, usually insulated.

leading edge The initial portion of a pulse or wave in which voltage or current rise rapidly from zero to a final value.

leading load A reactive load in which the reactance of capacitance is greater than that of inductance. Current through such a load *leads* the applied voltage causing the current.

leakage The loss of energy resulting from the flow of electricity past an insulating material, the escape of electromagnetic radiation beyond its shielding, or the extension of magnetic lines of force beyond their intended working area.

leakage resistance The resistance of a path through which leakage current flows.

level The strength or intensity of a given signal.

level alignment The adjustment of transmission levels of single links and links in tandem to prevent overloading of transmission subsystems.

life cycle The predicted useful life of a class of equipment, operating under normal (specified) working conditions.

life safety system A system designed to protect life and property, such as emergency lighting, fire alarms, smoke exhaust and ventilating fans, and site security.

life test A test in which random samples of a product are checked to see how long they can continue to perform their functions satisfactorily. A form of *stress testing* is used, including temperature, current, voltage, and/or vibration effects, cycled at many times the rate that would apply in normal usage.

limiter An electronic device in which some characteristic of the output is automatically prevented from exceeding a predetermined value.

limiter circuit A circuit of nonlinear elements that restricts the electrical excursion of a variable in accordance with some specified criteria.

limiting A process by which some characteristic at the output of a device is prevented from exceeding a predetermined value.

line loss The total end-to-end loss in decibels in a transmission line.

line-up The process of adjusting transmission parameters to bring a circuit to its specified values.

linear A circuit, device, or channel whose output is directly proportional to its input.

linear distortion A distortion mechanism that is independent of signal amplitude.

linearity A constant relationship, over a designated range, between the input and output characteristics of a circuit or device.

lines of force A group of imaginary lines indicating the direction of the electric or magnetic field at all points along it.

lissajous pattern The looping patterns generated by a CRT spot when the horizontal (X) and vertical (Y) deflection signals are sinusoids. The lissajous pattern is useful for evaluating the delay or phase of two sinusoids of the same frequency.

live A device or system connected to a source of electric potential.

load The work required of an electrical or mechanical system.

load factor The ratio of the average load over a designated period of time to the peak load occurring during the same period.

load line A straight line drawn across a grouping of plate current/plate voltage characteristic curves showing the relationship between grid voltage and plate current for a particular plate load resistance of an electron tube.

logarithm The power to which a base must be raised to produce a given number. Common logarithms are to base 10.

logarithmic scale A meter scale with displacement proportional to the logarithm of the quantity represented.

long persistence The quality of a cathode ray tube that has phosphorescent compounds on its screen (in addition to fluorescent compounds) so that the image continues to glow after the original electron beam has ceased to create it by producing the usual fluorescence effect. Long persistence is often used in radar screens or where photographic evidence is needed of a display. Most such applications, however, have been superseded through the use of digital storage techniques.

longitude The angular measurement of a point on the surface of the earth in relation to the meridian of Greenwich (London). The earth is divided into 360° of longitude, beginning at the Greenwich mean. As one travels west around the globe, the longitude increases.

longitudinal current A current that travels in the same direction on both wires of a pair. The return current either flows in another pair or via a ground return path.

loss The power dissipated in a circuit, usually expressed in decibels, that performs no useful work.

loss deviation The change of actual loss in a circuit or system from a designed value.

loss variation The change in actual measured loss over time.

lossy The condition when the line loss per unit length is significantly greater than some defined normal parameter.

lossy cable A coaxial cable constructed to have high transmission loss so it can be used as an artificial load or as an attenuator.

lot size A specific quantity of similar material or a collection of similar units from a common source; in inspection work, the quantity offered for inspection and acceptance at any one time. The **lot size** may be a collection of raw material, parts, subassemblies inspected during production, or a consignment of finished products to be sent out for service.

low tension A low voltage circuit.

low-pass filter A filter network that passes all frequencies below a specified frequency with little or no loss, but that significantly attenuates higher frequencies.

lug A tag or projecting terminal onto which a wire may be connected by wrapping, soldering, or crimping.

lumped constant A resistance, inductance, or capacitance connected at a point, and not distributed uniformly throughout the length of a route or circuit.

M

mA An abbreviation for *milliamperes* (0.001 A).

magnet A device that produces a magnetic field and can attract iron, and attract or repel other magnets.

magnetic field An energy field that exists around magnetic materials and current-carrying conductors. Magnetic fields combine with electric fields in light and radio waves.

magnetic flux The field produced in the area surrounding a magnet or electric current. The standard unit of flux is the *Weber*.

magnetic flux density A vector quantity measured by a standard unit called the *Tesla*. The *magnetic flux density* is the number of magnetic lines of force per unit area, at right angles to the lines.

magnetic leakage The magnetic flux that does not follow a useful path.

magnetic pole A point that appears from the outside to be the center of magnetic attraction or repulsion at or near one end of a magnet.

magnetic storm A violent local variation in the earth's magnetic field, usually the result of sunspot activity.

magnetism A property of iron and some other materials by which external magnetic fields are maintained, other magnets being thereby attracted or repelled.

magnetization The exposure of a magnetic material to a magnetizing current, field, or force.

magnetizing force The force producing magnetization.

magnetomotive force The force that tends to produce lines of force in a magnetic circuit. The *magnetomotive force* bears the same relationship to a magnetic circuit that voltage does to an electrical circuit.

magnetron A high-power, ultra high frequency electron tube oscillator that employs the interaction of a strong electric field between an anode and cathode with the field of a strong permanent magnet to cause oscillatory electron flow through multiple internal cavity resonators. The magnetron may operate in a continuous or pulsed mode.

maintainability The probability that a failure will be repaired within a specified time after the failure occurs.

maintenance Any activity intended to keep a functional unit in satisfactory working condition. The term includes the tests, measurements, replacements, adjustments, and repairs necessary to keep a device or system operating properly.

malfunction An equipment failure or a fault.

manometer A test device for measuring gas pressure.

margin The difference between the value of an operating parameter and the value that would result in unsatisfactory operation. Typical *margin* parameters include signal level, signal-to-noise ratio, distortion, crosstalk coupling, and/or undesired emission level.

Markov model A statistical model of the behavior of a complex system over time in which the probabilities of the occurrence of various future states depend only on the present state of the system, and not on the path by which the present state was achieved. This term was named for the Russian mathematician Andrei Andreevich Markov (1856–1922).

master clock An accurate timing device that generates a synchronous signal to control other clocks or equipment.

master oscillator A stable oscillator that provides a standard frequency signal for other hardware and/or systems.

matched termination A termination that absorbs all the incident power and so produces no reflected waves or mismatch loss.

matching The connection of channels, circuits, or devices in a manner that results in minimal reflected energy.

matrix A logical network configured in a rectangular array of intersections of input/output signals.

Maxwell's equations Four differential equations that relate electric and magnetic fields to electromagnetic waves. The equations are a basis of electrical and electronic engineering.

mean An arithmetic average in which values are added and divided by the number of such values.

mean time between failures (MTBF) For a particular interval, the total functioning life of a population of an item divided by the total number of failures within the population during the measurement interval.

mean time to failure (MTTF) The measured operating time of a single piece of equipment divided by the total number of failures during the measured period of time. This measurement is normally made during that period between early life and wear-out failures.

mean time to repair (MTTR) The total corrective maintenance time on a component or system divided by the total number of corrective maintenance actions during a given period of time.

measurement A procedure for determining the amount of a quantity.

median A value in a series that has as many readings or values above it as below.

medium An electronic pathway or mechanism for passing information from one point to another.

mega A prefix meaning one million.

megahertz (MHz) A quantity equal to one million Hertz (cycles per second).

megohm A quantity equal to one million ohms.

metric system A decimal system of measurement based on the meter, the kilogram, and the second.

micro A prefix meaning one millionth.

micron A unit of length equal to one millionth of a meter (1/25,000 of an inch).

microphonic(s) Unintended noise introduced into an electronic system by mechanical vibration of electrical components.

microsecond One millionth of a second (0.000001 s).

microvolt A quantity equal to one-millionth of a volt.

milli A prefix meaning one thousandth.

milliammeter A test instrument for measuring electrical current, often part of a *multimeter*.

millihenry A quantity equal to one-thousandth of a henry.

milliwatt A quantity equal to one thousandth of a watt.

minimum discernible signal The smallest input that will produce a discernible change in the output of a circuit or device.

mixer A circuit used to combine two or more signals to produce a third signal that is a function of the input waveforms.

mixing ratio The ratio of the mass of water vapor to the mass of dry air in a given volume of air. The *mixing ratio* affects radio propagation.

mode An electromagnetic field distribution that satisfies theoretical requirements for propagation in a waveguide or oscillation in a cavity.

modified refractive index The sum of the refractive index of the air at a given height above sea level, and the ratio of this height to the radius of the earth.

modular An equipment design in which major elements are readily separable, and which the user may replace, reducing the mean-time-to-repair.

modulation The process whereby the amplitude, frequency, or phase of a single-frequency wave (the *carrier*) is varied in step with the instantaneous value of, or samples of, a complex wave (the *modulating wave*).

modulator A device that enables the intelligence in an information-carrying modulating wave to be conveyed by a signal at a higher frequency. A *modulator* modifies a carrier wave by amplitude, phase, and/or frequency as a function of a control signal that carries intelligence. Signals are *modulated* in this way to permit more efficient and/or reliable transmission over any of several media.

module An assembly replaceable as an entity, often as an interchangeable plug-in item. A *module* is not normally capable of being disassembled.

monostable A device that is stable in one state only. An input pulse causes the device to change state, but it reverts immediately to its stable state.

motor A machine that converts electrical energy into mechanical energy.

motor effect The repulsion force exerted between adjacent conductors carrying currents in opposite directions.

moving coil Any device that utilizes a coil of wire in a magnetic field in such a way that the coil is made to move by varying the applied current, or itself produces a varying voltage because of its movement.

ms An abbreviation for *millisecond* (0.001 s).

multimeter A test instrument fitted with several ranges for measuring voltage, resistance, and current, and equipped with an analog meter or digital display readout. The *multimeter* is also known as a *volt-ohm-milliammeter*, or *VOM*.

multiplex (MUX) The use of a common channel to convey two or more channels. This is done either by splitting of the common channel frequency band into narrower bands, each of which is used to constitute a distinct channel (*frequency division multiplex*), or by allotting this common channel to multiple users in turn to constitute different intermittent channels (*time division multiplex*).

multiplexer A device or circuit that combines several signals onto a single signal.

multiplexing A technique that uses a single transmission path to carry multiple channels. In *time division multiplexing* (TDM), path time is shared. For *frequency division multiplexing* (FDM) or *wavelength division multiplexing* (WDM), signals are divided into individual channels sent along the same path but at different frequencies.

multiplication Signal mixing that occurs within a multiplier circuit.

multiplier A circuit in which one or more input signals are mixed under the direction of one or more control signals. The resulting output is a composite of the input signals, the characteristics of which are determined by the scaling specified for the circuit.

mutual induction The property of the magnetic flux around a conductor that induces a voltage in a nearby conductor. The voltage generated in the secondary conductor in turn induces a voltage in the primary conductor. The inductance of two conductors so coupled is referred to as *mutual inductance*.

mV An abbreviation for *millivolt* (0.001 V).

mW An abbreviation for *milliwatt* (0.001 W).

N

nano A prefix meaning one billionth.

nanometer 1×10^{-9} meter.

nanosecond (ns) One billionth of a second (1×10^{-9} s).

narrowband A communications channel of restricted bandwidth, often resulting in degradation of the transmitted signal.

narrowband emission An emission having a spectrum exhibiting one or more sharp peaks that are narrow in width compared to the nominal bandwidth of the measuring instrument, and are far enough apart in frequency to be resolvable by the instrument.

National Electrical Code (NEC) A document providing rules for the installation of electric wiring and equipment in public and private buildings, published by the National Fire Protection Association. The NEC has been adopted as law by many states and municipalities in the U.S.

National Institute of Standards and Technology (NIST) A nonregulatory agency of the Department of Commerce that serves as a national reference and measurement laboratory for the physical and engineering sciences. Formerly called the *National Bureau of Standards*, the agency was renamed in 1988 and given the additional responsibility of aiding U.S. companies in adopting new technologies to increase their international competitiveness.

negative In a conductor or semiconductor material, an excess of electrons or a deficiency of positive charge.

negative feedback The return of a portion of the output signal from a circuit to the input but 180° out of phase. This type of feedback decreases signal amplitude but stabilizes the amplifier and reduces distortion and noise.

negative impedance An impedance characterized by a decrease in voltage drop across a device as the current through the device is increased, or a decrease in current through the device as the voltage across it is increased.

neutral A device or object having no electrical charge.

neutral conductor A conductor in a power distribution system connected to a point in the system that is designed to be at neutral potential. In a balanced system, the neutral conductor carries no current.

neutral ground An intentional ground applied to the neutral conductor or neutral point of a circuit, transformer, machine, apparatus, or system.

newton The standard unit of force. One *newton* is the force that, when applied to a body having a mass of 1 kg, gives it an acceleration of 1 m/s^2.

nitrogen A gas widely used to pressurize radio frequency transmission lines. If a small puncture occurs in the cable sheath, the nitrogen keeps moisture out so that service is not adversely affected.

node The points at which the current is at minimum in a transmission system in which standing waves are present.

noise Any random disturbance or unwanted signal in a communication system that tends to obscure the clarity or usefulness of a signal in relation to its intended use.

noise factor (NF) The ratio of the noise power measured at the output of a receiver to the noise power that would be present at the output if the thermal noise resulting from the resistive component of the source impedance were the only source of noise in the system.

noise figure A measure of the noise in dB generated at the input of an amplifier, compared with the noise generated by an impedance-method resistor at a specified temperature.

noise filter A network that attenuates noise frequencies.

noise generator A generator of wideband random noise.

noise power ratio (NPR) The ratio, expressed in decibels, of signal power to intermodulation product power plus residual noise power, measured at the baseband level.

noise suppressor A filter or digital signal processing circuit in a receiver or transmitter that automatically reduces or eliminates noise.

noise temperature The temperature, expressed in Kelvin, at which a resistor will develop a particular noise voltage. The noise temperature of a radio receiver is the value by which the temperature of the resistive component of the source impedance should be increased—if it were the only source of noise in the system—to cause the noise power at the output of the receiver to be the same as in the real system.

nominal The most common value for a component or parameter that falls between the maximum and minimum limits of a tolerance range.

nominal value A specified or intended value independent of any uncertainty in its realization.

nomogram A chart showing three or more scales across which a straight edge may be held in order to read off a graphical solution to a three-variable equation.

nonionizing radiation Electromagnetic radiation that does not turn an atom into an ion. Examples of nonionizing radiation include visible light and radio waves.

nonconductor A material that does not conduct energy, such as electricity, heat, or sound.

noncritical technical load That part of the technical power load for a facility not required for minimum acceptable operation.

noninductive A device or circuit without significant inductance.

nonlinearity A distortion in which the output of a circuit or system does not rise or fall in direct proportion to the input.

nontechnical load The part of the total operational load of a facility used for such purposes as general lighting, air conditioning, and ventilating equipment during normal operation.

normal A line perpendicular to another line or to a surface.

normal-mode noise Unwanted signals in the form of voltages appearing in line-to-line and line-to-neutral signals.

normalized frequency The ratio between the actual frequency and its nominal value.

normalized frequency departure The frequency departure divided by the nominal frequency value.

normalized frequency difference The algebraic difference between two normalized frequencies.

normalized frequency drift The frequency drift divided by the nominal frequency value.

normally closed Switch contacts that are closed in their nonoperated state, or relay contacts that are closed when the relay is de-energized.

normally open Switch contacts that are open in their nonoperated state, or relay contacts that are open when the relay is de-energized.

north pole The pole of a magnet that seeks the north magnetic pole of the earth.

notch filter A circuit designed to attenuate a specific frequency band; also known as a *band stop filter.*

notched noise A noise signal in which a narrow band of frequencies has been removed.

ns An abbreviation for *nanosecond.*

null A zero or minimum amount or position.

O

octave Any frequency band in which the highest frequency is twice the lowest frequency.

off-line A condition wherein devices or subsystems are not connected into, do not form a part of, and are not subject to the same controls as an operational system.

offset An intentional difference between the realized value and the nominal value.

ohm The unit of electric resistance through which one ampere of current will flow when there is a difference of one volt. The quantity is named for the German physicist Georg Simon Ohm (1787–1854).

Ohm's law A law that sets forth the relationship between voltage (E), current (I), and resistance (R). The law states that $E = I^2 R$. *Ohm's Law* is named for the German physicist Georg Simon Ohm (1787–1854).

ohmic loss The power dissipation in a line or circuit caused by electrical resistance.

ohmmeter A test instrument used for measuring resistance, often part of a *multimeter.*

ohms-per-volt A measure of the sensitivity of a voltmeter.

on-line A device or system that is energized and operational, and ready to perform useful work.

open An interruption in the flow of electrical current, as caused by a broken wire or connection.

open-circuit A defined loop or path that closes on itself and contains an infinite impedance.

open-circuit impedance The input impedance of a circuit when its output terminals are open, that is, not terminated.

open-circuit voltage The voltage measured at the terminals of a circuit when there is no load and, hence, no current flowing.

operating lifetime The period of time during which the principal parameters of a component or system remain within a prescribed range.

optimize The process of adjusting for the best output or maximum response from a circuit or system.

orbit The path, relative to a specified frame of reference, described by the center of mass of a satellite or other object in space, subjected solely to natural forces (mainly gravitational attraction).

order of diversity The number of independently fading propagation paths or frequencies, or both, used in a diversity reception system.

original equipment manufacturer (OEM) A manufacturer of equipment that is used in systems assembled and sold by others.

oscillation A variation with time of the magnitude of a quantity with respect to a specified reference when the magnitude is alternately greater than and smaller than the reference.

oscillator A nonrotating device for producing alternating current, the output frequency of which is determined by the characteristics of the circuit.

oscilloscope A test instrument that uses a display, usually a cathode-ray tube, to show the instantaneous values and waveforms of a signal that varies with time or some other parameter.

out-of-band energy Energy emitted by a transmission system that falls outside the frequency spectrum of the intended transmission.

outage duration The average elapsed time between the start and the end of an outage period.

outage probability The probability that an outage state will occur within a specified time period. In the absence of specific known causes of outages, the *outage probability* is the sum of all outage durations divided by the time period of measurement.

outage threshold A defined value for a supported performance parameter that establishes the minimum operational service performance level for that parameter.

output impedance The impedance presented at the output terminals of a circuit, device, or channel.

output stage The final driving circuit in a piece of electronic equipment.

ovenized crystal oscillator (OXO) A crystal oscillator enclosed within a temperature regulated heater (oven) to maintain a stable frequency despite external temperature variations.

overcoupling A degree of coupling greater than the *critical coupling* between two resonant circuits. *Overcoupling* results in a wide bandwidth circuit with two peaks in the response curve.

overload In a transmission system, a power greater than the amount the system was designed to carry. In a power system, an overload could cause excessive heating. In a communications system, distortion of a signal could result.

overshoot The first maximum excursion of a pulse beyond the 100% level. Overshoot is the portion of the pulse that exceeds its defined level temporarily before settling to the correct level. Overshoot amplitude is expressed as a percentage of the defined level.

P

pentode An electron tube with five electrodes, the cathode, control grid, screen grid, suppressor grid, and plate.

photocathode An electrode in an electron tube that will emit electrons when bombarded by photons of light.

picture tube A cathode-ray tube used to produce an image by variation of the intensity of a scanning beam on a phosphor screen.

pin A terminal on the base of a component, such as an electron tube.

plasma (1—arc) An ionized gas in an arc-discharge tube that provides a conducting path for the discharge. **(2—solar)** The ionized gas at extremely high temperature found in the sun.

plate (1—electron tube) The anode of an electron tube. **(2—battery)** An electrode in a storage battery. **(3—capacitor)** One of the surfaces in a capacitor. **(4—chassis)** A mounting surface to which equipment may be fastened.

propagation time delay The time required for a signal to travel from one point to another.

protector A device or circuit that prevents damage to lines or equipment by conducting dangerously high voltages or currents to ground. Protector types include spark gaps, semiconductors, varistors, and gas tubes.

proximity effect A nonuniform current distribution in a conductor, caused by current flow in a nearby conductor.

pseudonoise In a spread-spectrum system, a seemingly random series of pulses whose frequency spectrum resembles that of continuous noise.

pseudorandom A sequence of signals that appears to be completely random but have, in fact, been carefully drawn up and repeat after a significant time interval.

pseudorandom noise A noise signal that satisfies one or more of the standard tests for statistical randomness. Although it seems to lack any definite pattern, there is a sequence of pulses that repeats after a long time interval.

pseudorandom number sequence A sequence of numbers that satisfies one or more of the standard tests for statistical randomness. Although it seems to lack any definite pattern, there is a sequence that repeats after a long time interval.

pulsating direct current A current changing in value at regular or irregular intervals but which has the same direction at all times.

pulse One of the elements of a repetitive signal characterized by the rise and decay in time of its magnitude. A *pulse* is usually short in relation to the time span of interest.

pulse decay time The time required for the trailing edge of a pulse to decrease from 90 percent to 10 percent of its peak amplitude.

pulse duration The time interval between the points on the leading and trailing edges of a pulse at which the instantaneous value bears a specified relation to the peak pulse amplitude.

pulse duration modulation (PDM) The modulation of a pulse carrier by varying the width of the pulses according to the instantaneous values of the voltage samples of the modulating signal (also called *pulse width modulation*).

pulse edge The leading or trailing edge of a pulse, defined as the 50 percent point of the pulse rise or fall time.

pulse fall time The interval of time required for the edge of a pulse to fall from 90 percent to 10 percent of its peak amplitude.

pulse interval The time between the start of one pulse and the start of the next.

pulse length The duration of a pulse (also called *pulse width*).

pulse level The voltage amplitude of a pulse.

pulse period The time between the start of one pulse and the start of the next.

pulse ratio The ratio of the length of any pulse to the total pulse period.

pulse repetition period The time interval from the beginning of one pulse to the beginning of the next pulse.

pulse repetition rate The number of times each second that pulses are transmitted.

pulse rise time The time required for the leading edge of a pulse to rise from 10 percent to 90 percent of its peak amplitude.

pulse train A series of pulses having similar characteristics.

pulse width The measured interval between the 50 percent amplitude points of the leading and trailing edges of a pulse.

puncture A breakdown of insulation or of a dielectric, such as in a cable sheath or in the insulant around a conductor.

pW An abbreviation for picowatt, a unit of power equal to 10^{-12} W (–90 dBm).

Q

Q (quality factor) A figure of merit that defines how close a coil comes to functioning as a pure inductor. *High Q* describes an inductor with little energy loss resulting from resistance. *Q* is found by dividing the inductive reactance of a device by its resistance.

quadrature A state of alternating current signals separated by one quarter of a cycle (90°).

quadrature amplitude modulation (QAM) A process that allows two different signals to modulate a single carrier frequency. The two signals of interest amplitude modulate two samples of the carrier that are of the same frequency, but differ in phase by 90°. The two resultant signals can be added and transmitted. Both signals may be recovered at a decoder when they are demodulated 90° apart.

quadrature component The component of a voltage or current at an angle of 90° to a reference signal, resulting from inductive or capacitive reactance.

quadrature phase shift keying (QPSK) A type of phase shift keying using four phase states.

quality The absence of objectionable distortion.

quality assurance (QA) All those activities, including surveillance, inspection, control, and documentation, aimed at ensuring that a given product will meet its performance specifications.

quality control (QC) A function whereby management exercises control over the quality of raw material or intermediate products in order to prevent the production of defective devices or systems.

quantum noise Any noise attributable to the discrete nature of electromagnetic radiation. Examples include shot noise, photon noise, and recombination noise.

quantum-limited operation An operation wherein the minimum detectable signal is limited by quantum noise.

quartz A crystalline mineral that when electrically excited vibrates with a stable period. Quartz is typically used as the frequency-determining element in oscillators and filters.

quasi-peak detector A detector that delivers an output voltage that is some fraction of the peak value of the regularly repeated pulses applied to it. The fraction increases toward unity as the pulse repetition rate increases.

quick-break fuse A fuse in which the fusible link is under tension, providing for rapid operation.

quiescent An inactive device, signal, or system.

quiescent current The current that flows in a device in the absence of an applied signal.

R

rack An equipment rack, usually measuring 19 in (48.26 cm) wide at the front mounting rails.

rack unit (RU) A unit of measure of vertical space in an equipment enclosure. One rack unit is equal to 1.75 in (4.45 cm).

radiate The process of emitting electromagnetic energy.

radiation The emission and propagation of electromagnetic energy in the form of waves. *Radiation* is also called *radiant energy.*

radiation scattering The diversion of thermal, electromagnetic, or nuclear radiation from its original path as a result of interactions or collisions with atoms, molecules, or large particles in the atmosphere or other media between the source of radiation and a point some distance away. As a result of scattering, radiation (especially gamma rays and neutrons) will be received at such a point from many directions, rather than only from the direction of the source.

radio The transmission of signals over a distance by means of electromagnetic waves in the approximate frequency range of 150 kHz to 300 GHz. The term may also be used to describe the equipment used to transmit or receive electromagnetic waves.

radio detection The detection of the presence of an object by radio location without precise determination of its position.

radio frequency interference (RFI) The intrusion of unwanted signals or electromagnetic noise into various types of equipment resulting from radio frequency transmission equipment or other devices using radio frequencies.

radio frequency spectrum Those frequency bands in the electromagnetic spectrum that range from several hundred thousand cycles per second (*very low frequency*) to several billion cycles per second (*microwave frequencies*).

radio recognition In military communications, the determination by radio means of the "friendly" or "unfriendly" character of an aircraft or ship.

random noise Electromagnetic signals that originate in transient electrical disturbances and have random time and amplitude patterns. Random noise is generally undesirable; however, it may also be generated for testing purposes.

rated output power The power available from an amplifier or other device under specified conditions of operation.

RC constant The time constant of a resistor-capacitor circuit. The *RC constant* is the time in seconds required for current in an RC circuit to rise to 63 percent of its final steady value or fall to 37 percent of its original steady value, obtained by multiplying the resistance value in ohms by the capacitance value in farads.

RC network A circuit that contains resistors and capacitors, normally connected in series.

reactance The part of the impedance of a network resulting from inductance or capacitance. The *reactance* of a component varies with the frequency of the applied signal.

reactive power The power circulating in an ac circuit. It is delivered to the circuit during part of the cycle and is returned during the other half of the cycle. The *reactive power* is obtained by multiplying the voltage, current, and the sine of the phase angle between them.

reactor A component with inductive reactance.

received signal level (RSL) The value of a specified bandwidth of signals at the receiver input terminals relative to an established reference.

receiver Any device for receiving electrical signals and converting them to audible sound, visible light, data, or some combination of these elements.

receptacle An electrical socket designed to receive a mating plug.

reception The act of receiving, listening to, or watching information-carrying signals.

rectification The conversion of alternating current into direct current.

rectifier A device for converting alternating current into direct current. A *rectifier* normally includes filters so that the output is, within specified limits, smooth and free of ac components.

rectify The process of converting alternating current into direct current.

redundancy A system design that provides a back-up for key circuits or components in the event of a failure. Redundancy improves the overall reliability of a system.

redundant A configuration when two complete systems are available at one time. If the online system fails, the backup will take over with no loss of service.

reference voltage A voltage used for control or comparison purposes.

reflectance The ratio of reflected power to incident power.

reflection An abrupt change, resulting from an impedance mismatch, in the direction of propagation of an electromagnetic wave. For light, at the interface of two dissimilar materials, the incident wave is returned to its medium of origin.

reflection coefficient The ratio between the amplitude of a reflected wave and the amplitude of the incident wave. For large smooth surfaces, the reflection coefficient may be near unity.

reflection gain The increase in signal strength that results when a reflected wave combines, in phase, with an incident wave.

reflection loss The apparent loss of signal strength caused by an impedance mismatch in a transmission line or circuit. The loss results from the reflection of part of the signal back toward the source from the point of the impedance discontinuity. The greater the mismatch, the greater the loss.

reflectometer A device that measures energy traveling in each direction in a waveguide, used in determining the standing wave ratio.

refraction The bending of a sound, radio, or light wave as it passes obliquely from a medium of one density to a medium of another density that varies its speed.

regulation The process of adjusting the level of some quantity, such as circuit gain, by means of an electronic system that monitors an output and feeds back a controlling signal to constantly maintain a desired level.

regulator A device that maintains its output voltage at a constant level.

relative envelope delay The difference in envelope delay at various frequencies when compared with a reference frequency that is chosen as having zero delay.

relative humidity The ratio of the quantity of water vapor in the atmosphere to the quantity that would cause saturation at the ambient temperature.

relative transmission level The ratio of the signal power in a transmission system to the signal power at some point chosen as a reference. The ratio is usually determined by applying a standard test signal at the input to the system and measuring the gain or loss at the location of interest.

relay A device by which current flowing in one circuit causes contacts to operate that control the flow of current in another circuit.

relay armature The movable part of an electromechanical relay, usually coupled to spring sets on which contacts are mounted.

relay bypass A device that, in the event of a loss of power or other failure, routes a critical signal around the equipment that has failed.

release time The time required for a pulse to drop from steady-state level to zero, also referred to as the *decay time*.

reliability The ability of a system or subsystem to perform within the prescribed parameters of quality of service. *Reliability* is often expressed as the probability that a system or subsystem will perform its intended function for a specified interval under stated conditions.

reliability growth The action taken to move a hardware item toward its reliability potential, during development or subsequent manufacturing or operation.

reliability predictions The compiled failure rates for parts, components, subassemblies, assemblies, and systems. These generic failure rates are used as basic data to predict the reliability of a given device or system.

remote control A system used to control a device from a distance.

remote station A station or terminal that is physically remote from a main station or center but can gain access through a communication channel.

repeater The equipment between two circuits that receives a signal degraded by normal factors during transmission and amplifies the signal to its original level for retransmission.

repetition rate The rate at which regularly recurring pulses are repeated.

reply A transmitted message that is a direct response to an original message.

repulsion The mechanical force that tends to separate like magnetic poles, like electric charges, or conductors carrying currents in opposite directions.

reset The act of restoring a device to its default or original state.

residual flux The magnetic flux that remains after a magnetomotive force has been removed.

residual magnetism The magnetism or flux that remains in a core after current ceases to flow in the coil producing the magnetomotive force.

residual voltage The vector sum of the voltages in all the phase wires of an unbalanced polyphase power system.

resistance The opposition of a material to the flow of electrical current. Resistance is equal to the voltage drop through a given material divided by the current flow through it. The standard unit of resistance is the *ohm*, named for the German physicist Georg Simon Ohm (1787–1854).

resistance drop The fall in potential (volts) between two points, the product of the current and resistance.

resistance-grounded A circuit or system grounded for safety through a resistance, which limits the value of the current flowing through the circuit in the event of a fault.

resistive load A load in which the voltage is in phase with the current.

resistivity The resistance per unit volume or per unit area.

resistor A device the primary function of which is to introduce a specified resistance into an electrical circuit.

resonance A tuned condition conducive to oscillation, when the reactance resulting from capacitance in a circuit is equal in value to the reactance resulting from inductance.

resonant frequency The frequency at which the inductive reactance and capacitive reactance of a circuit are equal.

resonator A resonant cavity.

return A return path for current, sometimes through ground.

reversal A change in magnetic polarity, in the direction of current flow.

reverse current A small current that flows through a diode when the voltage across it is such that normal forward current does not flow.

reverse voltage A voltage in the reverse direction from that normally applied.

rheostat A two-terminal variable resistor, usually constructed with a sliding or rotating shaft that can be used to vary the resistance value of the device.

ripple An ac voltage superimposed on the output of a dc power supply, usually resulting from imperfect filtering.

rise time The time required for a pulse to rise from 10 percent to 90 percent of its peak value.

roll-off A gradual attenuation of gain-frequency response at either or both ends of a transmission pass band.

root-mean-square (RMS) The square root of the average value of the squares of all the instantaneous values of current or voltage during one half-cycle of an alternating current. For an alternating current, the RMS voltage or current is equal to the amount of direct current or voltage that would produce the same heating effect in a purely resistive circuit. For a sinewave, the root-mean-square value is equal to 0.707 times the peak value. RMS is also called the *effective value*.

rotor The rotating part of an electric generator or motor.

RU An abbreviation for *rack unit*.

S

scan One sweep of the target area in a camera tube, or of the screen in a picture tube.

screen grid A grid in an electron tube that improves performance of the device by shielding the control grid from the plate.

self-bias The provision of bias in an electron tube through a voltage drop in the cathode circuit.

shot noise The noise developed in a vacuum tube or photoconductor resulting from the random number and velocity of emitted charge carriers.

slope The rate of change, with respect to frequency, of transmission line attenuation over a given frequency spectrum.

slope equalizer A device or circuit used to achieve a specified slope in a transmission line.

smoothing circuit A filter designed to reduce the amount of ripple in a circuit, usually a dc power supply.

snubber An electronic circuit used to suppress high frequency noise.

solar wind Charged particles from the sun that continuously bombard the surface of the earth.

solid A single wire conductor, as contrasted with a stranded, braided, or rope-type wire.

solid-state The use of semiconductors rather than electron tubes in a circuit or system.

source The part of a system from which signals or messages are considered to originate.

source terminated A circuit whose output is terminated for correct impedance matching with standard cable.

spare A system that is available but not presently in use.

spark gap A gap between two electrodes designed to produce a spark under given conditions.

specific gravity The ratio of the weight of a volume, liquid, or solid to the weight of the same volume of water at a specified temperature.

spectrum A continuous band of frequencies within which waves have some common characteristics.

spectrum analyzer A test instrument that presents a graphic display of signals over a selected frequency bandwidth. A cathode-ray tube is often used for the display.

spectrum designation of frequency A method of referring to a range of communication frequencies. In American practice, the designation is a two or three letter acronym for the name. The ranges are: below 300 Hz, ELF (extremely low frequency); 300 Hz–3000 Hz, ILF (infra low frequency); 3 kHz–30 kHz, VLF (very low frequency); 30 kHz–300 kHz, LF (low frequency); 300 kHz–3000 kHz, MF (medium frequency); 3 MHz–30 MHz, HF (high frequency); 30 MHz–300 MHz, VHF (very high frequency); 300 MHz–3000 MHz, UHF (ultra high frequency); 3 GHz–30 GHz, SHF (super high frequency); 30 GHz–300 GHz, EHF (extremely high frequency); 300 GHz–3000 GHz, THF (tremendously high frequency).

spherical antenna A type of satellite receiving antenna that permits more than one satellite to be accessed at any given time. A spherical antenna has a broader angle of acceptance than a parabolic antenna.

spike A high amplitude, short duration pulse superimposed on an otherwise regular waveform.

split-phase A device that derives a second phase from a single phase power supply by passing it through a capacitive or inductive reactor.

splitter A circuit or device that accepts one input signal and distributes it to several outputs.

splitting ratio The ratio of the power emerging from the output ports of a coupler.

sporadic An event occurring at random and infrequent intervals.

spread spectrum A communications technique in which the frequency components of a narrow-band signal are spread over a wide band. The resulting signal resembles white noise. The technique is used to achieve signal security and privacy, and to enable the use of a common band by many users.

spurious signal Any portion of a given signal that is not part of the fundamental waveform. Spurious signals include transients, noise, and hum.

square wave A square or rectangular-shaped periodic wave that alternately assumes two fixed values for equal lengths of time, the transition being negligible in comparison with the duration of each fixed value.

square wave testing The use of a square wave containing many odd harmonics of the fundamental frequency as an input signal to a device. Visual examination of the output signal on an oscilloscope indicates the amount of distortion introduced.

stability The ability of a device or circuit to remain stable in frequency, power level, and/or other specified parameters.

standard The specific signal configuration, reference pulses, voltage levels, and other parameters that describe the input/output requirements for a particular type of equipment.

standard time and frequency signal A time-controlled radio signal broadcast at scheduled intervals on a number of different frequencies by government-operated radio stations to provide a method for calibrating instruments.

standing wave ratio (SWR) The ratio of the maximum to the minimum value of a component of a wave in a transmission line or waveguide, such as the maximum voltage to the minimum voltage.

static charge An electric charge on the surface of an object, particularly a dielectric.

station One of the input or output points in a communications system.

stator The stationary part of a rotating electric machine.

status The present condition of a device.

statute mile A unit of distance equal to 1,609 km or 5,280 ft.

steady-state A condition in which circuit values remain essentially constant, occurring after all initial transients or fluctuating conditions have passed.

steady-state condition A condition occurring after all initial transient or fluctuating conditions have damped out in which currents, voltages, or fields remain essentially constant or oscillate uniformly without changes in characteristics such as amplitude, frequency, or wave shape.

steep wavefront A rapid rise in voltage of a given signal, indicating the presence of high frequency odd harmonics of a fundamental wave frequency.

step up (or down) The process of increasing (or decreasing) the voltage of an electrical signal, as in a step-up (or step-down) transformer.

straight-line capacitance A capacitance employing a variable capacitor with plates so shaped that capacitance varies directly with the angle of rotation.

stray capacitance An unintended—and usually undesired—capacitance between wires and components in a circuit or system.

stray current A current through a path other than the intended one.

stress The force per unit of cross-sectional area on a given object or structure.

subassembly A functional unit of a system.

subcarrier (SC) A carrier applied as modulation on another carrier, or on an intermediate subcarrier.

subharmonic A frequency equal to the fundamental frequency of a given signal divided by a whole number.

submodule A small circuit board or device that mounts on a larger module or device.

subrefraction A refraction for which the refractivity gradient is greater than standard.

subsystem A functional unit of a system.

superheterodyne receiver A radio receiver in which all signals are first converted to a common frequency for which the intermediate stages of the receiver have been optimized, both for tuning and filtering. Signals are converted by mixing them with the output of a local oscillator whose output is varied in accordance with the frequency of the received signals so as to maintain the desired *intermediate frequency.*

suppressor grid The fifth grid of a pentode electron tube, which provides screening between plate and screen grid.

surface leakage A leakage current from line to ground over the face of an insulator supporting an open wire route.

surface refractivity The refractive index, calculated from observations of pressure, temperature, and humidity at the surface of the earth.

surge A rapid rise in current or voltage, usually followed by a fall back to the normal value.

susceptance The reciprocal of reactance, and the imaginary component of admittance, expressed in siemens.

sweep The process of varying the frequency of a signal over a specified bandwidth.

sweep generator A test oscillator, the frequency of which is constantly varied over a specified bandwidth.

switching The process of making and breaking (connecting and disconnecting) two or more electrical circuits.

synchronization The process of adjusting the corresponding significant instants of signals—for example, the zero-crossings—to make them synchronous. The term *synchronization* is often abbreviated as *sync*.

synchronize The process of causing two systems to operate at the same speed.

synchronous In step or in phase, as applied to two or more devices; a system in which all events occur in a predetermined timed sequence.

synchronous detection A demodulation process in which the original signal is recovered by multiplying the modulated signal by the output of a synchronous oscillator locked to the carrier.

synchronous system A system in which the transmitter and receiver are operating in a fixed time relationship.

system standards The minimum required electrical performance characteristics of a specific collection of hardware and/or software.

systems analysis An analysis of a given activity to determine precisely what must be accomplished and how it is to be done.

T

tetrode A four element electron tube consisting of a cathode, control grid, screen grid, and plate.

thyratron A gas-filled electron tube in which plate current flows when the grid voltage reaches a predetermined level. At that point, the grid has no further control over the current, which continues to flow until it is interrupted or reversed.

tolerance The permissible variation from a standard.

torque A moment of force acting on a body and tending to produce rotation about an axis.

total harmonic distortion (THD) The ratio of the sum of the amplitudes of all signals harmonically related to the fundamental versus the amplitude of the fundamental signal. THD is expressed in percent.

trace The pattern on an oscilloscope screen when displaying a signal.

tracking The locking of tuned stages in a radio receiver so that all stages are changed appropriately as the receiver tuning is changed.

trade-off The process of weighing conflicting requirements and reaching a compromise decision in the design of a component or a subsystem.

transceiver Any circuit or device that receives and transmits signals.

transconductance The mutual conductance of an electron tube expressed as the change in plate current divided by the change in control grid voltage that produced it.

transducer A device that converts energy from one form to another.

transfer characteristics The intrinsic parameters of a system, subsystem, or unit of equipment which, when applied to the input of the system, subsystem, or unit of equipment, will fully describe its output.

transformer A device consisting of two or more windings wrapped around a single core or linked by a common magnetic circuit.

transformer ratio The ratio of the number of turns in the secondary winding of a transformer to the number of turns in the primary winding, also known as the *turns ratio*.

transient A sudden variance of current or voltage from a steady-state value. A transient normally results from changes in load or effects related to switching action.

transient disturbance A voltage pulse of high energy and short duration impressed upon the ac waveform. The overvoltage pulse can be one to 100 times the normal ac potential (or more) and can last up to 15 ms. Rise times measure in the nanosecond range.

transient response The time response of a system under test to a stated input stimulus.

transition A sequence of actions that occurs when a process changes from one state to another in response to an input.

transmission The transfer of electrical power, signals, or an intelligence from one location to another by wire, fiber optic, or radio means.

transmission facility A transmission medium and all the associated equipment required to transmit information.

transmission loss The ratio, in decibels, of the power of a signal at a point along a transmission path to the power of the same signal at a more distant point along the same path. This value is often used as a measure of the quality of the transmission medium for conveying signals. Changes in power level are normally expressed in decibels by calculating ten times the logarithm (base 10) of the ratio of the two powers.

transmission mode One of the field patterns in a waveguide in a plane transverse to the direction of propagation.

transmission system The set of equipment that provides single or multichannel communications facilities capable of carrying audio, video, or data signals.

transmitter The device or circuit that launches a signal into a passive medium, such as the atmosphere.

transparency The property of a communications system that enables it to carry a signal without altering or otherwise affecting the electrical characteristics of the signal.

tray The metal cabinet that holds circuit boards.

tremendously high frequency (THF) The frequency band from 300 GHz to 3000 GHz.

triangular wave An oscillation, the values of which rise and fall linearly, and immediately change upon reaching their peak maximum and minimum. A graphical representation of a triangular wave resembles a triangle.

trim The process of making fine adjustments to a circuit or a circuit element.

trimmer A small mechanically-adjustable component connected in parallel or series with a major component so that the net value of the two can be finely adjusted for tuning purposes.

triode A three-element electron tube, consisting of a cathode, control grid, and plate.

triple beat A third-order beat whose three beating carriers all have different frequencies, but are spaced at equal frequency separations.

troposphere The layer of the earth's atmosphere, between the surface and the stratosphere, in which about 80 percent of the total mass of atmospheric air is concentrated and in which temperature normally decreases with altitude.

trouble A failure or fault affecting the service provided by a system.

troubleshoot The process of investigating, localizing, and (if possible) correcting a fault.

tube (1—electron) An evacuated or gas-filled tube enclosed in a glass or metal case in which the electrodes are maintained at different voltages, giving rise to a controlled flow of electrons from the cathode to the anode. **(2—cathode ray, CRT)** An electron beam tube used for the display of changing electrical phenomena, generally similar to a television picture tube. **(3—cold-cathode)** An electron tube whose cathode emits electrons without the need of a heating filament. **(4—gas)** A gas-filled electron tube in which the gas plays an essential role in operation of the device. **(5—mercury-vapor)** A tube filled with mercury vapor at low pressure, used as a rectifying device. **(6—metal)** An electron tube enclosed in a metal case. **(7—traveling wave, TWT)** A wide band microwave amplifier in which a stream of electrons interacts with a guided electromagnetic wave moving substantially in synchronism with the electron stream, resulting in a net transfer of energy from the electron stream to the wave. **(8—velocity-modulated)** An electron tube in which the velocity of the electron stream is continually changing, as in a klystron.

tune The process of adjusting the frequency of a device or circuit, such as for resonance or for maximum response to an input signal.

tuned trap A series resonant network bridged across a circuit that eliminates ("traps") the frequency of the resonant network.

tuner The radio frequency and intermediate frequency parts of a radio receiver that produce a low level output signal.

tuning The process of adjusting a given frequency; in particular, to adjust for resonance or for maximum response to a particular incoming signal.

turns ratio In a transformer, the ratio of the number of turns on the secondary to the number of turns on the primary.

tweaking The process of adjusting an electronic circuit to optimize its performance.

twin-line A feeder cable with two parallel, insulated conductors.

two-phase A source of alternating current circuit with two sinusoidal voltages that are 90° apart.

U

ultra high frequency (UHF) The frequency range from 300 MHz to 3000 MHz.

ultraviolet radiation Electromagnetic radiation in a frequency range between visible light and high-frequency X-rays.

unattended A device or system designed to operate without a human attendant.

unattended operation A system that permits a station to receive and transmit messages without the presence of an attendant or operator.

unavailability A measure of the degree to which a system, subsystem, or piece of equipment is not operable and not in a committable state at the start of a mission, when the mission is called for at a random point in time.

unbalanced circuit A two-wire circuit with legs that differ from one another in resistance, capacity to earth or to other conductors, leakage, or inductance.

unbalanced line A transmission line in which the magnitudes of the voltages on the two conductors are not equal with respect to ground. A coaxial cable is an example of an unbalanced line.

unbalanced modulator A modulator whose output includes the carrier signal.

unbalanced output An output with one leg at ground potential.

unbalanced wire circuit A circuit whose two sides are inherently electrically unlike.

uncertainty An expression of the magnitude of a possible deviation of a measured value from the true value. Frequently, it is possible to distinguish two components: the *systematic uncertainty* and the *random uncertainty.* The random uncertainty is expressed by the standard deviation or by a multiple of the standard deviation. The systematic uncertainty is generally estimated on the basis of the parameter characteristics.

undamped wave A signal with constant amplitude.

underbunching A condition in a traveling wave tube wherein the tube is not operating at its optimum bunching rate.

Underwriters Laboratories, Inc. A laboratory established by the National Board of Fire Underwriters which tests equipment, materials, and systems that may affect insurance risks, with special attention to fire dangers and other hazards to life.

ungrounded A circuit or line not connected to ground.

unicoupler A device used to couple a balanced circuit to an unbalanced circuit.

unidirectional A signal or current flowing in one direction only.

uniform transmission line A transmission line with electrical characteristics that are identical, per unit length, over its entire length.

unit An assembly of equipment and associated wiring that together forms a complete system or independent subsystem.

unity coupling In a theoretically perfect transformer, complete electromagnetic coupling between the primary and secondary windings with no loss of power.

unity gain An amplifier or active circuit in which the output amplitude is the same as the input amplitude.

unity power factor A power factor of 1.0, which means that the load is—in effect—a pure resistance, with ac voltage and current completely in phase.

unterminated A device or system that is not terminated.

up-converter A frequency translation device in which the frequency of the output signal is greater than that of the input signal. Such devices are commonly found in microwave radio and satellite systems.

uplink A transmission system for sending radio signals from the ground to a satellite or aircraft.

upstream A device or system placed ahead of other devices or systems in a signal path.

useful life The period during which a low, constant failure rate can be expected for a given device or system. The *useful life* is the portion of a product life cycle between break-in and wear out.

user A person, organization, or group that employs the services of a system for the transfer of information or other purposes.

V

VA An abbreviation for *volt-amperes*, volts times amperes.

vacuum relay A relay whose contacts are enclosed in an evacuated space, usually to provide reliable long-term operation.

vacuum switch A switch whose contacts are enclosed in an evacuated container so that spark formation is discouraged.

vacuum tube An electron tube. The most common vacuum tubes include the diode, triode, tetrode, and pentode.

validity check A test designed to ensure that the quality of transmission is maintained over a given system.

varactor A semiconductor that behaves like a capacitor under the influence of an external control voltage.

varactor diode A semiconductor device whose capacitance is a function of the applied voltage. A varactor diode, also called a *variable reactance diode* or simply a *varactor*, is often used to tune the operating frequency of a radio circuit.

variable frequency oscillator (VFO) An oscillator whose frequency can be set to any required value in a given range of frequencies.

variable impedance A capacitor, inductor, or resistor that is adjustable in value.

variable-gain amplifier An amplifier whose gain can be controlled by an external signal source.

variable-reluctance A transducer in which the input (usually a mechanical movement) varies the magnetic reluctance of a device.

variation monitor A device used for sensing a deviation in voltage, current, or frequency, which is capable of providing an alarm and/or initiating transfer to another power source when programmed limits of voltage, frequency, current, or time are exceeded.

varicap A diode used as a variable capacitor.

VCXO (voltage controlled crystal oscillator) A device whose output frequency is determined by an input control voltage.

vector A quantity having both magnitude and direction.

vector diagram A diagram using vectors to indicate the relationship between voltage and current in a circuit.

vector sum The sum of two vectors which, when they are at right angles to each other, equal the length of the hypotenuse of the right triangle so formed. In the general case, the vector sum of the two vectors equals the diagonal of the parallelogram formed on the two vectors.

velocity of light The speed of propagation of electromagnetic waves in a vacuum, equal to 299,792,458 m/s, or approximately 186,000 mi/s. For rough calculations, the figure of 300,000 km/s is used.

velocity of propagation The velocity of signal transmission. In free space, electromagnetic waves travel at the speed of light. In a cable, the velocity is substantially lower.

vernier A device that enables precision reading of a measuring set or gauge, or the setting of a dial with precision.

very low frequency (VLF) A radio frequency in the band 3 kHz to 30 kHz.

vestigial sideband A form of transmission in which one sideband is significantly attenuated. The carrier and the other sideband are transmitted without attenuation.

vibration testing A testing procedure whereby subsystems are mounted on a test base that vibrates, thereby revealing any faults resulting from badly soldered joints or other poor mechanical design features.

volt The standard unit of electromotive force, equal to the potential difference between two points on a conductor that is carrying a constant current of one ampere when the power dissipated between the two points is equal to one watt. One *volt* is equivalent to the potential difference across a resistance of one ohm when one ampere is flowing through it. The volt is named for the Italian physicist Alessandro Volta (1745–1827).

volt-ampere (VA) The apparent power in an ac circuit (volts times amperes).

volt-ohm-milliammeter (VOM) A general purpose multirange test meter used to measure voltage, resistance, and current.

voltage The potential difference between two points.

voltage drop A decrease in electrical potential resulting from current flow through a resistance.

voltage gradient The continuous drop in electrical potential, per unit length, along a uniform conductor or thickness of a uniform dielectric.

voltage level The ratio of the voltage at a given point to the voltage at an arbitrary reference point.

voltage reference circuit A stable voltage reference source.

voltage regulation The deviation from a nominal voltage, expressed as a percentage of the nominal voltage.

voltage regulator A circuit used for controlling and maintaining a voltage at a constant level.

voltage stabilizer A device that produces a constant or substantially constant output voltage despite variations in input voltage or output load current.

voltage to ground The voltage between any given portion of a piece of equipment and the ground potential.

voltmeter An instrument used to measure differences in electrical potential.

vox A voice-operated relay circuit that permits the equivalent of push-to-talk operation of a transmitter by the operator.

VSAT (very small aperture terminal) A satellite Ku-band earth station intended for fixed or portable use. The antenna diameter of a VSAT is on the order of 1.5 m or less.

W

watt The unit of power equal to the work done at one joule per second, or the rate of work measured as a current of one ampere under an electric potential of one volt. Designated by the symbol *W*, the watt is named after the Scottish inventor James Watt (1736–1819).

watt meter A meter indicating in watts the rate of consumption of electrical energy.

watt-hour The work performed by one watt over a one hour period.

wave A disturbance that is a function of time or space, or both, and is propagated in a medium or through space.

wave number The reciprocal of wavelength; the number of wave lengths per unit distance in the direction of propagation of a wave.

waveband A band of wavelengths defined for some given purpose.

waveform The characteristic shape of a periodic wave, determined by the frequencies present and their amplitudes and relative phases.

wavefront A continuous surface that is a locus of points having the same phase at a given instant. A *wavefront* is a surface at right angles to rays that proceed from the wave source. The surface passes through those parts of the wave that are in the same phase and travel in the same direction. For parallel rays the wavefront is a plane; for rays that radiate from a point, the wavefront is spherical.

waveguide Generally, a rectangular or circular pipe that constrains the propagation of an acoustic or electromagnetic wave along a path between two locations. The dimensions of a waveguide determine the frequencies for optimum transmission.

wavelength For a sinusoidal wave, the distance between points of corresponding phase of two consecutive cycles.

weber The unit of magnetic flux equal to the flux that, when linked to a circuit of one turn, produces an electromotive force of one volt as the flux is reduced at a uniform rate to zero in one second. The *weber* is named for the German physicist Wilhelm Eduard Weber (1804–1891).

weighted The condition when a correction factor is applied to a measurement.

weighting The adjustment of a measured value to account for conditions that would otherwise be different or appropriate during a measurement.

weighting network A circuit, used with a test instrument, that has a specified amplitude-versus-frequency characteristic.

wideband The passing or processing of a wide range of frequencies. The meaning varies with the context.

Wien bridge An ac bridge used to measure capacitance or inductance.

winding A coil of wire used to form an inductor.

wire A single metallic conductor, usually solid-drawn and circular in cross section.

working range The permitted range of values of an analog signal over which transmitting or other processing equipment can operate.

working voltage The rated voltage that may safely be applied continuously to a given circuit or device.

X

x-band A microwave frequency band from 5.2 GHz to 10.9 GHz.

x-cut A method of cutting a quartz plate for an oscillator, with the x-axis of the crystal perpendicular to the faces of the plate.

X ray An electromagnetic radiation of approximately 100 nm to 0.1 nm, capable of penetrating nonmetallic materials.

Y

y-cut A method of cutting a quartz plate for an oscillator, with the y-axis of the crystal perpendicular to the faces of the plate.

yield strength The magnitude of mechanical stress at which a material will begin to deform. Beyond the *yield strength* point, extension is no longer proportional to stress and rupture is possible.

yoke A material that interconnects magnetic cores. *Yoke* can also refer to the deflection windings of a CRT.

yttrium-iron garnet (YIG) A crystalline material used in microwave devices.

12.3 References

1. *ATSC Digital Television Standard*, Doc. A/53, Advanced Television Systems Committee, Washington, D.C., 1996.

2. *Digital Audio Compression (AC-3) Standard*, Doc. A/52, Advanced Television Systems Committee, Washington, D.C., 1996.
3. *Guide to the Use of the ATSC Digital Television Standard*, Doc. A/54, Advanced Television Systems Committee, Washington, D.C., 1996.
4. *Program Guide for Digital Television*, Doc. A/55, Advanced Television Systems Committee, Washington, D.C., 1996.
5. *System Information for Digital Television*, Doc. A/56, Advanced Television Systems Committee, Washington, D.C., 1996.
6. "Advanced Television Enhancement Forum Specification," Draft, Version 1.1r26 updated 2/2/99, ATVEF, Portland, Ore., 1999.

12.4 Bibliography

Whitaker, Jerry C.: *Power Vacuum Tubes Handbook*, CRC Press, Boca Raton, Fla., 1999.

General Services Administration, Information Technology Service, National Communications System: *Glossary of Telecommunication Terms*, Technology and Standards Division, Federal Standard 1037C, General Services Administration, Washington, D.C., August, 7, 1996.

Chapter

13

Acronyms and Abbreviations

Jerry C. Whitaker, editor

13.1 Acronyms and Abbreviations Relating to Digital Television

The following definitions apply to digital television in general, and the ATSC DTV system in particular [1–6]:

A/D analog to digital converter

ACATS Advisory Committee on Advanced Television Service

AES Audio Engineering Society

ANSI American National Standards Institute

ATEL Advanced Television Evaluation Laboratory

ATM asynchronous transfer mode

ATSC Advanced Television Systems Committee

ATTC Advanced Television Test Center

ATV advanced television

bps bits per second

bslbf bit serial, leftmost bit first

CAT conditional access table

CDT carrier definition table

CDTV conventional definition television

CRC cyclic redundancy check

DCT discrete cosine transform

DSM digital storage media

DSM-CC digital storage media command and control

DTS decoding time-stamp

DVCR digital video cassette recorder

ECM entitlement control message

EMM entitlement management message

ES elementary stream

ESCR elementary stream clock reference

FPLL frequency- and phase-locked loop

GA Grand Alliance

GMT Greenwich mean time

GOP group of pictures

GPS global positioning system

HDTV high-definition television

IEC International Electrotechnical Commission

IRD integrated receiver-decoder

ISO International Organization for Standardization

ITU International Telecommunication Union

JEC Joint Engineering Committee of EIA and NCTA

MCPT multiple carriers per transponder

MMT modulation mode table

MP@HL main profile at high level

MP@ML main profile at main level

MPEG Moving Picture Experts Group

NAB National Association of Broadcasters

NTSC National Television System Committee

NVOD near video on demand

PAL phase alternation each line

PAT program association table

PCR program clock reference

pel pixel

PES packetized elementary stream

PID packet identifier

PMT program map table

PSI program specific information

PTS presentation time stamp

PU presentation unit

SCR system clock reference

SDTV standard definition television

SECAM sequential couleur avec mémoire (sequential color with memory)

SIT satellite information table

SMPTE Society of Motion Picture and Television Engineers

STD system target decoder

TAI international atomic time

TDT transponder data table

TNT transponder name table

TOV threshold of visibility

TS transport stream

UTC universal coordinated time

VBV video buffering verifier

VCN virtual channel number

VCT virtual channel table

13.2 General Electronics Acronyms and Abbreviations[1]

A

a	atto (10^{-18})
Å	angstrom
A	ampere
AAR	automatic alternate routing
AARTS	automatic audio remote test set
ac	alternating current
ACA	automatic circuit assurance
ACC	automatic callback calling
ACD	automatic call distributor

1. This section adapted from: General Services Administration, Information Technology Service, National Communications System: *Glossary of Telecommunication Terms*, Technology and Standards Division, Federal Standard 1037C, General Services Administration, Washington, D.C., August, 7, 1996.

ac-dc	alternating current - direct current
ACK	acknowledge character
ACTS	Advanced Communications Technology Satellite
ACU	automatic calling unit
A-D	analog-to-digital
ADC	analog-to-digital converter; analog-to-digital conversion
ADCCP	Advanced Data Communication Control Procedures
ADH	automatic data handling
ADP	automatic data processing
ADPCM	adaptive differential pulse-code modulation
ADPE	automatic data processing equipment
ADU	automatic dialing unit
ADX	automatic data exchange
AECS	Aeronautical Emergency Communications System [Plan]
AF	audio frequency
AFC	area frequency coordinator; automatic frequency control
AFRS	Armed Forces Radio Service
AGC	automatic gain control
AGE	aerospace ground equipment
AI	artificial intelligence
AIG	address indicator group; address indicating group
AIM	amplitude intensity modulation
AIN	advanced intelligent network
AIOD	automatic identified outward dialing
AIS	automated information system
AJ	anti-jamming
ALC	automatic level control; automatic load control
ALE	automatic link establishment
ALU	arithmetic and logic unit
AM	amplitude modulation
AMA	automatic message accounting
AMC	administrative management complex
AME	amplitude modulation equivalent; automatic message exchange

AMI	alternate mark inversion [signal]
AM/PM/VSB	amplitude modulation/phase modulation/vestigial sideband
AMPS	automatic message processing system
AMPSSO	automated message processing system security officer
AMSC	American Mobile Satellite Corporation
AMTS	automated maritime telecommunications system
ANI	automatic number identification
ANL	automatic noise limiter
ANMCC	Alternate National Military Command Center
ANS	American National Standard
ANSI	American National Standards Institute
AP	anomalous propagation
APC	adaptive predictive coding
API	application program interface
APK	amplitude phase-shift keying
APL	average picture level
ARP	address resolution protocol
ARPA	Advanced Research Projects Agency [now DARPA]
ARPANET	Advanced Research Projects Agency Network
ARQ	automatic repeat-request
ARS	automatic route selection
ARSR	air route surveillance radar
ARU	audio response unit
ASCII	American Standard Code for Information Interchange
ASIC	application-specific integrated circuits
ASP	Aggregated Switch Procurement; adjunct service point
ASR	automatic send and receive; airport surveillance radar
AT	access tandem
ATACS	Army Tactical Communications System
ATB	all trunks busy
ATCRBS	air traffic control radar beacon system
ATDM	asynchronous time-division multiplexing
ATE	automatic test equipment

ATM	asynchronous transfer mode
ATV	advanced television
au	astronomical unit
AUI	attachment unit interface
AUTODIN	Automatic Digital Network
AUTOVON	Automatic Voice Network
AVD	alternate voice/data
AWG	American wire gauge
AWGN	additive white Gaussian noise
AZ	azimuth

B

b	bit
B	bel; byte
balun	balanced to unbalanced
basecom	base communications
BASIC	beginners' all-purpose symbolic instruction code
BCC	block check character
BCD	binary coded decimal; binary-coded decimal notation
BCI	bit-count integrity
Bd	baud
B8ZS	bipolar with eight-zero substitution
bell	BEL character
BER	bit error ratio
BERT	bit error ratio tester
BETRS	basic exchange telecommunications radio service
BEX	broadband exchange
BIH	International Time Bureau
B-ISDN	broadband ISDN
bi-sync	binary synchronous [communication]
bit	binary digit
BIT	built-in test
BITE	built-in test equipment

BIU	bus interface unit
BNF	Backus Naur form
BOC	Bell Operating Company
bpi	bits per inch; bytes per inch
BPSK	binary phase-shift keying
b/s	bits per second
b/in	bits per inch
BPSK	binary phase-shift keying
BR	bit rate
BRI	basic rate interface
BSA	basic serving arrangement
BSE	basic service element
BSI	British Standards Institution
B6ZS	bipolar with six-zero substitution
B3ZS	bipolar with three-zero substitution
BTN	billing telephone number
BW	bandwidth

C

c	centi (10^{-2})
CACS	centralized alarm control system
CAM	computer-aided manufacturing
CAMA	centralized automatic message accounting
CAN	cancel character
CAP	competitive access provider; customer administration panel
CARS	cable television relay service [station]
CAS	centralized attendant services
CASE	computer-aided software engineering; computer aided system engineering; computer-assisted software engineering
CATV	cable TV; cable television; community antenna television
CBX	computer branch exchange
C^2	command and control
C^3	command, control, and communications

C^3CM	C^3 countermeasures
C^3I	command, control, communications and intelligence
CCA	carrier-controlled approach
CCD	charge-coupled device
CCH	connections per circuit hour
CCIF	International Telephone Consultative Committee
CCIR	International Radio Consultative Committee
CCIS	common-channel interoffice signaling
CCIT	International Telegraph Consultative Committee
CCITT	International Telegraph and Telephone Consultative Committee
CCL	continuous communications link
CCS	hundred call-seconds
CCSA	common control switching arrangement
CCTV	closed-circuit television
CCW	cable cutoff wavelength
cd	candela
CD	collision detection; compact disk
CDF	combined distribution frame; cumulative distribution function
CDMA	code-division multiple access
CDPSK	coherent differential phase-shift keying
CDR	call detail recording
CD ROM	compact disk read-only memory
CDT	control data terminal
CDU	central display unit
C-E	communications-electronics
CEI	comparably efficient interconnection
CELP	code-excited linear prediction
CEP	circular error probable
CFE	contractor-furnished equipment
cgs	centimeter-gram-second
ChR	channel reliability
CIAS	circuit inventory and analysis system
CIC	content indicator code

CIF	common intermediate format
CIFAX	ciphered facsimile
CiR	circuit reliability
C/kT	carrier-to-receiver noise density
CLASS	custom local area signaling service
cm	centimeter
CMI	coded mark inversion
CMIP	Common Management Information Protocol
CMIS	common management information service
CMOS	complementary metal oxide substrate
CMRR	common-mode rejection ratio
CNR	carrier-to-noise ratio; combat-net radio
CNS	complementary network service
C.O.	central office
COAM	customer owned and maintained equipment
COBOL	common business oriented language
codec	coder-decoder
COG	centralized ordering group
COMINT	communications intelligence
COMJAM	communications jamming
compandor	compressor-expander
COMPUSEC	computer security
COMSAT	Communications Satellite Corporation
COMSEC	communications security
CONEX	connectivity exchange
CONUS	Continental United States
COP	Committee of Principals
COR	Council of Representatives
COT	customer office terminal
CPAS	cellular priority access services
CPE	customer premises equipment
cpi	characters per inch
cpm	counts per minute

cps	characters per second
CPU	central processing unit; communications processor unit
CR	channel reliability; circuit reliability
CRC	cyclic redundancy check
CRITICOM	Critical Intelligence Communications
CROM	control read-only memory
CRT	cathode ray tube
c/s	cycles per second
CSA	Canadian Standards Association
CSC	circuit-switching center; common signaling channel
CSMA	carrier sense multiple access
CSMA/CA	carrier sense multiple access with collision avoidance
CSMA/CD	carrier sense multiple access with collision detection
CSU	channel service unit; circuit switching unit; customer service unit
CTS	clear to send
CTX	Centrex® [service]; clear to transmit
CVD	chemical vapor deposition
CVSD	continuously variable slope delta [modulation]
cw	carrier wave; composite wave; continuous wave
CX	composite signaling
cxr	carrier

D

d	deci (10^{-1})
da	deka (10)
D-A	digital-to-analog; digital-to-analog converter
D/L	downlink
DACS	digital access and cross-connect system
DAMA	demand assignment multiple access
DARPA	Defense Advanced Research Projects Agency
dB	decibel
dBa	decibels adjusted
dBa0	noise power measured at zero transmission level point

dBc	dB relative to carrier power
dBm	dB referred to 1 milliwatt
dBm(psoph)	noise power in dBm measured by a set with psophometric weighting
DBMS	database management system
dBmV	dB referred to 1 millivolt across 75 ohms
dBm0	noise power in dBm referred to or measured at 0TLP
dBm0p	noise power in dBm0 measured by a psophometric or noise measuring set having psophometric weighting
dBr	power difference in dB between any point and a reference point
dBrn	dB above reference noise
dBrnC	noise power in dBrn measured by a set with C-message weighting
dBrnC0	noise power in dBrnC referred to or measured at 0TLP
dBrn(f_1-f_2)	flat noise power in dBrn
dBrn(144)	noise power in dBrn measured by a set with 144-line weighting
dBv	dB relative to 1 V (volt) peak-to-peak
dBW	dB referred to 1 W (watt)
dBx	dB above reference coupling
dc	direct current
DCA	Defense Communications Agency
DCE	data circuit-terminating equipment
DCL	direct communications link
DCPSK	differentially coherent phase-shift keying
DCS	Defense Communications System
DCTN	Defense Commercial Telecommunications Network
DCWV	direct-current working volts
DDD	direct distance dialing
DDN	Defense Data Network
DDS	digital data service
DEL	delete character
demarc	demarcation point
demux	demultiplex; demultiplexer; demultiplexing
dequeue	double-ended queue
DES	Data Encryption Standard

detem	detector/emitter
DFSK	double-frequency shift keying
DIA	Defense Intelligence Agency
DID	direct inward dialing
DIN	Deutsches Institut für Normung
DIP	dual in-line package
DISA	Defense Information Systems Agency
DISC	disconnect command
DISN	Defense Information System Network
DISNET	Defense Integrated Secure Network
DLA	Defense Logistic Agency
DLC	digital loop carrier
DLE	data link escape character
DM	delta modulation
DMA	Defense Mapping Agency; direct memory access
DME	distance measuring equipment
DMS	Defense Message System
DNA	Defense Nuclear Agency
DNIC	data network identification code
DNPA	data numbering plan area
DNS	Domain Name System
DO	design objective
DoC	Department of Commerce
DOD	Department of Defense; direct outward dialing
DODD	Department of Defense Directive
DODISS	Department of Defense Index of Specifications and Standards
DOD-STD	Department of Defense Standard
DOS	Department of State
DPCM	differential pulse-code modulation
DPSK	differential phase-shift keying
DQDB	distributed-queue dual-bus [network]
DRAM	dynamic random access memory
DRSI	destination station routing indicator

DS	digital signal; direct support
DS0	digital signal 0
DS1	digital signal 1
DS1C	digital signal 1C
DS2	digital signal 2
DS3	digital signal 3
DS4	digital signal 4
DSA	dial service assistance
DSB	double sideband (transmission); Defense Science Board
DSB-RC	double-sideband reduced carrier transmission
DSB-SC	double-sideband suppressed-carrier transmission
DSC	digital selective calling
DSCS	Defense Satellite Communications System
DSE	data switching exchange
DSI	digital speech interpolation
DSL	digital subscriber line
DSN	Defense Switched Network
DSR	data signaling rate
DSS	direct station selection
DSSCS	Defense Special Service Communications System
DSTE	data subscriber terminal equipment
DSU	data service unit
DTE	data terminal equipment
DTG	date-time group
DTMF	dual-tone multifrequency (signaling)
DTN	data transmission network
DTS	Diplomatic Telecommunications Service
DTU	data transfer unit; data tape unit; digital transmission unit; direct to user
DVL	direct voice link
DX signaling	direct current signaling; duplex signaling

E

E	exa (10^{18})

E-MAIL	electronic mail
EAS	extended area service
EBCDIC	extended binary coded decimal interchange code
E_b/N_0	signal energy per bit per hertz of thermal noise
EBO	embedded base organization
EBS	Emergency Broadcast System
EBX	electronic branch exchange
EC	Earth coverage; Earth curvature
ECC	electronically controlled coupling; enhance call completion
ECCM	electronic counter-countermeasures
ECM	electronic countermeasures
EDC	error detection and correction
EDI	electronic data interchange
EDTV	extended-definition television
EHF	extremely high frequency
EIA	Electronic Industries Association
eirp	effective isotropically radiated power; equivalent isotopically radiated power
EIS	Emergency Information System
el	elevation
ELF	extremely low frequency
ELINT	electronics intelligence; electromagnetic intelligence
ELSEC	electronics security
ELT	emergency locator transmitter
EMC	electromagnetic compatibility
EMCON	emission control
EMD	equilibrium mode distribution
EME	electromagnetic environment
emf	electromotive force
EMI	electromagnetic interference; electromagnetic interference control
EMP	electromagnetic pulse
EMR	electromagnetic radiation
e.m.r.p.	effective monopole radiated power; equivalent monopole radiated power
EMS	electronic message system

EMSEC	emanations security
emu	electromagnetic unit
EMV	electromagnetic vulnerability
EMW	electromagnetic warfare; electromagnetic wave
ENQ	enquiry character
EO	end office
E.O.	Executive Order
EOD	end of data
EOF	end of file
EOL	end of line
EOM	end of message
EOP	end of program; end output
EOS	end-of-selection character
EOT	end-of-transmission character; end of tape
EOW	engineering orderwire
EPROM	erasable programmable read-only memory
EPSCS	enhanced private switched communications system
ERL	echo return loss
ERLINK	emergency response link
ERP, e.r.p.	effective radiated power
ES	end system; expert system
ESC	escape character; enhanced satellite capability
ESF	extended superframe
ESM	electronic warfare support measures
ESP	enhanced service provider
ESS	electronic switching system
ETB	end-of-transmission-block character
ETX	end-of-text character
EW	electronic warfare
EXCSA	Exchange Carriers Standards Association

F

f	femto (10^{-15})

f	frequency
FAA	Federal Aviation Administration
FAQ file	Frequently Asked Questions file
FAX	facsimile
FC	functional component
FCC	Federal Communications Commission
FCS	frame check sequence
FDDI	fiber distributed data interface
FDDI-2	fiber distributed data interface-2
FDHM	full duration at half maximum
FDM	frequency-division multiplexing
FDMA	frequency-division multiple access
FDX	full duplex
FEC	forward error correction
FECC	Federal Emergency Communications Coordinators
FED-STD	Federal Standard
FEMA	Federal Emergency Management Agency
FEP	front-end processor
FET	field effect transistor
FIFO	first-in first-out
FIP	Federal Information Processing
FIPS	Federal Information Processing Standards
FIR	finite impulse response
FIRMR	Federal Information Resources Management Regulations
FISINT	foreign instrumentation signals intelligence
flops	floating-point operations per second
FM	frequency modulation
FO	fiber optics
FOC	final operational capability; full operational capability
FOT	frequency of optimum traffic; frequency of optimum transmission
FPIS	forward propagation ionospheric scatter
fps	foot-pound-second
FPS	frames per second; focus projection and scanning

FRP	Federal Response Plan
FSDPSK	filtered symmetric differential phase-shift keying
FSK	frequency-shift keying
FSS	fully separate subsidiary
FT	fiber optic T-carrier
FTAM	file transfer, access, and management
FTF	Federal Telecommunications Fund
ft/min	feet per minute
FTP	file transfer protocol
ft/s	feet per second
FTS	Federal Telecommunications System
FTS2000	Federal Telecommunications System 2000
FTSC	Federal Telecommunications Standards Committee
FWHM	full width at half maximum
FX	fixed service; foreign exchange service
FYDP	Five Year Defense Plan

G

g	profile parameter
G	giga (10^9)
GBH	group busy hour
GCT	Greenwich Civil Time
GDF	group distribution frame
GETS	Government Emergency Telecommunications Service
GFE	Government-furnished equipment
GGCL	government-to-government communications link
GHz	gigahertz
GII	Global Information Infrastructure
GMT	Greenwich Mean Time
GOS	grade of service
GOSIP	Government Open Systems Interconnection Profile
GSA	General Services Administration
GSTN	general switched telephone network

G/T	antenna gain-to-noise-temperature
GTP	Government Telecommunications Program
GTS	Government Telecommunications System
GUI	graphical user interface

H

h	hecto (10^2); hour; Planck's constant
HCS	hard clad silica (fiber)
HDLC	high-level data link control
HDTV	high-definition television
HDX	half-duplex (operation)
HE_{11} mode	the fundamental hybrid mode (of an optical fiber)
HEMP	high-altitude electromagnetic pulse
HERF	hazards of electromagnetic radiation to fuel
HERO	hazards of electromagnetic radiation to ordnance
HERP	hazards of electromagnetic radiation to personnel
HF	high frequency
HFDF	high-frequency distribution frame
HLL	high-level language
HPC	high probability of completion
HV	high voltage
Hz	hertz

I

IA	International Alphabet
I&C	installation and checkout
IC	integrated circuit
ICI	incoming call identification
ICNI	Integrated Communications, Navigation, and Identification
ICW	interrupted continuous wave
IDDD	International Direct Distance Dialing
IDF	intermediate distribution frame
IDN	integrated digital network
IDTV	improved-definition television

IEC	International Electrotechnical Commission
IEEE	Institute of Electrical and Electronics Engineers
IES	Industry Executive Subcommittee
IF	intermediate frequency
I/F	interface
IFF	identification, friend or foe
IFRB	International Frequency Registration Board
IFS	ionospheric forward scatter
IIR	infinite impulse response
IITF	Information Infrastructure Task Force
ILD	injection laser diode
ILS	instrument landing system
IM	intensity modulation; intermodulation
I&M	installation and maintenance
IMD	intermodulation distortion
IMP	interface message processor
IN	intelligent network
INFOSEC	information systems security
INS	inertial navigation system
INTELSAT	International Telecommunications Satellite Consortium
INWATS	Inward Wide-Area Telephone Service
I/O	input/output (device)
IOC	integrated optical circuit; initial operational capability; input-output controller
IP	Internet protocol; intelligent peripheral
IPA	intermediate power amplifier
IPC	information processing center
IPM	impulses per minute; interference prediction model; internal polarization modulation; interruptions per minute
in/s	inches per second
ips	interruptions per second
IPX	Internet Packet Exchange
IQF	intrinsic quality factor
IR	infrared

IRAC	Interdepartment Radio Advisory Committee
IRC	international record carrier; Interagency Radio Committee
ISB	independent-sideband (transmission)
ISDN	Integrated Services Digital Network
ISM	industrial, scientific, and medical (applications)
ISO	International Organization for Standardization
ITA	International Telegraph Alphabet
ITA-5	International Telegraph Alphabet Number 5
ITC	International Teletraffic Congress
ITS	Institute for Telecommunication Sciences
ITSO	International Telecommunications Satellite Organization
ITU	International Telecommunication Union
IVDT	integrated voice data terminal
IXC	interexchange carrier

J

JANAP	Joint Army-Navy-Air Force Publication(s)
JCS	Joint Chiefs of Staff
JPL	Jet Propulsion Laboratory
JSC	Joint Steering Committee; Joint Spectrum Center
JTC^3A	Joint Tactical Command, Control and Communications Agency
JTIDS	Joint Tactical Information Distribution System
JTRB	Joint Telecommunications Resources Board
JTSSG	Joint Telecommunications Standards Steering Group
JWID	Joint Warrior Interoperability Demonstration

K

k	kilo (10^3); Boltzmann's constant
K	coefficient of absorption; kelvin
KDC	key distribution center
KDR	keyboard data recorder
KDT	keyboard display terminal
kg	kilogram
kg·m·s	kilogram-meter-second

kHz	kilohertz
km	kilometer
kΩ, k	kilohm
KSR	keyboard send/receive device
kT	noise power density
KTS	key telephone system
KTU	key telephone unit

L

LAN	local area network
LAP-B	Data Link Layer protocol (CCITT Recommendation X.25 [1989])
LAP-D	link access procedure D
laser	light amplification by stimulated emission of radiation
LASINT	laser intelligence
LATA	local access and transport area
LBO	line buildout
LC	limited capability
LCD	liquid crystal display
LD	long distance
LDM	limited distance modem
LEC	local exchange carrier
LED	light-emitting diode
LF	low frequency
LFB	look-ahead-for-busy (information)
LIFO	last-in first-out
LLC	logical link control (sublayer)
l/m	lines per minute
LMF	language media format
LMR	land mobile radio
LNA	launch numerical aperture
LOF	lowest operating frequency
loran	long-range aid to navigation system; long-range radio navigation; long-range radio aid to navigation system

LOS	line of sight, loss of signal
LP	linearly polarized (mode); linear programming; linking protection; log-periodic (antenna); log-periodic (array)
LPA	linear power amplifier
LPC	linear predictive coding
LPD	low probability of detection
LPI	low probability of interception
lpi	lines per inch
lpm	lines per minute
LP_{01}	the fundamental mode (of an optical fiber)
LQA	link quality analysis
LRC	longitudinal redundancy check
LSB	lower sideband, least significant bit
LSI	large scale integrated (circuit); large scale integration; line status indication
LTC	line traffic coordinator
LUF	lowest usable high frequency
LULT	line-unit-line termination
LUNT	line-unit-network termination
LV	low voltage

M

m	meter
M	mega (10^6)
MAC	medium access control [sublayer]
MACOM	major command
MAN	metropolitan area network
MAP	manufacturers' automation protocol
maser	microwave amplification by the stimulated emission of radiation
MAU	medium access unit
MCC	maintenance control circuit
MCEB	Military Communications-Electronics Board
MCM	multicarrier modulation
MCS	Master Control System

MCW	modulated continuous wave
MCXO	microcomputer compensated crystal oscillator
MDF	main distribution frame
MDT	mean downtime
MEECN	Minimum Essential Emergency Communications Network
MERCAST	merchant-ship broadcast system
MF	medium frequency; multifrequency (signaling)
MFD	mode field diameter
MFJ	Modification of Final Judgment
MFSK	multiple frequency-shift keying
MHF	medium high frequency
MHS	message handling service; message handling system
MHz	megahertz
mi	mile
MIC	medium interface connector; microphone; microwave integrated circuit; minimum ignition current; monolithic integrated circuit; mutual interface chart
MILNET	military network
MIL-STD	Military Standard
min	minute
MIP	medium interface point
MIPS, mips	million instructions per second
MIS	management information system
MKS	meter-kilogram-second
MLPP	multilevel precedence and preemption
MMW	millimeter wave
modem	modulator-demodulator
mol	mole
ms	millisecond (10^{-3} second)
MSB	most significant bit
MSK	minimum-shift keying
MTBF	mean time between failures
MTBM	mean time between maintenance
MTBO	mean time between outages

MTBPM	mean time between preventive maintenance
MTF	modulation transfer function
MTSO	mobile telephone switching office
MTSR	mean time to service restoration
MTTR	mean time to repair
μ	micro (10^{-6})
μs	microsecond
MUF	maximum usable frequency
MUX	multiplex; multiplexer
MUXing	multiplexing
mw	microwave
MWI	message waiting indicator
MWV	maximum working voltage

N

n	nano (10^{-9}); refractive index
N_0	sea level refractivity; spectral noise density
NA	numerical aperture
NACSEM	National Communications Security Emanation Memorandum
NACSIM	National Communications Security Information Memorandum
NAK	negative-acknowledge character
NASA	National Aeronautics and Space Administration
NATA	North American Telecommunications Association
NATO	North Atlantic Treaty Organization
NAVSTAR	Navigational Satellite Timing and Ranging
NBFM	narrowband frequency modulation
NBH	network busy hour
NBRVF	narrowband radio voice frequency
NBS	National Bureau of Standards
NBSV	narrowband secure voice
NCA	National Command Authorities
NCC	National Coordinating Center for Telecommunications
NCS	National Communications System; net control station

NCSC	National Communications Security Committee
NDCS	network data control system
NDER	National Defense Executive Reserve
NEACP	National Emergency Airborne Command Post
NEC	National Electric Code®
NEP	noise equivalent power
NES	noise equivalent signal
NF	noise figure
NFS	Network File System
NIC	network interface card
NICS	NATO Integrated Communications System
NID	network interface device; network inward dialing; network information database
NII	National Information Infrastructure
NIOD	network inward/outward dialing
NIST	National Institute of Standards and Technology
NIU	network interface unit
NLP	National Level Program
nm	nanometer
NMCS	National Military Command System
nmi	nautical mile
NOD	network outward dialing
Np	neper
NPA	numbering plan area
NPR	noise power ratio
NRI	net radio interface
NRM	network resource manager
NRRC	Nuclear Risk Reduction Center
NRZ	non-return-to-zero
NRZI	non-return-to-zero inverted
NRZ-M	non-return-to-zero mark
NRZ-S	non-return-to-zero space
NRZ1	non-return-to-zero, change on ones
NRZ-1	non-return-to-zero mark

ns	nanosecond
NSA	National Security Agency
NSC	National Security Council
NS/EP	National Security or Emergency Preparedness telecommunications
NSTAC	National Security Telecommunications Advisory Committee
NTCN	National Telecommunications Coordinating Network
NTDS	Naval Tactical Data System
NT1	Network termination 1
NT2	Network termination 2
NTI	network terminating interface
NTIA	National Telecommunications and Information Administration
NTMS	National Telecommunications Management Structure
NTN	network terminal number
NTSC	National Television Standards Committee; National Television Standards Committee (standard)
NUL	null character
NVIS	near vertical incidence skywave

O

O&M	operations and maintenance
OC	operations center
OCC	other common carrier
OCR	optical character reader; optical character recognition
OCU	orderwire control unit
OCVCXO	oven controlled-voltage controlled crystal oscillator
OCXO	oven controlled crystal oscillator
OD	optical density; outside diameter
OFC	optical fiber, conductive
OFCP	optical fiber, conductive, plenum
OFCR	optical fiber, conductive, riser
OFN	optical fiber, nonconductive
OFNP	optical fiber, nonconductive, plenum
OFNR	optical fiber, nonconductive, riser

OMB	Office of Management and Budget
ONA	open network architecture
opm	operations per minute
OPMODEL	operations model
OPSEC	operations security
OPX	off-premises extension
OR	off-route service; off-route aeronautical mobile service
OSHA	Occupational Safety and Health Administration
OSI	open switching interval; Open Systems Interconnection
OSI-RM	Open Systems Interconnection—Reference Model
OSRI	originating stations routing indicator
OSSN	originating stations serial number
OTAM	over-the-air management of automated HF network nodes
OTAR	over-the-air rekeying
OTDR	optical time domain reflectometer; optical time domain reflectometry
OW	orderwire [circuit]

P

p	pico (10^{-12})
P	peta (10^{15})
PABX	private automatic branch exchange
PAD	packet assembler/disassembler
PAL	phase alternation by line
PAL-M	phase alternation by line—modified
PAM	pulse-amplitude modulation
PAMA	pulse-address multiple access
p/a r	peak-to-average ratio
PAR	performance analysis and review
PARAMP	parametric amplifier
par meter	peak-to-average ratio meter
PAX	private automatic exchange
PBER	pseudo-bit-error-ratio
PBX	private branch exchange

PC	carrier power (of a radio transmitter); personal computer
PCB	power circuit breaker; printed circuit board
PCM	pulse-code modulation; plug compatible module; process control module
PCS	Personal Communications Services; personal communications system; plastic-clad silica (fiber)
PCSR	parallel channels signaling rate
PD	photodetector
PDM	pulse delta modulation; pulse-duration modulation
PDN	public data network
PDS	protected distribution system; power distribution system; program data source
PDT	programmable data terminal
PDU	protocol data unit
PE	phase-encoded (recording)
PEP	peak envelope power (of a radio transmitter)
pF	picofarad
PF	power factor
PFM	pulse-frequency modulation
PI	protection interval
PIC	plastic insulated cable
ping	packet Internet groper
PIV	peak inverse voltage
PLA	programmable logic array
PL/I	programming language 1
PLL	phase-locked loop
PLN	private line network
PLR	pulse link repeater
PLS	physical signaling sublayer
pm	phase modulation
PM	mean power; polarization-maintaining (optical fiber); preventive maintenance; pulse modulation
PMB	pilot-make-busy (circuit)
PMO	program management office
POI	point of interface

POP	point of presence
POSIX	portable operating system interface for computer environments
POTS	plain old telephone service
PP	polarization-preserving (optical fiber)
P-P	peak-to-peak (value)
P/P	point-to-point
PPM	pulse-position modulation
pps	pulses per second
PR	pulse rate
PRF	pulse-repetition frequency
PRI	primary rate interface
PRM	pulse-rate modulation
PROM	programmable read-only memory
PRR	pulse repetition rate
PRSL	primary area switch locator
PS	permanent signal
psi	pounds (force) per square inch
PSK	phase-shift keying
PSN	public switched network
p-static	precipitation static
PSTN	public switched telephone network
PTF	patch and test facility
PTM	pulse-time modulation
PTT	postal, telephone, and telegraph; push-to-talk (operation)
PTTC	paper tape transmission code
PTTI	precise time and time interval
PU	power unit
PUC	public utility commission; public utilities commission
PVC	permanent virtual circuit; polyvinyl chloride (insulation)
pW	picowatt
PWM	pulse-width modulation
PX	private exchange

Q

QA	quality assurance
QAM	quadrature amplitude modulation
QC	quality control
QCIF	quarter common intermediate format
QMR	qualitative material requirement
QOS	quality of service
QPSK	quadrature phase-shift keying
QRC	quick reaction capability

R

racon	radar beacon
rad	radian; radiation absorbed dose
radar	radio detection and ranging
RADHAZ	electromagnetic radiation hazards
RADINT	radar intelligence
RAM	random access memory; reliability, availability, and maintainability
R&D	research and development
RATT	radio teletypewriter system
RBOC	Regional Bell Operating Company
RbXO	rubidium-crystal oscillator
RC	reflection coefficient; resource controller
RCC	radio common carrier
RCVR	receiver
RDF	radio-direction finding
REA	Rural Electrification Administration
REN	ringer equivalency number
RF	radio frequency; range finder
RFI	radio frequency interference
RFP	request for proposal
RFQ	request for quotation
RGB	red-green-blue
RH	relative humidity

RHR	radio horizon range
RI	routing indicator
RISC	reduced instruction set chip
RJ	registered jack
RJE	remote job entry
rms	root-mean-square (deviation)
RO	read only; receive only
ROA	recognized operating agency
ROC	required operational capability
ROM	read-only memory
ROSE	remote operations service element protocol
rpm	revolutions per minute
RPM	rate per minute
RPOA	recognized private operating agency
rps	revolutions per second
RQ	repeat-request
RR	repetition rate
RSL	received signal level
rss	root-sum-square
R/T	real time
RTA	remote trunk arrangement
RTS	request to send
RTTY	radio teletypewriter
RTU	remote terminal unit
RTX	request to transmit
RVA	reactive volt-ampere
RVWG	Reliability and Vulnerability Working Group
RWI	radio and wire integration
RX	receive; receiver
RZ	return-to-zero

S

s	second

SCC	specialized common carrier
SCE	service creation environment
SCF	service control facility
SCP	service control point
SCPC	single channel per carrier
SCR	semiconductor-controlled rectifier; silicon-controlled rectifier
SCSR	single channel signaling rate
SDLC	synchronous data link control
SDM	space-division multiplexing
SDN	software-defined network
SECDEF	Secretary of Defense
SECORD	secure voice cord board
SECTEL	secure telephone
SETAMS	systems engineering, technical assistance, and management services
SEVAS	Secure Voice Access System
S-F	store-and-forward
SF	single-frequency (signaling)
SGDF	supergroup distribution frame
S/H	sample and hold
SHA	sidereal hour angle
SHARES	Shared Resources (SHARES) HF Radio Program
SHF	super high frequency
SI	International System of Units
SID	sudden ionospheric disturbance
SIGINT	signals intelligence
SINAD	signal-plus-noise-plus-distortion to noise-plus-distortion ratio
SLD	superluminescent diode
SLI	service logic interpreter
SLP	service logic program
SMDR	station message-detail recording
SMSA	standard metropolitan statistical area
SNR	signal-to-noise ratio
SOH	start-of-heading character

SOM	start of message
sonar	sound navigation and ranging
SONET	synchronous optical network
SOP	standard operating procedure
SOR	start of record
SOW	statement of work
(S+N)/N	signal-plus-noise-to-noise ratio
sr	steradian
S/R	send and receive
SSB	single-sideband (transmission)
SSB-SC	single-sideband suppressed carrier (transmission)
SSN	station serial number
SSP	service switching point
SSUPS	solid-state uninterruptible power system
STALO	stabilized local oscillator
STD	subscriber trunk dialing
STFS	standard time and frequency signal (service); standard time and frequency service
STL	standard telegraph level; studio-to-transmitter link
STP	standard temperature and pressure; signal transfer point
STU	secure telephone unit
STX	start-of-text character
SUB	substitute character
SWR	standing wave ratio
SX	simplex signaling
SXS	step-by-step switching system
SYN	synchronous idle character
SYSGEN	system generation

T

T	tera (10^{12})
TADIL	tactical data information link
TADIL-A	tactical data information link-A
TADIL-B	tactical data information link-B

TADS	teletypewriter automatic dispatch system
TADSS	Tactical Automatic Digital Switching System
TAI	International Atomic Time
TASI	time-assignment speech interpolation
TAT	trans-Atlantic telecommunication (cable)
TC	toll center
TCB	trusted computing base
TCC	telecommunications center
TCCF	Tactical Communications Control Facility
TCF	technical control facility
TCP	transmission control protocol
TCS	trusted computer system
TCU	teletypewriter control unit
TCVXO	temperature compensated-voltage controlled crystal oscillator
TCXO	temperature controlled crystal oscillator
TD	time delay; transmitter distributor
TDD	Telecommunications Device for the Deaf
TDM	time-division multiplexing
TDMA	time-division multiple access
TE	transverse electric [mode]
TED	trunk encryption device
TEK	traffic encryption key
TEM	transverse electric and magnetic [mode]
TEMPEST	compromising emanations
TEMS	telecommunications management system
TGM	trunk group multiplexer
THD	total harmonic distortion
THF	tremendously high frequency
THz	terahertz
TIA	Telecommunications Industry Association
TIE	time interval error
TIFF	tag image file format
TIP	terminal interface processor

T_K	response timer
TLP	transmission level point
TM	transverse magnetic [mode]
TP	toll point
TRANSEC	transmission security
TRC	transverse redundancy check
TRF	tuned radio frequency
TRI-TAC	tri-services tactical [equipment]
TSK	transmission security key
TSP	Telecommunications Service Priority [system]
TSPS	traffic service position system
TSR	telecommunications service request; terminate and stay resident
TTL	transistor-transistor logic
TTTN	tandem tie trunk network
TTY	teletypewriter
TTY/TDD	Telecommunications Device for the Deaf
TV	television
TW	traveling wave
TWT	traveling wave tube
TWTA	traveling wave tube amplifier
TWX®	teletypewriter exchange service
TX	transmit; transmitter

U

UDP	User Datagram Protocol
UHF	ultra high frequency
U/L	uplink
ULF	ultra low frequency
UPS	uninterruptible power supply
UPT	Universal Personal Telecommunications service
USB	upper sideband
USDA	U.S. Department of Agriculture
USFJ	U.S. Forces, Japan

USFK	U.S. Forces, Korea
USNO	U.S. Naval observatory
USTA	U.S. Telephone Association
UT	Universal Time
UTC	Coordinated Universal Time
uv	ultraviolet

V

V	volt
VA	volt-ampere
VAN	value-added network
VAR	value added reseller
VARISTAR	variable resistor
vars	volt-amperes reactive
VC	virtual circuit
VCO	voltage-controlled oscillator
VCXO	voltage-controlled crystal oscillator
V/D	voice/data
Vdc	volts direct current
VDU	video display unit; visual display unit
VF	voice frequency
VFCT	voice frequency carrier telegraph
VFDF	voice frequency distribution frame
VFO	variable-frequency oscillator
VFTG	voice-frequency telegraph
VHF	very high frequency
VLF	very low frequency
V/m	volts per meter
VNL	via net loss
VNLF	via net loss factor
vocoder	voice-coder
vodas	voice-operated device anti-sing
vogad	voice-operated gain-adjusting device

volcas	voice-operated loss control and echo/signaling suppression
vox	voice-operated relay circuit; voice operated transmit
VRC	vertical redundancy check
VSAT	very small aperture terminal
VSB	vestigial sideband [transmission]
VSM	vestigial sideband modulation
VSWR	voltage standing wave ratio
VT	virtual terminal
VTU	video teleconferencing unit
vu	volume unit

W

WADS	wide area data service
WAIS	Wide Area Information Servers
WAN	wide area network
WARC	World Administrative Radio Conference
WATS	Wide Area Telecommunications Service; Wide Area Telephone Service
WAWS	Washington Area Wideband System
WDM	wavelength-division multiplexing
WHSR	White House Situation Room
WIN	WWMCCS Intercomputer Network
WITS	Washington Integrated Telecommunications System
WORM	write once, read many times
wpm	words per minute
wps	words per second
wv	working voltage
WVDC	working voltage direct current
WWDSA	worldwide digital system architecture
WWMCCS	Worldwide Military Command and Control System
WWW	World Wide Web

X

XMIT	transmit
XMSN	transmission

XMTD	transmitted
XMTR	transmitter
XO	crystal oscillator
XOFF	transmitter off
XON	transmitter on
XT	crosstalk
XTAL	crystal

Z

Z	Zulu time
ZD	zero defects
Z_o	characteristic impedance
0TLP	zero transmission level point

13.3 References

1. *ATSC Digital Television Standard*, Doc. A/53, Advanced Television Systems Committee, Washington, D.C., 1996.
2. *Digital Audio Compression (AC-3) Standard*, Doc. A/52, Advanced Television Systems Committee, Washington, D.C., 1996.
3. *Guide to the Use of the ATSC Digital Television Standard*, Doc. A/54, Advanced Television Systems Committee, Washington, D.C., 1996.
4. *Program Guide for Digital Television*, Doc. A/55, Advanced Television Systems Committee, Washington, D.C., 1996.
5. *System Information for Digital Television*, Doc. A/56, Advanced Television Systems Committee, Washington, D.C., 1996.
6. "Advanced Television Enhancement Forum Specification," Draft, Version 1.1r26 updated 2/2/99, ATVEF, Portland, Ore., 1999.

13.4 Bibliography

Whitaker, Jerry C.: *Power Vacuum Tubes Handbook*, CRC Press, Boca Raton, Fla., 1999.

General Services Administration, Information Technology Service, National Communications System: *Glossary of Telecommunication Terms*, Technology and Standards Division, Federal Standard 1037C, General Services Administration, Washington, D.C., August, 7, 1996.

Chapter

14

Reference Data and Tables

Jerry C. Whitaker, editor

14.1 Standard Units

Name	Symbol	Quantity
ampere	A	electric current
ampere per meter	A/m	magnetic field strength
ampere per square meter	A/m^2	current density
becquerel	Bg	activity (of a radionuclide)
candela	cd	luminous intensity
coulomb	C	electric charge
coulomb per kilogram	C/kg	exposure (x and gamma rays)
coulomb per sq. meter	C/m^2	electric flux density
cubic meter	m^3	volume
cubic meter per kilogram	m^3/kg	specific volume
degree Celsius	°C	Celsius temperature
farad	F	capacitance
farad per meter	F/m	permittivity
henry	H	inductance
henry per meter	H/m	permeability
hertz	Hz	frequency
joule	J	energy, work, quantity of heat
joule per cubic meter	J/m^3	energy density
joule per kelvin	J/K	heat capacity
joule per kilogram K	J/(kg•K)	specific heat capacity
joule per mole	J/mol	molar energy
kelvin	K	thermodynamic temperature

kilogram	kg	mass
kilogram per cubic meter	kg/m^3	density, mass density
lumen	lm	luminous flux
lux	lx	luminance
meter	m	length
meter per second	m/s	speed, velocity
meter per second sq.	m/s^2	acceleration
mole	mol	amount of substance
newton	N	force
newton per meter	N/m	surface tension
ohm	Ω	electrical resistance
pascal	Pa	pressure, stress
pascal second	Pa•s	dynamic viscosity
radian	rad	plane angle
radian per second	rad/s	angular velocity
radian per second squared	rad/s^2	angular acceleration
second	s	time
siemens	S	electrical conductance
square meter	m^2	area
steradian	sr	solid angle
tesla	T	magnetic flux density
volt	V	electrical potential
volt per meter	V/m	electric field strength
watt	W	power, radiant flux
watt per meter kelvin	W/(m•K)	thermal conductivity
watt per square meter	W/m^2	heat (power) flux density
weber	Wb	magnetic flux

14.1a Standard Prefixes

Multiple	Prefix	Symbol
10^{18}	exa	E
10^{15}	peta	P
10^{12}	tera	T
10^{9}	giga	G
10^{6}	mega	M
10^{3}	kilo	k
10^{2}	hecto	h
10	deka	da
10^{-1}	deci	d
10^{-2}	centi	c
10^{-3}	milli	m
10^{-6}	micro	μ
10^{-9}	nano	n
10^{-12}	pico	p
10^{-15}	femto	f
10^{-18}	atto	a

14.1b Common Standard Units

Unit	Symbol
centimeter	cm
cubic centimeter	cm^3
cubic meter per second	m^3/s
gigahertz	GHz
gram	g
kilohertz	kHz
kilohm	kΩ
kilojoule	kJ
kilometer	km
kilovolt	kV
kilovoltampere	kVA
kilowatt	kW

megahertz	MHz
megavolt	MV
megawatt	MW
megohm	MΩ
microampere	μA
microfarad	μF
microgram	μg
microhenry	μH
microsecond	μs
microwatt	μW
milliampere	mA
milligram	mg
millihenry	mH
millimeter	mm
millisecond	ms
millivolt	mV
milliwatt	mW
nanoampere	nA
nanofarad	nF
nanometer	nm
nanosecond	ns
nanowatt	nW
picoampere	pA
picofarad	pF
picosecond	ps
picowatt	pW

14.2 Conversion Reference Data

To Convert	Into	Multiply By
abcoulomb	statcoulombs	2.998×10^{10}
acre	sq. chain (Gunters)	10
acre	rods	160
acre	square links (Gunters)	1×10^{5}
acre	Hectare or sq. hectometer	0.4047
acre-feet	cubic feet	43,560.0
acre-feet	gallons	3.259×10^{5}
acres	sq. feet	43,560.0
acres	sq. meters	4,047
acres	sq. miles	1.562×10^{-3}
acres	sq. yards	4,840
ampere-hours	coulombs	3,600.0
ampere-hours	faradays	0.03731
amperes/sq. cm	amps/sq. in	6.452
amperes/sq. cm	amps/sq. meter	10^{4}
amperes/sq. in	amps/sq. cm	0.1550
amperes/sq. in	amps/sq. meter	1,550.0
amperes/sq. meter	amps/sq. cm	10^{-4}
amperes/sq. meter	amps/sq. in	6.452×10^{-4}
ampere-turns	gilberts	1.257
ampere-turns/cm	amp-turns/in	2.540
ampere-turns/cm	amp-turns/meter	100.0
ampere-turns/cm	gilberts/cm	1.257
ampere-turns/in	amp-turns/cm	0.3937
ampere-turns/in	amp-turns/m	39.37
ampere-turns/in	gilberts/cm	0.4950
ampere-turns/meter	amp-turns/cm	0.01
ampere-turns/meter	amp-turns/in	0.0254
ampere-turns/meter	gilberts/cm	0.01257
Angstrom unit	inch	3937×10^{-9}
Angstrom unit	meter	1×10^{-10}
Angstrom unit	micron or (Mu)	1×10^{-4}
are	acre (U.S.)	0.02471
ares	sq. yards	119.60

ares	acres	0.02471
ares	sq. meters	100.0
astronomical unit	kilometers	1.495×10^8
atmospheres	ton/sq. in	0.007348
atmospheres	cm of mercury	76.0
atmospheres	ft of water (at 4°C)	33.90
atmospheres	in of mercury (at 0°C)	29.92
atmospheres	kg/sq. cm	1.0333
atmospheres	kg/sq. m	10,332
atmospheres	pounds/sq. in	14.70
atmospheres	tons/sq. ft	1.058
barrels (U.S., dry)	cubic inches	7056
barrels (U.S., dry)	quarts (dry)	105.0
barrels (U.S., liquid)	gallons	31.5
barrels (oil)	gallons (oil)	42.0
bars	atmospheres	0.9869
bars	dynes/sq. cm	10^4
bars	kg/sq. m	1.020×10^4
bars	pounds/sq. ft	2,089
bars	pounds/sq. in	14.50
baryl	dyne/sq. cm	1.000
bolt (U.S. cloth)	meters	36.576
Btu	liter-atmosphere	10.409
Btu	ergs	1.0550×10^{10}
Btu	foot-lb	778.3
Btu	gram-calories	252.0
Btu	horsepower-hr	3.931×10^{-4}
Btu	joules	1,054.8
Btu	kilogram-calories	0.2520
Btu	kilogram-meters	107.5
Btu	kilowatt-hr	2.928×10^{-4}
Btu/hr	foot-pounds/s	0.2162
Btu/hr	gram-calories/s	0.0700
Btu/hr	horsepower-hr	3.929×10^{-4}
Btu/hr	watts	0.2931
Btu/min	foot-lbs/s	12.96
Btu/min	horsepower	0.02356

Btu/min	kilowatts	0.01757
Btu/min	watts	17.57
Btu/sq. ft/min	watts/sq. in	0.1221
bucket (br. dry)	cubic cm	1.818×10^{4}
bushels	cubic ft	1.2445
bushels	cubic in	2,150.4
bushels	cubic m	0.03524
bushels	liters	35.24
bushels	pecks	4.0
bushels	pints (dry)	64.0
bushels	quarts (dry)	32.0
calories, gram (mean)	Btu (mean)	3.9685×10^{-3}
candle/sq. cm	Lamberts	3.142
candle/sq. in	Lamberts	0.4870
centares (centiares)	sq. meters	1.0
Centigrade	Fahrenheit	(C° x 9/5) + 32
centigrams	grams	0.01
centiliter	ounce fluid (U.S.)	0.3382
centiliter	cubic inch	0.6103
centiliter	drams	2.705
centiliter	liters	0.01
centimeter	feet	3.281×10^{-2}
centimeter	inches	0.3937
centimeter	kilometers	10^{-5}
centimeter	meters	0.01
centimeter	miles	6.214×10^{-6}
centimeter	millimeters	10.0
centimeter	mils	393.7
centimeter	yards	1.094×10^{-2}
centimeter-dynes	cm-grams	1.020×10^{-3}
centimeter-dynes	meter-kg	1.020×10^{-8}
centimeter-dynes	pound-ft	7.376×10^{-8}
centimeter-grams	cm-dynes	980.7
centimeter-grams	meter-kg	10^{-5}
centimeter-grams	pound-ft	7.233×10^{-5}
centimeters of mercury	atmospheres	0.01316

centimeters of mercury	feet of water	0.4461
centimeters of mercury	kg/sq. meter	136.0
centimeters of mercury	pounds/sq. ft	27.85
centimeters of mercury	pounds/sq. in	0.1934
centimeters/sec	feet/min	1.9686
centimeters/sec	feet/sec	0.03281
centimeters/sec	kilometers/hr	0.036
centimeters/sec	knots	0.1943
centimeters/sec	meters/min	0.6
centimeters/sec	miles/hr	0.02237
centimeters/sec	miles/min	3.728×10^{-4}
centimeters/sec/sec	feet/sec/sec	0.03281
centimeters/sec/sec	km/hr/sec	0.036
centimeters/sec/sec	meters/sec/sec	0.01
centimeters/sec/sec	miles/hr/sec	0.02237
chain	inches	792.00
chain	meters	20.12
chains (surveyor's or Gunter's)	yards	22.00
circular mils	sq. cm	5.067×10^{-6}
circular mils	sq. mils	0.7854
circular mils	sq. inches	7.854×10^{-7}
circumference	Radians	6.283
cord feet	cubic feet	16
cords	cord feet	8
coulomb	statcoulombs	2.998×10^{9}
coulombs	faradays	1.036×10^{-5}
coulombs/sq. cm	coulombs/sq. in	64.52
coulombs/sq. cm	coulombs/sq. meter	10^{4}
coulombs/sq. in	coulombs/sq. cm	0.1550
coulombs/sq. in	coulombs/sq. meter	1,550
coulombs/sq. meter	coulombs/sq. cm	10^{-4}
coulombs/sq. meter	coulombs/sq. in	6.452×10^{-4}
cubic centimeters	cubic feet	3.531×10^{-5}
cubic centimeters	cubic inches	0.06102
cubic centimeters	cubic meters	10^{-6}
cubic centimeters	cubic yards	1.308×10^{-6}

cubic centimeters	gallons (U.S. liq.)	2.642×10^{-4}
cubic centimeters	liters	0.001
cubic centimeters	pints (U.S. liq.)	2.113×10^{-3}
cubic centimeters	quarts (U.S. liq.)	1.057×10^{-3}
cubic feet	bushels (dry)	0.8036
cubic feet	cubic cm	28,320.0
cubic feet	cubic inches	1,728.0
cubic feet	cubic meters	0.02832
cubic feet	cubic yards	0.03704
cubic feet	gallons (U.S. liq.)	7.48052
cubic feet	liters	28.32
cubic feet	pints (U.S. liq.)	59.84
cubic feet	quarts (U.S. liq.)	29.92
cubic feet/min	cubic cm/sec	472.0
cubic feet/min	gallons/sec	0.1247
cubic feet/min	liters/sec	0.4720
cubic feet/min	pounds of water/min	62.43
cubic feet/sec	million gal/day	0.646317
cubic feet/sec	gallons/min	448.831
cubic inches	cubic cm	16.39
cubic inches	cubic feet	5.787×10^{-4}
cubic inches	cubic meters	1.639×10^{-5}
cubic inches	cubic yards	2.143×10^{-5}
cubic inches	gallons	4.329×10^{-3}
cubic inches	liters	0.01639
cubic inches	mil-feet	1.061×10^{5}
cubic inches	pints (U.S. liq.)	0.03463
cubic inches	quarts (U.S. liq.)	0.01732
cubic meters	bushels (dry)	28.38
cubic meters	cubic cm	10^{6}
cubic meters	cubic feet	35.31
cubic meters	cubic inches	61,023.0
cubic meters	cubic yards	1.308
cubic meters	gallons (U.S. liq.)	264.2
cubic meters	liters	1,000.0
cubic meters	pints (U.S. liq.)	2,113.0
cubic meters	quarts (U.S. liq.)	1,057.

cubic yards	cubic cm	7.646×10^5
cubic yards	cubic feet	27.0
cubic yards	cubic inches	46,656.0
cubic yards	cubic meters	0.7646
cubic yards	gallons (U.S. liq.)	202.0
cubic yards	liters	764.6
cubic yards	pints (U.S. liq.)	1,615.9
cubic yards	quarts (U.S. liq.)	807.9
cubic yards/min	cubic ft/sec	0.45
cubic yards/min	gallons/sec	3.367
cubic yards/min	liters/sec	12.74
Dalton	gram	1.650×10^{-24}
days	seconds	86,400.0
decigrams	grams	0.1
deciliters	liters	0.1
decimeters	meters	0.1
degrees (angle)	quadrants	0.01111
degrees (angle)	radians	0.01745
degrees (angle)	seconds	3,600.0
degrees/sec	radians/sec	0.01745
degrees/sec	revolutions/min	0.1667
degrees/sec	revolutions/sec	2.778×10^{-3}
dekagrams	grams	10.0
dekaliters	liters	10.0
dekameters	meters	10.0
drams (apothecaries or troy)	ounces (avoirdupois)	0.1371429
drams (apothecaries or troy)	ounces (troy)	0.125
drams (U.S., fluid or apothecaries)	cubic cm	3.6967
drams	grams	1.7718
drams	grains	27.3437
drams	ounces	0.0625
dyne/cm	erg/sq. millimeter	0.01
dyne/sq. cm	atmospheres	9.869×10^{-7}
dyne/sq. cm	inch of mercury at 0°C	2.953×10^{-5}
dyne/sq. cm	inch of water at 4°C	4.015×10^{-4}
dynes	grams	1.020×10^{-3}
dynes	joules/cm	10^{-7}

dynes	joules/meter (newtons)	10^{-5}
dynes	kilograms	1.020×10^{-6}
dynes	poundals	7.233×10^{-5}
dynes	pounds	2.248×10^{-6}
dynes/sq. cm	bars	10^{-6}
ell	cm	114.30
ell	inches	45
em, pica	inch	0.167
em, pica	cm	0.4233
erg/sec	Dyne-cm/sec	1.000
ergs	Btu	9.480×10^{-11}
ergs	dyne-centimeters	1.0
ergs	foot-pounds	7.367×10^{-8}
ergs	gram-calories	0.2389×10^{-7}
ergs	gram-cm	1.020×10^{-3}
ergs	horsepower-hr	3.7250×10^{-14}
ergs	joules	10^{-7}
ergs	kg-calories	2.389×10^{-11}
ergs	kg-meters	1.020×10^{-8}
ergs	kilowatt-hr	0.2778×10^{-13}
ergs	watt-hours	0.2778×10^{-10}
ergs/sec	Btu/min	$5,688 \times 10^{-9}$
ergs/sec	ft-lb/min	4.427×10^{-6}
ergs/sec	ft-lb/sec	7.3756×10^{-8}
ergs/sec	horsepower	1.341×10^{-10}
ergs/sec	kg-calories/min	1.433×10^{-9}
ergs/sec	kilowatts	10^{-10}
farad	microfarads	10^{6}
Faraday/sec	ampere (absolute)	9.6500×10^{4}
faradays	ampere-hours	26.80
faradays	coulombs	9.649×10^{4}
fathom	meter	1.828804
fathoms	feet	6.0
feet	centimeters	30.48

feet	kilometers	3.048×10^{-4}
feet	meters	0.3048
feet	miles (naut.)	1.645×10^{-4}
feet	miles (stat.)	1.894×10^{-4}
feet	millimeters	304.8
feet	mils	1.2×10^{4}
feet of water	atmospheres	0.02950
feet of water	in of mercury	0.8826
feet of water	kg/sq. cm	0.03048
feet of water	kg/sq. meter	304.8
feet of water	pounds/sq. ft	62.43
feet of water	pounds/sq. in	0.4335
feet/min	cm/sec	0.5080
feet/min	feet/sec	0.01667
feet/min	km/hr	0.01829
feet/min	meters/min	0.3048
feet/min	miles/hr	0.01136
feet/sec	cm/sec	30.48
feet/sec	km/hr	1.097
feet/sec	knots	0.5921
feet/sec	meters/min	18.29
feet/sec	miles/hr	0.6818
feet/sec	miles/min	0.01136
feet/sec/sec	cm/sec/sec	30.48
feet/sec/sec	km/hr/sec	1.097
feet/sec/sec	meters/sec/sec	0.3048
feet/sec/sec	miles/hr/sec	0.6818
feet/100 feet	per centigrade	1.0
foot-candle	lumen/sq. meter	10.764
foot-pounds	Btu	1.286×10^{-3}
foot-pounds	ergs	1.356×10^{7}
foot-pounds	gram-calories	0.3238
foot-pounds	hp-hr	5.050×10^{-7}
foot-pounds	joules	1.356
foot-pounds	kg-calories	3.24×10^{-4}
foot-pounds	kg-meters	0.1383
foot-pounds	kilowatt-hr	3.766×10^{-7}

foot-pounds/min	Btu/min	1.286×10^{-3}
foot-pounds/min	foot-pounds/sec	0.01667
foot-pounds/min	horsepower	3.030×10^{-5}
foot-pounds/min	kg-calories/min	3.24×10^{-4}
foot-pounds/min	kilowatts	2.260×10^{-5}
foot-pounds/sec	Btu/hr	4.6263
foot-pounds/sec	Btu/min	0.07717
foot-pounds/sec	horsepower	1.818×10^{-3}
foot-pounds/sec	kg-calories/min	0.01945
foot-pounds/sec	kilowatts	1.356×10^{-3}
Furlongs	miles (U.S.)	0.125
furlongs	rods	40.0
furlongs	feet	660.0
gallons	cubic cm	3,785.0
gallons	cubic feet	0.1337
gallons	cubic inches	231.0
gallons	cubic meters	3.785×10^{-3}
gallons	cubic yards	4.951×10^{-3}
gallons	liters	3.785
gallons (liq. Br. Imp.)	gallons (U.S. liq.)	1.20095
gallons (U.S.)	gallons (Imp.)	0.83267
gallons of water	pounds of water	8.3453
gallons/min	cubic ft/sec	2.228×10^{-3}
gallons/min	liters/sec	0.06308
gallons/min	cubic ft/hr	8.0208
gausses	lines/sq. in	6.452
gausses	webers/sq. cm	10^{-8}
gausses	webers/sq. in	6.452×10^{-8}
gausses	webers/sq. meter	10^{-4}
gilberts	ampere-turns	0.7958
gilberts/cm	amp-turns/cm	0.7958
gilberts/cm	amp-turns/in	2.021
gilberts/cm	amp-turns/meter	79.58
gills	liters	0.1183
gills	pints (liq.)	0.25
gills (British)	cubic cm	142.07

grade	radian	0.01571
grains	drams (avoirdupois)	0.03657143
grains (troy)	grains (avdp.)	1.0
grains (troy)	grams	0.06480
grains (troy)	ounces (avdp.)	2.0833×10^{-3}
grains (troy)	pennyweight (troy)	0.04167
grains/Imp. gal	parts/million	14.286
grains/U.S. gal	parts/million	17.118
grains/U.S. gal	pounds/million gal	142.86
gram-calories	Btu	3.9683×10^{-3}
gram-calories	ergs	4.1868×10^{7}
gram-calories	foot-pounds	3.0880
gram-calories	horsepower-hr	1.5596×10^{-6}
gram-calories	kilowatt-hr	1.1630×10^{-6}
gram-calories	watt-hr	1.1630×10^{-3}
gram-calories/sec	Btu/hr	14.286
gram-centimeters	Btu	9.297×10^{-8}
gram-centimeters	ergs	980.7
gram-centimeters	joules	9.807×10^{-5}
gram-centimeters	kg-calories	2.343×10^{-8}
gram-centimeters	kg-meters	10^{-5}
grams	dynes	980.7
grams	grains	15.43
grams	joules/cm	9.807×10^{-5}
grams	joules/meter (newtons)	9.807×10^{-3}
grams	kilograms	0.001
grams	milligrams	1,000
grams	ounces (avdp.)	0.03527
grams	ounces (troy)	0.03215
grams	poundals	0.07093
grams	pounds	2.205×10^{-3}
grams/cm	pounds/inch	5.600×10^{-3}
grams/cubic cm	pounds/cubic ft	62.43
grams/cubic cm	pounds/cubic in	0.03613
grams/cubic cm	pounds/mil-foot	3.405×10^{-7}
grams/liter	grains/gal	58.417

grams/liter	pounds/1,000 gal	8.345
grams/liter	pounds/cubic ft	0.062427
grams/liter	parts/million	1,000.0
grams/sq. cm	pounds/sq. ft	2.0481
hand	cm	10.16
hectares	acres	2.471
hectares	sq. feet	1.076×10^5
hectograms	grams	100.0
hectoliters	liters	100.0
hectometers	meters	100.0
hectowatts	watts	100.0
henries	millihenries	1,000.0
horsepower	Btu/min	42.44
horsepower	foot-lb/min	33,000
horsepower	foot-lb/sec	550.0
horsepower	kg-calories/min	10.68
horsepower	kilowatts	0.7457
horsepower	watts	745.7
horsepower (boiler)	Btu/hr	33.479
horsepower (boiler)	kilowatts	9.803
horsepower, metric (542.5 ft lb./sec)	horsepower (550 ft lb./sec)	0.9863
horsepower (550 ft lb./sec)	horsepower, metric (542.5 ft lb./sec)	1.014
horsepower-hr	Btu	2,547
horsepower-hr	ergs	2.6845×10^{13}
horsepower-hr	foot-lb	1.98×10^6
horsepower-hr	gram-calories	641,190
horsepower-hr	joules	2.684×10^6
horsepower-hr	kg-calories	641.1
horsepower-hr	kg-meters	2.737×10^5
horsepower-hr	kilowatt-hr	0.7457
hours	days	4.167×10^{-2}
hours	weeks	5.952×10^{-3}
hundredweights (long)	pounds	112
hundredweights (long)	tons (long)	0.05
hundredweights (short)	ounces (avoirdupois)	1,600
hundredweights (short)	pounds	100
hundredweights (short)	tons (metric)	0.0453592

hundredweights (short)	tons (long)	0.0446429
inches	centimeters	2.540
inches	meters	2.540×10^{-2}
inches	miles	1.578×10^{-5}
inches	millimeters	25.40
inches	mils	1,000.0
inches	yards	2.778×10^{-2}
inches of mercury	atmospheres	0.03342
inches of mercury	feet of water	1.133
inches of mercury	kg/sq. cm	0.03453
inches of mercury	kg/sq. meter	345.3
inches of mercury	pounds/sq. ft	70.73
inches of mercury	pounds/sq. in	0.4912
inches of water (at 4°C)	atmospheres	2.458×10^{-3}
inches of water (at 4°C)	inches of mercury	0.07355
inches of water (at 4°C)	kg/sq. cm	2.540×10^{-3}
inches of water (at 4°C)	ounces/sq. in	0.5781
inches of water (at 4°C)	pounds/sq. ft	5.204
inches of water (at 4°C)	pounds/sq. in	0.03613
international ampere	ampere (absolute)	0.9998
international Volt	volts (absolute)	1.0003
international volt	joules (absolute)	1.593×10^{-19}
international volt	joules	9.654×10^{4}
joules	Btu	9.480×10^{-4}
joules	ergs	10^{7}
joules	foot-pounds	0.7376
joules	kg-calories	2.389×10^{-4}
joules	kg-meters	0.1020
joules	watt-hr	2.778×10^{-4}
joules/cm	grams	1.020×10^{4}
joules/cm	dynes	10^{7}
joules/cm	joules/meter (newtons)	100.0
joules/cm	poundals	723.3
joules/cm	pounds	22.48
kilogram-calories	Btu	3.968
kilogram-calories	foot-pounds	3,088

kilogram-calories	hp-hr	1.560×10^{-3}
kilogram-calories	joules	4,186
kilogram-calories	kg-meters	426.9
kilogram-calories	kilojoules	4.186
kilogram-calories	kilowatt-hr	1.163×10^{-3}
kilogram meters	Btu	9.294×10^{-3}
kilogram meters	ergs	9.804×10^{7}
kilogram meters	foot-pounds	7.233
kilogram meters	joules	9.804
kilogram meters	kg-calories	2.342×10^{-3}
kilogram meters	kilowatt-hr	2.723×10^{-6}
kilograms	dynes	980,665
kilograms	grams	1,000.0
kilograms	joules/cm	0.09807
kilograms	joules/meter (newtons)	9.807
kilograms	poundals	70.93
kilograms	pounds	2.205
kilograms	tons (long)	9.842×10^{-4}
kilograms	tons (short)	1.102×10^{-3}
kilograms/cubic meter	grams/cubic cm	0.001
kilograms/cubic meter	pounds/cubic ft	0.06243
kilograms/cubic meter	pounds/cubic in	3.613×10^{-5}
kilograms/cubic meter	pounds/mil-foot	3.405×10^{-10}
kilograms/meter	pounds/ft	0.6720
kilograms/sq. cm	dynes	980,665
kilograms/sq. cm	atmospheres	0.9678
kilograms/sq. cm	feet of water	32.81
kilograms/sq. cm	inches of mercury	28.96
kilograms/sq. cm	pounds/sq. ft	2,048
kilograms/sq. cm	pounds/sq. in	14.22
kilograms/sq. meter	atmospheres	9.678×10^{-5}
kilograms/sq. meter	bars	98.07×10^{-6}
kilograms/sq. meter	feet of water	3.281×10^{-3}
kilograms/sq. meter	inches of mercury	2.896×10^{-3}
kilograms/sq. meter	pounds/sq. ft	0.2048
kilograms/sq. meter	pounds/sq. in	1.422×10^{-3}

kilograms/sq. mm	kg/sq. meter	10^6
kilolines	maxwells	1,000.0
kiloliters	liters	1,000.0
kilometers	centimeters	10^5
kilometers	feet	3,281
kilometers	inches	3.937×10^4
kilometers	meters	1,000.0
kilometers	miles	0.6214
kilometers	millimeters	10^4
kilometers	yards	1,094
kilometers/hr	cm/sec	27.78
kilometers/hr	feet/min	54.68
kilometers/hr	feet/sec	0.9113
kilometers/hr	knots	0.5396
kilometers/hr	meters/min	16.67
kilometers/hr	miles/hr	0.6214
kilometers/hr/sec	cm/sec/sec	27.78
kilometers/hr/sec	feet/sec/sec	0.9113
kilometers/hr/sec	meters/sec/sec	0.2778
kilometers/hr/sec	miles/hr/sec	0.6214
kilowatt-hr	Btu	3,413
kilowatt-hr	ergs	3.600×10^{13}
kilowatt-hr	foot-lb	2.655×10^6
kilowatt-hr	gram-calories	859,850
kilowatt-hr	horsepower-hr	1.341
kilowatt-hr	joules	3.6×10^6
kilowatt-hr	kg-calories	860.5
kilowatt-hr	kg-meters	3.671×10^5
kilowatt-hr	pounds of water raised from 62° to 212°F	22.75
kilowatts	Btu/min	56.92
kilowatts	foot-lb/min	4.426×10^4
kilowatts	foot-lb/sec	737.6
kilowatts	horsepower	1.341
kilowatts	kg-calories/min	14.34
kilowatts	watts	1,000.0
knots	feet/hr	6,080

knots	kilometers/hr	1.8532
knots	nautical miles/hr	1.0
knots	statute miles/hr	1.151
knots	yards/hr	2,027
knots	feet/sec	1.689
league	miles (approx.)	3.0
light year	miles	5.9×10^{12}
light year	kilometers	9.4637×10^{12}
lines/sq. cm	gausses	1.0
lines/sq. in	gausses	0.1550
lines/sq. in	webers/sq. cm	1.550×10^{-9}
lines/sq. in	webers/sq. in	10^{-8}
lines/sq. in	webers/sq. meter	1.550×10^{-5}
links (engineer's)	inches	12.0
links (surveyor's)	inches	7.92
liters	bushels (U.S. dry)	0.02838
liters	cubic cm	1,000.0
liters	cubic feet	0.03531
liters	cubic inches	61.02
liters	cubic meters	0.001
liters	cubic yards	1.308×10^{-3}
liters	gallons (U.S. liq.)	0.2642
liters	pints (U.S. liq.)	2.113
liters	quarts (U.S. liq.)	1.057
liters/min	cubic ft/sec	5.886×10^{-4}
liters/min	gal/sec	4.403×10^{-3}
lumen	spherical candle power	0.07958
lumen	watt	0.001496
lumens/sq. ft	foot-candles	1.0
lumens/sq. ft	lumen/sq. meter	10.76
lux	foot-candles	0.0929
maxwells	kilolines	0.001
maxwells	webers	10^{-8}
megalines	maxwells	10^{6}
megohms	microhms	10^{12}
megohms	ohms	10^{6}

meter-kilograms	cm-dynes	9.807×10^{7}
meter-kilograms	cm-grams	10^{5}
meter-kilograms	pound-feet	7.233
meters	centimeters	100.0
meters	feet	3.281
meters	inches	39.37
meters	kilometers	0.001
meters	miles (naut.)	5.396×10^{-4}
meters	miles (stat.)	6.214×10^{-4}
meters	millimeters	1,000.0
meters	yards	1.094
meters	varas	1.179
meters/min	cm/sec	1,667
meters/min	feet/min	3.281
meters/min	feet/sec	0.05468
meters/min	km/hr	0.06
meters/min	knots	0.03238
meters/min	miles/hr	0.03728
meters/sec	feet/min	196.8
meters/sec	feet/sec	3.281
meters/sec	kilometers/hr	3.6
meters/sec	kilometers/min	0.06
meters/sec	miles/hr	2.237
meters/sec	miles/min	0.03728
meters/sec/sec	cm/sec/sec	100.0
meters/sec/sec	ft/sec/sec	3.281
meters/sec/sec	km/hr/sec	3.6
meters/sec/sec	miles/hr/sec	2.237
microfarad	farads	10^{-6}
micrograms	grams	10^{-6}
microhms	megohms	10^{-12}
microhms	ohms	10^{-6}
microliters	liters	10^{-6}
microns	meters	1×10^{-6}
miles (naut.)	feet	6,080.27
miles (naut.)	kilometers	1.853
miles (naut.)	meters	1,853

miles (naut.)	miles (statute)	1.1516
miles (naut.)	yards	2,027
miles (statute)	centimeters	1.609×10^5
miles (statute)	feet	5,280
miles (statute)	inches	6.336×10^4
miles (statute)	kilometers	1.609
miles (statute)	meters	1,609
miles (statute)	miles (naut.)	0.8684
miles (statute)	yards	1,760
miles/hr	cm/sec	44.70
miles/hr	feet/min	88
miles/hr	feet/sec	1.467
miles/hr	km/hr	1.609
miles/hr	km/min	0.02682
miles/hr	knots	0.8684
miles/hr	meters/min	26.82
miles/hr	miles/min	0.1667
miles/hr/sec	cm/sec/sec	44.70
miles/hr/sec	feet/sec/sec	1.467
miles/hr/sec	km/hr/sec	1.609
miles/hr/sec	meters/sec/sec	0.4470
miles/min	cm/sec	2,682
miles/min	feet/sec	88
miles/min	km/min	1.609
miles/min	knots/min	0.8684
miles/min	miles/hr	60
mil-feet	cubic inches	9.425×10^{-6}
milliers	kilograms	1,000
milligrams	grains	0.01543236
milligrams	grams	0.001
milligrams/liter	parts/million	1.0
millihenries	henries	0.001
milliliters	liters	0.001
millimeters	centimeters	0.1
millimeters	feet	3.281×10^{-3}
millimeters	inches	0.03937
millimeters	kilometers	10^{-6}

millimeters	meters	0.001
millimeters	miles	6.214×10^{-7}
millimeters	mils	39.37
millimeters	yards	1.094×10^{-3}
millimicrons	meters	1×10^{-9}
million gal/day	cubic ft/sec	1.54723
mils	centimeters	2.540×10^{-3}
mils	feet	8.333×10^{-5}
mils	inches	0.001
mils	kilometers	2.540×10^{-8}
mils	yards	2.778×10^{-5}
miner's inches	cubic ft/min	1.5
minims (British)	cubic cm	0.059192
minims (U.S., fluid)	cubic cm	0.061612
minutes (angles)	degrees	0.01667
minutes (angles)	quadrants	1.852×10^{-4}
minutes (angles)	radians	2.909×10^{-4}
minutes (angles)	seconds	60.0
myriagrams	kilograms	10.0
myriameters	kilometers	10.0
myriawatts	kilowatts	10.0
nepers	decibels	8.686
Newton	dynes	1 x 105
ohm (international)	ohm (absolute)	1.0005
ohms	megohms	10^{-6}
ohms	microhms	10^{6}
ounces	drams	16.0
ounces	grains	437.5
ounces	grams	28.349527
ounces	pounds	0.0625
ounces	ounces (troy)	0.9115
ounces	tons (long)	2.790×10^{-5}
ounces	tons (metric)	2.835×10^{-5}
ounces (fluid)	cubic inches	1.805
ounces (fluid)	liters	0.02957
ounces (troy)	grains	480.0

ounces (troy)	grams	31.103481
ounces (troy)	ounces (avdp.)	1.09714
ounces (troy)	pennyweights (troy)	20.0
ounces (troy)	pounds (troy)	0.08333
ounces/sq. inch	dynes/sq. cm	4,309
ounces/sq. in	pounds/sq. in	0.0625
parsec	miles	19×10^{12}
parsec	kilometers	3.084×10^{13}
parts/million	grains/U.S. gal	0.0584
parts/million	grains/Imp. gal	0.07016
parts/million	pounds/million gal	8.345
pecks (British)	cubic inches	554.6
pecks (British)	liters	9.091901
pecks (U.S.)	bushels	0.25
pecks (U.S.)	cubic inches	537.605
pecks (U.S.)	liters	8.809582
pecks (U.S.)	quarts (dry)	8
pennyweights (troy)	grains	24.0
pennyweights (troy)	ounces (troy)	0.05
pennyweights (troy)	grams	1.55517
pennyweights (troy)	pounds (troy)	4.1667×10^{-3}
pints (dry)	cubic inches	33.60
pints (liq.)	cubic cm	473.2
pints (liq.)	cubic feet	0.01671
pints (liq.)	cubic inches	28.87
pints (liq.)	cubic meters	4.732×10^{-4}
pints (liq.)	cubic yards	6.189×10^{-4}
pints (liq.)	gallons	0.125
pints (liq.)	liters	0.4732
pints (liq.)	quarts (liq.)	0.5
Planck's quantum	erg - second	6.624×10^{-27}
poise	gram/cm sec	1.00
poundals	dynes	13,826
poundals	grams	14.10
poundals	joules/cm	1.383×10^{-3}
poundals	joules/meter (newtons)	0.1383
poundals	kilograms	0.01410

poundals	pounds	0.03108
pound-feet	cm-dynes	1.356×10^7
pound-feet	cm-grams	13,825
pound-feet	meter-kg	0.1383
pounds	drams	256
pounds	dynes	44.4823×10^4
pounds	grains	7,000
pounds	grams	453.5924
pounds	joules/cm	0.04448
pounds	joules/meter (newtons)	4.448
pounds	kilograms	0.4536
pounds	ounces	16.0
pounds	ounces (troy)	14.5833
pounds	poundals	32.17
pounds	pounds (troy)	1.21528
pounds	tons (short)	0.0005
pounds (avoirdupois)	ounces (troy)	14.5833
pounds (troy)	grains	5,760
pounds (troy)	grams	373.24177
pounds (troy)	ounces (avdp.)	13.1657
pounds (troy)	ounces (troy)	12.0
pounds (troy)	pennyweights (troy)	240.0
pounds (troy)	pounds (avdp.)	0.822857
pounds (troy)	tons (long)	3.6735×10^{-4}
pounds (troy)	tons (metric)	3.7324×10^{-4}
pounds (troy)	tons (short)	4.1143×10^{-4}
pounds of water	cubic ft	0.01602
pounds of water	cubic inches	27.68
pounds of water	gallons	0.1198
pounds of water/min	cubic ft/sec	2.670×10^{-4}
pounds/cubic ft	grams/cubic cm	0.01602
pounds/cubic ft	kg/cubic meter	16.02
pounds/cubic ft	pounds/cubic in	5.787×10^{-4}
pounds/cubic ft	pounds/mil-foot	5.456×10^{-9}
pounds/cubic in	gm/cubic cm	27.68
pounds/cubic in	kg/cubic meter	2.768×10^4
pounds/cubic in	pounds/cubic ft	1,728

pounds/cubic in	pounds/mil-foot	9.425×10^{-6}
pounds/ft	kg/meter	1.488
pounds/in	gm/cm	178.6
pounds/mil-foot	gm/cubic cm	2.306×10^{6}
pounds/sq. ft	atmospheres	4.725×10^{-4}
pounds/sq. ft	feet of water	0.01602
pounds/sq. ft	inches of mercury	0.01414
pounds/sq. ft	kg/sq. meter	4.882
pounds/sq. ft	pounds/sq. in	6.944×10^{-3}
pounds/sq. in	atmospheres	0.06804
pounds/sq. in	feet of water	2.307
pounds/sq. in	inches of mercury	2.036
pounds/sq. in	kg/sq. meter	703.1
pounds/sq. in	pounds/sq. ft	144.0
quadrants (angle)	degrees	90.0
quadrants (angle)	minutes	5,400.0
quadrants (angle)	radians	1.571
quadrants (angle)	seconds	3.24×10^{5}
quarts (dry)	cubic inches	67.20
quarts (liq.)	cubic cm	946.4
quarts (liq.)	cubic feet	0.03342
quarts (liq.)	cubic inches	57.75
quarts (liq.)	cubic meters	9.464×10^{-4}
quarts (liq.)	cubic yards	1.238×10^{-3}
quarts (liq.)	gallons	0.25
quarts (liq.)	liters	0.9463
radians	degrees	57.30
radians	minutes	3,438
radians	quadrants	0.6366
radians	seconds	2.063×10^{5}
radians/sec	degrees/sec	57.30
radians/sec	revolutions/min	9.549
radians/sec	revolutions/sec	0.1592
radians/sec/sec	revolutions/min/min	573.0
radians/sec/sec	revolutions/min/sec	9.549
radians/sec/sec	revolutions/sec/sec	0.1592
revolutions	degrees	360.0

revolutions	quadrants	4.0
revolutions	radians	6.283
revolutions/min	degrees/sec	6.0
revolutions/min	radians/sec	0.1047
revolutions/min	revolutions/sec	0.01667
revolutions/min/min	radians/sec/sec	1.745×10^{-3}
revolutions/min/min	revolutions/min/sec	0.01667
revolutions/min/min	revolutions/sec/sec	2.778×10^{-4}
revolutions/sec	degrees/sec	360.0
revolutions/sec	radians/sec	6.283
revolutions/sec	revolutions/min	60.0
revolutions/sec/sec	radians/sec/sec	6.283
revolutions/sec/sec	revolutions/min/min	3,600.0
revolutions/sec/sec	revolutions/min/sec	60.0
rod	chain (Gunters)	0.25
rod	meters	5.029
rods	feet	16.5
rods (surveyors' meas.)	yards	5.5
scruples	grains	20
seconds (angle)	degrees	2.778×10^{-4}
seconds (angle)	minutes	0.01667
seconds (angle)	quadrants	3.087×10^{-6}
seconds (angle)	radians	4.848×10^{-6}
slug	kilogram	14.59
slug	pounds	32.17
sphere	steradians	12.57
square centimeters	circular mils	1.973×10^{5}
square centimeters	sq. feet	1.076×10^{-3}
square centimeters	sq. inches	0.1550
square centimeters	sq. meters	0.0001
square centimeters	sq. miles	3.861×10^{-11}
square centimeters	sq. millimeters	100.0
square centimeters	sq. yards	1.196×10^{-4}
square feet	acres	2.296×10^{-5}
square feet	circular mils	1.833×10^{8}
square feet	sq. cm	929.0

square feet	sq. inches	144.0
square feet	sq. meters	0.09290
square feet	sq. miles	3.587×10^{-8}
square feet	sq. millimeters	9.290×10^{4}
square feet	sq. yards	0.1111
square inches	circular mils	1.273×10^{6}
square inches	sq. cm	6.452
square inches	sq. feet	6.944×10^{-3}
square inches	sq. millimeters	645.2
square inches	sq. mils	10^{6}
square inches	sq. yards	7.716×10^{-4}
square kilometers	acres	247.1
square kilometers	sq. cm	10^{10}
square kilometers	sq. ft	10.76×10^{6}
square kilometers	sq. inches	1.550×10^{9}
square kilometers	sq. meters	10^{6}
square kilometers	sq. miles	0.3861
square kilometers	sq. yards	1.196×10^{6}
square meters	acres	2.471×10^{-4}
square meters	sq. cm	10^{4}
square meters	sq. feet	10.76
square meters	sq. inches	1,550
square meters	sq. miles	3.861×10^{-7}
square meters	sq. millimeters	10^{6}
square meters	sq. yards	1.196
square miles	acres	640.0
square miles	sq. feet	27.88×10^{6}
square miles	sq. km	2.590
square miles	sq. meters	2.590×10^{6}
square miles	sq. yards	3.098×10^{6}
square millimeters	circular mils	1,973
square millimeters	sq. cm	0.01
square millimeters	sq. feet	1.076×10^{-5}
square millimeters	sq. inches	1.550×10^{-3}
square mils	circular mils	1.273

square mils	sq. cm	6.452×10^{-6}
square mils	sq. inches	10^{-6}
square yards	acres	2.066×10^{-4}
square yards	sq. cm	8,361
square yards	sq. feet	9.0
square yards	sq. inches	1,296
square yards	sq. meters	0.8361
square yards	sq. miles	3.228×10^{-7}
square yards	sq. millimeters	8.361×10^{5}
temperature (°C)+273	absolute temperature (°C)	1.0
temperature (°C)+17.78	temperature (°F)	1.8
temperature (°F)+460	absolute temperature (°F)	1.0
temperature (°F)–32	temperature (°C)	5/9
tons (long)	kilograms	1,016
tons (long)	pounds	2,240
tons (long)	tons (short)	1.120
tons (metric)	kilograms	1,000
tons (metric)	pounds	2,205
tons (short)	kilograms	907.1848
tons (short)	ounces	32,000
tons (short)	ounces (troy)	29,166.66
tons (short)	pounds	2,000
tons (short)	pounds (troy)	2,430.56
tons (short)	tons (long)	0.89287
tons (short)	tons (metric)	0.9078
tons (short)/sq. ft	kg/sq. meter	9,765
tons (short)/sq. ft	pounds/sq. in	2,000
tons of water/24 hr	pounds of water/hr	83.333
tons of water/24 hr	gallons/min	0.16643
tons of water/24 hr	cubic ft/hr	1.3349
volt (absolute)	statvolts	0.003336
volt/inch	volt/cm	0.39370
watt-hours	Btu	3.413
watt-hours	ergs	3.60×10^{10}
watt-hours	foot-pounds	2,656
watt-hours	gram-calories	859.85
watt-hours	horsepower-hr	1.341×10^{-3}

watt-hours	kilogram-calories	0.8605
watt-hours	kilogram-meters	367.2
watt-hours	kilowatt-hr	0.001
watt (international)	watt (absolute)	1.0002
watts	Btu/hr	3.4129
watts	Btu/min	0.05688
watts	ergs/sec	107
watts	foot-lb/min	44.27
watts	foot-lb/sec	0.7378
watts	horsepower	1.341×10^{-3}
watts	horsepower (metric)	1.360×10^{-3}
watts	kg-calories/min	0.01433
watts	kilowatts	0.001
watts (Abs.)	Btu (mean)/min	0.056884
watts (Abs.)	joules/sec	1
webers	maxwells	10^{8}
webers	kilolines	10^{5}
webers/sq. in	gausses	1.550×10^{7}
webers/sq. in	lines/sq. in	10^{8}
webers/sq. in	webers/sq. cm	0.1550
webers/sq. in	webers/sq. meter	1,550
webers/sq. meter	gausses	10^{4}
webers/sq. meter	lines/sq. in	6.452×10^{4}
webers/sq. meter	webers/sq. cm	10^{-4}
webers/sq. meter	webers/sq. in	6.452×10^{-4}
yards	centimeters	91.44
yards	kilometers	9.144×10^{-4}
yards	meters	0.9144
yards	miles (naut.)	4.934×10^{-4}
yards	miles (stat.)	5.682×10^{-4}
yards	millimeters	914.4

14.3 Reference Tables

14.3a Power Conversion Factors

dBm	dBw	Watts	Multiple	Prefix
+150	+120	1,000,000,000,000	10^{12}	1 Terawatt
+140	+110	100,000,000,000	10^{11}	100 Gigawatts
+130	+100	10,000,000,000	10^{10}	10 Gigawatts
+120	+90	1,000,000,000	10^{9}	1 Gigawatt
+110	+80	100,000,000	10^{8}	100 Megawatts
+100	+70	10,000,000	10^{7}	10 Megawatts
+90	+60	1,000,000	10^{6}	1 Megawatt
+80	+50	100,000	10^{5}	100 Kilowatts
+70	+40	10,000	10^{4}	10 Kilowatts
+60	+30	1,000	10^{3}	1 Kilowatt
+50	+20	100	10^{2}	1 Hectrowatt
+40	+10	10	10	1 Decawatt
+30	0	1	1	1 Watt
+20	–10	0.1	10^{-1}	1 Deciwatt
+10	–20	0.01	10^{-2}	1 Centiwatt
0	–30	0.001	10^{-3}	1 Milliwatt
–10	–40	0.0001	10^{-4}	100 Microwatts
–20	–50	0.00001	10^{-5}	10 Microwatts
–30	–60	0.000,001	10^{-6}	1 Microwatt
–40	–70	0.0,000,001	10^{-7}	100 Nanowatts
–50	–80	0.00,000,001	10^{-8}	10 Nanowatts
–60	–90	0.000,000,001	10^{-9}	1 Nanowatt
–70	–100	0.0,000,000,001	10^{-10}	100 Picowatts
–80	–110	0.00,000,000,001	10^{-11}	10 Picowatts
–90	–120	0.000,000,000,001	10^{-12}	1 Picowatt

14.3b Standing Wave Ratio

SWR	Reflection Coefficient	Return Loss	Power Ratio	Percent Reflected
1.01:1	0.0050	46.1 dB	0.00002	0.002
1.02:1	0.0099	40.1 dB	0.00010	0.010
1.04:1	0.0196	34.2 dB	0.00038	0.038
1.06:1	0.0291	30.7 dB	0.00085	0.085
1.08:1	0.0385	28.3 dB	0.00148	0.148
1.10:1	0.0476	26.4 dB	0.00227	0.227
1.20:1	0.0909	20.8 dB	0.00826	0.826
1.30:1	0.1304	17.7 dB	0.01701	1.7
1.40:1	0.1667	15.6 dB	0.02778	2.8
1.50:1	0.2000	14.0 dB	0.04000	4.0
1.60:1	0.2308	12.7 dB	0.05325	5.3
1.70:1	0.2593	11.7 dB	0.06722	6.7
1.80:1	0.2857	10.9 dB	0.08163	8.2
1.90:1	0.3103	10.2 dB	0.09631	9.6
2.00:1	0.3333	9.5 dB	0.11111	11.1
2.20:1	0.3750	8.5 dB	0.14063	14.1
2.40:1	0.4118	7.7 dB	0.16955	17.0
2.60:1	0.4444	7.0 dB	0.19753	19.8
2.80:1	0.4737	6.5 dB	0.22438	22.4
3.00:1	0.5000	6.0 dB	0.25000	25.0
3.50:1	0.5556	5.1 dB	0.30864	30.9
4.00:1	0.6000	4.4 dB	0.36000	36.0
4.50:1	0.6364	3.9 dB	0.40496	40.5
5.00:1	0.6667	3.5 dB	0.44444	44.4
6.00:1	0.7143	2.9 dB	0.51020	51.0
7.00:1	0.7500	2.5 dB	0.56250	56.3
8.00:1	0.7778	2.2 dB	0.60494	60.5
9.00:1	0.8000	1.9 dB	0.64000	64.0
10.00:1	0.8182	1.7 dB	0.66942	66.9
15.00:1	0.8750	1.2 dB	0.76563	76.6
20.00:1	0.9048	0.9 dB	0.81859	81.9
30.00:1	0.9355	0.6 dB	0.87513	97.5
40.00:1	0.9512	0.4 dB	0.90482	90.5
50.00:1	0.9608	0.3 dB	0.92311	92.3

14.3c Specifications of Standard Copper Wire Sizes

Wire Size (AWG)	Diameter (mils)	Circular mil Area	Turns per Linear Inch[1]			Ohms per 100 ft[2]	Current Carrying Capacity[3]	Diameter (mm)
			Enamel	SCE	DCC			
1	289.3	83810	-	-	-	0.1239	119.6	7.348
2	257.6	05370	-	-	-	0.1563	94.8	6.544
3	229.4	62640	-	-	-	0.1970	75.2	5.827
4	204.3	41740	-	-	-	0.2485	59.6	5.189
5	181.9	33100	-	-	-	0.3133	47.3	4.621
6	162.0	26250	-	-	-	0.3951	37.5	4.115
7	144.3	20820	-	-	-	0.4982	29.7	3.665
8	128.5	16510	7.6	-	7.1	0.6282	23.6	3.264
9	114.4	13090	8.6	-	7.8	0.7921	18.7	2.906
10	101.9	10380	9.6	9.1	8.9	0.9989	14.8	2.588
11	90.7	8234	10.7	-	9.8	1.26	11.8	2.305
12	80.8	6530	12.0	11.3	10.9	1.588	9.33	2.063
13	72.0	5178	13.5	-	12.8	2.003	7.40	1.828
14	64.1	4107	15.0	14.0	13.8	2.525	5.87	1.628
15	57.1	3257	16.8	-	14.7	3.184	4.65	1.450
16	50.8	2583	18.9	17.3	16.4	4.016	3.69	1.291
17	45.3	2048	21.2	-	18.1	5.064	2.93	1.150
18	40.3	1624	23.6	21.2	19.8	6.386	2.32	1.024
19	35.9	1288	26.4	-	21.8	8.051	1.84	0.912
20	32.0	1022	29.4	25.8	23.8	10.15	1.46	0.812
21	28.5	810	33.1	-	26.0	12.8	1.16	0.723
22	25.3	642	37.0	31.3	30.0	16.14	0.918	0.644
23	22.6	510	41.3	-	37.6	20.36	0.728	0.573
24	20.1	404	46.3	37.6	35.6	25.67	0.577	0.511
25	17.9	320	51.7	-	38.6	32.37	0.458	0.455
26	15.9	254	58.0	46.1	41.8	40.81	0.363	0.406
27	14.2	202	64.9	-	45.0	51.47	0.288	0.361
28	12.6	160	72.7	54.6	48.5	64.9	0.228	0.321
29	11.3	127	81.6	-	51.8	81.83	0.181	0.286
30	10.0	101	90.5	64.1	55.5	103.2	0.144	0.255
31	8.9	50	101	-	59.2	130.1	0.114	0.227
32	8.0	63	113	74.1	61.6	164.1	0.090	0.202
33	7.1	50	127	-	66.3	206.9	0.072	0.180
34	6.3	40	143	86.2	70.0	260.9	0.057	0.160

35	5.6	32	158	-	73.5	329.0	0.045	0.143
36	5.0	25	175	103.1	T7.0	414.8	0.036	0.127
37	4.5	20	198	-	80.3	523.1	0.028	0.113
38	4.0	16	224	116.3	83.6	659.6	0.022	0.101
39	3.5	12	248	-	86.6	831.8	0.018	0.090

1. Based on 25.4 mm.
2. Ohms per 1,000 ft measured at 20°C.
3. Current-carrying capacity at 700 cm/amp.

14.3d Celcius-to-Fahrenheit Conversion Table

°Celsius	°Fahrenheit	°Celsius	°Fahrenheit
–50	–58	125	257
–45	–49	130	266
–40	–40	135	275
–35	–31	140	284
–30	–22	145	293
–25	–13	150	302
–20	4	155	311
–15	5	160	320
–10	14	165	329
–5	23	170	338
0	32	175	347
5	41	180	356
10	50	185	365
15	59	190	374
20	68	195	383
25	77	200	392
30	86	205	401
35	95	210	410
40	104	215	419
45	113	220	428
50	122	225	437
55	131	230	446
60	140	235	455
65	149	240	464
70	158	245	473
75	167	250	482
80	176	255	491
85	185	260	500
90	194	265	509
95	203	270	518
100	212	275	527
105	221	280	536
110	230	285	545
115	239	290	554
120	248	295	563

14.3e Inch-to-Millimeter Conversion Table

Inch	0	1/8	1/4	3/8	1/2	5/8	3/4	7/8	Inch
0	0.0	3.18	6.35	9.52	12.70	15.88	19.05	22.22	0
1	25.40	28.58	31.75	34.92	38.10	41.28	44.45	47.62	1
2	50.80	53.98	57.15	60.32	63.50	66.68	69.85	73.02	2
3	76.20	79.38	82.55	85.72	88.90	92.08	95.25	98.42	3
4	101.6	104.8	108.0	111.1	114.3	117.5	120.6	123.8	4
5	127.0	130.2	133.4	136.5	139.7	142.9	146.0	149.2	5
6	152.4	155.6	158.8	161.9	165.1	168.3	171.4	174.6	6
7	177.8	181.0	184.2	187.3	190.5	193.7	196.8	200.0	7
8	203.2	206.4	209.6	212.7	215.9	219.1	222.2	225.4	8
9	228.6	231.8	235.0	238.1	241.3	244.5	247.6	250.8	9
10	254.0	257.2	260.4	263.5	266.7	269.9	273.0	276.2	10
11	279	283	286	289	292	295	298	302	11
12	305	308	311	314	317	321	324	327	12
13	330	333	337	340	343	346	349	352	13
14	356	359	362	365	368	371	375	378	14
15	381	384	387	391	394	397	400	403	15
16	406	410	413	416	419	422	425	429	16
17	432	435	438	441	445	448	451	454	17
18	457	460	464	467	470	473	476	479	18
19	483	486	489	492	495	498	502	505	19
20	508	511	514	518	521	524	527	530	20

14.3f Conversion of Millimeters to Decimal Inches

mm	Inches	mm	Inches	mm	Inches
1	0.039370	46	1.811020	91	3.582670
2	0.078740	47	1.850390	92	3.622040
3	0.118110	48	1.889760	93	3.661410
4	0.157480	49	1.929130	94	3.700780
5	0.196850	50	1.968500	95	3.740150
6	0.236220	51	2.007870	96	3.779520
7	0.275590	52	2.047240	97	3.818890
8	0.314960	53	2.086610	98	3.858260
9	0.354330	54	2.125980	99	3.897630
10	0.393700	55	2.165350	100	3.937000
11	0.433070	56	2.204720	105	4.133848

12	0.472440	57	2.244090	110	4.330700
13	0.511810	58	2.283460	115	4.527550
14	0.551180	59	2.322830	120	4.724400
15	0.590550	60	2.362200	125	4.921250
16	0.629920	61	2.401570	210	8.267700
17	0.669290	62	2.440940	220	8.661400
18	0.708660	63	2.480310	230	9.055100
19	0.748030	64	2.519680	240	9.448800
20	0.787400	65	2.559050	250	9.842500
21	0.826770	66	2.598420	260	10.236200
22	0.866140	67	2.637790	270	10.629900
23	0.905510	68	2.677160	280	11.032600
24	0.944880	69	2.716530	290	11.417300
25	0.984250	70	2.755900	300	11.811000
26	1.023620	71	2.795270	310	12.204700
27	1.062990	72	2.834640	320	12.598400
28	1.102360	73	2.874010	330	12.992100
29	1.141730	74	2.913380	340	13.385800
30	1.181100	75	2.952750	350	13.779500
31	1.220470	76	2.992120	360	14.173200
32	1.259840	77	3.031490	370	14.566900
33	1.299210	78	3.070860	380	14.960600
34	1.338580	79	3.110230	390	15.354300
35	1.377949	80	3.149600	400	15.748000
36	1.417319	81	3.188970	500	19.685000
37	1.456689	82	3.228340	600	23.622000
38	1.496050	83	3.267710	700	27.559000
39	1.535430	84	3.307080	800	31.496000
40	1.574800	85	3.346450	900	35.433000
41	1.614170	86	3.385820	1000	39.370000
42	1.653540	87	3.425190	2000	78.740000
43	1.692910	88	3.464560	3000	118.110000
44	1.732280	89	3.503903	4000	157.480000
45	1.771650	90	3.543300	5000	196.850000

14.3g Convertion of Common Fractions to Decimal and Millimeter Units

Common Fractions	Decimal Fractions	mm (approx.)	Common Fractions	Decimal Fractions	mm (appox.)
1/128	0.008	0.20	1/2	0.500	12.70
1/64	0.016	0.40	33/64	0.516	13.10
1/32	0.031	0.79	17/32	0.531	13.49
3/64	0.047	1.19	35/64	0.547	13.89
1/16	0.063	1.59	9/16	0.563	14.29
5/64	0.078	1.98	37/64	0.578	14.68
3/32	0.094	2.38	19/32	0.594	15.08
7/64	0.109	2.78	39/64	0.609	15.48
1/8	0.125	3.18	5/8	0.625	15.88
9/64	0.141	3.57	41/64	0.641	16.27
5/32	0.156	3.97	21/32	0.656	16.67
11/64	0.172	4.37	43/64	0.672	17.07
3/16	0.188	4.76	11/16	0.688	17.46
13/64	0.203	5.16	45/64	0.703	17.86
7/32	0.219	5.56	23/32	0.719	18.26
15/64	0.234	5.95	47/64	0.734	18.65
1/4	0.250	6.35	3/4	0.750	19.05
17/64	0.266	6.75	49/64	0.766	19.45
9/32	0.281	7.14	25/32	0.781	19.84
19/64	0.297	7.54	51/64	0.797	20.24
5/16	0.313	7.94	13/16	0.813	20.64
21/64	0.328	8.33	53/64	0.828	21.03
11/32	0.344	8.73	27/32	0.844	21.43
23/64	0.359	9.13	55/64	0.859	21.83
3/8	0.375	9.53	7/8	0.875	22.23
25/64	0.391	9.92	57/64	0.891	22.62
13/32	0.406	10.32	29/32	0.906	23.02
27/64	0.422	10.72	59/64	0.922	23.42
7/16	0.438	11.11	15/16	0.938	23.81
29/64	0.453	11.51	61/64	0.953	24.21
15/32	0.469	11.91	31/32	0.969	24.61
31/64	0.484	12.30	63/64	0.984	25.00

14.3h Decimal Equivalent Size of Drill Numbers

Drill Number	Decimal Equivalent	Drill Number	Decimal Equivalent	Drill Number	Decimal Equivalent
80	0.0135	53	0.0595	26	0.1470
79	0.0145	52	0.0635	25	0.1495
78	0.0160	51	0.0670	24	0.1520
77	0.0180	50	0.0700	23	0.1540
76	0.0200	49	0.0730	22	0.1570
75	0.0210	48	0.0760	21	0.1590
74	0.0225	47	0.0785	20	0.1610
73	0.0240	46	0.0810	19	0.1660
72	0.0250	45	0.0820	18	0.1695
71	0.0260	44	0.0860	17	0.1730
70	0.0280	43	0.0890	16	0.1770
69	0.0292	42	0.0935	15	0.1800
68	0.0310	41	0.0960	14	0.1820
67	0.0320	40	0.0980	13	0.1850
66	0.0330	39	0.0995	12	0.1890
65	0.0350	38	0.1015	11	0.1910
64	0.0360	37	0.1040	10	0.1935
63	0.0370	36	0.1065	9	0.1960
62	0.0380	35	0.1100	8	0.1990
61	0.0390	34	0.1110	7	0.2010
60	0.0400	33	0.1130	6	0.2040
59	0.0410	32	0.1160	5	0.2055
58	0.0420	31	0.1200	4	0.2090
57	0.0430	30	0.1285	3	0.2130
56	0.0465	29	0.1360	2	0.2210
55	0.0520	28	0.1405	1	0.2280
54	0.0550	27	0.1440		

14.3i Decimal Equivalent Size of Drill Letters

Letter Drill	Decimal Equivalent	Letter Drill	Decimal Equivalent	Letter Drill	Decimal Equivalent
A	0.234	J	0.277	S	0.348
B	0.238	K	0.281	T	0.358
C	0.242	L	0.290	U	0.368
D	0.246	M	0.295	V	0.377
E	0.250	N	0.302	W	0.386
F	0.257	O	0.316	X	0.397
G	0.261	P	0.323	Y	0.404
H	0.266	Q	0.332	Z	0.413
I	0.272	R	0.339		

14.3j Conversion Ratios for Length

Known Quantity	Multiply by	Quantity to Find
inches (in)	2.54	centimeters (cm)
feet (ft)	30	centimeters (cm)
yards (yd)	0.9	meters (m)
miles (mi)	1.6	kilometers (km)
millimeters (mm)	0.04	inches (in)
centimeters (cm)	0.4	inches (in)
meters (m)	3.3	feet (ft)
meters (m)	1.1	yards (yd)
kilometers (km)	0.6	miles (mi)
centimeters (cm)	10	millimeters (mm)
decimeters (dm)	10	centimeters (cm)
decimeters (dm)	100	millimeters (mm)
meters (m)	10	decimeters (dm)
meters (m)	1000	millimeters (mm)
dekameters (dam)	10	meters (m)
hectometers (hm)	10	dekameters (dam)
hectometers (hm)	100	meters (m)
kilometers (km)	10	hectometers (hm)
kilometers (km)	1000	meters (m)

14.3k Conversion Ratios for Area

Known Quantity	Multiply by	Quantity to Find
square inches (in^2)	6.5	square centimeters (cm^2)
square feet (ft^2)	0.09	square meters (m^2)
square yards (yd^2)	0.8	square meters (m^2)
square miles (mi^2)	2.6	square kilometers (km^2)
acres	0.4	hectares (ha)
square centimeters (cm^2)	0.16	square inches (in^2)
square meters (m^2)	1.2	square yards (yd^2)
square kilometers (km^2)	0.4	square miles (mi^2)
hectares (ha)	2.5	acres
square centimeters (cm^2)	100	square millimeters (mm^2)
square meters (m^2)	10,000	square centimeters (cm^2)
square meters (m^2)	1,000,000	square millimeters (mm^2)
ares (a)	100	square meters (m^2)
hectares (ha)	100	ares (a)
hectares (ha)	10,000	square meters (m^2)
square kilometers (km^2)	100	hectares (ha)
square kilometers (km^2)	1,000	square meters (m^2)

14.3l Conversion Ratios for Mass

Known Quantity	Multiply by	Quantity to Find
ounces (oz)	28	grams (g)
pounds (lb)	0.45	kilograms (kg)
tons	0.9	tonnes (t)
grams (g)	0.035	ounces (oz)
kilograms (kg)	2.2	pounds (lb)
tonnes (t)	100	kilograms (kg)
tonnes (t)	1.1	tons
centigrams (cg)	10	milligrams (mg)
decigrams (dg)	10	centigrams (cg)
decigrams (dg)	100	milligrams (mg)
grams (g)	10	decigrams (dg)
grams (g)	1000	milligrams (mg)
dekagram (dag)	10	grams (g)
hectogram (hg)	10	dekagrams (dag)
hectogram (hg)	100	grams (g)
kilograms (kg)	10	hectograms (hg)
kilograms (kg)	1000	grams (g)

14.3m Conversion Ratios for Volume

Known Quantity	Multiply by	Quantity to Find
milliliters (mL)	0.03	fluid ounces (fl oz)
liters (L)	2.1	pints (pt)
liters (L)	1.06	quarts (qt)
liters (L)	0.26	gallons (gal)
gallons (gal)	3.8	liters (L)
quarts (qt)	0.95	liters (L)
pints (pt)	0.47	liters (L)
cups (c)	0.24	liters (L)
fluid ounces (fl oz)	30	milliliters (mL)
teaspoons (tsp)	5	milliliters (mL)
tablespoons (tbsp)	15	milliliters (mL)
liters (L)	100	milliliters (mL)

14.3n Conversion Ratios for Cubic Measure

Known Quantity	Multiply by	Quantity to Find
cubic meters (m^3)	35	cubic feet (ft^3)
cubic meters (m^3)	1.3	cubic yards (yd^3)
cubic yards (yd^3)	0.76	cubic meters (m^3)
cubic feet (ft^3)	0.028	cubic meters (m^3)
cubic centimeters (cm^3)	1000	cubic millimeters (mm^3)
cubic decimeters (dm^3)	1000	cubic centimeters (cm^3)
cubic decimeters (dm^3)	1,000,000	cubic millimeters (mm^3)
cubic meters (m^3)	1000	cubic decimeters (dm^3)
cubic meters (m^3)	1	steres
cubic feet (ft^3)	1728	cubic inches (in^3)
cubic feet (ft^3)	28.32	liters (L)
cubic inches (in^3)	16.39	cubic centimeters (cm^3)
cubic meters (m^3	264	gallons (gal)
cubic yards (yd^3)	27	cubic feet (ft^3)
cubic yards (yd^3)	202	gallons (gal)
gallons (gal)	231	cubic inches (in^3)

14.3o Conversion Ratios for Electrical Quantities

Known Quantity	Multiply by	Quantity to Find
Btu per minute	0.024	horsepower (hp)
Btu per minute	17.57	watts (W)
horsepower (hp)	33,000	foot-pounds per min (ft-lb/min)
horsepower (hp)	746	watts (W)
kilowatts (kW)	57	Btu per minute
kilowatts (kW)	1.34	horsepower (hp)

Chapter

15

Informative Documents by Subject

Jerry C. Whitaker, editor

15.1 Introduction

The following documents are listed as a means of finding additional information on specific aspects of audio and video engineering.

15.2 Audio

15.2a Principles and Sound and Hearing

Backus, John: *The Acoustical Foundations of Music,* Norton, New York, N.Y., 1969.

Batteau, D. W.: "The Role of the Pinna in Human Localization," *Proc. R. Soc. London*, B168, pp. 158–180, 1967.

Benade, A. H.: *Fundamentals of Musical Acoustics,* Oxford University Press, New York, N.Y., 1976.

Beranek, Leo L: *Acoustics,* McGraw-Hill, New York, N.Y., 1954.

Blauert, J., and W. Lindemann: "Auditory Spaciousness: Some Further Psychoacoustic Studies," *J. Acoust. Soc. Am.*, vol. 80, 533–542, 1986.

Blauert, J: *Spatial Hearing*, translation by J. S. Allen, M.I.T., Cambridge. Mass., 1983.

Bloom, P. J.: "Creating Source Elevation Illusions by Spectral Manipulations," *J. Audio Eng. Soc.*, vol. 25, pp. 560–565, 1977.

Bose, A. G.: "On the Design, Measurement and Evaluation of Loudspeakers," presented at the 35th convention of the Audio Engineering Society, preprint 622, 1962.

Buchlein, R.: "The Audibility of Frequency Response Irregularities" (1962), reprinted in English translation in *J. Audio Eng. Soc.*, vol. 29, pp. 126–131, 1981.

Denes, Peter B., and E. N. Pinson: *The Speech Chain,* Bell Telephone Laboratories, Waverly, 1963.

Durlach, N. I., and H. S. Colburn: "Binaural Phenemena," in *Handbook of Perception*, E. C. Carterette and M. P. Friedman (eds.), vol. 4, Academic, New York, N.Y., 1978.

Ehara, Shiro: "Instantaneous Pressure Distributions of Orchestra Sounds," *J. Acoust. Soc. Japan*, vol. 22, pp. 276–289, 1966.

Fletcher, H., and W. A. Munson: "Loudness, Its Definition, Measurement and Calculation," *J. Acoust. Soc. Am.*, vol. 5, pp. 82–108, 1933.

Fryer, P.: "Loudspeaker Distortions—Can We Rear Them?," *Hi-Fi News Record Rev.*, vol. 22, pp. 51–56, 1977.

Gabrielsson, A., and B. Lindstrom: "Perceived Sound Quality of High-Fidelity Loudspeakers." *J. Audio Eng. Soc.*, vol. 33, pp. 33–53, 1985.

Gabrielsson, A., and H. Siogren: "Perceived Sound Quality of Sound-Reproducing Systems," *J. Aoust. Soc. Am.*, vol. 65, pp. 1019–1033, 1979.

Haas, H.: "The Influence of a Single Echo on the Audibility of Speech," *Acustica*, vol. I, pp. 49–58, 1951; English translation reprinted in *J. Audio Eng. Soc.*, vol. 20, pp. 146–159, 1972.

Hall, Donald: *Musical Acoustics—An Introduction*, Wadsworth, Belmont, Calif., 1980.

International Electrotechnical Commission: *Sound System Equipment*, part 10, *Programme Level Meters*, Publication 268-1 0A, 1978.

International Organization for Standardization: *Normal Equal-Loudness Contours for Pure Tones and Normal Threshold for Hearing under Free Field Listening Conditions*, Recommendation R226, December 1961.

Jones, B. L., and E. L. Torick: "A New Loudness Indicator for Use in Broadcasting," *J. SMPTE*, Society of Motion Picture and Television Engineers, White Plains, N.Y., vol. 90, pp. 772–777, 1981.

Kuhl, W., and R. Plantz: "The Significance of the Diffuse Sound Radiated from Loudspeakers for the Subjective Hearing Event," *Acustica*, vol. 40, pp. 182–190, 1978.

Kuhn, G. F.: "Model for the Interaural Time Differences in the Azimuthal Plane," *J. Acoust. Soc. Am.*, vol. 62, pp. 157–167, 1977.

Kurozumi, K., and K. Ohgushi: "The Relationship between the Cross-Correlation Coefficient of Two-Channel Acoustic Signals and Sound Image Quality," *J. Acoust. Soc. Am.*, vol. 74, pp. 1726–1733, 1983.

Main, Ian G.: *Vibrations and Waves in Physics,* Cambridge, London, 1978.

Mankovsky, V. S.: *Acoustics of Studios and Auditoria,* Focal Press, London, 1971.

Meyer, J.: *Acoustics and the Performance of Music*, Verlag das Musikinstrument, Frankfurt am Main, 1987.

Morse, Philip M.: *Vibrations and Sound,* 1964, reprinted by the Acoustical Society of America, New York, N.Y., 1976.

Olson, Harry F.: *Acoustical Engineering*, Van Nostrand, New York, N.Y., 1957.

Pickett, J. M.: *The Sounds of Speech Communications*, University Park Press, Baltimore, MD, 1980.

Pierce, John R.: *The Science of Musical Sound*, Scientific American Library, New York, N.Y., 1983.

Piercy, J. E., and T. F. W. Embleton: "Sound Propagation in the Open Air," in *Handbook* of *Noise Control*, 2d ed., C. M. Harris (ed.), McGraw-Hill, New York, N.Y., 1979.

Plomp, R.: *Aspects* of *Tone Sensation—A Psychophysical Study*," Academic, New York, N.Y., 1976.

Rakerd, B., and W. M. Hartmann: "Localization of Sound in Rooms, II—The Effects of a Single Reflecting Surface," *J. Acoust. Soc. Am.*, vol. 78, pp. 524–533, 1985.

Rasch, R. A., and R. Plomp: "The Listener and the Acoustic Environment," in D. Deutsch (ed.), *The Psychology of Music*, Academic, New York, N.Y., 1982.

Robinson, D. W., and R. S. Dadson: "A Redetermination of the Equal-Loudness Relations for Pure Tones," *Br. J. Appl. Physics*, vol. 7, pp. 166–181, 1956.

Scharf, B.: "Loudness," in E. C. Carterette and M. P. Friedman (eds.), *Handbook* of *Perception*, vol. 4, *Hearing*, chapter 6, Academic, New York, N.Y., 1978.

Shaw, E. A. G., and M. M. Vaillancourt: "Transformation of Sound-Pressure Level from the Free Field to the Eardrum Presented in Numerical Form," *J. Acoust. Soc. Am.*, vol. 78, pp. 1120–1123, 1985.

Shaw, E. A. G., and R. Teranishi: "Sound Pressure Generated in an External-Ear Replica and Real Human Ears by a Nearby Sound Source," *J. Acoust. Soc. Am.*, vol. 44, pp. 240–249, 1968.

Shaw, E. A. G.: "Aural Reception," in A. Lara Saenz and R. W. B. Stevens (eds.), *Noise Pollution*, Wiley, New York, N.Y., 1986.

Shaw, E. A. G.: "External Ear Response and Sound Localization," in R. W. Gatehouse (ed.), *Localization of Sound: Theory and Applications*, Amphora Press, Groton, Conn., 1982.

Shaw, E. A. G.: "Noise Pollution—What Can be Done?" *Phys. Today*, vol. 28, no. 1, pp. 46–58, 1975.

Shaw, E. A. G.: "The Acoustics of the External Ear," in W. D. Keidel and W. D. Neff (eds.), *Handbook* of *Sensory Physiology*, vol. V/I, *Auditory System,* Springer-Verlag, Berlin, 1974.

Shaw, E. A. G.: "Transformation of Sound Pressure Level from the Free Field to the Eardrum in the Horizontal Plane," *J. Acoust. Soc. Am.*, vol. 56, pp. 1848–1861, 1974.

Stephens, R. W. B., and A. E. Bate: *Acoustics and Vibrational Physics,* 2nd ed., E. Arnold (ed.), London, 1966.

Stevens, W. R.: "Loudspeakers—Cabinet Effects," *Hi-Fi News Record Rev.*, vol. 21, pp. 87–93, 1976.

Sundberg, Johan: "The Acoustics of the Singing Voice," in *The Physics of Music,* introduction by C. M. Hutchins, Scientific American/Freeman, San Francisco, Calif., 1978.

Tonic, F. E.: "Loudness—Applications and Implications to Audio," *dB,* Part 1, vol. 7, no. 5, pp. 27–30; Part 2, vol. 7, no. 6, pp. 25–28, 1973.

Toole, F. E., and B. McA. Sayers: "Lateralization Judgments and the Nature of Binaural Acoustic Images," *J. Acoust. Soc. Am.*, vol. 37, pp. 319–324, 1965.

Toole, F. E.: "Loudspeaker Measurements and Their Relationship to Listener Preferences," *J. Audio Eng. Soc.*, vol. 34, part 1, pp. 227–235, part 2, pp. 323–348, 1986.

Toole, F. E.: "Subjective Measurements of Loudspeaker Sound Quality and Listener Performance," *J. Audio Eng. Soc.*, vol. 33, pp. 2–32, 1985.

Voelker, E. J.: "Control Rooms for Music Monitoring," *J. Audio Eng. Soc.*, vol. 33, pp. 452–462, 1985.

Ward, W. D.: "Subjective Musical Pitch," *J. Acoust. Soc. Am.*, vol. 26, pp. 369–380, 1954.

Waterhouse, R. V., and C. M. Harris: "Sound in Enclosed Spaces," in *Handbook of Noise Control*, 2d ed., C. M. Harris (ed.), McGraw-Hill, New York, N.Y., 1979.

Wong, G. S. K.: "Speed of Sound in Standard Air," *J. Acoust. Soc. Am.*, vol. 79, pp. 1359–1366, 1986.

Zurek, P. M.: "Measurements of Binaural Echo Suppression," *J. Acoust. Soc. Am.*, vol. 66, pp. 1750–1757, 1979.

Zwislocki, J. J.: "Masking—Experimental and Theoretical Aspects of Simultaneous, For-ward, Backward and Central Masking," in E. C. Carterette and M. P. Friedman (eds.), *Handbook of Perception,* vol. 4, *Hearing*, chapter 8, Academic, New York, N.Y., 1978.

15.2b The Audio Spectrum

Bendat, J. S., and A. G. Riersol: *Engineering Applications of Correlation and Spectral Analysis*, Wiley, New York, 1980.

Bendat, J. S., and A. G. Piersol: *Random Data: Analysis and Measurement Procedures,* Wiley-Interscience, New York, N.Y., 1971.

Blinchikoff, H. J., and A. I. Zverev: *Filtering in the Time and Frequency Domains*, Wiley, New York, N.Y., 1976.

Bloom, P. J., and D. Preis: "Perceptual Identification and Discrimination of Phase Distortions," *IEEE ICASSP Proc.*, pp. 1396–1399, April 1983.

Bode, H. W.: *Network Analysis and Feedback Amplifier Design,* Van Nostrand, New York, N.Y., 1945.

Bracewell, R.: *The Fourier Integral and Its Applications,* McGraw-Hill, New York, N.Y., 1965.

Cheng, D. K.: *Analysis of Linear Systems*, Addison-Wesley, Reading, Mass., 1961.

Childers, D. G.: *Modern Spectral Analysis,* IEEE, New York, N.Y., 1978.

Connor, F. R.: *Signals,* Arnold, London, 1972.

Deer, J. A., P. J. Bloom, and D. Preis: "Perception of Phase Distortion in All-Pass Filters," *J. Audio Eng. Soc.*, vol. 33, no. 10, pp. 782–786, October 1985.

Di Toro. M. J.: "Phase and Amplitude Distortion in Linear Networks," *Proc. IRE*, vol. 36, pp. 24–36, January 1948.

Guillemin, E. A.: *Communication Networks*, vol. 11, Wiley, New York, N.Y., 1935.

Henderson. K. W., and W. H. Kautz: "Transient Response of Conventional Filters," *IRE Trans. Circuit Theory*, CT-5, pp. 333–347, December 1958.

Hewlett-Packard: "Application Note 63—Section II, Appendix A, "Table of Important Transforms," Hewlett-Packard, Palo Alto, Calif, pp. 37, 38, 1954.

Jenkins, G. M., and D. G. Watts: *Spectral Analysis and Its Applications,* Holden-Day, San Francisco, Calif., 1968.

Kharkevich, A. A.: *Spectra and Analysis,* English translation, Consultants Bureau, New York, N.Y., 1960.

Kupfmuller, K.: *Die Systemtheorie der elektrischen Nachrichtenuhertragung,* S. Hirzel Verlag, Stuttgart, 1968.

Lane, C. E.: "Phase Distortion in Telephone Apparatus," *Bell* Syst. *Tech. J.*, vol. 9, pp. 493–521, July 1930.

Lathi, B. P.: *Signals, Systems and Communications,* Wiley, New York, N.Y., 1965.

Lynn, P. A.: *An Introduction to the Analysis and Processing of Signals,* 2nd ed. Macmillan, London, 1982.

Mallinson, J. C.: "Tutorial Review of Magnetic Recording." *Proc. IEEE*, vol. 62, pp. 196–208, February 1976.

Members of the Technical Staff of Bell Telephone Laboratories: *Transmission Systems for Communications,* 4th ed., Western Electric Company, Technical Publications, Winston-Salem, N.C., 197 I.

Oppenheim, A. V., and R. W. Schafer: *Digital Signal Processing*, Prentice-Hall, Englewood Cliffs, N.J., 1975.

Panter, P. F.: *Modulation, Noise and Spectral Analysis,* McGraw-Hill, New York, N.Y., 1965.

Papoulis, A.: *Signal Analysis*, McGraw-Hill, New York, N.Y., 1977.

Papoulis, A.: *The Fourier Integral and Its Applications,* McGraw-Hill, New York, N.Y., 1962.

Peus, S.: "Microphones and Transients," *db Mag.*, translated from *Radio Mentor* by S. Temmer, vol. 11, pp. 35–38, May 1977.

Preis, D: "A Catalog of Frequency and Transient Responses," *J. Audio Eng. Soc.*, vol. 25, no. 12, pp. 990–1007, December 1977.

Pries, D.: "Audio Signal Processing with Transversal Filters," *IEEE Conf. Proc.*, 1979 ICASSP, pp. 310–313, April 1979.

Preis, D.: "Hilbert-Transformer Side-Chain Phase Equalizer for Analogue Magnetic Recording," *Electron. Lett.*, vol. 13, pp. 616–617, September 1977.

Preis, D.: "Impulse Testing and Peak Clipping," *J. Audio Eng. Soc.*, vol. 25, no. 1, pp. 2–14, January 1977.

Preis, D.: "Least-Squares Time-Domain Deconvolution for Transversal-Filter Equalizers," *Electron. Lett.*, vol. 13, no. 12, pp. 356–357, June 1977.

Preis, D.: "Linear Distortion," *J. Audio Eng. Soc.*, vol. 24, no. 5, pp. 346–367, June 1976.

Pries, D.: "Measures and Perception of Phase Distortion in Electroacoustical Systems," *IEEE Conf. Proc.*, 1980 ICASSP, pp. 490–493, 1980.

Pries, D.: "Phase Equalization for Analogue Magnetic Recorders by Transversal Filtering," *Electron. Lett.*, vol. 13, pp. 127–128, March 1977.

Pries, D.: "Phase Equalization for Magnetic Recording," *IEEE Conf. Proc.*, 1981 ICASSP, pp. 790–795, March 1981.

Preis, D.: "Phase Distortion and Phase Equalization in Audio Signal Processing—A Tutorial Review," *J. Audio Eng Soc.*, vol. 30, no. 11, pp. 774–794, November 1982.

Pries, D., and C. Bunks: "Three Algorithms for the Design of Transversal-Filter Equalizers," *Proc. 1981 IEEE Int. Symp. Circuits Sys.*, pp. 536–539, 1981.

Pries, D., and P. J. Bloom: "Perception of Phase Distortion in Anti-Alias Filters," *J. Audio Eng. Soc.*, vol. 32, no. 11, pp. 842–848, November 1984.

Preis, D., F. Hlawatsch, P. J. Bloom, and J. A. Deer: "Wigner Distribution Analysis of Filters with Perceptible Phase Distortion," *J. Audio Eng. Soc.*, December 1987.

Rabiner, L. R., and C. M. Rader (eds.): *Digital Signal Processing,* IEEE, New York, N.Y., 1972.

Schwartz, M.: *Information Transmission, Modulation and Noise,* McGraw-Hill, New York, N.Y., 1970.

Small, R. H.: "Closed-Box Loudspeaker Systems, Part 1: Analysis," *J. Audio Eng. Soc.*, vol. 20, pp. 798–808, December 1972.

Totzek, U., and D. Press: "How to Measure and Interpret Coherence Loss in Magnetic Recording," *J. Audio Eng. Soc.*, December 1987.

Totzek, U., D. Preis, and J. F. Boebme: "A Spectral Model for Time-Base Distortions and Magnetic Recording," *Archiv. fur Elektronik und Ubertragungstechnik*, vol. 41, no. 4, pp. 223–231, July-August 1987.

Westman, H. P. (ed.): *ITT Reference Data for Radio Engineers,* Howard W. Sams, New York, N.Y., 1973.

Wheeler, H. A.: "The Interpretation of Amplitude and Phase Distortion in Terms of Paired Echoes," *Proc. IRE*, vol. 27, pp. 359–385, June 1939.

Williams, A. B.: *Active Filter Design,* Artech House. Dedham, Mass., 1975.

Zverev, A. 1.: *Handbook of Filter Synthesis*, Wiley, New York, N.Y., 1967.

15.2c Architectural Acoustic Principles and Design Techniques

ANSI: *American National Standard.for Rating Noise with Respect to Speech Interference,* ANSI S3.14-1977, American National Standards Institute, New York, N.Y., 1977.

ANSI: *Method for the Measurement of Monosyllabic Word Intelligibility,* ANSI S3.2-1960, rev. 1977, American National Standards Institute, New York, N.Y., 1976.

ASA Standards Index 2, Acoustical Society of America, New York, N.Y., 1980.

ASHRAE: *ASHRAE Handbook—1984 Systems,* American Society of Heating, Refrigerating and Air-Conditioning Engineers, Atlanta, Ga., 1984.7.

Beranek, L. L.: *Acoustics,* McGraw-Hill, New York, N.Y., 1954.

Beranek, L. L.: *Noise and Vibration Control,* McGraw-Hill. New York, N.Y., 1971.

Catalogue of STC and IIC Ratings for Wall and Floor/Ceiling Assemblies, Office of Noise Control, Berkeley, Calif.

Egan, M. D.: Co*ncepts in Architectural Acoustics,* McGraw-Hill, New York, N.Y., 1972.

Huntington, W. C., R. A. Mickadeit, and W. Cavanaugh: *Building Construction Materials,* 5th ed., Wiley, New York, N.Y., 1981.

Jones, Robert S.: *Noise and Vibration Control in Buildings,* McGraw-Hill, New York, N.Y., 1980.

Kryter, K. D.: *The Effects of Noise on Man,* Academic, New York, N.Y., 1985.

Lyon R. H., and R. G. Cann: *Acoustical Scale Modeling,* Grozier Technical Systems, Inc., Brookline, Mass.

Marris, Cyril M.: *Handbook of Noise Control*, 2nd ed., McGraw-Hill, New York, N.Y., 1979.

Marshall, Harold, and M. Barron: "Spatial Impression Due to Early Lateral Reflections in Concert Halls: The Derivation of the Physical Measure," *JSV*, vol.77, no. 2, pp. 211–232, 1981.

Morse, P. M.: *Vibration and Sound*, American Institute of Physics, New York, N.Y., 1981.

Siebein, Gary W.: *Prolect Design Phase Analysis Techniques for Predicting the Acoustical Qualities of Buildings,* research report to the National Science Foundation, grant CEE8307948, Florida Architecture and Building Research Center, Gainesville, Fla., 1986.

Talaske, Richard H., Ewart A. Wetherill, and William J. Cavanaugh (eds.): *Halls for Music Performance Two Decades of Experience, 1962-1982*, American Institute of Physics for the Acoustical Society of America, New York, N.Y., 1982.

15.2d Microphone Devices and Systems

"A Phased Array," *Hi Fi News Record Rev.,* July 1981.

Bevan, W. R., R. B. Schulein, and C. E. Seeler: "Design of a Studio-Quality Condenser Microphone Using Electret Technology," *J. Audio Eng Soc.* (*Engineering Reports*), vol. 26, pp. 947–957, December 1978.

Black, H. S.: U.S. Patent 2,102,671.

Blumlein, A.: British Patent 394,325, December 14, 1931; reprinted in *J. Audio Eng. Soc.*, vol. 6, April 1958.

Dooley, W. L., and R. D. Streicher: "M-S Stereo: A Powerful Technique for Working in Stereo," *J. Audio Eng. Soc.*, vol. 30, pp. 707–718, October 1982.

Eargle, J.: *Sound Recording,* Van Nostrand Reinhold, New York, 1976.

Eargle, J.: *The Microphone Handbook,* Elar Publishing, Plainview, N.Y., 1981.

Fewer, D. R.: "Design Principles for Junction Transistor Audio Power Amplifiers," *Trans. IRE PGA*, AU-3(6), November–December 1955.

Fredericksen, E., N. Firby, and H. Mathiasen: "Prepolarized Condenser Microphones for Measurement Purposes," *Tech. Rev.,* Bruel & Kjaer, Copenhagen, no.4, 1979.

Garner, L. H.: "High-Power Solid State Amplifiers," *Trans. IRE PGA*, 15(4), December 1967.

Gordon, J.: "Recording in 2 and 4 Channels," *Audio*, pp. 36–38, December 1973.

Harper, C. A. (ed.): *Handbook of Components for Electronics,* McGraw-Hill, New York, N.Y., 1977.

Instruction Book for RCA BK-16A Dynamic Microphone, IB-24898, Radio Corporation of America, Camden, N.J.

Killion, M. C., and E. V. Carlson: "A Subminiature Electret-Condenser Microphone of New Design," *J. Audio Eng. Soc.*, vol. 22, pg. 237–243, May 1974.

Kirchner, R. J.: "Properties of Junction Transistors," *Trans. IRE PGA*, AU-3(4), July-August 1955.

Kishi, K., N. Tsuchiya, and K. Shimura: "Unidirectional Microphone," U.S. Patent 3,581,012, May 25, 1971.

Kubota, H.: "Back Electret Microphones," presented at the 55th Convention of the Audio Engineering Society, *J. Audio Eng. Soc.* (*Abstracts*), vol. 24, no. 862, preprint 1157, December 1976.

Lipshitz, S. P.: "Stereo Microphone Techniques: Are the Purists Wrong?" presented at the 78th Convention of the Audio Engineering Society, *J. Audio Eng. Soc.* (*Abstracts*), vol. 33, pg. 594, preprint 2261, July-August 1985.

Long, E. M., and R. J. Wickersham: "Pressure Recording Process and Device," U.S. Patent 4,361,736, November 30, 1982.

Lynn, D. K., C. S. Meyer, and D. C. Hamilton (eds.): *Analysis and Design of Integrated Circuits,* McGraw-Hill, New York, N.Y., 1967.

Microphones—Anthology, Audio Engineering Society, New York, 1979.

"Miking with the 3-Point System," *Audio*, pp. 28–36, December 1975.

Nisbett, A.: *The Technique of the Sound Studio,* Hastings House, New York, N.Y., 1974.

Olson, H. F.: *Acoustical Engineering,* Van Nostrand, Princeton, N.J., 1957.

Olson, H. F.: "Directional Electrostatic Microphone," U.S. Patent 3,007,012, October 31, 1961.

Olson, H. F. (ed.): *McGraw-Hill Encyclopedia of Science and Technology,* 5th ed., vol. 18, McGraw-Hill, New York, N.Y., pg. 506, 1982.

Olson, H. F.: *Music, Physics, and Engineering,* 2d ed., Dover, New York, N.Y., 1967.

Olson, H. F.: "Ribbon Velocity Microphones," *J. Audio Eng. Soc.*, vol. 18, pp. 263–268, June 1970.

Petersen, A., and D. B. Sinclair: "A Singled-Ended Push-Pull Audio Amplifier," *Proc. IRE*, vol. 40, January 1952.

Rasmussen, G.: "A New Condenser Microphone," *Tech. Rev.,* Bruel & Kjaer, Copenhagen, no.1, 1959.

Sank, J. R.: "Equipment Profile-Nakamichi CM-700 Electret Condenser Microphone System." *Audio*, September 1978.

Shockley, W: "A Unipolar Field-Effect Transistor," *Proc. IRE*, vol. 40, November 1952.

Shockley, W: "The Theory of P-N Junctions in Semiconductors and P-N Junction Transistors," *Proc. JRE*, vol. 41, June 1953.

Trent, R. L.: "Design Principles for Transistor Audio Amplifiers," *Trans. IRE PGA*, AU-3(5), September–October 1955.

Walker, P. J.: "A Current Dumping Audio Power Amplifier," *Wireless World,* December 1975.

Weinberg, L.: *Network Analysis and Synthesis,* McGraw-Hill, New York, N.Y., 1967.

Widlar, R. J.: "A Unique Current Design for a High Performance Operational Amplifier Especially Suited to Monolithic Construction," *Proc. NEC*, 1965.

Woszczyk, W. R.: "A Microphone Technique Employing the Principle of Second-Order Gradient Unidirectionality," presented at the 69th Convention of the Audio Engineering Society., *J. Audio Eng. Soc.* (*Abstracts*), vol. 29, pg. 550, preprint 1800, July-August 1981.

15.2e Sound Reproduction Devices and Systems

Allison, R., et at.: "On the Magnitude and Audibility of FM Distortion in Loudspeakers," *J. Audio Eng Soc.*, vol. 30, no. 10, pg. 694, 1982.

Beranek, L. L.: *Acoustics*, McGraw-Hill, New York, N.Y., pg. 183–185, 1954.

Hayasaka, T., et al.: *Onkyo-Kogaku Gairon* (*An Introduction to Sound and Vibration*), Nikkan Kogyo Shinbunshya, pg. 67, 1973 (in Japanese).

Hayasaka, T., et al.: *Onkyo-Shindo Ron* (*Sound and Vibration*), Maruzen Kabushikigaishya, pg. 201, 1974 (in Japanese).

Hirata, Y.: "Study of Nonlinear Distortion in Audio Instruments," *J. Audio Eng. Soc.*, vol. 29, no. 9, pg. 607, 1981.

Kinsler, L. E., et al: *Fundamentals of Acoustics*, Wiley, New York, N.Y., 1982.

Melillo, L., et al.: "Ferrolluids as a Means of Controlling Woofer Design Parameters," presented at the 63d Convention of the Audio Engineering Society, vol. 1, pg. 177, 1979.

Morse, P. M.: *Vibration and Sound*, McGraw-Hill, New York, N.Y., pg. 326, 1948.

Morse, P. M., and K. U. Ingard: Theoretical Acoustics, McGraw-Hill, New York, N.Y., pg. 366, 1968.

Niguchi, H., et al.: "Reinforced Olefin Polymer Diaphragm for Loudspeakers," *J. Audio Eng. Soc.*, vol. 29, no. 11, pg. 808, 1981.

Okahara, M., et al: *Audio Handbook*, Ohm Sya, pg. 285, 1978 (in Japanese).

Olson, H. F.: *Elements of Acoustical Engineering*, Van Nostrand, Princeton, N.J., 1957.

Rayleigh, J. W. S.: *The Theory of Sound*, Dover, New York, N.Y., pg 162, 1945.

Sakamoto, N.: *Loudspeaker and Loudspeaker Systems*, Nikkan Kogyo Shinbunshya, pg. 36, 1967 (in Japanese).

Sakamotoct, N., et. al.: "Loudspeaker with Honeycomb Disk Diaphragm," *J. Audio Eng. Soc.*, vol. 29, no. 10, pg. 711, 1981.

Shindo, T., et al: "Effect of Voice-Coil and Surround on Vibration and Sound Pressure Response of Loudspeaker Cones," *J. Audio Eng. Soc.*, vol. 28, no 7–8, pg. 490, 1980.

Suwa, H., et al.: "Heat Pipe Cooling Enables Loudspeakers to Handle Higher Power," presented at the 63d Convention of the Audio Engineering Society, vol. 1, pg. 213, 1979.

Suzuki, H., et al.: "Radiation and Diffraction Effects by Convex and Concave Domes," *J. Audio Eng Soc.*, vol. 29, no. 12, pg. 873, 1981.

Takahashi, S., et al.: "Glass-Fiber and Graphite-Flake Reinforced Polyimide Composite Diaphragm for Loudspeakers," *J. Audio Eng. Soc.*, vol. 31, no. 10, pg. 723, 1983.

Tsuchiya, H., et al.: "Reducing Harmonic Distortion in Loudspeakers," presented at the 63d Convention of the Audio Engineering Society, vol. 2, pg. 1, 1979.

Yamamoto, T., et al.: "High-Fidelity Loudspeakers with Boronized Titanium Diaphragm," *J. Audio Eng. Soc.*, vol. 28, no. 12, pg. 868, 1980.

Yoshihisa, N., et al.: "Nonlinear Distortion in Cone Loudspeakers," Chuyu-Ou Univ. Rep., vol. 23, pg. 271, 1980.

15.2f Digital Coding of Audio Signals

Alkin, Oktay: "Digital Coding Schemes," *The Electronics Handbook*, Jerry C. Whitaker (ed.), CRC Press, Boca Raton, Fla., pp. 1252–1258, 1996.

Benson, K. B., and D. G. Fink: "Digital Operations in Video Systems," *HDTV: Advanced Television for the 1990s*, McGraw-Hill, New York, pp. 4.1–4.8, 1990.

Chambers, J. A., S. Tantaratana, and B. W. Bomar: "Digital Filters," *The Electronics Handbook*, Jerry C. Whitaker (ed.), CRC Press, Boca Raton, Fla., pp. 749–772, 1996.

Garrod, Susan A. R.: "D/A and A/D Converters," *The Electronics Handbook*, Jerry C. Whitaker (ed.), CRC Press, Boca Raton, Fla., pp. 723–730, 1996.

Garrod, Susan, and R. Borns: *Digital Logic: Analysis, Application, and Design*, Saunders College Publishing, Philadelphia, 1991.

Lee, E. A., and D. G. Messerschmitt: *Digital Communications*, 2nd ed., Kluwer, Norell, Mass., 1994.

Nyquist, H.: "Certain Factors Affecting Telegraph Speed," *Bell System Tech. J.*, vol. 3, pp. 324–346, March 1924.

Parks, T. W., and J. H. McClellan: "A Program for the Design of Linear Phase Infinite Impulse Response Filters," *IEEE Trans. Audio Electroacoustics*, AU-20(3), pp. 195–199, 1972.

Peterson, R., R. Ziemer, and D. Borth: *Introduction to Spread Spectrum Communications*, Prentice-Hall, Englewood Cliffs, N. J., 1995.

Pohlmann, Ken: *Principles of Digital Audio*, McGraw-Hill, New York, N.Y., 2000.

Sklar, B.: *Digital Communications: Fundamentals and Applications*, Prentice-Hall, Englewood Cliffs, N. J., 1988.

TMS320C55x DSP Functional Overview, Texas Instruments, Dallas, TX, literature No. SRPU312, June 2000.

Ungerboeck, G.: "Trellis-Coded Modulation with Redundant Signal Sets," parts I and II, *IEEE Comm. Mag.*, vol. 25 (Feb.), pp. 5-11 and 12-21, 1987.

Ziemer, R., and W. Tranter: *Principles of Communications: Systems, Modulation, and Noise*, 4th ed., Wiley, New York, 1995.

Ziemer, Rodger E.: "Digital Modulation," *The Electronics Handbook*, Jerry C. Whitaker (ed.), CRC Press, Boca Raton, Fla., pp. 1213–1236, 1996.

15.2g Compression Technologies for Audio

Brandenburg, K., and Gerhard Stoll: "ISO-MPEG-1 Audio: A Generic Standard for Coding of High Quality Digital Audio," *92nd AES Convention Proceedings*, Audio Engineering Society, New York, N.Y., 1992, revised 1994.

Ehmer, R. H.: "Masking Patterns of Tones," J. Acoust. Soc. Am., vol. 31, pp. 1115–1120, August 1959.

Fibush, David K.: "Testing MPEG-Compressed Signals," *Broadcast Engineering*, Overland Park, Kan., pp. 76–86, February 1996.

Herre, J., and B. Grill: "MPEG-4 Audio—Overview and Perspectives for the Broadcaster," *IBC 2000 Proceedings*, International Broadcast Convention, Amsterdam, September 2000.

IEEE Standard Dictionary of Electrical and Electronics Terms, ANSI/IEEE Standard 100-1984, Institute of Electrical and Electronics Engineers, New York, 1984.

ITU-R Recommendation BS-775, "Multi-channel Stereophonic Sound System with and Without Accompanying Picture."

Lyman, Stephen, "A Multichannel Audio Infrastructure Based on Dolby E Coding," *Proceedings of the NAB Broadcast Engineering Conference*, National Association of Broadcasters, Washington, D.C., 1999.

Moore, B. C. J., and B. R. Glasberg: "Formulae Describing Frequency Selectivity as a Function of Frequency and Level, and Their Use in Calculating Excitation Patterns," Hearing Research, vol. 28, pp. 209–225, 1987.

Robin, Michael, and Michel Poulin: *Digital Television Fundamentals*, McGraw-Hill, New York, N.Y., 1998.

SMPTE Standard for Television: "12-Channel Serial Interface for Digital Audio and Auxiliary Data," SMPTE 324M, SMPTE, White Plains, N.Y., 1999.

SMPTE Standard for Television: "Channel Assignments and Levels on Multichannel Audio Media," SMPTE 320M-1999, SMPTE, White Plains, N.Y., 1999.

Smyth, Stephen: "Digital Audio Data Compression," *Broadcast Engineering*, Intertec Publishing, Overland Park, Kan., February 1992.

Terry, K. B., and S. B. Lyman: "Dolby E—A New Audio Distribution Format for Digital Broadcast Applications," *International Broadcasting Convention Proceedings*, IBC, London, England, pp. 204–209, September 1999.

Todd, C., et. al.: "AC-3: Flexible Perceptual Coding for Audio Transmission and Storage," AES 96th Convention, Preprint 3796, Audio Engineering Society, New York, February 1994.

Vernon, S., and T. Spath: "Carrying Multichannel Audio in a Stereo Production and Distribution Infrastructure," *Proceedings of IBC 2000*, International Broadcasting Convention, Amsterdam, September 2000.

Wylie, Fred: "Audio Compression Techniques," *The Electronics Handbook*, Jerry C. Whitaker (ed.), CRC Press, Boca Raton, Fla., pp. 1260–1272, 1996.

Wylie, Fred: "Audio Compression Technologies," *NAB Engineering Handbook*, 9th ed., Jerry C. Whitaker (ed.), National Association of Broadcasters, Washington, D.C., 1998.

Zwicker, E.: "Subdivision of the Audible Frequency Range Into Critical Bands (Frequenzgruppen)," J. Acoust. Soc. of Am., vol. 33, p. 248, February 1961.

15.2h Audio Networking

ATSC, "Guide to the Use of the Digital Television Standard," Advanced Television Systems Committee, Washington, D.C., Doc. A/54, Oct. 4, 1995.

Craig, Donald: "Network Architectures: What does Isochronous Mean?," *IBC Daily News*, IBC, Amsterdam, September 1999.

Dahlgren, Michael W.: "Servicing Local Area Networks," *Broadcast Engineering*, Intertec Publishing, Overland Park, Kan., November 1989.

Fibush, David: *A Guide to Digital Television Systems and Measurement*, Tektronix, Beaverton, OR, 1994.

Gaggioni, H., M. Ueda, F. Saga, K. Tomita, and N. Kobayashi, "The Development of a High-Definition Serial Digital Interface," Sony Technical Paper, Sony Broadcast Group, San Jose, Calif., 1998.

Gallo and Hancock: *Networking Explained*, Digital Press, pp. 191–235, 1999.

Goldman, J: *Applied Data Communications: A Business Oriented Approach*, 2md ed., Wiley, New York, N.Y., 1998.

Goldman, J: *Local Area Networks: A Business Oriented Approach*, 2nd ed., Wiley, New York, N.Y., 2000.

Goldman, James E.: "Network Communication," in *The Electronics Handbook*, Jerry C. Whitaker (ed.), CRC Press, Boca Raton, Fla., 1996.

Held, G.: *Ethernet Networks: Design Implementation, Operation and Management*, Wiley, New York, N.Y., 1994.

Held, G.: *Internetworking LANs and WANs*, Wiley, New York, N.Y., 1993.

Held, G.: *Local Area Network Performance Issues and Answers*, Wiley, New York, N.Y., 1994.

Held, G.: *The Complete Modem Reference*, Wiley, New York, N.Y., 1994.

International Organization for Standardization: "Information Processing Systems—Open Systems Interconnection—Basic Reference Model," ISO 7498, 1984.

Legault, Alain, and Janet Matey: "Interconnectivity in the DTV Era—The Emergence of SDTI," *Proceedings of Digital Television '98*, Intertec Publishing, Overland Park, Kan., 1998.

Miller, Mark A.: "Servicing Local Area Networks," *Microservice Management*, Intertec Publishing, Overland Park, Kan., February 1990.

Miller, Mark A.: *LAN Troubleshooting Handbook*, M&T Books, Redwood City, Calif., 1990.

"Networking and Internet Broadcasting," Omneon Video Networks, Campbell, Calif, 1999.

"Networking and Production," Omneon Video Networks, Campbell, Calif., 1999.

Owen, Peter: "Gigabit Ethernet for Broadcast and Beyond," *Proceedings of DTV99*, Intertec Publishing, Overland Park, Kan., November 1999.

Piercy, John: "ATM Networked Video: Moving From Leased-Lines to Packetized Transmission," *Proceedings of the Transition to Digital Conference*, Intertec Publishing, Overland Park, Kan., 1996.

"SMPTE Recommended Practice—Error Detection Checkwords and Status Flags for Use in Bit-Serial Digital Interfaces for Television," RP 165-1994, SMPTE, White Plains, N.Y., 1994.

"SMPTE Recommended Practice—SDTI-CP MPEG Decoder Templates," RP 204, SMPTE, White Plains, N.Y., 1999.

"SMPTE Standard for Television—24-Bit Digital Audio Format for HDTV Bit-Serial Interface," SMPTE 299M-1997, SMPTE, White Plains, N.Y., 1997.

"SMPTE Standard for Television—Ancillary Data Packet and Space Formatting," SMPTE 291M-1998, SMPTE, White Plains, N.Y., 1998.

"SMPTE Standard for Television—Bit-Serial Digital Interface for High-Definition Television Systems," SMPTE 292M-1998, SMPTE, White Plains, N.Y., 1998.

"SMPTE Standard for Television—Element and Metadata Definitions for the SDTI-CP," SMPTE 331M-2000, SMPTE, White Plains, N.Y., 2000.

"SMPTE Standard for Television—Encapsulation of Data Packet Streams over SDTI (SDTI-PF)," SMPTE 332M-2000, SMPTE, White Plains, N.Y., 2000.

"SMPTE Standard for Television—Mapping of AES3 Data into MPEG-2 Transport Stream," SMPTE 302M-1998, SMPTE, White Plains, N.Y., 1998.

"SMPTE Standard for Television—SDTI Content Package Format (SDTI-CP)," SMPTE 326M-2000, SMPTE, White Plains, N.Y., 2000.

"SMPTE Standard for Television—Serial Data Transport Interface," SMPTE 305M-1998, SMPTE, White Plains, N.Y., 1998.

SMPTE 344M, "540 Mb/s Serial Digital Interface," SMPTE, White Plains, N.Y., 2000.

"SMPTE Standard for Television—High Data-Rate Serial Data Transport Interface (HD-SDTI)," SMPTE 348M, SMPTE, White Plains, N.Y., 2000.

"SMPTE Standard for Television—Signals and Generic Data over High-Definition Interfaces," SMPTE 346M, SMPTE, White Plains, N.Y., 2000.

"SMPTE Standard for Television—Vertical Ancillary Data Mapping for Bit-Serial Interface, SMPTE 334M, SMPTE, White Plains, N.Y., 2000.

"Technology Brief—Networking and Storage Strategies," Omneon Video Networks, Campbell, Calif., 1999.

Turow, Dan: "SDTI and the Evolution of Studio Interconnect," *International Broadcasting Convention Proceedings*, IBC, Amsterdam, September 1998.

Wilkinson, J. H., H. Sakamoto, and P. Horne: "SDDI as a Video Data Network Solution," *International Broadcasting Convention Proceedings*, IBC, Amsterdam, September 1997.

Wu, Tsong-Ho: "Network Switching Concepts," *The Electronics Handbook*, Jerry C. Whitaker (ed.), CRC Press, Boca Raton, Fla., p. 1513, 1996.

15.2i Audio Recording Systems

Anderson, D: "Fibre Channel-Arbitrated Loop: The Preferred Path to Higher I/O Performance, Flexibility in Design," Seagate Technology Paper #MN-24, Seagate, Scotts Valley, Calif., 1995.

Bate, G.: "Recent Developments in Magnetic Recording Materials," *J. Appl. Phys*. pg. 2447, 1981.

Bertram, H. N.: "Long Wavelength ac Bias Recording Theory," *IEEE Trans. Magnetics*, vol. MAG-10, pp. 1039–1048, 1974.

Bozorth, Richard M.: *Ferromagnetism*, Van Nostrand, Princeton, N.J., 1961.

Chikazumi, Soshin: *Physics of Magnetism*, Wiley, New York, N.Y., 1964.

Goldberg, Thomas: "New Storage Technology," *Proceedings of the Advanced Television Summit*, Intertec Publishing, Overland Park, Kan., 1996.

Grega, Joe: "Magnetic and Optical Recording Media," in *NAB Engineering Handbook*, 9th ed., Jerry C. Whitaker (ed.), National Association of Broadcasters, Washington, D.C., pp. 893–906, 1999.

Hawthorne, J. M., and C. J. Hefielinger: "Polyester Films," in *Encyclopedia of Polymer Science and Technology*, N. M. Bikales (ed.), vol. 11, Wiley, New York, N.Y., pg. 42, 1969.

Heyn, T.: "The RAID Advantage," Seagate Technology Paper, Seagate, Scotts Valley, Calif., 1995.

Jorgensen, F.: *The Complete Handbook of Magnetic Recording*, Tab Books, Blue Ridge Summit, Pa., 1980.

Kalil, F. (ed): *Magnetic Tape Recording for the Eighties*, NASA References Publication 1975, April 1982.

Kraus, John D.: *Electromagnetics*, McGraw-Hill, New York, N.Y., 1953.

Lehtinen, Rick, "Editing Systems," *Broadcast Engineering*, Intertec Publishing, Overland Park. Kan., pp. 26–36, May 1996.

Lueck, L. B. (ed): *Symposium Proceedings Textbook*, Symposium on Magnetic Media Manufacturing Methods, Honolulu, May 25–27, 1983.

McConathy, Charles F.: "A Digital Video Disk Array Primer," *SMPTE Journal*, SMPTE, New York, N.Y., pp. 220–223, April 1998.

McKnight, John G.: "Erasure of Magnetic Tape," *J. Audio Eng. Soc.*, Audio Engineering Society, New York, N.Y., vol. 11, no.3, pp. 223–232, 1963.

Nylen, P., and E. Sunderland: *Modern Surface Coatings*, Interscience Publishers Division, Wiley, London, 1965.

Pear, C. B.: *Magnetic Recording in Science and Industry*, Reinhold, New York, N.Y., 1967.

Perry, R. H., and A. A. Nishimura: "Magnetic Tape," in *Encyclopedia of Chemical Technology*, 3d ed., Kirk Othmer (ed.), vol. 14, Wiley, New York, N.Y., pp. 732–753, 1981.

Plank, Bob: "Video Disk and Server Operation," *International Broadcast Engineer*, September 1995.

Robin, Michael, and Michel Poulin: "Multimedia and Television," in *Digital Television Fundamentals*, McGraw-Hill, New York, N.Y., pp. 455–488, 1997.

Sharrock, Michael P., and D. P. Stubs: "Perpendicular Magnetic Recording Technology: A Review," *SMPTE J.*, SMPTE, White Plains, N.Y., vol. 93, pp. 1127–1133, December 1984.

Smit, J., and H. P. J. Wijn: *Ferrite*, Wiley, New York, N.Y., 1959.

Tochihara, S.: "Magnetic Coatings and Their Applications in Japan," *Prog. Organic Coatings*, vol. 10, pp. 195–204, 1982.

Tyson, H: "Barracuda and Elite: Disk Drive Storage for Professional Audio/Video," Seagate Technology Paper #SV-25, Seagate, Scotts Valley, Calif., 1995.

Whitaker, Jerry C.: "Data Storage Systems," in *The Electronics Handbook*, Jerry C. Whitaker (ed.), CRC Press, Boca Raton, Fla., pp. 1445–1459, 1996.

Whitaker, Jerry C.: "Tape Recording Technology," *Broadcast Engineering*, Intertec Publishing, Overland Park, Kan., vol. 31, no. 11, pp. 78–108, 1989.

15.2j Audio Production Facility Design

DeSantis, Gene, Jerry C. Whitaker, and C. Robert Paulson: *Interconnecting Electronic Systems*, CRC Press, Boca Raton, Fla., 1992.

Whitaker, Jerry C.: *Facility Design Handbook*, CRC Press, Boca Raton, Fla., 2000.

15.2k Radio Broadcast Transmission Systems

Bean, B. R., and E. J. Dutton: "Radio Meteorology," National Bureau of Standards Monograph 92, March 1, 1966.

Benson, B., and Whitaker, J.: *Television and Audio Handbook for Technicians and Engineers*, McGraw-Hill, New York, N.Y., 1989.

Bingeman, Grant: "AM Tower Impedance Matching," *Broadcast Engineering*, Intertec Publishing, Overland Park, Kan., July 1985.

Bixby, Jeffrey: "AM DAs—Doing it Right," *Broadcast Engineering*, Intertec Publishing, Overland Park, Kan., February 1984.

Bullington, K.: "Radio Propagation at Frequencies above 30 Mc," *Proc. IRE*, pg. 1122, October 1947.

Bullington, K.: "Radio Propagation Variations at VHF and UHF," *Proc. IRE*, pg. 27, January 1950.

Burrows, C. R., and M. C. Gray: "The Effect of the Earth's Curvature on Groundwave Propagation," *Proc. IRE*, pg. 16, January 1941.

Chick, Elton B.: "Monitoring Directional Antennas," *Broadcast Engineering*, Intertec Publishing, Overland Park, Kan., July 1985.

Collocott, T. C., A. B. Dobson, and W. R. Chambers (eds.): *Dictionary of Science & Technology.*

DeComier, Bill: "Inside FM Multiplexer Systems," *Broadcast Engineering*, Intertec Publishing, Overland Park, Kan., May 1988.

de Lisle, E. W.: "Computations of VHF and UHF Propagation for Radio Relay Applications," RCA, Report by International Division, New York, N.Y.

Dickson, F. H., J. J. Egli, J. W. Herbstreit, and G. S. Wickizer: "Large Reductions of VHF Transmission Loss and Fading by the Presence of a Mountain Obstacle in Beyond-Line-of-Sight Paths," *Proc. IRE*, vol. 41, no. 8, pg. 96, August 1953.

"Documents of the XVth Plenary Assembly," CCIR Report 238, vol. 5, Geneva, 1982.

"Documents of the XVth Plenary Assembly," CCIR Report 563, vol. 5, Geneva, 1982.

"Documents of the XVth Plenary Assembly," CCIR Report 881, vol. 5, Geneva, 1982.

Dougherty, H. T., and E. J. Dutton: "The Role of Elevated Ducting for Radio Service and Interference Fields," NTIA Report 81–69, March 1981.

Eckersley, T. L.: "Ultra-Short-Wave Refraction and Diffraction," *J. Inst. Elec. Engrs.*, pg. 286, March 1937.

Epstein, J., and D. Peterson: "An Experimental Study of Wave Propagation at 850 Mc," *Proc. IRE*, pg. 595, May 1953.

Fink, D. G., (ed.): *Television Engineering Handbook*, McGraw-Hill, New York, N.Y., 1957.

Fink, D., and D. Christiansen (eds.): Electronics Engineer's Handbook, 3rd ed., McGraw-Hill, New York, N.Y., 1989.

Handbook of Physics, McGraw-Hill, New York, N.Y., 1958.

Harrison, Cecil: "Passive Filters," in *The Electronics Handbook*, Jerry C. Whitaker (ed.), CRC Press, Boca Raton, Fla., pp. 279–290, 1996.

Hauptstuek, Jim: "Interconnecting the Digital Chain," *NAB 1996 Broadcast Engineering Conference Proceedings*, National Association of Broadcasters, Washington, D.C., pp. 360–358, 1996.

Heymans, Dennis: "Hot Switches and Combiners," *Broadcast Engineering*, Overland Park, Kan., December 1987.

Jordan, Edward C. (ed.): "*Reference Data for Engineers—Radio, Electronics, Computer and Communications*, 7th ed., Howard W. Sams, Indianapolis, Ind., 1985.

Judd, D. B., and G. Wyszecki: *Color in Business, Science and Industry*, 3rd ed., John Wiley and Sons, New York, N.Y.

Kaufman, Ed: *IES Illumination Handbook*, Illumination Engineering Society.

Lapedes, D. N. (ed.): *The McGraw-Hill Encyclopedia of Science & Technology*, 2nd ed., McGraw-Hill, New York, N.Y.

Longley, A. G., and P. L. Rice: "Prediction of Tropospheric Radio Transmission over Irregular Terrain—A Computer Method," ESSA (Environmental Science Services Administration), U.S. Dept. of Commerce, Report ERL (Environment Research Laboratories) 79-ITS 67, July 1968.

McClanahan, M. E.: "Aural Broadcast Auxiliary Links," in *NAB Engineering Handbook*, 8th ed., E. B. Cructhfield (ed.), National Association of Broadcasters, Washington, D.C, pp. 671–678, 1992.

McPetrie, J. S., and L. H. Ford: "An Experimental Investigation on the Propagation of Radio Waves over Bare Ridges in the Wavelength Range 10 cm to 10 m," J. *Inst. Elec. Engrs.*, pt. 3, vol. 93, pg. 527, 1946.

Megaw, E. C. S.: "Some Effects of Obstacles on the Propagation of Very Short Radio Waves," J. *Inst. Elec. Engrs.*, pt. 3, vol. 95, no. 34, pg. 97, March 1948.

Mullaney, John H.: "The Folded Unipole Antenna for AM Broadcast," *Broadcast Engineering*, Intertec Publishing, Overland Park, Kan., January 1960.

Mullaney, John H.: "The Folded Unipole Antenna," *Broadcast Engineering*, Intertec Publishing, Overland Park, Kan., July 1986.

National Bureau of Standards Circular 462, "Ionospheric Radio Propagation," June 1948.

NIST: *Manual of Regulations and Procedures for Federal Radio Frequency Management*, September 1995 edition, revisions for September 1996, January and May 1997, NTIA, Washington, D.C., 1997.

Norgard, John: "Electromagnetic Spectrum," *NAB Engineering Handbook*, 9th ed., Jerry C. Whitaker (ed.), National Association of Broadcasters, Washington, D.C., 1999.

Norgard, John: "Electromagnetic Spectrum," *The Electronics Handbook*, Jerry C. Whitaker (ed.), CRC Press, Boca Raton, Fla., 1996.

Norton, K. A.: "Ground Wave Intensity over a Finitely Conducting Spherical Earth," *Proc. IRE*, pg. 622, December 1941.

Norton, K. A.: "The Propagation of Radio Waves over a Finitely Conducting Spherical Earth," *Phil. Mag*., June 1938.

Parker, Darryl: "TFT DMM92 Meets STL Requirements," *Radio World*, Falls Church, VA, October 21, 1992.

"Radio Wave Propagation," Summary Technical Report of the Committee on Propagation of the National Defense Research Committee, Academic Press, New York, N.Y., 1949.

"Report of the Ad Hoc Committee, Federal Communications Commission," vol. 1, May 1949; vol. 2, July 1950.

Rollins, William W., and Robert L. Band: "T1 Digital STL: Discrete vs. Composite Transmission," *NAB 1996 Broadcast Engineering Conference Proceedings*, National Association of Broadcasters, Washington, D.C., pp. 356–359, 1996

Salek, Stanley: "Analysis of FM Booster System Configurations," *Proceedings of the 1992 NAB Broadcast Engineering Conference*, National Association of Broadcasters, Washington, DC, April 1992.

Selvidge, H.:"Diffraction Measurements at Ultra High Frequencies," *Proc. IRE*, pg. 10, January 1941.

Smith, E. E., and E. W. Davis: "Wind-induced Ions Thwart TV Reception," *IEEE Spectrum*, pp. 52—55, February 1981.

Stemson, A: *Photometry and Radiometry for Engineers*, John Wiley and Sons, New York, N.Y.

Stenberg, James T.: "Using Super Power Isolators in the Broadcast Plant," *Proceedings of the Broadcast Engineering Conference*, Society of Broadcast Engineers, Indianapolis, IN, 1988.

Surette, Robert A.: "Combiners and Combining Networks," in *The Electronics Handbook*, Jerry C. Whitaker (ed.), CRC Press, Boca Raton, Fla., pp. 1368–1381, 1996.

The Cambridge Encyclopedia, Cambridge University Press, 1990.

The Columbia Encyclopedia, Columbia University Press, 1993.

"The Propagation of Radio Waves through the Standard Atmosphere," *Summary Technical Report of the Committee on Propagation*, vol. 3, National Defense Research Council, Washington, D.C., 1946, published by Academic Press, New York, N.Y.

van der Pol, Balth, and H. Bremmer: "The Diffraction of Electromagnetic Waves from an Electrical Point Source Round a Finitely Conducting Sphere, with Applications to Radiotelegraphy and to Theory of the Rainbow," pt. 1, *Phil. Mag.*, July, 1937; pt. 2, *Phil. Mag.*, November 1937.

Vaughan, T., and E. Pivit: "High Power Isolator for UHF Television," *Proceedings of the NAB Engineering Conference*, National Association of Broadcasters, Washington, D.C., 1989.

Webster's New World Encyclopedia, Prentice Hall, 1992.

Westberg, J. M.: "Effect of 90° Stub on Medium Wave Antennas," *NAB Engineering Handbook*, 7the ed., National Association of Broadcasters, Washington, D.C., 1985.

Whitaker, Jerry C., (ed.): *A Primer: Digital Aural Studio to Transmitter Links*, TFT, Santa Clara, CA, 1994.

Whitaker, Jerry C., and Skip. Pizzi: "Radio Electronic News Gathering and Field Production," in *NAB Engineering Handbook*, 8th ed., E. B. Cructhfield (ed.), National Association of Broadcasters, Washington, D.C, pp. 1051–1072, 1992.

Wyszecki, G., and W. S. Stiles: *Color Science, Concepts and Methods, Quantitative Data and Formulae*, 2nd ed., John Wiley and Sons, New York, N.Y.

15.21 Radio Receivers

Amos, S. W.: "FM Detectors," *Wireless World*, vol. 87, no. 1540, pg. 77, January 1981.

Benson, K. Blair, and Jerry C. Whitaker: *Television and Audio Handbook for Engineers and Technicians*, McGraw-Hill, New York, N.Y., 1990.

Engelson, M., and J. Herbert: "Effective Characterization of CDMA Signals," *Microwave Journal*, pg. 90, January 1995.

Howald, R.: "Understand the Mathematics of Phase Noise," *Microwaves & RF*, pg. 97, December 1993.

Johnson, J. B:, "Thermal Agitation of Electricity in Conduction," *Phys. Rev.*, vol. 32, pg. 97, July 1928.

Nyquist, H.: "Thermal Agitation of Electrical Charge in Conductors," *Phys. Rev.*, vol. 32, pg. 110, July 1928.

Pleasant, D.: "Practical Simulation of Bit Error Rates," *Applied Microwave and Wireless*, pg. 65, Spring 1995.

Rohde, Ulrich L.: *Digital PLL Frequency Synthesizers*, Prentice-Hall, Englewood Cliffs, N.J., 1983.

Rohde, Ulrich L.: "Key Components of Modern Receiver Design—Part 1," *QST*, pg. 29, May 1994.

Rohde, Ulrich L. Rohde and David P. Newkirk: *RF/Microwave Circuit Design for Wireless Applications*, John Wiley & Sons, New York, N.Y., 2000.

Rohde, Ulrich L, and Jerry C. Whitaker: *Communications Receivers*, 3rd ed., McGraw-Hill, New York, N.Y., 2000.

"Standards Testing: Bit Error Rate," application note 3SW-8136-2, Tektronix, Beaverton, OR, July 1993.

Using Vector Modulation Analysis in the Integration, Troubleshooting and Design of Digital RF Communications Systems, Product Note HP89400-8, Hewlett-Packard, Palo Alto, Calif., 1994.

Watson, R.: "Receiver Dynamic Range; Pt. 1, Guidelines for Receiver Analysis," *Microwaves & RF*, vol. 25, pg. 113, December 1986.

"Waveform Analysis: Noise and Jitter," application note 3SW8142-2, Tektronix, Beaverton, OR, March 1993.

Wilson, E.: "Evaluate the Distortion of Modular Cascades," *Microwaves*, vol. 20, March 1981.

Whitaker, Jerry C. (ed.): *NAB Engineering Handbook*, 9th ed., National Association of Broadcasters, Washington, D.C., 1999.

15.2m Standards and Practices

Baumgartner, Fred, and Terrence Baun: "Broadcast Engineering Documentation," in *NAB Engineering Handbook*, 9th ed., Jerry C. Whitaker (ed.), National Association of Broadcasters, Washington, D.C., 1999.

Baumgartner, Fred, and Terrence Baun: "Engineering Documentation," in *The Electronics Handbook*, Jerry C. Whitaker (ed.), CRC Press, Boca Raton, Fla., 1996.

"Current Intelligence Bulletin #45," National Institute for Occupational Safety and Health, Division of Standards Development and Technology Transfer, February 24, 1986.

Code of Federal Regulations, 40, Part 761.

Delatore, J. P., E. M. Prell, and M. K. Vora: "Translating Customer Needs Into Product Specifications", *Quality Progress*, January 1989.

DeSantis, Gene: "Systems Engineering Concepts," in *NAB Engineering Handbook*, 9th ed., Jerry C. Whitaker (ed.), National Association of Broadcasters, Washington, D.C., 1999.

DeSantis, Gene: "Systems Engineering," in *The Electronics Handbook*, Jerry C. Whitaker (ed.), CRC Press, Boca Raton, Fla., 1996.

"Electrical Standards Reference Manual," U.S. Department of Labor, Washington, D.C.

Finkelstein, L.: "Systems Theory", *IEE Proceedings*, vol. 135, Part A, no. 6, July 1988.

Hammett, William F.: "Meeting IEEE C95.1-1991 Requirements," *NAB 1993 Broadcast Engineering Conference Proceedings*, National Association of Broadcasters, Washington, D.C., pp. 471–476, April 1993.

Hammar, Willie: *Occupational Safety Management and Engineering*, Prentice Hall, New York, N.Y.

Hoban, F. T., and W. M. Lawbaugh: *Readings In Systems Engineering*, NASA, Washington, D.C., 1993.

Markley, Donald: "Complying with RF Emission Standards," *Broadcast Engineering*, Intertec Publishing, Overland Park, Kan., May 1986.

"Occupational Injuries and Illnesses in the United States by Industry," OSHA Bulletin 2278, U.S. Department of Labor, Washington, D.C, 1985.

OSHA, "Electrical Hazard Fact Sheets," U.S. Department of Labor, Washington, D.C, January 1987.

OSHA, "Handbook for Small Business," U.S. Department of Labor, Washington, D.C.

Pfrimmer, Jack, "Identifying and Managing PCBs in Broadcast Facilities," *1987 NAB Engineering Conference Proceedings*, National Association of Broadcasters, Washington, D.C, 1987.

"Safety Precautions," Publication no. 3386A, Varian Associates, Palo Alto, Calif., March 1985.

Shinners, S. M.: *A Guide to Systems Engineering and Management*, Lexington, 1976.

Smith, Milford K., Jr., "RF Radiation Compliance," *Proceedings of the Broadcast Engineering Conference*, Society of Broadcast Engineers, Indianapolis, IN, 1989.

System Engineering Management Guide, Defense Systems Management College, Virginia, 1983.

"Toxics Information Series," Office of Toxic Substances, July 1983.

Tuxal, J. G.: *Introductory System Engineering*, McGraw-Hill, New York, N.Y., 1972.

Whitaker, Jerry C.: *AC Power Systems*, 2nd Ed., CRC Press, Boca Raton, Fla., 1998.

Whitaker, Jerry C.: G. DeSantis, and C. Paulson: *Interconnecting Electronic Systems*, CRC Press, Boca Raton, Fla., 1993.

Whitaker, Jerry C.: *Maintaining Electronic Systems*, CRC Press, Boca Raton, Fla. 1991.

Whitaker, Jerry C.: *Power Vacuum Tubes Handbook*, 2nd Ed., CRC Press, Boca Raton, Fla., 1999.

Whitaker, Jerry C.: *Radio Frequency Transmission Systems: Design and Operation*, McGraw-Hill, New York, N.Y., 1990.

15.3 Video

15.3a Light, Vision, and Photometry

Barten, Peter G. J.: "Physical Model for the Contrast Sensitivity of the Human Eye," *Human Vision, Visual Processing, and Digital Display III*, Bernice E. Rogowitz ed., Proc. SPIE 1666, SPIE, Bellingham, Wash., pp. 57–72, 1992.

Boynton, R. M.: *Human Color Vision*, Holt, New York, 1979.

Committee on Colorimetry, Optical Society of America: *The Science of Color*, Optical Society of America, New York, N.Y., 1953.

Daly, Scott: "The Visible Differences Predictor: An Algorithm for the Assessment of Image Fidelity," *Human Vision, Visual Processing, and Digital Display III*, Bernice E. Rogowitz ed., Proc. SPIE 1666, SPIE, Bellingham, Wash., pp. 2–15, 1992.

Davson, H.: *Physiology of the Eye*, 4th ed., Academic, New York, N.Y., 1980.

Evans, R. M., W. T. Hanson, Jr., and W. L. Brewer: *Principles of Color Photography*, Wiley, New York, N.Y., 1953.

Fink, D. G.: *Television Engineering Handbook*, McGraw-Hill, New York, N.Y., 1957.

Fink, D. G: *Television Engineering*, 2nd ed., McGraw-Hill, New York, N.Y., 1952.

Grogan, T. A.: "Image Evaluation with a Contour-Based Perceptual Model," *Human Vision, Visual Processing, and Digital Display III*, Bernice E. Rogowitz ed., Proc. SPIE 1666, SPIE, Bellingham, Wash., pp. 188–197, 1992.

Grogan, Timothy A.: "Image Evaluation with a Contour-Based Perceptual Model," *Human Vision, Visual Processing, and Digital Display III*, Bernice E. Rogowitz ed., Proc. SPIE 1666, SPIE, Bellingham, Wash., pp. 188–197, 1992.

Hecht, S., S. Shiaer, and E. L. Smith: "Intermittent Light Stimulation and the Duplicity Theory of Vision," Cold Spring Harbor Symposia on Quantitative Biology, vol. 3, pg. 241, 1935.

Hecht, S.: "The Visual Discrimination of Intensity and the Weber-Fechner Law," *J. Gen Physiol.*, vol. 7, pg. 241, 1924.

IES Lighting Handbook, Illuminating Engineering Society of North America, New York, N.Y., 1981.

Kingslake, R. (ed.): *Applied Optics and Optical Engineering*, vol. 1, Academic, New York, N.Y., 1965.

Martin, Russel A., Albert J. Ahumanda, Jr., and James O. Larimer: "Color Matrix Display Simulation Based Upon Luminance and Chromatic Contrast Sensitivity of Early Vision," in *Human Vision, Visual Processing, and Digital Display III*, Bernice E. Rogowitz ed., Proc. SPIE 1666, SPIE, Bellingham, Wash., pp. 336–342, 1992.

Polysak, S. L.: *The Retina*, University of Chicago Press, Chicago, Ill., 1941.

Reese, G.: "Enhancing Images with Intensity-Dependent Spread Functions," *Human Vision, Visual Processing, and Digital Display III*, Bernice E. Rogowitz ed., Proc. SPIE 1666, SPIE, Bellingham, Wash., pp. 253–261, 1992.

Reese, Greg: "Enhancing Images with Intensity-Dependent Spread Functions," *Human Vision, Visual Processing, and Digital Display III*, Bernice E. Rogowitz ed., Proc. SPIE 1666, SPIE, Bellingham, Wash., pp. 253–261, 1992.

Schade, O. H.: "Electro-optical Characteristics of Television Systems," *RCA Review*, vol. 9, pp. 5–37, 245–286, 490–530, 653–686, 1948.

Wright, W. D.: *The Measurement of Colour*, 4th ed., Adam Hilger, London, 1969.

15.3b Color Vision, Representation, and Reproduction

Baldwin, M., Jr.: "The Subjective Sharpness of Simulated Television Images," *Proceedings of the IRE*, vol. 28, July 1940.

Belton, J.: "The Development of the CinemaScope by Twentieth Century Fox," *SMPTE Journal*, vol. 97, SMPTE, White Plains, N.Y., September 1988.

Benson, K. B., and D. G. Fink: *HDTV: Advanced Television for the 1990s*, McGraw-Hill, New York, N.Y., 1990.

Bingley, F. J.: "Colorimetry in Color Television—Pt. I," *Proc. IRE*, vol. 41, pp. 838–851, 1953.

Bingley, F. J.: "Colorimetry in Color Television—Pts. II and III," *Proc. IRE*, vol. 42, pp. 48–57, 1954.

Bingley, F. J.: "The Application of Projective Geometry to the Theory of Color Mixture," *Proc. IRE*, vol. 36, pp. 709–723, 1948.

Boynton, R.M.: *Human Color Vision*, Holt, New York, N.Y., p. 404, 1979.

"Colorimetry," Publication no. 15, Commission Internationale de l'Eclairage, Paris, 1971.

DeMarsh, L. E.: "Colorimetric Standards in US Color Television," *J. SMPTE*, vol. 83, pp. 1–5, 1974.

Epstein, D. W.: "Colorimetric Analysis of RCA Color Television System," *RCA Review*, vol. 14, pp. 227–258, 1953.

Fink, D. G., et. al.: "The Future of High Definition Television," *SMPTE Journal*, vol. 89, SMPTE, White Plains, N.Y., February/March 1980.

Fink, D. G.: "Perspectives on Television: The Role Played by the Two NTSCs in Preparing Television Service for the American Public," *Proceedings of the IEEE*, vol. 64, IEEE, New York, N.Y., September 1976.

Fink, D. G.: *Color Television Standards*, McGraw-Hill, New York, N.Y., 1986.

Foley, James D., et al.: *Computer Graphics: Principles and Practice*, 2nd ed., Addison-Wesley, Reading, Mass., pp. 584–592, 1991.

Fujio, T., J. Ishida, T. Komoto and T. Nishizawa: "High Definition Television Systems—Signal Standards and Transmission," *SMPTE Journal*, vol. 89, SMPTE, White Plains, N.Y., August 1980.

Guild, J.: "The Colorimetric Properties of the Spectrum," *Phil. Trans. Roy. Soc. A.*, vol. 230, pp. 149–187, 1931.

Herman, S.: "The Design of Television Color Rendition," *J. SMPTE*, SMPTE, White Plains, N.Y., vol. 84, pp. 267–273, 1975.

Hubel, David H.: *Eye, Brain and Vision*, Scientific American Library, New York, N.Y., 1988.

Hunt, R. W. G.: *The Reproduction of Colour*, 3d ed., Fountain Press, England, 1975.

Isnardi, M. A.: "Exploring and Exploiting Subchannels in the NTSC Spectrum," *SMPTE J.*, SMPTE, White Plains, N.Y., vol. 97, pp. 526–532, July 1988.

Isnardi, M. A.: "Multidimensional Interpretation of NTSC Encoding and Decoding," *IEEE Transactions on Consumer Electronics*, vol. 34, pp. 179–193, February 1988.

Judd, D. B., and G. Wyszencki: *Color in Business, Science, and Industry,*. 3rd ed., Wiley, New York, N.Y., pp. 44-45, 1975.

Judd, D. B.: "The 1931 C.I.E. Standard Observer and Coordinate System for Colorimetry," *Journal of the Optical Society of America*, vol. 23, 1933.

Kaufman, J. E. (ed.): *IES Lighting Handbook-1981 Reference Volume*, Illuminating Engineering Society of North America, New York, N.Y., 1981.

Kelly, K. L.: "Color Designation of Lights," *Journal of the Optical Society of America*, vol. 33, 1943.

Kelly, R. D., A. V. Bedbord and M. Trainer: "Scanning Sequence and Repetition of Television Images," *Proceedings of the IRE*, vol. 24, April 1936.

Miller, Howard: "Options in Advanced Television Broadcasting in North America," *Proceedings of the ITS*, International Television Symposium, Montreux, Switzerland, 1991.

Morizono, M.: "Technological Trends in High-Resolution Displays Including HDTV," *SID International Symposium Digest*, paper 3.1, May 1990.

Munsell Book of Color, Munsell Color Co., 2441 No. Calvert Street, Baltimore, MD 21218.

Neal, C. B.: "Television Colorimetry for Receiver Engineers," *IEEE Trans. BTR*, vol. 19, pp. 149–162, 1973.

Newhall, S. M., D. Nickerson, and D. B. Judd: "Final Report of the OSA Subcommittee on the Spacing of the Munsell Colors," *Journal of the Optical Society of America*, vol. 33, pp. 385–418, 1943.

Nickerson, D.: "History of the Munsell Color System, Company and Foundation, I," *Color Res. Appl.*, vol. 1, pp. 7–10, 1976.

Nickerson, D.: "History of the Munsell Color System, Company and Foundation, II: Its Scientific Application," *Color Res. Appl.*, vol. 1, pp. 69–77, 1976.

Nickerson, D.: "History of the Munsell Color System, Company and Foundation, III," *Color Res. Appl.,* vol. 1, pp. 121–130, 1976.

Nyquist, H.: "Certain Factors Affecting Telegraph Speed," *Bell System Tech. J.*, vol. 3, pp. 324–346, March 1924.

Pearson, M. (ed.): Proc. ISCC Conf. on Optimum Reproduction of Color, Williamsburg, Va., 1971, Graphic Arts Research Center, Rochester, N.Y., 1971.

Pitts, K. and N. Hurst: "How Much Do People Prefer Widescreen (16 × 9) to Standard NTSC (4 × 3)?," *IEEE Transactions on Consumer Electronics*, IEEE, New York, N.Y., August 1989.

Pointer, M. R.: "The Gamut of Real Surface Colours," *Color Res. Appl.*, vol. 5, pp. 145–155, 1980.

Pointer, R. M.: "The Gamut of Real Surface Colors, *Color Res. App.*, vol. 5, 1945.

Pritchard, D. H.: "US Color Television Fundamentals—A Review," *IEEE Trans. CE*, vol. 23, pp. 467–478, 1977.

Smith, A. R.: "Color Gamut Transform Pairs," *SIGGRAPH 78*, 12–19, 1978.

Smith, V. C., and J. Pokorny: "Spectral Sensitivity of the Foveal Cone Pigments Between 400 and 500 nm," *Vision Res.*, vol. 15, pp. 161–171, 1975.

Sproson, W. N.: *Colour Science in Television and Display Systems*, Adam Hilger, Bristol, England, 1983.

Tektronix application note #21W-7165: "Colorimetry and Television Camera Color Measurement," Tektronix, Beaverton, Ore., 1992.

Uba, T., K. Omae, R. Ashiya, and K. Saita: "16:9 Aspect Ratio 38V-High Resolution Trinitron for HDTV," *IEEE Transactions on Consumer Electronics*, IEEE, New York, N.Y., February 1988.

van Raalte, John A.: "CRT Technologies for HDTV Applications," *1991 HDTV World Conference Proceedings*, National Association of Broadcasters, Washington, D.C., April 1991.

Wentworth, J. W.: *Color Television Engineering*, McGraw-Hill, New York, N.Y., 1955.

Wintringham, W. T.: "Color Television and Colorimetry," *Proc. IRE*, vol. 39, pp. 1135– 1172, 1951.

Wright, W. D.: "A Redetermination of the Trichromatic Coefficients of the Spectral Colours," *Trans. Opt. Soc.*, vol. 30, pp. 141–164, 1928–1929.

Wright, W. D.: *The Measurement of Colour*, 4th ed., Adam Hilger, London, 1969.

Wyszecki, G., and W. S. Stiles: *Color Science*, 2nd ed., Wiley, New York, N.Y., 1982.

15.3c Optical Components and Systems

Fink, D. G. (ed.): *Television Engineering Handbook*, McGraw-Hill, New York, N.Y., 1957.

Hardy, A. C., and F. H. Perrin: *The Principles of Optics*, McGraw-Hill, New York, N.Y., 1932.

Kingslake, Rudolf (ed.): *Applied Optics and Optical Engineering*, vol. 1, Chapter 6, Academic, New York, N.Y., 1965.

Sears, F. W.: *Principles of Physics*, III, Optics, Addison-Wesley, Cambridge, Mass., 1946.

Williams, Charles S., and Becklund, Orville A.: *Optics: A Short Course for Engineers and Scientists*, Wiley Interscience, New York, N.Y., 1972.

15.3d Digital Coding of Video Signals

Alkin, Oktay: "Digital Coding Schemes," *The Electronics Handbook*, Jerry C. Whitaker (ed.), CRC Press, Boca Raton, Fla., pp. 1252–1258, 1996.

Benson, K. B., and D. G. Fink: "Digital Operations in Video Systems," *HDTV: Advanced Television for the 1990s*, McGraw-Hill, New York, pp. 4.1–4.8, 1990.

Chambers, J. A., S. Tantaratana, and B. W. Bomar: "Digital Filters," *The Electronics Handbook*, Jerry C. Whitaker (ed.), CRC Press, Boca Raton, Fla., pp. 749–772, 1996.

DeMarsh, LeRoy E.: "Displays and Colorimetry for Future Television," *SMPTE Journal*, SMPTE, White Plains, N.Y., pp. 666–672, October 1994.

Garrod, Susan A. R.: "D/A and A/D Converters," *The Electronics Handbook*, Jerry C. Whitaker (ed.), CRC Press, Boca Raton, Fla., pp. 723–730, 1996.

Garrod, Susan, and R. Borns: *Digital Logic: Analysis, Application, and Design*, Saunders College Publishing, Philadelphia, 1991.

Hunold, Kenneth: "4:2:2 or 4:1:1—What are the Differences?," *Broadcast Engineering*, Intertec Publishing, Overland Park, Kan., pp. 62–74, October 1997.

Lee, E. A., and D. G. Messerschmitt: *Digital Communications*, 2nd ed., Kluwer, Norell, Mass., 1994.

Mazur, Jeff: "Video Special Effects Systems," *NAB Engineering Handbook*, 9th ed., Jerry C. Whitaker (ed.), National Association of Broadcasters, Washington, D.C., to be published 1998.

Nyquist, H.: "Certain Factors Affecting Telegraph Speed," *Bell System Tech. J.*, vol. 3, pp. 324–346, March 1924.

Parks, T. W., and J. H. McClellan: "A Program for the Design of Linear Phase Infinite Impulse Response Filters," *IEEE Trans. Audio Electroacoustics*, AU-20(3), pp. 195–199, 1972.

Peterson, R., R. Ziemer, and D. Borth: *Introduction to Spread Spectrum Communications*, Prentice-Hall, Englewood Cliffs, N. J., 1995.

Pohlmann, Ken: *Principles of Digital Audio*, McGraw-Hill, New York, N.Y., 2000.

Sklar, B.: *Digital Communications: Fundamentals and Applications*, Prentice-Hall, Englewood Cliffs, N. J., 1988.

TMS320C55x DSP Functional Overview, Texas Instruments, Dallas, TX, literature No. SRPU312, June 2000.

"SMPTE C Color Monitor Colorimetry," SMPTE Recommended Practice RP 145-1994, SMPTE, White Plains, N.Y., June 1, 1994.

Ungerboeck, G.: "Trellis-Coded Modulation with Redundant Signal Sets," parts I and II, *IEEE Comm. Mag.*, vol. 25 (Feb.), pp. 5-11 and 12-21, 1987.

Ziemer, R., and W. Tranter: *Principles of Communications: Systems, Modulation, and Noise*, 4th ed., Wiley, New York, 1995.

Ziemer, Rodger E.: "Digital Modulation," *The Electronics Handbook*, Jerry C. Whitaker (ed.), CRC Press, Boca Raton, Fla., pp. 1213–1236, 1996.

15.3e Electron Optics and Deflection

Aiken, W. R.: "A Thin Cathode Ray Tube," *Proc. IRE*, vol. 45, no. 12, pp. 1599–1604, December 1957.

Barkow, W. H., and J. Gross: "The RCA Large Screen 110° Precision In-Line System," ST-5015, RCA Entertainment, Lancaster, Pa.

Boers, J.: "Computer Simulation of Space Charge Flows," Rome Air Development Command RADC-TR-68-175, University of Michigan, 1968.

Casteloano, Joseph A.: *Handbook of Display Technology*, Academic, New York, N.Y., 1992.

Cathode Ray Tube Displays, MIT Radiation Laboratory Series, vol. 22, McGraw-Hill, New York, N.Y., 1953.

Cloz, R., et al.: "Mechanism of Thin Film Electroluminescence," Conference Record, *SID Proceedings*, Society for Information Display, San Jose, Calif., vol. 20, no. 3, 1979.

Dasgupta, B. B.: "Recent Advances in Deflection Yoke Design," *SID International Symposium Digest of Technical Papers*, Society for Information Display, San Jose, Calif., pp. 248–252, May 1999.

Fink, Donald, (ed.): *Television Engineering Handbook*, McGraw-Hill, New York, N.Y., 1957.

Fink, Donald, and Donald Christiansen (eds.): *Electronics Engineers Handbook*, 3rd ed., McGraw-Hill, New York, N.Y., 1989.

Hutter, Rudolph G. E., "The Deflection of Electron Beams," in *Advances in Image Pickup and Display*, B. Kazan (ed.), vol. 1, pp. 212–215, Academic, New York, N.Y., 1974.

IEEE Standard Dictionary of Electrical and Electronics Terms, 2nd ed., Wiley, New York, N.Y., 1977.

Jordan, Edward C. (ed.): *Reference Data for Engineers: Radio, Electronics, Computer, and Communications*, 7th ed., Howard W. Sams, Indianapolis, IN, 1985.

Langmuir, D.: "Limitations of Cathode Ray Tubes," *Proc. IRE*, vol. 25, pp. 977–991, 1937.

Luxenberg, H. R., and R. L. Kuehn (eds.): *Display Systems Engineering*, McGraw-Hill, New York, N.Y. 1968.

Morell, A. M., et al.: "Color Television Picture Tubes," in *Advances in Image Pickup and Display*, vol. 1, B. Kazan (ed.), pg. 136, Academic, New York,N.Y., 1974.

Moss, Hilary: *Narrow Angle Electron Guns and Cathode Ray Tubes*, Academic, New York, N.Y., 1968.

Nix, L.: "Spot Growth Reduction in Bright, Wide Deflection Angle CRTs," *SID Proc.*, Society for Information Display, San Jose, Calif., vol. 21, no. 4, pg. 315, 1980.

Pender, H., and K. McIlwain (eds.), *Electrical Engineers Handbook*, Wiley, New York, N.Y., 1950.

Poole, H. H.: *Fundamentals of Display Systems*, Spartan, Washington, D.C., 1966.

Popodi, A. E., "Linearity Correction for Magnetically Deflected Cathode Ray Tubes," *Elect. Design News*, vol. 9, no. 1, January 1964.

Sadowski, M.: *RCA Review*, vol 95, 1957.

Sherr, S.: *Electronic Displays*, Wiley, New York, N.Y., 1979.

Sherr, S.: *Fundamentals of Display Systems Design*, Wiley, New York, N.Y., 1970.

Sinclair, Clive, "Small Flat Cathode Ray Tube," *SID Digest*, Society for Information Display, San Jose, Calif., pp. 138–139, 1981.

Spangenberg, K. R., *Vacuum Tubes*, McGraw-Hill, New York, N.Y., 1948.

True, R.: "Space Charge Limited Beam Forming Systems Analyzed by the Method of Self-Consistent Fields with Solution of Poisson's Equation on a Deformable Relaxation Mesh," Ph.D. thesis, University of Connecticut, Storrs, 1968.

Zworykin, V. K., and G. Morton: *Television*, 2d ed., Wiley, New York, N.Y., 1954.

15.3f Video Cameras

Bendel, Sidney L., and C. A. Johnson: "Matching the Performance of a New Pickup Tube to the TK-47 Camera," *SMPTE J.*, SMPTE, White Plains, N.Y., vol. 86, no. 11, pp. 838–841, November 1980.

Crutchfield, E. B., (ed.): *NAB Engineering Handbook*, 8th ed., National Association of Broadcasters, Washington, D.C., 1993.

Crutchfield, E. B., (ed.); *NAB Engineering Handbook*, 7th ed., National Association of Broadcasters, Washington, D.C., 1988.

Favreau, M., S. Soca, J. Bajon, and M. Cattoen: "Adaptive Contrast Corrector Using Real-Time Histogram Modification," *SMPTE J.*, SMPTE, White Plains, N.Y., vol. 93, pp. 488–491, May 1984.

Fink, D. G., and D. Christiansen (eds.): *Electronics Engineers' Handbook*, 2nd ed., McGraw-Hill, New York, N.Y., pp. 20–30, 1982.

Gloeggler, Peter: "Video Pickup Devices and Systems," in *NAB Engineering Handbook*, 9th Ed., Jerry C. Whitaker (ed.), National Association of Broadcasters, Washington, D.C., 1999.

Inglis, A.F.: *Behind the Tube*, Focal Press, London, 1990.

Levitt, R.S.: *Operating Characteristics of the Plumbicon*,1968.

Mathias, H.: "Gamma and Dynamic Range Needs for an HDTV Electronic Cinematography System," *SMPTE J.*, SMPTE, White Plains, N.Y., vol. 96, pp. 840–845, September 1987.

Ogomo, M., T. Yamada, K. Ando, and E. Yamazaki: "Considerations on Required Property for HDTV Displays," *Proc. of HDTV 90 Colloquium*, vols. 1, 2B, 1990.

Philips Components: "Plumbicon Application Bulletin 43," Philips, Slatersville, R.I., January 1985.

Rao, N.V.: "Development of High-Resolution Camera Tube for 2000 line TV System," 1968.

"SMPTE Standard for Television—Broadcast Cameras: Hybrid Electrical and Fiber-Optic Connector," SMPTE 304M-1998, SMPTE, White Plains, N.Y., 1998.

"SMPTE Standard for Television—Camera Positioning Information Conveyed by Ancillary Data Packets," SMPTE 315M-1999, SMPTE, White Plains, N.Y., 1999.

"SMPTE Standard for Television—Hybrid Electrical and Fiber-Optic Camera Cable," SMPTE 311M-1998, SMPTE, White Plains, N.Y., 1998.

SPIE: *Electron Image Tubes and Image Intensifiers*, SPIE, Bellingham, Wash., vol. 1243, pp. 80–86.

Steen, R.: "CCDs vs. Camera Tubes: A Comparison," *Broadcast Engineering*, Intertec Publishing, Overland Park, Kan., May,1991.

Stupp, E.H.: Physical Properties of the Plumbicon, 1968.

Tanaka, H., and L. J. Thorpe: "The Sony PCL HDVS Production Facility," *SMPTE J.*, SMPTE, White Plains, N.Y., vol. 100, pp. 404–415, June 1991.

Thorpe, L. J., E. Tamura, and T. Iwasaki: "New Advances in CCD Imaging," *SMPTE J.*, SMPTE, White Plains, N.Y., vol. 97, pp. 378–387, May 1988.

Thorpe, L., et. al.: "New High Resolution CCD Imager," *NAB Engineering Conference Proceedings*, National Association of Broadcasters, Washington, D.C., pp. 334– 345, 1988.

Thorpe, Laurence J.: "HDTV and Film—Digitization and Extended Dynamic Range," *133rd SMPTE Technical Conference*, Paper no. 133-100, SMPTE, White Plains, N.Y., October 1991.

Thorpe, Laurence J.: "Television Cameras," in *Electronic Engineers' Handbook*, 4th ed., Donald Christiansen (ed.), McGraw-Hill, New York, N.Y., pp. 24.58–24.74, 1997.

Thorpe, Laurence J.: "The HDTV Camcorder and the March to Marketplace Reality," *SMPTE Journal*, SMPTE, White Plains, N.Y., pp. 164–177, March 1998.

15.3g Monochrome and Color Image Display Devices

Aiken, J. A.: "A Thin Cathode Ray Tube," *Proc. IRE*, vol. 45, pg. 1599, 1957.

Allison, J.: *Electronic Engineering Semiconductors and Devices*, 2nd ed., McGraw-Hill, London, pg. 308–309, 1990.

Amm, D. T., and R. W. Corrigan: "Optical Performance of the Grating Light Valve Technology," Projection Displays V Symposium, SPIE Proceedings, SPIE, San Jose, Calif., vol. EI 3634-10, February 1999.

Amm, D. T., and R.W. Corrigan: "Grating Light Valve Technology: Update and Novel Applications," SID Symposium—Anaheim, SID, San Jose, Calif., May 1998.

Ashizaki, S., Y. Suzuki, K. Mitsuda, and H. Omae: "Direct-View and Projection CRTs for HDTV," *IEEE Transactions on Consumer Electronics*, vol. 34, no. 1, pp. 91–98, February 1988.

Barbin, R., and R. Hughes: "New Color Picture Tube System for Portable TV Receivers," *IEEE Trans. Broadcast TV Receivers*, vol. BTR-18, no. 3, pp. 193–200, August 1972.

Barkow, W. H., and J. Gross: "The RCA Large Screen 110° Precision In-Line System," ST-5015, RCA Entertainment, Lancaster, Pa.

Bates, W., P. Gelinas, and P. Recuay: "Light Valve Projection System for HDTV," *Proceedings of the ITS*, International Television Symposium, Montreux, Switzerland, 1991.

Baur, G.: *The Physics and Chemistry of Liquid Crystal Devices*, G. J. Sprokel (ed.), Plenum, New York, N.Y., pg. 62, 1980.

Bauman, E.: "The Fischer Large-Screen Projection System," *SMPTE Journal*, SMPTE, White Plains, N.Y., vol. 60, pg. 351, 1953.

Benson, K. B., and D. G. Fink: *HDTV: Advanced Television for the 1990s*, McGraw-Hill, New York, N.Y., 1990.

Blacker, A., et al.: "A New Form of Extended Field Lens for Use in Color Television Picture Tube Guns," *IEEE Trans. Consumer Electronics*, pp. 238–246, August 1966.

Blaha, Richard J.: "Degaussing Circuits for Color TV Receivers," *IEEE Trans. Broadcast TV Receivers*, vol. BTR-18, no. 1, pp. 7–10, February 1972.

Blaha, Richard J.: "Large Screen Display Technology Assessment for Military Applications," *Large-Screen Projection Displays II*, William P. Bleha, Jr., (ed.), Proc. SPIE 1255, SPIE, Bellingham, Wash., pp. 80–92, 1990.

Bleha, Wiliam. P.: "Image Light Amplifier (ILA) Technology for Large-Screen Projection," *SMPTE Journal*, SMPTE, White Plains, N.Y., pp. 710–717, October 1997.

Bleha, William P., Jr., (ed.): *Large-Screen Projection Displays II*, Proc. SPIE 1255, SPIE, Bellingham, Wash., 1990.

Burgmans, A., et. al.: *Information Display*, pg. 14, April/May 1998.

Buzak, Thomas S.: "Recent Advances in PALC Technology," in *Proceedings of the 18th International Display Research Conference*, Society for Information Display, San Jose, Calif., Asia Display '98, pp. 273–276, 1998.

Carpenter, C.: et al., "An Analysis of Focusing and Deflection in the Post-Deflection Focus Color Kinescope," *IRE Trans. Electron Devices*, vol. 2, pp. 1–7, 1955.

Casteloano, Joseph A.: *Handbook of Display Technology*, Academic, New York, N.Y., 1992.

Chang, I.: "Recent Advances in Display Technologies," *Proc. SID*, Society for Information Display, San Jose, Calif., vol. 21, no. 2, pg. 45, 1980.

Chen, H., and R. Hughes: "A High Performance Color CRT Gun with an Asymmetrical Beam Forming Region," *IEEE Trans. Consumer Electronics*, vol. CE-26, pp. 459–465, August, 1980.

Chen, K. C., W. Y. Ho and C. H. Tseng: "Invar Mask for Color Cathode Ray Tubes," in *Display Technologies*, Shu-Hsia Chen and Shin-Tson Wu (eds.), Proc. SPIE 1815, SPIE, Bellingham, Wash., pp.42–48, 1992.

Cheng, J. B., and Q. H. Wang: "Studies on YAG Phosphor Screen for HDTV Projector," *Proc SPIE 2892*, SPIE, Bellingham, Wash., pg. 36, 1996.

Cheng, Jia-Shyong, et. al.: "The Optimum Design of LCD Parameters in Projection and Direct View Applications," *Display Technologies*, Shu-Hsia Chen and Shin-Tson Wu (eds.), Proc. SPIE 1815, SPIE, Bellingham, Wash., pp. 69–80, 1992.

Clapp, R., et al.: "A New Beam Indexing Color Television Display System," *Proc. IRE*, vol. 44, no. 9, pp. 1108–1114, September 1956.

Cohen, C.: "Sony's Pocket TV Slims Down CRT Technology," *Electronics*, pg. 81, February 10, 1982.

Corrigan, R. W., B. R. Lang, D.A. LeHoty, and P.A. Alioshin: "An Alternative Architecture for High Performance Display," Silicon Light Machines, Sunnyvale, Calif., 1999. Presented at the 141st SMPTE Technical Conference (paper 141-25).

Credelle, T. L., et al.: "Cathodoluminescent Flat Panel TV Using Electron Beam Guides," *SID Int. Symp. Digest*, Society for Information Display, San Jose, Calif., pg. 26, 1980.

Credelle, T. L., et al.: *Japan Display '83*, pg. 26, 1983.

Credelle, T. L.: "Modular Flat Display Device with Beam Convergence," U.S. Patent 4,131,823.

"CRT Control Grid Having Orthogonal Openings on Opposite Sides," U.S. Patent 4,242,613, Dec. 30, 1980.

"CRTs: Glossary of Terms and Definitions," Publication TEP92, Electronic Industries Association, Washington, 1975.

Davis, C., and D. Say: "High Performance Guns for Color TV—A Comparison of Recent Designs," *IEEE Trans. Consumer Electronics*, vol. CE-25, August 1979.

Donofrio, R.: "Image Sharpness of a Color Picture Tube by MTF Techniques," *IEEE Trans. Broadcast TV Receivers*, vol. BTR-18, no. 1, pp. 1–6, February 1972.

Dressler, R.: "The PDF Chromatron—A Single or Multi-Gun CRT," *Proc. IRE*, vol. 41, no. 7, July 1953.

Dworsky, Lawrence N., and Babu R. Chalamala: *Field-Emission Displays*, Society for Information Display, San Jose, Calif., pp. F-1/3–F1/66, 1999.

Eccles, D. A., and Y. Zhang: "Digital-Television Signal Processing and Display Technology," *SID 99 Digest*, Society for Information Display, San Jose, Calif., pp. 108–111, 1999.

"Electron Gun with Astigmatic Flare-Reducing Beam Forming Region," U.S. Patent 4,234,814, Nov. 18, 1980.

Fink, D. G., et al.: "The Future of High-Definition Television," *SMPTE Journal*, SMPTE, White Plains, N.Y., vol. 89, February/March 1980.

Fink, D. G.: *Color Television Standards*, McGraw-Hill, New York, N.Y., 1986.

Fink, Donald, (ed.): *Television Engineering Handbook*, McGraw-Hill, New York, N.Y., 1957.

Fink, Donald, and Donald Christiansen (eds.): *Electronics Engineers Handbook*, 3rd ed., McGraw-Hill, New York, N.Y., 1989.

Fiore, J., and S. Kaplin: "A Second Generation Color Tube Providing More Than Twice the Brightness and Improved Contrast," *IEEE Trans. Consumer Electronics*, vol. CE-28, no. 1, pp. 65–73, February 1982.

Flechsig, W.: "CRT for the Production of Multicolored Pictures on a Luminescent Screen," French Patent 866,065, 1939.

Florence, J., and L. Yoder: "Display System Architectures for Digital Micromirror Device (DMD) Based Projectors," *Proc. SPIE*, SPIE, Bellingham, Wash., vol. 2650, Projection Displays II, pp. 193–208, 1996.

Fritz, Victor J.: "Full-Color Liquid Crystal Light Valve Projector for Shipboard Use," *Large Screen Projection Displays II*, William P. Bleha, Jr. (ed.), Proc. SPIE 1255, SPIE, Bellingham, Wash., pp. 59–68, 1990.

Fujio, T., J. Ishida, T. Komoto, and T. Nishizawa: "High-Definition Television Systems—Signal Standards and Transmission," *SMPTE Journal*, SMPTE, White Plains, N.Y., vol. 89, August 1980.

Gerhard-Multhaupt, R.: "Light Valve Technologies for HDTV Projection Displays: A Summary," *Proceedings of the ITS*, International Television Symposium, Montreux, Switzerland, 1991.

Glenn, W. E., C. E. Holton, G. J. Dixon, and P. J. Bos: "High-Efficiency Light Valve Projectors and High-Efficiency Laser Light Sources," *SMPTE Journal*, SMPTE, White Plains, N.Y., pp. 210–216, April 1997.

Glenn, William E.: "Large Screen Displays for Consumer and Theater Use," *Large Screen Projection Displays II*, William P. Bleha, Jr., (ed.), Proc. SPIE 1255, SPIE, Bellingham, Wash., pp. 36–43, 1990.

Glenn, William. E.: "Principles of Simultaneous Color Projection Using Fluid Deformation," *SMPTE Journal*, SMPTE, White Plains, N.Y., vol. 79, pg. 788, 1970.

Godfrey, R., et al.: "Development of the Permachrome Color Picture Tube," *IEEE Trans. Broadcast TV Receivers*, vol. BTR-14, no. 1, 1968.

Goede, Walter F: "Electronic Information Display Perspective," *SID Seminar Lecture Notes*, Society for Information Display, San Jose, Calif., vol. 1, pp. M-1/3–M1/49, May 17, 1999.

Good, W.: "Projection Television," *IEEE Trans.*, vol. CE-21, no. 3, pp. 206–212, August 1975.

Good, W.: "Recent Advances in the Single-Gun Color Television Light-Valve Projector," *Soc. Photo-Optical Instrumentation Engrs.*, vol. 59, 1975.

Gove, R. J., V. Markandey, S. Marshall, D. Doherty, G. Sextro, and M. DuVal: "High-Definition Display System Based on Digital Micromirror Device," International Workshop on HDTV (HDTV '94), International Institute for Communications, Turin, Italy (October 1994).

Gow, J., and R. Door: "Compatible Color Picture Presentation with the Single-Gun Tri Color Chromatron," *Proc. IRE*, vol. 42, no. 1, pp. 308–314, January 1954.

Gretag AG: "What You May Want to Know about the Technique of Eidophor," Regensdorf, Switzerland.

Grinberg, J. et al.: "Photoactivated Birefringent Liquid-Crystal Light Valve for Color Symbology Display," *IEEE Trans. Electron Devices*, vol. ED-22, no. 9, pp. 775–783, September 1975.

Hardy, A. C., and F. H. Perrin: *The Principles of Optics*, McGraw-Hill, New York, N.Y., 1932.

Hasker, J.: "Astigmatic Electron Gun for the Beam Indexing Color TV Display," *IEEE Trans. Electron Devices*, vol. ED-18, no. 9, pg. 703, September 1971.

Hayashi, M., N. Yamada, and B. Sastra: "Development of a 42-in. High-Definition Plasma-Addressed LCD," in *SID International Symposium Digest of Technical Papers*, Society for Information Display, San Jose, Calif., pp.280–284, 1999.

Herold, E.: "A History of Color TV Displays," *Proc. IEEE*, vol. 64, no. 9, pp. 1331–1337, September 1976.

Hockenbrock, Richard: "New Technology Advances for Brighter Color CRT Displays," *Display System Optics II*, Harry M. Assenheim (ed.), Proc. SPIE 1117, SPIE, Bellingham, Wash., pp. 219-226, 1989.

Hornbeck, Larry J.: "Digital Light Processing for High-Brightness, High-Resolution Applications," *Projection Displays III*, Electronic Imaging '97 Conference, SPIE, Bellingham, Wash., February 1997.

Hoskoshi, K., et al.: "A New Approach to a High Performance Electron Gun Design for Color Picture Tubes," 1980 IEEE Chicago Spring Conf. Consumer Electronics.

Howe, R., and B. Welham: "Developments in Plastic Optics for Projection Television Systems," *IEEE Trans*., vol. CE-26, no. 1, pp. 44–53, February 1980.

Hu, C., Y. Yu and K. Wang: "Antiglare/Antistatic Coatings for Cathode Ray Tube Based on Polymer System," in *Display Technologies*, Shu-Hsia Chen and Shin-Tson Wu (eds)., Proc. SPIE 1815, SPIE, Bellingham, Wash., pp.42–48, 1992.

Hubel, David H.: *Eye, Brain and Vision*, Scientific American Library, New York, N.Y., 1988.

Hutter, Rudolph G. E.: "The Deflection of Electron Beams," *Advances in Image Pickup and Display*, B. Kazen (ed.), vol. 1, Academic Press, New York, pp. 212-215, 1974.

Ilcisen, K. J., et. al.: *Eurodisplay '96*, pg. 595, 1996.

Itah, N., et al.: "New Color Video Projection System with Glass Optics and Three Primary Color Tubes for Consumer Use," *IEEE Trans. Consumer Electronics*, vol. CE-25, no. 4, pp. 497–503, August 1979.

Johnson, A.: "Color Tubes for Data Display—A System Study," Philips ECG, Electronic Tube Division.

Judd, D. B.: "The 1931 C.I.E. Standard Observer and Coordinate System for Colorimetry," *Journal of the Optical Society of America*, vol. 23, 1933.

Kikuchi, M., et al.: "A New Coolant-Sealed CRT for Projection Color TV," *IEEE Trans*., vol. CE-27, IEEE, New York, no. 3, pp. 478-485, August 1981.

Kingslake, Rudolf (ed.): *Applied Optics and Optical Engineering*, vol. 1, Chap. 6, Academic, New York, N.Y., 1965.

"Kodak Filters for Scientific and Technical Uses," Eastman Kodak Co., Rochester, N.Y.

Kurahashi, K., et al.: "An Outdoor Screen Color Display System," *SID Int. Symp. Digest 7*, Technical Papers, vol. XII, Society for Information Display, San Jose, Calif., pp. 132– 133, April 1981.

Lakatos, A. I., and R. F. Bergen: "Projection Display Using an Amorphous-Se-Type Ruticon Light Valve," *IEEE Trans. Electron Devices*, vol. ED-24, no. 7, pp. 930–934, July 1977.

Law, H.: "A Three-Gun Shadowmask Color Kinescope," *Proc. IRE*, vol. 39, pp. 1186–1194, October 1951.

Lim, G. S., et. al.: "New Driving Method for Improvement of Picture Quality in 40-in. AC PDP," in *Asia Display '98—Proceedings of the 18th International Display Research Conference*, Society for Information Display, San Jose, Calif., pp. 591–594, 1998.

Lucchesi, B., and M. Carpenter: "Pictures of Deflected Electron Spots from a Computer," *IEEE Trans. Consumer Electronics*, vol. CE-25, no. 4, pp. 468–474, 1979.

Luxenberg, H., and R. Kuehn: *Display Systems Engineering*, McGraw-Hill, New York, N.Y., 1968.

Maeda, M.: *Japan Display '83*, pg. 2, 1971.

Maseo, Imai, et. al.: "High-Brightness Liquid Crystal Light Valve Projector Using a New Polarization Converter," *Large Screen Projection Displays II*, William P. Bleha, Jr., (ed.), Proc. SPIE 1255, SPIE, Bellingham, Wash., pp. 52–58, 1990.

Masterson, W., and R. Barbin: "Designing Out the Problems of Wide-Angle Color TV Tube," *Electronics*, pp. 60–63, April 26, 1971.

Masuda, T., et. al: *Conference Record—International Display Resolution Conference*, pg. 357, 1994.

McKechnie, S.: Philips Laboratories (NA) report, 1981, unpublished.

Mears, N., "Method and Apparatus for Producing Perforated Metal Webs," U.S. Patent 2,762,149, 1956.

Mikoshiba, Shigeo: *Color Plasma Displays*, Society for Information Display, San Jose, Calif., pp M-4/3–M-4/68, 1999.

Mitsuhashi, Tetsuo: "HDTV and Large Screen Display," *Large-Screen Projection Displays II*, William P. Bleha, Jr., (ed.), Proc. SPIE 1255, SPIE, Bellingham, Wash., pp. 2– 12, 1990.

Mokhoff, N.: "A Step Toward Perfect Resolution," *IEEE Spectrum*, IEEE, New York, N.Y., vol. 18, no. 7, pp. 56–58, July 1981.

Morizono, M.: "Technological Trends in High Resolution Displays Including HDTV," *SID International Symposium Digest*, paper 3.1, Society for Information Display, San Jose, Calif., May 1990.

Morrell, A., et al.: *Color Picture Tubes*, Academic Press, New York, pp. 91-98, 1974.

Morrell, A.: "Color Picture Tube Design Trends," *Proc. SID*, Society for Information Display, San Jose, Calif., vol. 22, no. 1, pp. 3–9, 1981.

Morris, James E.: "Liquid-Crystal Displays," in *The Electrical Engineering Handbook*, Richard C. Dorf (ed.), CRC Press, Boca Raton, Fla., 1993.

Moss, H.: *Narrow Angle Electron Guns and Cathode Ray Tubes*, Academic, New York, N.Y., 1968.

Na, Young-Sun, et. al.: "A New Data Driver Circuit for Field Emission Display," in *Asia Display '98—Proceedings of the 18th International Display Research Conference*, Society for Information Display, San Jose, Calif., pp. 137–140, 1998.

Nasibov, A., et al.: "Electron-Beam Tube with a Laser Screen," *Sov. J. Quant. Electron.*, vol. 4, no. 3, pp. 296–300, September 1974.

Nishio, T., and K. Amemiya: "High-Luminance and High-Definition 50-in.-Diagonal Co-Planar Color PDPs with T-Shaped Electrodes," in *SID International Symposium Digest of Technical Papers*, Society for Information Display, San Jose, Calif., pp.268–272, 1999.

Oess, F.: "CRT Considerations for Raster Dot Alpha Numeric Presentations," *Proc. SID*, Society for Information Display, San Jose, Calif., vol. 20, no. 2, pp. 81–88, second quarter, 1979.

Ohkoshi, A., et al.: "A New 30V" Beam Index Color Cathode Ray Tube," *IEEE Trans. Consumer Electronics*, vol. CE-27, p. 433, August 1981.

Palac, K.: Method for Manufacturing a Color CRT Using Mask and Screen Masters, U.S. Patent 3,989,524, 1976.

Pease, Richard W.: "An Overview of Technology for Large Wall Screen Projection Using Lasers as a Light Source," *Large Screen Projection Displays II*, William P. Bleha, Jr., (ed.), Proc. SPIE 1255, SPIE, Bellingham, Wash., pp. 93–103, 1990.

Pfahnl, A.: "Aging of Electronic Phosphors in Cathode Ray Tubes," *Advances in Electron Tube Techniques*, Pergamon, New York, N.Y., pp. 204–208.

Phillips, Thomas E., et. al.: "1280 × 1024 Video Rate Laser-Addressed Liquid Crystal Light Valve Color Projection Display," *Optical Engineering*, Society of Photo-Optical Instrumentation Engineers, vol. 31, no. 11, pp. 2300–2312, November 1992.

Pitts, K., and N. Hurst: "How Much do People Prefer Widescreen (16 × 9) to Standard NTSC (4 × 3)?," *IEEE Transactions on Consumer Electronics*, vol. 35, no. 3, pp. 160–169, August 1989.

Pointer, R. M.: "The Gamut of Real Surface Colors," *Color Res. App.*, vol. 5, 1945.

Poorter, T., and F. W. deVrijer: "The Projection of Color Television Pictures," *SMPTE Journal*, SMPTE, White Plains, N.Y., vol. 68, pg. 141, 1959.

"Recommended Practice for Measurement of X-Radiation from Direct View TV Picture Tubes," Publication TEP 164, Electronics Industries Association, Washington, D.C., 1981.

Robbins, J., and D. Mackey: "Moire Pattern in Color TV," *IEEE Trans. Consumer Electronics*, vol. CE-28, no. 1, pp. 44–55, February 1982.

Robertson, A.: "Projection Television—1 Review of Practice," *Wireless World*, vol. 82, no. 1489, pp. 47–52, September 1976.

Robinder, R., D. Bates, P. Green: "A High Brightness Shadow Mask Color CRT for Cockpit Displays," *SID Digest*, Society for Information Display, vol. 14, pp. 72-73, 1983.

Rublack, W.: "In-Line Plural Beam CRT with an Aspherical Mask," U.S. Patent 3,435,668, 1969.

Sakamoto, Y.: and E. Miyazaki, *Japan Display '83*, pg. 30, 1983.

Sarma, Kalluri R.: "Active-Matrix LCDs," in *Seminar Lecture Notes*, Society for Information Display, San Jose, Calif., vol. 1, pp. M3/3–M3/45, 1999.

Say, D.: "Picture Tube Spot Analysis Using Direct Photography," *IEEE Trans. Consumer Electronics*, vol. CE-23, pp. 32–37, February 1977.

Say, D.: "The High Voltage Bipotential Approach to Enhanced Color Tube Performance," *IEEE Trans. Consumer Electronics*, vol. CE-24, no. 1, pg. 75, February 1978.

Schiecke, K.: "Projection Television: Correcting Distortions," *IEEE Spectrum*, IEEE, New York, N.Y., vol. 18, no. 11, pp. 40–45, November 1981.

Schwartz, J.: "Electron Beam Cathodoluminescent Panel Display," U.S. Patent 4,137,486.

Sears, F. W.: *Principles of Physics, III, Optics*, Addison-Wesley, Cambridge, Mass., 1946.

Sextro, G., I. Ballew, and J. Lwai: "High-Definition Projection System Using DMD Display Technology," *SID 95 Digest*, Society for Information Display, San Jose, Calif., pp. 70–73, 1995.

Sherr, S.: *Electronic Displays*, Wiley, New York, N.Y., 1979.

Sherr, S.: *Fundamentals of Display System Design*, Wiley-Interscience, New York, N.Y., 1970.

Sinclair, C.: "Small Flat Cathode Ray Tube," *SID Digest*, Society for Information Display, San Jose, Calif., pg. 138, 1981.

Stanley, T.: "Flat Cathode Ray Tube," U.S. Patent 4,031,427.

Swartz, J.: "Beam Index Tube Technology," *SID Proceedings*, Society for Information Display, San Jose, Calif., vol. 20, no. 2, p. 45, 1979.

Takeuchi, Kazuhiko, et. al.: "A 750-TV-Line Resolution Projector using 1.5 Megapixel a-Si TFT LC Modules," *SID 91 Digest*, Society for Information Display, San Jose, Calif., pp. 415–418, 1991.

Taneda, T., et al.: "A 1125-Scanning Line Laser Color TV Display," *SID 1973 Symp. Digest Technical Papers*, Society for Information Display, San Jose, Calif., vol. IV, pp. 86– 87, May 1973.

Tomioka, M., and Y. Hayshi: "Liquid Crystal Projection Display for HDTV," *Proceedings of the International Television Symposium*, ITS, Montreux, Switzerland, 1991.

Tong, Hua-Sou: "HDTV Display—A CRT Approach," *Display Technologies*, Shu-Hsia Chen and Shin-Tson Wu (eds.), Proc. SPIE 1815, SPIE, Bellingham, Wash., pp. 2–8, 1992.

Tsuruta, Masahiko, and Neil Neubert: "An Advanced High Resolution, High Brightness LCD Color Video Projector," *SMPTE Journal*, SMPTE, White Plains, N.Y., pp. 399– 403, June 1992.

Uba, T., K. Omae, R. Ashiya, and K. Saita: "16:9 Aspect Ratio 38V-High Resolution Trinitron for HDTV," *IEEE Transactions on Consumer Electronics*, vol. 34, no. 1., pp. 85–89, February 1988.

Van, M. W., and J. Von Esdonk: "A High Luminance High-Resolution Cathode-Ray Tube for Special Purposes," *IEEE Trans Electron Dev.*, IEEE, New York, N.Y., ED-30, pg. 193, 1983.

Wang, Q. H., J. B. Cheng, and Z. L. Lin: "A New YAG Phosphor Screen for Projection CRT," *Electron Lett.*, vol. 34, no. 14, pg. 1420, 1998.

Wang, Qionghua, Jianbo Cheng, Zulun Lin, and Gang Yang: "A High-Luminance and High-Resolution CRT for Projection HDTV Display," *Journal of the SID*, Society for Information Display, San Jose, Calif., vol. 7, no. 3, pg 183–186, 1999.

Wang, S., et al.: "Spectral and Spatial Distribution of X-Rays from Color Television Receivers," *Proc. Conf. Detection and Measurement of X-radiation from Color Television Receivers*, Washington, D.C., pp. 53–72, March 28–29, 1968.

Weber, Larry F.: "Plasma Displays," in *The Electrical Engineering Handbook*, Richard C. Dorf (ed.), CRC Press, Boca Raton, Fla., 1786–1798, 1993.

Wedding, Donald K., Sr.: "Large Area Full Color ac Plasma Display Monitor," *Large Screen Projection Displays II*, William P. Bleha, Jr., (ed.), Proc. SPIE 1255, SPIE, Bellingham, Wash., pp. 29–35, 1990.

Williams, Charles S., and Becklund, Orville A.: *Optics: A Short Course for Engineers and Scientists*, Wiley Interscience, New York, N.Y., 1972.

Wilson, J., and J. F. B. Hawkes: *Optoelectronics: An Introduction*, Prentice-Hall, London, pg. 145, 1989.

Woodhead, A., et al.: *1982 SID Digest*, Society for Information Display, San Jose, Calif., pg. 206, 1982.

"X-Radiation Measurement Procedures for Projection Tubes," TEPAC Publication 102, Electronic Industries Association, Washington, D. C.

Yamamoto, Y., Y. Nagaoka, Y. Nakajima, and T. Murao: "Super-compact Projection Lenses for Projection Television," *IEEE Transactions on Consumer Electronics*, IEEE, New York, N.Y., August 1986.

Yoshida, S., et al.: "25-V Inch 114-Degree Trinitron Color Picture Tube and Associated New Development," *Trans. BTR*, pp. 193-200, August 1974.

Yoshida, S., et al.: "A Wide Deflection Angle (114°) Trinitron Color Picture Tube," *IEEE Trans. Electron Devices*, vol. 19, no. 4, pp. 231–238, 1973.

Yoshida. S.: et al., "The Trinitron—A New Color Tube," *IEEE Trans. Consumer Electronics*, vol. CE-28, no. 1, pp. 56–64, February 1982.

Yoshizawa, T., S. Hatakeyama, A. Ueno, M. Tsukahara, K. Matsumi, K. Hirota: "A 61-in High-Definition Projection TV for the ATSC Standard," *SID 99 Digest*, Society for Information Display, San Jose, Calif., pp. 112–115, 1999.

Younse, J. M.: "Projection Display Systems Based on the Digital Micromirror Device (DMD)," *SPIE Conference on Microelectronic Structures and Microelectromechanical Devices for Optical Processing and Multimedia Applications*, Austin, Tex., SPIE Proceedings, SPIE, Bellingham, Wash., vol. 2641, pp. 64–75, Oct. 24, 1995.

15.3h Video Recording Systems

Ampex: *General Information, Volume 1*, Training Department, Ampex Corporation, Redwood City, Calif., 1983.

Anderson, D: "Fibre Channel-Arbitrated Loop: The Preferred Path to Higher I/O Performance, Flexibility in Design," Seagate Technology Paper #MN-24, Seagate, Scotts Valley, Calif., 1995.

Bate, G.: "Recent Developments in Magnetic Recording Materials," *J. Appl. Phys*. pg. 2447, 1981.

Bertram, H. N.: "Long Wavelength ac Bias Recording Theory," *IEEE Trans. Magnetics*, vol. MAG-10, pp. 1039–1048, 1974.

Bozorth, Richard M.: *Ferromagnetism*, Van Nostrand, Princeton, N.J., 1961.

Chikazumi, Soshin: *Physics of Magnetism*, Wiley, New York, N.Y., 1964.

Epstein, Steve: "Video Recording Principles," in *NAB Engineering Handbook*, 9th ed., Jerry C. Whitaker (ed.), National Association of Broadcasters, Washington, D.C., pp. 923–935, 1999.

Epstein, Steve: "Videotape Storage Systems," in *The Electronics Handbook*, Jerry C. Whitaker (ed.), CRC Press, Boca Raton, Fla., pp. 1412–1433, 1996.

Felix, Michael O.: "Video Recording Systems," in *Electronics Engineers' Handbook*, 4th ed., Donald Christiansen (ed.), McGraw-Hill, New York, N.Y., pp. 24.81–24.92, 1996.

Fink, D. G., and D. Christiansen (eds.): *Electronic Engineers' Handbook*, 2nd ed., McGraw-Hill, New York, N.Y., 1982.

Ginsburg, C. P.: *The Birth of Videotape Recording*, Ampex Corporation, Redwood City, Calif., 1981. Reprinted from notes of paper delivered to Society of Motion Picture and Television Engineers, October 5, 1957.

Goldberg, Thomas: "New Storage Technology," *Proceedings of the Advanced Television Summit*, Intertec Publishing, Overland Park, Kan., 1996.

Grega, Joe: "Magnetic and Optical Recording Media," in *NAB Engineering Handbook*, 9th ed., Jerry C. Whitaker (ed.), National Association of Broadcasters, Washington, D.C., pp. 893–906, 1999.

Hammer, P.: "The Birth of the VTR," *Broadcast Engineering*, Intertec Publishing, Overland Park, Kan., vol. 28, no. 6, pp. 158–164, 1986.

Hawthorne, J. M., and C. J. Hefielinger: "Polyester Films," in *Encyclopedia of Polymer Science and Technology*, N. M. Bikales (ed.), vol. 11, Wiley, New York, N.Y., pg. 42, 1969.

Heyn, T.: "The RAID Advantage," Seagate Technology Paper, Seagate, Scotts Valley, Calif., 1995.

Jorgensen, F.: *The Complete Handbook of Magnetic Recording*, Tab Books, Blue Ridge Summit, Pa., 1980.

Kalil, F. (ed): *Magnetic Tape Recording for the Eighties*, NASA References Publication 1975, April 1982.

Kraus, John D.: *Electromagnetics*, McGraw-Hill, New York, N.Y., 1953.

Lehtinen, Rick, "Editing Systems," *Broadcast Engineering*, Intertec Publishing, Overland Park. Kan., pp. 26–36, May 1996.

Lueck, L. B. (ed): *Symposium Proceedings Textbook*, Symposium on Magnetic Media Manufacturing Methods, Honolulu, May 25–27, 1983.

McConathy, Charles F.: "A Digital Video Disk Array Primer," *SMPTE Journal*, SMPTE, New York, N.Y., pp. 220–223, April 1998.

McKnight, John G.: "Erasure of Magnetic Tape," *J. Audio Eng. Soc.*, Audio Engineering Society, New York, N.Y., vol. 11, no.3, pp. 223–232, 1963.

Mee, C. D., and E. D. Daniel: *Magnetic Recording Handbook*, McGraw-Hill, New York, N.Y., 1990.

Nylen, P., and E. Sunderland: *Modern Surface Coatings*, Interscience Publishers Division, Wiley, London, 1965.

Pear, C. B.: *Magnetic Recording in Science and Industry*, Reinhold, New York, N.Y., 1967.

Perry, R. H., and A. A. Nishimura: "Magnetic Tape," in *Encyclopedia of Chemical Technology*, 3d ed., Kirk Othmer (ed.), vol. 14, Wiley, New York, N.Y., pp. 732–753, 1981.

Plank, Bob: "Video Disk and Server Operation," *International Broadcast Engineer*, September 1995.

Robin, Michael, and Michel Poulin: "Multimedia and Television," in *Digital Television Fundamentals*, McGraw-Hill, New York, N.Y., pp. 455–488, 1997.

Roizen, J.: *Magnetic Video Recording Techniques*, Ampex Training Manual—General Information, vol. 1, Ampex Corporation, Redwood City, Calif., 1964.

Roters, Herbert C.: *Electromagnetic Devices*, Wiley, New York, N.Y., 1961.

Sanders, M.: "Technology Report: AST," *Video Systems*, Intertec Publishing, Overland Park, Kan., April 1980.

Sharrock, Michael P., and D. P. Stubs: "Perpendicular Magnetic Recording Technology: A Review," *SMPTE J.*, SMPTE, White Plains, N.Y., vol. 93, pp. 1127–1133, December 1984.

Smit, J., and H. P. J. Wijn: *Ferrite*, Wiley, New York, N.Y., 1959.

Tochihara, S.: "Magnetic Coatings and Their Applications in Japan," *Prog. Organic Coatings*, vol. 10, pp. 195–204, 1982.

Tyson, H: "Barracuda and Elite: Disk Drive Storage for Professional Audio/Video," Seagate Technology Paper #SV-25, Seagate, Scotts Valley, Calif., 1995.

Whitaker, Jerry C.: "Data Storage Systems," in *The Electronics Handbook*, Jerry C. Whitaker (ed.), CRC Press, Boca Raton, Fla., pp. 1445–1459, 1996.

Whitaker, Jerry C.: "Tape Recording Technology," *Broadcast Engineering*, Intertec Publishing, Overland Park, Kan., vol. 31, no. 11, pp. 78–108, 1989.

15.3i Video Production Standards, Equipment, and System Design

Ajemian, Ronald G.: "Fiber Optic Connector Considerations for Professional Audio," *Journal of the Audio Engineering Society*, Audio Engineering Society, New York, N.Y., June 1992.

ATSC: "Implementation Subcommittee Report on Findings," Draft Version 0.4, ATSC, Washington, D.C., September 21, 1998.

Baldwin, M. Jr.: "The Subjective Sharpness of Simulated Television Images," *Proceedings of the IRE*, vol. 28, July 1940.

Belton, J.: "The Development of the CinemaScope by Twentieth Century Fox," *SMPTE Journal*, vol. 97, SMPTE, White Plains, N.Y., September 1988.

Benson, K. B., and D. G. Fink: *HDTV: Advanced Television for the 1990s*, McGraw-Hill, New York, N.Y., 1990.

Course notes, "DTV Express," PBS/Harris, Alexandria, Va., 1998.

Crutchfield, E. B.: *NAB Engineering Handbook*, 8th ed., National Association of Broadcasters, Washington, D.C., 1992.

DeSantis, Gene, Jerry C. Whitaker, and C. Robert Paulson: *Interconnecting Electronic Systems*, CRC Press, Boca Raton, Fla., 1992.

Dick, Bradley: "Building Fiber-Optic Transmission Systems," *Broadcast Engineering*, Intertec Publishing, Overland Park, Kan., November 1991 and December 1991.

Fink, D. G.: "Perspectives on Television: The Role Played by the Two NTSCs in Preparing Television Service for the American Public," *Proceedings of the IEEE*, vol. 64, IEEE, New York, September 1976.

Fink, D. G: *Color Television Standards*, McGraw-Hill, New York, 1986.

Fink, D. G, et. al.: "The Future of High Definition Television," *SMPTE Journal*, vol. 9, SMPTE, White Plains, N.Y., February/March 1980.

Fujio, T., J. Ishida, T. Komoto, and T. Nishizawa: "High-Definition Television Systems—Signal Standards and Transmission," *SMPTE Journal*, vol. 89, SMPTE, White Plains, N.Y., August 1980.

Hamasaki, Kimio: "How to Handle Sound with Large Screen," *Proceedings of the ITS*, International Television Symposium, Montreux, Switzerland, 1991.

Hopkins, Robert: "What We've Learned from the DTV Experience," *Proceedings of Digital Television '98*, Intertec Publishing, Overland Park, Kan., 1998.

Holman, Tomlinson: "Psychoacoustics of Multi-Channel Sound Systems for Television," *Proceedings of HDTV World*, National Association of Broadcasters, Washington, D.C., 1992.

Holman, Tomlinson: "The Impact of Multi-Channel Sound on Conversion to ATV," *Perspectives on Wide Screen and HDTV Production*, National Association of Broadcasters, Washington, D.C., 1995.

Hubel, David H.: *Eye, Brain and Vision*, Scientific American Library, New York, 1988.

Judd, D. B.: "The 1931 C.I.E. Standard Observer and Coordinate System for Colorimetry," *Journal of the Optical Society of America*, vol. 23, 1933.

Keller, Thomas B.: "Proposal for Advanced HDTV Audio," *1991 HDTV World Conference Proceedings*, National Association of Broadcasters, Washington, D.C., April 1991.

Kelly, R. D., A. V. Bedbord, and M. Trainer: "Scanning Sequence and Repetition of Television Images," *Proceedings of the IRE*, vol. 24, April 1936.

Kelly, K. L.: "Color Designation of Lights," *Journal of the Optical Society of America*, vol. 33, 1943.

Lagadec, Roger, Ph.D.: "Audio for Television: Digital Sound in Production and Transmission," *Proceedings of the ITS*, International Television Symposium, Montreux, Switzerland, 1991.

Leathers, David: "Production Considerations for DTV," in *NAB Engineering Handbook*, 9th ed., Jerry C. Whitaker (ed.), National Association of Broadcasters, Washington, D.C., pp. 1067–1072, 1999.

Mendrala, Jim: "Mastering at 24P," *Broadcast Engineering*, Intertec Publishing, Overland Park, Kan., pp. 92–94, February 1999.

Miller, Howard: "Options in Advanced Television Broadcasting in North America," *Proceedings of the ITS*, International Television Symposium, Montreux, Switzerland, 1991.

Pearson, Eric: How to Specify and Choose Fiber-Optic Cables, Pearson Technologies, Acworth, GA, 1991.

Pitts, K. and N. Hurst: "How Much Do People Prefer Widescreen (16 × 9) to Standard NTSC (4 × 3)?," *IEEE Transactions on Consumer Electronics*, IEEE, New York, August 1989.

Pointer, R. M.: "The Gamut of Real Surface Colors, *Color Res. App.*, vol. 5, 1945.

Robin, Michael: "Digital Resolution," *Broadcast Engineering*, Intertec Publishing, Overland Park, Kan., pp. 44–48, April 1998.

Slamin, Brendan: "Sound for High Definition Television," *Proceedings of the ITS*, International Television Symposium, Montreux, Switzerland, 1991.

SMPTE Recommended Practice RP 199-1999, "Mapping of Pictures in Wide-Screen (16:9) Scanning Structure to Retain Original Aspect Ratio of the Work," SMPTE, White Plains, N.Y., 1999.

SMPTE Recommended Practice—Implementation of 24P, 25P, and 30P Segmented Frames for 1920 × 1080 Production Format, RP 211-2000, SMPTE, White Plains, N.Y., 2000.

"SMPTE Standard for Television—1280 × 720 Scanning, Analog and Digital Representation and Analog Interface," SMPTE 296M-2001, SMPTE, White Plains, N.Y., 1997.

"SMPTE Standard for Television—1920 × 1080 50 Hz Scanning and Interfaces," SMPTE 295M-1997, SMPTE, White Plains, N.Y., 1997.

"SMPTE Standard for Television—1920 × 1080 Scanning and Analog and Parallel Digital Interfaces for Multiple-Picture Rates," SMPTE 274-1998, SMPTE, White Plains, N.Y., 1998.

"SMPTE Standard for Television—720 × 483 Active Line at 59.94 Hz Progressive Scan Production—Digital Representation," SMPTE 293M-1996, SMPTE, White Plains, N.Y., 1996.

"SMPTE Standard for Television—720 × 483 Active Line at 59.94-Hz Progressive Scan Production Bit-Serial Interfaces," SMPTE 294M-1997, SMPTE, White Plains, N.Y., 1997.

"SMPTE Standard for Television—Composite Analog Video Signal NTSC for Studio Applications," SMPTE 170M-1999, SMPTE, White Plains, N.Y., 1999.

"SMPTE Standard for Television—Digital Representation and Bit-Parallel Interface—1125/60 High-Definition Production System," SMPTE 260M-1992, SMPTE, White Plains, N.Y., 1992.

"SMPTE Standard for Television—MPEG-2 4:2:2 Profile at High Level," SMPTE 308M-1998, SMPTE, White Plains, N.Y., 1998.

"SMPTE Standard for Television—Signal Parameters—1125-Line High-Definition Production Systems," SMPTE 240M-1995, SMPTE, White Plains, N.Y., 1995.

"SMPTE Standard for Television—Synchronous Serial Interface for MPEG-2 Digital Transport Stream," SMPTE 310M-1998, SMPTE, White Plains, N.Y., 1998.

SMPTE: "System Overview—Advanced System Control Architecture, S22.02, Revision 2.0," S22.02 Advanced System Control Architectures Working Group, SMPTE, White Plains, N.Y., March 27, 2000.

Suitable Sound Systems to Accompany High-Definition and Enhanced Television Systems: Report 1072. Recommendations and Reports to the CCIR, 1986. Broadcast Service—Sound. International Telecommunications Union, Geneva, 1986.

Thorpe, Larry: "The Great Debate: Interlaced Versus Progressive Scanning," *Proceedings of the Digital Television '97 Conference*, Intertec Publishing, Overland Park, Kan., December 1997.

Thorpe, Laurence: "A New Global HDTV Program Origination Standard: Implications for Production and Technology," *Proceedings of DTV99*, Intertec Publishing, Overland Park, Kan., 1999.

Thorpe, Laurence J.: "Applying High-Definition Television," *Television Engineering Handbook*, rev. ed., K. B. Benson and Jerry C. Whitaker (eds.), McGraw-Hill, New York, N.Y., pg. 23.4, 1991.

Torick, Emil L.: "HDTV: High Definition Video—Low Definition Audio?," *1991 HDTV World Conference Proceedings*, National Association of Broadcasters, Washington, D.C., April 1991.

Venkat, Giri, "Understanding ATSC Datacasting—A Driver for Digital Television," *Proceedings of the NAB Broadcast Engineering Conference*, National Association of Broadcasters, Washington, D.C., pp. 113–116, 1999.

Whitaker, Jerry C.: *The Communications Facility Design Handbook*, CRC Press, Boca Raton, Fla., 2000.

15.3j Film for Video Applications

Bauer, Richard W.: "Film for Television," *NAB Engineering Handbook*, 9th ed., Jerry C. Whitaker (ed.), National Association of Broadcasters, Washington, D.C., 1998.

Belton, J.: "The Development of the CinemaScope by Twentieth Century Fox," *SMPTE Journal*, vol. 97, SMPTE, White Plains, N.Y., September 1988.

Benson, K. B., and D. G. Fink: *HDTV: Advanced Television for the 1990s*, McGraw-Hill, New York, 1990.

Evans, R. N., W. T. Hanson, Jr., and W. L. Brewer: *Principles of Color Photography*, Wiley, New York, N.Y., 1953.

Fink, D. G, et. al.: "The Future of High Definition Television," *SMPTE Journal*, vol. 9, SMPTE, White Plains, N.Y., February/March 1980.

Fink, D. G: *Color Television Standards*, McGraw-Hill, New York, 1986.

Hubel, David H.: *Eye, Brain and Vision*, Scientific American Library, New York, 1988.

James, T. H., (ed.): *The Theory of the Photographic Process*, 4th ed., MacMillan, New York, N.Y., 1977.

James, T. H., and G. C. Higgins: *Fundamentals of Photographic Theory*, Morgan and Morgan, Inc., Dobbs Ferry, N.Y., 1968.

Judd, D. B.: "The 1931 C.I.E. Standard Observer and Coordinate System for Colorimetry," *Journal of the Optical Society of America*, vol. 23, 1933.

Kelly, K. L.: "Color Designation of Lights," *Journal of the Optical Society of America*, vol. 33, 1943.

Pointer, R. M.: "The Gamut of Real Surface Colors," *Color Res. App.*, vol. 5, 1945.

Woodlief, Thomas, Jr. (ed.): *SPSE Handbook of Photographic Science and Engineering*, Wiley, New York, N.Y., 1973.

Wyszecki, G., and W. S. Stiles: *Color Science*, Wiley, New York, N.Y., 1967.

15.3k Compression Technologies for Video and Audio

Arvind, R., et al.: "Images and Video Coding Standards," *AT&T Technical J.*, p. 86, 1993.

ATSC, "Guide to the Use of the ATSC Digital Television Standard," Advanced Television Systems Committee, Washington, D.C., doc. A/54, Oct. 4, 1995.

Bennett, Christopher: "Three MPEG Myths," *Proceedings of the 1996 NAB Broadcast Engineering Conference*, National Association of Broadcasters, Washington, D.C., pp. 129–136, 1996.

Bonomi, Mauro: "The Art and Science of Digital Video Compression," *NAB Broadcast Engineering Conference Proceedings*, National Association of Broadcasters, Washington, D.C., pp. 7–14, 1995.

Brandenburg, K., and Gerhard Stoll: "ISO-MPEG-1 Audio: A Generic Standard for Coding of High Quality Digital Audio," *92nd AES Convention Proceedings*, Audio Engineering Society, New York, N.Y., 1992, revised 1994.

Cugnini, Aldo G.: "MPEG-2 Bitstream Splicing," *Proceedings of the Digital Television '97 Conference*, Intertec Publishing, Overland Park, Kan., December 1997.

Dare, Peter: "The Future of Networking," *Broadcast Engineering*, Intertec Publishing, Overland Park, Kan., p. 36, April 1996.

DeWith, P. H. N.: "Motion-Adaptive Intraframe Transform Coding of Video Signals," *Philips J. Res.*, vol. 44, pp. 345–364, 1989.

Epstein, Steve: "Editing MPEG Bitstreams," *Broadcast Engineering*, Intertec Publishing, Overland Park, Kan., pp. 37–42, October 1997.

Fibush, David K.: "Testing MPEG-Compressed Signals," *Broadcast Engineering*, Overland Park, Kan., pp. 76–86, February 1996.

Freed, Ken: "Video Compression," *Broadcast Engineering*, Overland Park, Kan., pp. 46– 77, January 1997.

Gilge, M.: "Region-Oriented Transform Coding in Picture Communication," VDI-Verlag, Advancement Report, Series 10, 1990.

Herre, J., and B. Grill: "MPEG-4 Audio—Overview and Perspectives for the Broadcaster," *IBC 2000 Proceedings*, International Broadcast Convention, Amsterdam, September 2000.

IEEE Standard Dictionary of Electrical and Electronics Terms, ANSI/IEEE Standard 100-1984, Institute of Electrical and Electronics Engineers, New York, 1984.

"IEEE Standard Specifications for the Implementation of 8 × 8 Inverse Discrete Cosine Transform," std. 1180-1990, Dec. 6, 1990.

Jones, Ken: "The Television LAN," *Proceedings of the 1995 NAB Engineering Conference*, National Association of Broadcasters, Washington, D.C., p. 168, April 1995.

Lakhani, Gopal: "Video Compression Techniques and Standards," *The Electronics Handbook*, Jerry C. Whitaker (ed.), CRC Press, Boca Raton, Fla., pp. 1273–1282, 1996.

Lyman, Stephen: "A Multichannel Audio Infrastructure Based on Dolby E Coding," *Proceedings of the NAB Broadcast Engineering Conference*, National Association of Broadcasters, Washington, D.C., 1999.

Nelson, Lee J.: "Video Compression," *Broadcast Engineering*, Intertec Publishing, Overland Park, Kan., pp. 42–46, October 1995.

Netravali, A. N., and B. G. Haskell: *Digital Pictures, Representation, and Compression*, Plenum Press, 1988.

Robin, Michael, and Michel Poulin: *Digital Television Fundamentals*, McGraw-Hill, New York, N.Y., 1998.

Smith, Terry: "MPEG-2 Systems: A Tutorial Overview," *Transition to Digital Conference*, Broadcast Engineering, Overland Park, Kan., Nov. 21, 1996.

SMPTE Recommended Practice: RP 202-2000, *Video Alignment for MPEG-2 Coding*, SMPTE, White Plains, N.Y., 1999.

SMPTE Standard: SMPTE 308M, *MPEG-2 4:2:2 Profile at High Level*, SMPTE, White Plains, N.Y., 1998.

SMPTE Standard: SMPTE 319M-2000, *Transporting MPEG-2 Recoding Information Through 4:2:2 Component Digital Interfaces*, SMPTE, White Plains, N.Y., 2000.

SMPTE Standard: SMPTE 312M, *Splice Points for MPEG-2 Transport Streams*, SMPTE, White Plains, N.Y., 1999.

SMPTE Standard: SMPTE 327M-2000, *MPEG-2 Video Recoding Data Set*, SMPTE, White Plains, N.Y., 2000.

SMPTE Standard: SMPTE 328M-2000, *MPEG-2 Video Elementary Stream Editing Information*, SMPTE, White Plains, N.Y., 2000.

SMPTE Standard: SMPTE 329M-2000, *MPEG-2 Video Recoding Data Set—Compressed Stream Format*, SMPTE, White Plains, N.Y., 2000.

SMPTE Standard: SMPTE 351M-2000, *Transporting MPEG-2 Recoding Information through High-Definition Digital Interfaces*, SMPTE, White Plains, N.Y., 2000.

SMPTE Standard: SMPTE 353M-2000, *Transport of MPEG-2 Recoding Information as Ancillary Data Packets*, SMPTE, White Plains, N.Y., 2000.

Smyth, Stephen: "Digital Audio Data Compression," *Broadcast Engineering*, Intertec Publishing, Overland Park, Kan., February 1992.

Solari, Steve. J.: *Digital Video and Audio Compression*, McGraw-Hill, New York, N.Y., 1997.

Stallings, William: *ISDN and Broadband ISDN*, 2nd Ed., MacMillan, New York.

Symes, Peter D.: "Video Compression Systems," *NAB Engineering Handbook*, 9th Ed., Jerry C. Whitaker (ed.), National Association of Broadcasters, Washington, D.C., pp. 907–922, 1999.

Symes, Peter D.: *Video Compression*, McGraw-Hill, N.Y., 1998.

Taylor, P.: "Broadcast Quality and Compression," *Broadcast Engineering*, Intertec Publishing, Overland Park, Kan., p. 46, October 1995.

Terry, K. B., and S. B. Lyman: "Dolby E—A New Audio Distribution Format for Digital Broadcast Applications," *International Broadcasting Convention Proceedings*, IBC, London, England, pp. 204–209, September 1999.

Vernon, S., and T. Spath: "Carrying Multichannel Audio in a Stereo Production and Distribution Infrastructure," *Proceedings of IBC 2000*, International Broadcasting Convention, Amsterdam, September 2000.

Ward, Christopher, C. Pecota, X. Lee and G. Hughes: "Seamless Splicing for MPEG-2 Transport Stream Video Servers," *Proceedings, 33rd SMPTE Advanced Motion Imaging Conference*, SMPTE, White Plains, N.Y., 2000.

Whitaker, Jerry C., and Harold Winard (eds.): *The Information Age Dictionary*, Intertec Publishing/Bellcore, Overland Park, Kan., 1992.

Wylie, Fred: "Audio Compression Techniques," *The Electronics Handbook*, Jerry C. Whitaker (ed.), CRC Press, Boca Raton, Fla., pp. 1260–1272, 1996.

Wylie, Fred: "Audio Compression Technologies," *NAB Engineering Handbook*, 9th ed., Jerry C. Whitaker (ed.), National Association of Broadcasters, Washington, D.C., 1998.

15.31 Video Networking

ATSC, "Guide to the Use of the Digital Television Standard," Advanced Television Systems Committee, Washington, D.C., Doc. A/54, Oct. 4, 1995.

Craig, Donald: "Network Architectures: What does Isochronous Mean?," *IBC Daily News*, IBC, Amsterdam, September 1999.

Dahlgren, Michael W.: "Servicing Local Area Networks," *Broadcast Engineering*, Intertec Publishing, Overland Park, Kan., November 1989.

Fibush, David: *A Guide to Digital Television Systems and Measurement*, Tektronix, Beaverton, OR, 1994.

Gaggioni, H., M. Ueda, F. Saga, K. Tomita, and N. Kobayashi, "The Development of a High-Definition Serial Digital Interface," Sony Technical Paper, Sony Broadcast Group, San Jose, Calif., 1998.

Gallo and Hancock: *Networking Explained*, Digital Press, pp. 191–235, 1999.

Goldman, J: *Applied Data Communications: A Business Oriented Approach*, 2md ed., Wiley, New York, N.Y., 1998.

Goldman, J: *Local Area Networks: A Business Oriented Approach*, 2nd ed., Wiley, New York, N.Y., 2000.

Goldman, James E.: "Network Communication," in *The Electronics Handbook*, Jerry C. Whitaker (ed.), CRC Press, Boca Raton, Fla., 1996.

Held, G.: *Ethernet Networks: Design Implementation, Operation and Management*, Wiley, New York, N.Y., 1994.

Held, G.: *Internetworking LANs and WANs*, Wiley, New York, N.Y., 1993.

Held, G.: *Local Area Network Performance Issues and Answers*, Wiley, New York, N.Y., 1994.

Held, G.: *The Complete Modem Reference*, Wiley, New York, N.Y., 1994.

International Organization for Standardization: "Information Processing Systems—Open Systems Interconnection—Basic Reference Model," ISO 7498, 1984.

Legault, Alain, and Janet Matey: "Interconnectivity in the DTV Era—The Emergence of SDTI," *Proceedings of Digital Television '98*, Intertec Publishing, Overland Park, Kan., 1998.

Miller, Mark A.: "Servicing Local Area Networks," *Microservice Management*, Intertec Publishing, Overland Park, Kan., February 1990.

Miller, Mark A.: *LAN Troubleshooting Handbook*, M&T Books, Redwood City, Calif., 1990.

"Networking and Internet Broadcasting," Omneon Video Networks, Campbell, Calif, 1999.

"Networking and Production," Omneon Video Networks, Campbell, Calif., 1999.

Owen, Peter: "Gigabit Ethernet for Broadcast and Beyond," *Proceedings of DTV99*, Intertec Publishing, Overland Park, Kan., November 1999.

Piercy, John: "ATM Networked Video: Moving From Leased-Lines to Packetized Transmission," *Proceedings of the Transition to Digital Conference*, Intertec Publishing, Overland Park, Kan., 1996.

"SMPTE Recommended Practice—Error Detection Checkwords and Status Flags for Use in Bit-Serial Digital Interfaces for Television," RP 165-1994, SMPTE, White Plains, N.Y., 1994.

"SMPTE Recommended Practice—SDTI-CP MPEG Decoder Templates," RP 204, SMPTE, White Plains, N.Y., 1999.

"SMPTE Standard for Television—540 Mb/s Serial Digital Interface, SMPTE 344M-2000, SMPTE, White Plains, N.Y., 2000.

"SMPTE Standard for Television—540 Mb/s Serial Digital Interface: Source Image Format Mapping," SMPTE 347M, SMPTE, White Plains, N.Y., 2001.

"SMPTE Standard for Television—24-Bit Digital Audio Format for HDTV Bit-Serial Interface," SMPTE 299M-1997, SMPTE, White Plains, N.Y., 1997.

"SMPTE Standard for Television—Ancillary Data Packet and Space Formatting," SMPTE 291M-1998, SMPTE, White Plains, N.Y., 1998.

"SMPTE Standard for Television—Bit-Serial Digital Interface for High-Definition Television Systems," SMPTE 292M-1998, SMPTE, White Plains, N.Y., 1998.

"SMPTE Standard for Television—Element and Metadata Definitions for the SDTI-CP," SMPTE 331M-2000, SMPTE, White Plains, N.Y., 2000.

"SMPTE Standard for Television—Encapsulation of Data Packet Streams over SDTI (SDTI-PF)," SMPTE 332M-2000, SMPTE, White Plains, N.Y., 2000.

"SMPTE Standard for Television—General Exchange Format (GXF)," SMPTE 360M, SMPTE, White Plains, N.Y., 2001.

"SMPTE Standard for Television—High Data-Rate Serial Data Transport Interface (HD-SDTI)," SMPTE 348M-2000, SMPTE, White Plains, N.Y., 2000.

"SMPTE Standard for Television—Mapping of AES3 Data into MPEG-2 Transport Stream," SMPTE 302M-1998, SMPTE, White Plains, N.Y., 1998.

"SMPTE Standard for Television—SDTI Content Package Format (SDTI-CP)," SMPTE 326M-2000, SMPTE, White Plains, N.Y., 2000.

"SMPTE Standard for Television—Serial Data Transport Interface," SMPTE 305M-1998, SMPTE, White Plains, N.Y., 1998.

"SMPTE Standard for Television—Signals and Generic Data over High-Definition Interfaces," SMPTE 346M-2000, SMPTE, White Plains, N.Y., 2000.

"SMPTE Standard for Television—Transport of Alternate Source Image Formats through SMPTE 292M, SMPTE 349M, SMPTE, White Plains, N.Y., 2001.

"SMPTE Standard for Television—Vertical Ancillary Data Mapping for Bit-Serial Interface, SMPTE 334M-2000, SMPTE, White Plains, N.Y., 2000.

"SMPTE Standard for Television—Video Payload Identification for Digital Television Interfaces," SMPTE 352M, SMPTE, White Plains, N.Y., 2001.

"Technology Brief—Networking and Storage Strategies," Omneon Video Networks, Campbell, Calif., 1999.

Turow, Dan: "SDTI and the Evolution of Studio Interconnect," *International Broadcasting Convention Proceedings*, IBC, Amsterdam, September 1998.

Wilkinson, J. H., H. Sakamoto, and P. Horne: "SDDI as a Video Data Network Solution," *International Broadcasting Convention Proceedings*, IBC, Amsterdam, September 1997.

Wu, Tsong-Ho: "Network Switching Concepts," *The Electronics Handbook*, Jerry C. Whitaker (ed.), CRC Press, Boca Raton, Fla., p. 1513, 1996.

15.3m Digital Television Transmission Systems

Advanced Television Enhancement Forum Specification," Draft, Version 1.1r26 updated 2/2/99, ATVEF, Portland, Ore., 1999.

Allision, Arthur: "PSIP 101: What You Need to Know," *Broadcast Engineering*, Intertec Publishing, Overland Park, Kan., pp. 140–144, June 2001.

Arragon, J. P., J. Chatel, J. Raven, and R. Story: "Instrumentation for a Compatible HD-MAC Coding System Using DATV," *Conference Record*, International Broadcasting Conference, Brighton, Institution of Electrical Engineers, London, 1989.

ATSC: "Amendment No. 1 to ATSC Standard: Program and System Information Protocol for Terrestrial Broadcast and Cable," Doc. A/67, ATSC, Washington, D.C, December 17, 1999.

ATSC: "ATSC Data Broadcast Standard," Advanced Television Systems Committee, Washington, D.C., Doc. A/90, July 26, 2000.

ATSC: "ATSC Digital Television Standard," Advanced Television Systems Committee, Washington, D.C., Doc. A/53, September 16, 1995.

ATSC: "Conditional Access System for Terrestrial Broadcast," Advanced Television Systems Committee, Washington, D.C., Doc. A/70, July 1999.

ATSC: "Digital Audio Compression Standard (AC-3), Annex A: AC-3 Elementary Streams in an MPEG-2 Multiplex," Advanced Television Systems Committee, Washington, D.C., Doc. A/52, December 20, 1995.

ATSC: "Digital Audio Compression Standard (AC-3)," Advanced Television Systems Committee, Washington, D.C., Doc. A/52, Dec. 20, 1995.

ATSC: "Guide to the Use of the Digital Television Standard," Advanced Television Systems Committee, Washington, D.C., Doc. A/54, October 4, 1995.

ATSC: "Implementation of Data Broadcasting in a DTV Station," Advanced Television Systems Committee, Washington, D.C., Doc. IS/151, November 1999.

ATSC: "Performance Assessment of the ATSC Transmission System, Equipment, and Future Directions," ATSC Task Force on RF System Performance, Advanced Television Systems Committee, Washington, DC, revision: 1.0, April 12, 2001.

ATSC: "Program and System Information Protocol for Terrestrial Broadcast and Cable," Advanced Television Systems Committee, Washington, D.C., Doc. A/65, February 1998.

ATSC: "Technical Corrigendum No.1 to ATSC Standard: Program and System Information Protocol for Terrestrial Broadcast and Cable," Doc. A/66, ATSC, Washington, D.C., December 17, 1999.

ATTC: "Digital HDTV Grand Alliance System Record of Test Results," Advanced Television Test Center, Alexandria, Virginia, October 1995.

Basile, C.: "An HDTV MAC Format for FM Environments," International Conference on Consumer Electronics, IEEE, New York, June 1989.

Cadzow, James A.: *Discrete Time Systems*, Prentice-Hall, Inc., Englewood Cliffs, N.J., 1973.

Chairman, ITU-R Task Group 11/3, "Report of the Second Meeting of ITU-R Task Group 11/3, Geneva, Oct. 13-19, 1993," p. 40, Jan. 5, 1994.

Chernock, Richard: "Implementation Recommendations for Data Broadcast," *NAB Broadcast Engineering Conference Proceedings*, National Association of Broadcasters, Washington, D.C., pp. 315–322, April 2000.

Chini, A., Y. Wu, M. El-Tanany, and S. Mahmoud: "An OFDM-based Digital ATV Terrestrial Broadcasting System with a Filtered Decision Feedback Channel Estimator," *IEEE Trans. Broadcasting*, IEEE, New York, N.Y., vol. 44, no. 1, pp. 2–11, March 1998.

Chini, A., Y. Wu, M. El-Tanany, and S. Mahmoud: "Hardware Nonlinearities in Digital TV Broadcasting Using OFDM Modulation," *IEEE Trans. Broadcasting*, IEEE, New York, N.Y., vol. 44, no. 1, March 1998.

Clement, Pierre, and Eric Gourmelen: "Internet and Television Convergence: IP and MPEG-2 Implementation Issues," *Proceedings of the 33rd SMPTE Advanced Motion Imaging Conference*, SMPTE, White Plains, N.Y., February 1999.

"Conclusions of the Extraordinary Meeting of Study Group 11 on High-Definition Television," Doc. 11/410-E, International Radio Consultative Committee (CCIR), Geneva, Switzerland, June 1989.

Ehmer, R. H.: "Masking of Tones Vs. Noise Bands," *J. Acoust. Soc. Am.*, vol. 31, pp. 1253–1256, September 1959.

Ehmer, R. H.: "Masking Patterns of Tones," *J. Acoust. Soc. Am.*, vol. 31, pp. 1115–1120, August 1959.

Erez, Beth: "Protecting content in the Digital Home," *Proceedings of the 33rd SMPTE Advanced Motion Imaging Conference*, SMPTE, White Plains, N.Y., pp. 231–238, February 1999.

ETS-300-421, "Digital Broadcasting Systems for Television, Sound, and Data Services; Framing Structure, Channel Coding and Modulation for 11–12 GHz Satellite Services," DVB Project technical publication.

ETS-300-429, "Digital Broadcasting Systems for Television, Sound, and Data Services; Framing Structure, Channel Coding and Modulation for Cable Systems," DVB Project technical publication.

ETS-300-468, "Digital Broadcasting Systems for Television, Sound, and Data Services; Specification for Service Information (SI) in Digital Video Broadcasting (DVB) Systems," DVB Project technical publication.

ETS-300-472, "Digital Broadcasting Systems for Television, Sound, and Data Services; Specification for Carrying ITU-R System B Teletext in Digital Video Broadcasting (DVB) Bitstreams," DVB Project technical publication.

ETS-300-473, "Digital Broadcasting Systems for Television, Sound, and Data Services; Satellite Master Antenna Television (SMATV) Distribution Systems," DVB Project technical publication.

ETS 300-744: "Digital Broadcasting Systems for Television, Sound and Data Services: Framing Structure, Channel Coding and Modulation for Digital Terrestrial Television," ETS 300 744, 1997.

Eureka 95 HDTV Directorate, Progressing Towards the Real Dimension, Eureka 95 Communications Committee, Eindhoven, Netherlands, June 1991.

European Telecommunications Standards Institute: "Digital Video Broadcasting; Framing Structure, Channel Coding and Modulation for Digital Terrestrial Television (DVB-T)", March 1997.

FCC Office of Engineering and Technology: "DTV Report on COFDM and 8-VSB Performance," Federal Communications Commission, Washington, D.C., OET Report FCC/OET 99-2, September 30, 1999.

FCC Report and Order: "Closed Captioning Requirements for Digital Television Receivers," Federal Communications Commission, Washington, D.C., ET Docket 99-254 and MM Docket 95-176, adopted July 21, 2000.

Fibush, David K.: *A Guide to Digital Television Systems and Measurements*, Tektronix, Beaverton, Ore., 1997.

"HD-MAC Bandwidth Reduction Coding Principles," Draft Report AZ-11, International Radio Consultative Committee (CCIR), Geneva, Switzerland, January 1989.

Husak, Walt, et. al.: "On-channel Repeater for Digital Television Implementation and Field Testing," *Proceedings 1999 Broadcast Engineering Conference*, NAB'99, Las Vegas, National Association of Broadcasters, Washington, D.C., pp. 397–403, April 1999.

ITU Radiocommunication Study Groups, Special Rapporteur's Group: "Guide for the Use of Digital Television Terrestrial Broadcasting Systems Based on Performance Comparison of ATSC 8-VSB and DVB-T COFDM Transmission Systems," International Telecommunications Union, Geneva, Document 11A/65-E, May 11, 1999.

ITU-R Document TG11/3-2, "Outline of Work for Task Group 11/3, Digital Terrestrial Television Broadcasting," June 30, 1992.

ITU-R Recommendation BS-775, "Multi-channel Stereophonic Sound System with and Without Accompanying Picture."

ITU-R SG 11, Special Rapporteur—Region 1, "Protection Ratios and Reference Receivers for DTTB Frequency Planning," ITU-R Doc. 11C/46-E, March 18, 1999.

Jacklin, Martin: "The Multimedia Home Platform: On the Critical Path to Convergence," DVB Project technical publication, 1998.

Joint ERC/EBU: "Planning and Introduction of Terrestrial Digital Television (DVB-T) in Europe," Izmir, Dec. 1997.

Ligeti, A., and J. Zander: "Minimal Cost Coverage Planning for Single Frequency Networks", *IEEE Trans. Broadcasting*, IEEE, New York, N.Y., vol. 45, no. 1, March 1999.

Lucas, K.: "B-MAC: A Transmission Standard for Pay DBS," *SMPTE Journal*, SMPTE, White Plains, N.Y., November 1984.

Luetteke, Georg: "The DVB Multimedia Home Platform," DVB Project technical publication, 1998.

Mignone, V., and A. Morello: "CD3-OFDM: A Novel Demodulation Scheme for Fixed and Mobile Receivers," *IEEE Trans. Commu.*, IEEE, New York, N.Y., vol. 44, pp. 1144–1151, September 1996.

Moore, B. C. J., and B. R. Glasberg: "Formulae Describing Frequency Selectivity as a Function of Frequency and Level, and Their Use in Calculating Excitation Patterns," *Hearing Research*, vol. 28, pp. 209–225, 1987.

Morello, Alberto, et. al.: "Performance Assessment of a DVB-T Television System," *Proceedings of the International Television Symposium 1997*, Montreux, Switzerland, June 1997.

Muschallik, C.: "Improving an OFDM Reception Using an Adaptive Nyquist Windowing," *IEEE Trans. on Consumer Electronics,* no. 03, 1996.

Muschallik, C.: "Influence of RF Oscillators on an OFDM Signal", IEEE Trans. Consumer Electronics, IEEE, New York, N.Y., vol. 41, no. 3, pp. 592–603, August 1995.

NAB TV TechCheck: National Association of Broadcasters, Washington, D.C., February 1, 1999.

NAB TV TechCheck: National Association of Broadcasters, Washington, D.C., January 4, 1999.

NAB: "An Introduction to DTV Data Broadcasting," *NAB TV TechCheck*, National Association of Broadcasters, Washington, D.C., August 2, 1999.

NAB: "Digital TV Closed Captions," *NAB TV TechCheck*, National Association of Broadcasters, Washington, D.C., August 7, 2000.

NAB: "Navigation of DTV Data Broadcasting Services," *NAB TV TechCheck*, National Association of Broadcasters, Washington, D.C., November 1, 1999.

NAB: "Pay TV Services for DTV," *NAB TV TechCheck*, National Association of Broadcasters, Washington, D.C., October 4, 1999.

Pickford, N.: "Laboratory Testing of DTTB Modulation Systems," Laboratory Report 98/01, Australia Department of Communications and Arts, June 1998.

Pollet, T., M. van Bladel, and M. Moeneclaey. "BER Sensitivity of OFDM Systems to Carrier Frequency Offset and Wiener Phase Noise," *IEEE Trans. on Communications*, vol. 43, 1995.

Raven, J. G.: "High-Definition MAC: The Compatible Route to HDTV," *IEEE Transactions on Consumer Electronics*, vol. 34, pp. 61–63, IEEE, New York, February 1988.

Robertson, P., and S.Kaiser: "Analysis of the Effects of Phase-Noise in Orthogonal Frequency Division Multiplex (OFDM) Systems," ICC 1995, pp. 1652–1657, 1995.

Sabatier, J., D. Pommier, and M. Mathiue: "The D2-MAC-Packet System for All Transmission Channels," *SMPTE Journal*, SMPTE, White Plains, N.Y., November 1984.

Salter, J. E.: "Noise in a DVB-T System," BBC R&D Technical Note, R&D 0873(98), February 1998.

Sari, H., G. Karam, and I. Jeanclaude: "Channel Equalization and Carrier Synchronization in OFDM Systems," *IEEE Proc.*, 6th Tirrenia Workshop on Digital Communications, Tirrenia, Italy, pp. 191–202, September 1993.

Sariowan, H.: "Comparative Studies Of Data Broadcasting," *International Broadcasting Convention Proceedings*, IBC, London, England, pp. 115–119, 1999.

Schachlbauer, Horst: "European Perspective on Advanced Television for Terrestrial Broadcasting," *Proceedings of the ITS, International Television Symposium*, Montreux, Switzerland, 1991.

Sheth, Amit, and Wolfgang Klas (eds.): *Multimedia Data Management*. McGraw-Hill, New York, N.Y., 1996.

SMPTE Recommended Practice RP 203: "Real Time Opportunistic Data Flow Control in an MPEG-2 Transport Emission Multiplex," SMPTE, White Plains, N.Y., 1999.

SMPTE Standard: SMPTE 320M-1999, "Channel Assignments and Levels on Multichannel Audio Media," SMPTE, White Plains, N.Y., 1999.

SMPTE Standard: SMPTE 324M, "12-Channel Serial Interface for Digital Audio and Auxiliary Data," SMPTE, White Plains, N.Y., 1999.

SMPTE Standard: SMPTE 325M-1999, "Opportunistic Data Broadcast Flow Control," SMPTE, White Plains, N.Y., 1999.

SMPTE Standard: SMPTE 333M-1999, "DTV Closed-Caption Server to Encoder Interface," SMPTE, White Plains, N.Y., 1999.

Story, R.: "HDTV Motion-Adaptive Bandwidth Reduction Using DATV," BBC Research Department Report, RD 1986/5.

Story, R.: "Motion Compensated DATV Bandwidth Compression for HDTV," International Radio Consultative Committee (CCIR), Geneva, Switzerland, January 1989.

Stott, J. H.: "Explaining Some of the Magic of COFDM", *Proceedings of the International TV Symposium 1997*, Montreux, Switzerland, June 1997.

Stott, J. H: "The Effect of Phase Noise in COFDM", *EBU Technical Review*, Summer 1998.

Teichmann, Wolfgang: "HD-MAC Transmission on Cable," *Proceedings of the ITS, International Television Symposium*, Montreux, Switzerland, 1991.

Thomas, Gomer: "ATSC Datacasting—Opportunities and Challenges," *Proceedings of the 33rd SMPTE Advanced Motion Imaging Conference*, SMPTE, White Plains, N.Y., pp. 307–314, February 1999.

Todd, C., et. al.: "AC-3: Flexible Perceptual Coding for Audio Transmission and Storage," AES 96th Convention, Preprint 3796, Audio Engineering Society, New York, February 1994.

Tvede, Lars, Peter Pircher, and Jens Bodenkamp: *Data Broadcasting: The Technology and the Business*, John Wiley & Sons, New York, N.Y., 1999.

van Klinken, N., and W. Renirie: "Receiving DVB: Technical Challenges," *Proceedings of the International Broadcasting Convention*, IBC, Amsterdam, September 2000.

Vreeswijk, F., F. Fonsalas, T. Trew, C. Carey-Smith, and M. Haghiri: "HD-MAC Coding for High-Definition Television Signals," International Radio Consultative Committee (CCIR), Geneva, Switzerland, January 1989.

VSB/COFDM Project: "8-VSBCOFDM Comparison Report," National Association of Broadcasters and Maximum Service Television, Washington, D.C., December 2000.

Wu, Y., and M. El-Tanany: "OFDM System Performance Under Phase Noise Distortion and Frequency Selective Channels," *Proceedings of Int'l Workshop of HDTV 1997*, Montreux Switzerland, June 10–11, 1997.

Wu, Y., et. al.: "Canadian Digital Terrestrial Television System Technical Parameters," *IEEE Transactions on Broadcasting*, IEEE, New York, N.Y., to be published in 1999.

Wu, Y., M. Guillet, B. Ledoux, and B. Caron: "Results of Laboratory and Field Tests of a COFDM Modem for ATV Transmission over 6 MHz Channels," *SMPTE Journal*, SMPTE, White Plains, N.Y., vol. 107, February 1998.

Wugofski, T. W.: "A Presentation Engine for Broadcast Digital Television," *International Broadcasting Convention Proceedings*, IBC, London, England, pp. 451–456, 1999.

Zwicker, E.: "Subdivision of the Audible Frequency Range Into Critical Bands (Frequenzgruppen)," *J. Acoust. Soc. of Am.*, vol. 33, p. 248, February 1961.

15.3n Frequency Bands and Propagation

Bean, B. R., and E. J. Dutton: "Radio Meteorology," National Bureau of Standards Monograph 92, March 1, 1966.

Bullington, K.: "Radio Propagation at Frequencies above 30 Mc," *Proc. IRE*, pg. 1122, October 1947.

Bullington, K.: "Radio Propagation Variations at VHF and UHF," *Proc. IRE*, pg. 27, January 1950.

Burrows, C. R., and M. C. Gray: "The Effect of the Earth's Curvature on Groundwave Propagation," *Proc. IRE*, pg. 16, January 1941.

Collocott, T. C., A. B. Dobson, and W. R. Chambers (eds.): *Dictionary of Science & Technology.*

de Lisle, E. W.: "Computations of VHF and UHF Propagation for Radio Relay Applications," RCA, Report by International Division, New York, N.Y.

Dickson, F. H., J. J. Egli, J. W. Herbstreit, and G. S. Wickizer: "Large Reductions of VHF Transmission Loss and Fading by the Presence of a Mountain Obstacle in Beyond-Line-of-Sight Paths," *Proc. IRE*, vol. 41, no. 8, pg. 96, August 1953.

"Documents of the XVth Plenary Assembly," CCIR Report 238, vol. 5, Geneva, 1982.

"Documents of the XVth Plenary Assembly," CCIR Report 563, vol. 5, Geneva, 1982.

"Documents of the XVth Plenary Assembly," CCIR Report 881, vol. 5, Geneva, 1982.

Dougherty, H. T., and E. J. Dutton: "The Role of Elevated Ducting for Radio Service and Interference Fields," NTIA Report 81–69, March 1981.

Dye, D. W.: *Proc. Phys. Soc.*, Vol. 38, pp. 399–457, 1926.

Eckersley, T. L.: "Ultra-Short-Wave Refraction and Diffraction," *J. Inst. Elec. Engrs.*, pg. 286, March 1937.

Epstein, J., and D. Peterson: "An Experimental Study of Wave Propagation at 850 Mc," *Proc. IRE*, pg. 595, May 1953.

Fink, D. G., (ed.): *Television Engineering Handbook*, McGraw-Hill, New York, N.Y., 1957.

Frerking, M. E.: *Crystal Oscillator Design and Temperature Compensation*, Van Nostrand Reinhold, New York, N. Y., 1978.

Handbook of Physics, McGraw-Hill, New York, N.Y., 1958.

Hietala, Alexander W., and Duane C. Rabe: "Latched Accumulator Fractional-*N* Synthesis With Residual Error Reduction," United States Patent, Patent No. 5,093,632, March 3, 1992.

Judd, D. B., and G. Wyszecki: *Color in Business, Science and Industry*, 3rd ed., John Wiley and Sons, New York, N.Y.

Kaufman, Ed: *IES Illumination Handbook*, Illumination Engineering Society.

King, Nigel J. R.: "Phase Locked Loop Variable Frequency Generator," United States Patent, Patent No. 4,204,174, May 20, 1980.

Lapedes, D. N. (ed.): *The McGraw-Hill Encyclopedia of Science & Technology*, 2nd ed., McGraw-Hill, New York, N.Y.

Longley, A. G., and P. L. Rice: "Prediction of Tropospheric Radio Transmission over Irregular Terrain—A Computer Method," ESSA (Environmental Science Services Administration), U.S. Dept. of Commerce, Report ERL (Environment Research Laboratories) 79-ITS 67, July 1968.

McPetrie, J. S., and L. H. Ford: "An Experimental Investigation on the Propagation of Radio Waves over Bare Ridges in the Wavelength Range 10 cm to 10 m," J. *Inst. Elec. Engrs.*, pt. 3, vol. 93, pg. 527, 1946.

Megaw, E. C. S.: "Some Effects of Obstacles on the Propagation of Very Short Radio Waves," J. *Inst. Elec. Engrs.*, pt. 3, vol. 95, no. 34, pg. 97, March 1948.

National Bureau of Standards Circular 462, "Ionospheric Radio Propagation," June 1948.

NIST: *Manual of Regulations and Procedures for Federal Radio Frequency Management*, September 1995 edition, revisions for September 1996, January and May 1997, NTIA, Washington, D.C., 1997.

Norgard, John: "Electromagnetic Spectrum," *NAB Engineering Handbook*, 9th ed., Jerry C. Whitaker (ed.), National Association of Broadcasters, Washington, D.C., 1999.

Norgard, John: "Electromagnetic Spectrum," *The Electronics Handbook*, Jerry C. Whitaker (ed.), CRC Press, Boca Raton, Fla., 1996.

Norton, K. A.: "Ground Wave Intensity over a Finitely Conducting Spherical Earth," *Proc. IRE*, pg. 622, December 1941.

Norton, K. A.: "The Propagation of Radio Waves over a Finitely Conducting Spherical Earth," *Phil. Mag.*, June 1938.

"Radio Wave Propagation," Summary Technical Report of the Committee on Propagation of the National Defense Research Committee, Academic Press, New York, N.Y., 1949.

"Report of the Ad Hoc Committee, Federal Communications Commission," vol. 1, May 1949; vol. 2, July 1950.

Riley, Thomas A. D.: "Frequency Synthesizers Having Dividing Ratio Controlled Sigma-Delta Modulator," United States Patent, Patent No. 4,965,531, October 23, 1990.

Rohde, Ulrich L.: *Digital PLL Frequency Synthesizers*, Prentice-Hall, Englewood Cliffs, N.J., 1983.

Rohde, Ulrich L.: *Microwave and Wireless Synthesizers: Theory and Design*, John Wiley & Sons, New York, N.Y., pg. 209, 1997.

Selvidge, H.:"Diffraction Measurements at Ultra High Frequencies," *Proc. IRE*, pg. 10, January 1941.

Smith, E. E., and E. W. Davis: "Wind-induced Ions Thwart TV Reception," *IEEE Spectrum*, pp. 52—55, February 1981.

Stemson, A: *Photometry and Radiometry for Engineers*, John Wiley and Sons, New York, N.Y.

Tate, Jeffrey P., and Patricia F. Mead: "Crystal Oscillators," in *The Electronics Handbook*, Jerry C. Whitaker (ed.), CRC Press, Boca Raton, Fla., pp. 185–199, 1996.

The Cambridge Encyclopedia, Cambridge University Press, 1990.

The Columbia Encyclopedia, Columbia University Press, 1993.

"The Propagation of Radio Waves through the Standard Atmosphere," *Summary Technical Report of the Committee on Propagation*, vol. 3, National Defense Research Council, Washington, D.C., 1946, published by Academic Press, New York, N.Y.

van der Pol, Balth, and H. Bremmer: "The Diffraction of Electromagnetic Waves from an Electrical Point Source Round a Finitely Conducting Sphere, with Applications to Radiotelegraphy and to Theory of the Rainbow," pt. 1, *Phil. Mag.*, July, 1937; pt. 2, *Phil. Mag.*, November 1937.

Webster's New World Encyclopedia, Prentice Hall, 1992.

Wells, John Norman: "Frequency Synthesizers," European Patent, Patent No. 012579OB2, July 5, 1995.

Wyszecki, G., and W. S. Stiles: *Color Science, Concepts and Methods, Quantitative Data and Formulae*, 2nd ed., John Wiley and Sons, New York, N.Y.

15.3o Television Transmission Systems

ACATS, "ATV System Description: ATV-into-NTSC Co-channel Test #016," Grand Alliance Advisory Committee on Advanced Television, p. I-14-10, Dec. 7, 1994.

Aitken, S., D. Carr, G. Clayworth, R. Heppinstall, and A. Wheelhouse: "A New, Higher Power, IOT System for Analogue and Digital UHF Television Transmission," *Proceedings of the 1997 NAB Broadcast Engineering Conference*, National Association of Broadcasters, Washington, D.C., p. 531, 1997.

Andrew Corporation: "Broadcast Transmission Line Systems," Technical Bulletin 1063H, Orland Park, Ill., 1982.

Andrew Corporation: "Circular Waveguide: System Planning, Installation and Tuning," Technical Bulletin 1061H, Orland Park, Ill., 1980.

ATSC Standard: "Modulation And Coding Requirements For Digital TV (DTV) Applications Over Satellite," Doc. A/80, ATSC, Washington, D.C., July, 17, 1999.

ATSC, "Guide to the Use of the Digital Television Standard," Advanced Television Systems Committee, Washington, D.C., Doc. A/54, Oct. 4, 1995.

Ben-Dov, O., and C. Plummer: "Doubly Truncated Waveguide," *Broadcast Engineering*, Intertec Publishing, Overland Park, Kan., January 1989.

Benson, K. B., and J. C. Whitaker: *Television and Audio Handbook for Technicians and Engineers*, McGraw-Hill, New York, N.Y., 1989.

Cablewave Systems: "Rigid Coaxial Transmission Lines," Cablewave Systems Catalog 700, North Haven, Conn., 1989.

Cablewave Systems: "The Broadcaster's Guide to Transmission Line Systems," Technical Bulletin 21A, North Haven, Conn., 1976.

Carnt, P. S., and G. B. Townsend: *Colour Television—Volume 1: NTSC*; *Volume 2: PAL and SECAM*, ILIFFE Bookes Ltd. (Wireless World), London, 1969.

"CCIR Characteristics of Systems for Monochrome and Colour Television—Recommendations and Reports," Recommendations 470-1 (1974–1978) of the Fourteenth Plenary Assembly of CCIR in Kyoto, Japan, 1978.

CCIR Report 122-4, 1990.

Crutchfield, E. B. (ed.), *NAB Engineering Handbook*, 8th Ed., National Association of Broadcasters, Washington, D.C., 1992.

DeComier, Bill: "Inside FM Multiplexer Systems," *Broadcast Engineering*, Intertec Publishing, Overland Park, Kan., May 1988.

Ericksen, Dane E.: "A Review of IOT Performance," *Broadcast Engineering*, Intertec Publishing, Overland Park, Kan., pg. 36, July 1996.

Fink, D., and D. Christiansen (eds.), *Electronics Engineers' Handbook*, 2nd Ed., McGraw-Hill, New York, N.Y., 1982.

Fink, D., and D. Christiansen (eds.): *Electronics Engineers' Handbook*, 3rd Ed., McGraw-Hill, New York, N.Y., 1989.

Fink, Donald G. (ed.): *Color Television Standards*, McGraw-Hill, New York, N.Y., 1955.

Gilmore, A. S.: *Microwave Tubes*, Artech House, Dedham, Mass., pp. 196–200, 1986.

Harrison, Cecil: "Passive Filters," in *The Electronics Handbook*, Jerry C. Whitaker (ed.), CRC Press, Boca Raton, Fla., pp. 279–290, 1996.

Herbstreit, J. W., and J. Pouliquen: "International Standards for Color Television," *IEEE Spectrum*, IEEE, New York, N.Y., March 1967.

Heymans, Dennis: "Hot Switches and Combiners," *Broadcast Engineering*, Overland Park, Kan., December 1987.

Hirsch, C. J.: "Color Television Standards for Region 2," *IEEE Spectrum*, IEEE, New York, N.Y., February 1968.

Hulick, Timothy P.: "60 kW Diacrode UHF TV Transmitter Design, Performance and Field Report," *Proceedings of the 1996 NAB Broadcast Engineering Conference*, National Association of Broadcasters, Washington, D.C., p. 442, 1996.

Hulick, Timothy P.: "Very Simple Out-of-Band IMD Correctors for Adjacent Channel NTSC/DTV Transmitters," *Proceedings of the Digital Television '98 Conference*, Intertec Publishing, Overland Park, Kan., 1998.

Jordan, Edward C.: *Reference Data for Engineers: Radio, Electronics, Computer and Communications*, 7th ed., Howard W. Sams, Indianapolis, IN, 1985.

Krohe, Gary L.: "Using Circular Waveguide," *Broadcast Engineering*, Intertec Publishing, Overland Park, Kan., May 1986.

Ostroff, Nat S.: "A Unique Solution to the Design of an ATV Transmitter," *Proceedings of the 1996 NAB Broadcast Engineering Conference*, National Association of Broadcasters, Washington, D.C., p. 144, 1996.

Perelman, R., and T. Sullivan: "Selecting Flexible Coaxial Cable," *Broadcast Engineering*, Intertec Publishing, Overland Park, Kan., May 1988.

Plonka, Robert J.: "Planning Your Digital Television Transmission System," *Proceedings of the 1997 NAB Broadcast Engineering Conference*, National Association of Broadcasters, Washington, D.C., p. 89, 1997.

Priest, D. H., and M. B. Shrader: "The Klystrode—An Unusual Transmitting Tube with Potential for UHF-TV," *Proc. IEEE*, vol. 70, no. 11, pp. 1318–1325, November 1982.

Pritchard, D. H.: "U.S. Color Television Fundamentals—A Review," *SMPTE Journal*, SMPTE, White Plains, N.Y., vol. 86, pp. 819–828, November 1977.

Rhodes, Charles W.: "Terrestrial High-Definition Television," *The Electronics Handbook*, Jerry C. Whitaker (ed.), CRC Press, Boca Raton, Fla., pp. 1599–1610, 1996.

Roizen, J.: "Universal Color Television: An Electronic Fantasia," *IEEE Spectrum*, IEEE, New York, N.Y., March 1967.

Stenberg, James T.: "Using Super Power Isolators in the Broadcast Plant," *Proceedings of the Broadcast Engineering Conference*, Society of Broadcast Engineers, Indianapolis, IN, 1988.

Surette, Robert A.: "Combiners and Combining Networks," in *The Electronics Handbook*, Jerry C. Whitaker (ed.), CRC Press, Boca Raton, Fla., pp. 1368–1381, 1996.

Symons, R., M. Boyle, J. Cipolla, H. Schult, and R. True: "The Constant Efficiency Amplifier—A Progress Report," *Proceedings of the NAB Broadcast Engineering Conference*, National Association of Broadcasters, Washington, D.C., pp. 77–84, 1998.

Symons, Robert S.: "The Constant Efficiency Amplifier," *Proceedings of the NAB Broadcast Engineering Conference*, National Association of Broadcasters, Washington, D.C., pp. 523–530, 1997.

Tardy, Michel-Pierre: "The Experience of High-Power UHF Tetrodes," *Proceedings of the 1993 NAB Broadcast Engineering Conference*, National Association of Broadcasters, Washington, D.C., p. 261, 1993.

Terman, F. E.: *Radio Engineering*, 3rd ed., McGraw-Hill, New York, N.Y., 1947.

Vaughan, T., and E. Pivit: "High Power Isolator for UHF Television," *Proceedings of the NAB Engineering Conference*, National Association of Broadcasters, Washington, D.C., 1989.

Whitaker, Jerry C., G. DeSantis, and C. Paulson: *Interconnecting Electronic Systems*, CRC Press, Boca Raton, Fla., 1993.

Whitaker, Jerry C.: *Radio Frequency Transmission Systems: Design and Operation*, McGraw-Hill, New York, N.Y., 1990.

Whitaker, Jerry C.: "Microwave Power Tubes," *Power Vacuum Tubes Handbook*, Van Nostrand Reinhold, New York, p. 259, 1994.

15.3p Television Antenna Systems

Allnatt, J. W., and R. D. Prosser: "Subjective Quality of Television Pictures Impaired by Long Delayed Echoes," *Proc. IEEE*, vol. 112, no. 3, March 1965.

Ben-Dov, O.: "Measurement of Circularly Polarized Broadcast Antennas," *IEEE Trans. Broadcasting*, vol. BC-19, no. 1, pp. 28–32, March 1972.

Bendov, Oded: "Coverage Contour Optimization of HDTV and NTSC Antennas," *Proceedings of the 1996 NAB Broadcast Engineering Conference*, National Association of Broadcasters, Washington, D.C., p. 69, 1996.

Brawn, D. A., and B. F. Kellom: "Butterfly VHF Panel Antenna," *RCA Broadcast News*, vol. 138, pp. 8–12, March 1968.

Clark, R. N., and N. A. L. Davidson: "The V-Z Panel as a Side Mounted Antenna," *IEEE Trans. Broadcasting*, vol. BC-13, no. 1, pp. 3–136, January 1967.

DeVito, G. G., and L. Mania: "Improved Dipole Panel for Circular Polarization," *IEEE Trans. Broadcasting*, vol. BC-28, no. 2, pp. 65–72, June 1982.

DeVito, G: "Considerations on Antennas with no Null Radiation Pattern and Pre-established Maximum-Minimum Shifts in the Vertical Plane," *Alta Frequenza*, vol. XXXVIII, no.6, 1969.

Dudzinsky, S. J., Jr.: "Polarization Discrimination for Satellite Communications," *Proc. IEEE*, vol. 57, no. 12, pp. 2179–2180, December 1969.

Fisk, R. E., and J. A. Donovan: "A New CP Antenna for Television Broadcast Service," *IEEE Trans. Broadcasting*, vol. BC-22, no. 3, pp. 91–96, September 1976.

Fowler, A. D., and H. N. Christopher: "Effective Sum of Multiple Echoes in Television," J. *SMPTE*, SMPTE, White Plains, N. Y., vol. 58, June 1952.

Fumes, N., and K. N. Stokke: "Reflection Problems in Mountainous Areas: Tests with Circular Polarization for Television and VHF/FM Broadcasting in Norway," *EBU Review*, Technical Part, no. 184, pp. 266–271, December 1980.

Heymans, Dennis: "Channel Combining in an NTSC/ATV Environment," *Proceedings of the 1996 NAB Broadcast Engineering Conference*, National Association of Broadcasters, Washington, D.C., p. 165, 1996.

Hill, P. C. J.: "Measurements of Reradiation from Lattice Masts at VHF," *Proc. IEEE*, vol. III, no. 12, pp. 1957–1968, December 1964.

Hill, P. C. J.: "Methods for Shaping Vertical Pattern of VHF and UHF Transmitting Aerials," *Proc. IEEE*, vol. 116, no. 8, pp. 1325–1337, August 1969.

Johns, M. R., and M. A. Ralston: "The First Candelabra for Circularly Polarized Broadcast Antennas," *IEEE Trans. Broadcasting*, vol. BC-27, no. 4, pp. 77–82, December 1981.

Johnson, R. C, and H. Jasik: *Antenna Engineering Handbook*, 2d ed., McGraw-Hill, New York, N.Y., 1984.

Knight, P.: "Reradiation from Masts and Similar Objects at Radio Frequencies," *Proc. IEEE*, vol. 114, pp. 30–42, January 1967.

Kraus, J. D.: *Antennas*, McGraw-Hill, New York, N.Y., 1950.

Lessman, A. M.: "The Subjective Effect of Echoes in 525-Line Monochrome and NTSC Color Television and the Resulting Echo Time Weighting," *J. SMPTE*, SMPTE, White Plains, N.Y., vol. 1, December 1972.

Mertz, P.: "Influence of Echoes on Television Transmission," *J. SMPTE*, SMPTE, White Plains, N.Y., vol. 60, May 1953.

Moreno, T.: *Microwave Transmission Design Data*, Dover, New York, N.Y.

Perini, J., and M. H. Ideslis: "Radiation Pattern Synthesis for Broadcast Antennas," *IEEE Trans. Broadcasting*, vol. BC-18, no. 3, pg. 53, September 1972.

Perini, J.: "Echo Performance of TV Transmitting Systems," *IEEE Trans. Broadcasting*, vol. BC-16, no. 3, September 1970.

Perini, J.: "Improvement of Pattern Circularity of Panel Antenna Mounted on Large Towers," *IEEE Trans. Broadcasting*, vol. BC-14, no. 1, pp. 33–40, March 1968.

Plonka, Robert J.: "Can ATV Coverage Be Improved With Circular, Elliptical, or Vertical Polarized Antennas?" *Proceedings of the 1996 NAB Broadcast Engineering Conference*, National Association of Broadcasters, Washington, D.C., p. 155, 1996.

Praba, K.: "Computer-aided Design of Vertical Patterns for TV Antenna Arrays," *RCA Engineer*, vol. 18-4, January–February 1973.

Praba, K.: "R. F. Pulse Measurement Techniques and Picture Quality," *IEEE Trans. Broadcasting*, vol. BC-23, no. 1, pp. 12–17, March 1976.

"Predicting Characteristics of Multiple Antenna Arrays," *RCA Broadcast News*, vol. 97, pp. 63–68, October 1957.

Sargent, D. W.: "A New Technique for Measuring FM and TV Antenna Systems," *IEEE Trans. Broadcasting*, vol. BC-27, no. 4, December 1981.

Siukola, M. S.: "Size and Performance Trade Off Characteristics of Horizontally and Circularly Polarized TV Antennas," *IEEE Trans. Broadcasting*, vol. BC-23, no. 1, March 1976.

Siukola, M. S.: "The Traveling Wave VHF Television Transmitting Antenna," *IRE Trans. Broadcasting*, vol. BTR-3, no. 2, pp. 49-58, October 1957.

Siukola, M. S.: "TV Antenna Performance Evaluation with RF Pulse Techniques," *IEEE Trans. Broadcasting*, vol. BC-16, no. 3, September 1970.

Smith, Paul D.: "New Channel Combining Devices for DTV, *Proceedings of the 1997 NAB Broadcast Engineering Conference*, National Association of Broadcasters, Washington, D.C., p. 218, 1996.

"WBAL, WJZ and WMAR Build World's First Three-Antenna Candelabra," *RCA Broadcast News*, vol. 106, pp. 30–35, December 1959.

Wescott, H. H.: "A Closer Look at the Sutro Tower Antenna Systems," *RCA Broadcast News*, vol. 152, pp. 35–41, February 1944.

Whythe, D. J.: "Specification of the Impedance of Transmitting Aerials for Monochrome and Color Television Signals," Tech. Rep. E-115, BBC, London, 1968.

15.3q Television Receivers and Cable/Satellite Distribution Systems

A. DeVries et al, "Characteristics of Surface-Wave Integratable Filters (SWIFS)," *IEEE Trans.*, vol. BTR-17, no. 1, p. 16.

"Advanced Television Enhancement Forum Specification," Draft, Version 1.1r26 updated 2/2/99, ATVEF, Portland, Ore., 1999.

Altman, F. J., and W. Sichak: "A Simplified Diversity Communication System," *IRE Trans.*, vol. CS-4, March 1956.

Applebaum. S. P.: "Adaptive Arrays," *IEEE Trans.*, vol. AP-24, pg. 585, September 1976.

ATSC: "Guide to the Use of the ATSC Digital Television Standard," Advanced Television Systems Committee, Washington, D.C., Doc. A/54, October 4, 1995.

ATSC: "Performance Assessment of the ATSC Transmission System, Equipment, and Future Directions," ATSC Task Force on RF System Performance, Advanced Television Systems Committee, Washington, DC, revision: 1.0, April 12, 2001.

Baldwin, T. F., and D. S. McVoy: *Cable Communications*, Prentice-Hall, Englewood Cliffs, N.J., 1983.

Bendov, O.: "On the Validity of the Longley-Rice (50,90/10) Propagation Model for HDTV Coverage and Interference Analysis," *Proceedings of the Broadcast Engineering Conference*, National Association of Broadcasters, Washington, D.C., 1999.

Beverage, H. H., and H. O. Peterson: "Diversity Receiving System of RCA Communications, Inc., for Radiotelegraphy," *Proc. IRE*, vol. 19, pg. 531, April 1931.

Bonang, C., and C. Auvray-Kander: "Next Generation Broadband Networks for Interactive Services," in Proceedings of IBC 2000, International Broadcasting Convention, Amsterdam, 2000.

Brennan, D. G.: "Linear Diversity Combining Techniques," *Proc. IRE*, vol. 47, pg. 1075, June 1959.

"Cable Television Information Bulletin," Federal Communications Commission, Washington, D.C., June 2000.

Ciciora, Walter S.: "Cable Television," in *NAB Engineering Handbook*, 9th ed., Jerry C. Whitaker (ed.), National Association of Broadcasters, Washington, D.C., pp. 1339– 1363, 1999.

Ciciora, Walter, et. al.: "A Tutorial on Ghost Canceling in Television Systems," *IEEE Transactions on Consumer Electronics*, IEEE, New York, vol. CE-25, no. 1, pp. 9–44, February 1979.

Compton, R. T., Jr., R. J. Huff, W. G. Swarner, and A. A. Ksienski: "Adaptive Arrays for Communication Systems: An Overview of Research at the Ohio State University," *IEEE Trans.*, vol. AP-24, pg. 599, September 1976.

Cook, James H., Jr., Gary Springer, Jorge B. Vespoli: "Satellite Earth Stations," in *NAB Engineering Handbook*, Jerry C. Whitaker (ed.), National Association of Broadcasters, Washington, D.C., pp. 1285–1322, 1999.

Digital HDTV Grand Alliance System: *Record of Test Results*, October 1995.

Di Toro, M. J.: "Communications in Time-Frequency Spread Media," *Proc. IEEE*, vol. 56, October 1968.

Einolf, Charles: "DTV Receiver Performance in the Real World," *2000 Broadcast Engineering Conference Proceedings*, National Association of Broadcasters, Washington, D.C., pp. 478–482, 2000.

Elliot, R. S.: *Antenna Theory and Design*, Prentice-Hall, Englewood Cliffs, N.J., pg. 64, 1981.

Engelson, M., and J. Herbert: "Effective Characterization of CDMA Signals," *Microwave Journal*, pg. 90, January 1995.

FCC/ACATS SSWP2-1306: *Grand Alliance System Test Procedures*, May 18, 1994.

FCC Regulations, 47 CFR, 15.65, Washington, D.C.

Feinstein, J.: "Passive Microwave Components," *Electronic Engineers' Handbook*, D. Fink and D. Christiansen (eds.), Handbook, McGraw-Hill, New York, N.Y., 1982.

Fink, D. G., and D. Christiansen (eds.): *Electronic Engineer's Handbook*, 2nd ed., McGraw-Hill, New York, 1982.

Fockens, P., and C. G. Eilers: "Intercarrier Buzz Phenomena Analysis and Cures," *IEEE Trans. Consumer Electronics*, vol. CE-27, no. 3, pg. 381, August 1981.

Gibson, E. D.: "Automatic Equalization Using Time-Domain Equalizers," *Proc. IEEE*, vol. 53, pg. 1140, August 1965.

Grossner, N.: *Transformers for Electronic Circuits*, 2nd ed., McGraw-Hill, New York, pp. 344–358, 1983.

Hoff, L. E., and A. R. King: "Skywave Communication Techniques," Tech. Rep. 709, Naval Ocean Systems Center, San Diego, Calif, March 30,1981.

Hoffman, Gary A.: "IEEE 1394: The A/V Digital Interface of Choice," 1394 Technology Association Technical Brief, 1394 Technology Association, Santa Clara, Calif., 1999.

Howald, R.: "Understand the Mathematics of Phase Noise," *Microwaves & RF*, pg. 97, December 1993.

Hufford, G. A: "A Characterization of the Multipath in the HDTV Channel," *IEEE Trans. on Broadcasting*, vol. 38, no. 4, December 1992.

Hulst, G. D.: "Inverse Ionosphere," *IRE Trans.*, vol. CS-8, pg. 3, March 1960.

IEEE Guide for Surge Withstand Capability, (SWC) Tests, ANSI C37.90a-1974/IEEE Std. 472-1974, IEEE, New York, 1974.

Jasik, H., *Antenna Engineering Handbook*, McGraw-Hill, New York, Chapter 24, 1961.

Johnson, J. B:, "Thermal Agitation of Electricity in Conduction," *Phys. Rev.*, vol. 32, pg. 97, July 1928.

Kahn, L. R.: "Ratio Squarer," *Proc. IRE*, vol. 42, pg. 1704, November 1954.

Kase, C. A., and W. L. Pritchard: "Getting Set for Direct-Broadcast Satellites," *IEEE Spectrum*, vol. 18, no. 8, pp. 22–28, 1981.

Kraus, J. D.: *Antennas*, McGraw-Hill, New York, N. Y., Chapter 12, 1950.

Ledoux, B, P. Bouchard, S. Laflèche, Y. Wu, and B. Caron: "Performance of 8-VSB Digital Television Receivers," *Proceedings of the International Broadcasting Convention*, IBC, Amsterdam, September 2000.

Lo, Y. T.: "TV Receiving Antennas," in *Antenna Engineering Handbook*, H. Jasik (ed.), McGraw-Hill, New York, N.Y., pp. 24–25, 1961.

Lucky, R. W.: "Automatic Equalization for Digital Communication," *Bell Sys.Tech. J.*, vol. 44, pg. 547, April 1965.

Mack, C. L.: "Diversity Reception in UHF Long-Range Communications," *Proc. IRE*, vol. 43, October 1955.

Megawave Corp.: "Megawave/NAB Joint Technology Development Improved Antennas For NTSC Off-The-Air TV Reception," *NAB Engineering Conference Proceedings*, National Association of Broadcasters, Washington, D.C., 1996.

Miyazawa, H.: "Evaluation and Measurement of Airplane Flutter Interference," *IEEE Trans. on Broadcasting*, vol. 35, no. 4, pp 362–367, December 1989.

Monsen, P.: "Fading Channel Communications," *IEEE Commun..*, vol. 18, pg. 16, January 1980.

NAB TV TechCheck: "Canadians Perform Indoor DTV Reception Tests," National Association of Broadcasters, Washington, D.C., October 9, 2000..

NAB TV TechCheck: "CEA Establishes Definitions for Digital Television Products," National Association of Broadcasters, Washington, D.C., September 1, 2000.

NAB TV TechCheck: "Consumer Electronic Consortium Publishes Updated Specifications for Home Audio Video Interoperability," National Association of Broadcasters, Washington, D.C., May 21, 2001.

NAB TV TechCheck: "Digital Cable–DTV Receiver Compatibility Standard Announced," National Association of Broadcasters, Washington, D.C., November 15, 1999.

NAB TV TechCheck: "FCC Adopts Rules for Labeling of DTV Receivers," National Association of Broadcasters, Washington, D.C., September 25, 2000.

NAB TV TechCheck: "New Digital Receiver Technology Announced," National Association of Broadcasters, Washington, D.C., August 30, 1999.

Neal, C. B., and S. Goyal, "Frequency and Amplitude Phase Effects in Television Broadcast Systems," *IEEE Trans*., vol. CE-23, no. 3, pg. 241, August 1977.

Nyquist, H.: "Thermal Agitation of Electrical Charge in Conductors," *Phys. Rev.*, vol. 32, pg. 110, July 1928.

Peterson, O. H., H. H. Beverage, and J. B. Moore: "Diversity Telephone Receiving System of RCA Communications, Inc.," *Proc. IRE*, vol. 19, pg. 562, April 1931.

Pleasant, D.: "Practical Simulation of Bit Error Rates," *Applied Microwave and Wireless*, pg. 65, Spring 1995.

Plonka, Robert: "Can ATV Coverage be Improved with Circular, Elliptical or Vertical Polarized Antennas?," *NAB Engineering Conference Proceedings*, National Association of Broadcasters, Washington, D.C., 1996.

Price, R., and P. E. Green, Jr.: "A Communication Technique for Multipath Channels," *Proc. IRE*, vol. 46, ph. 555, March 1958.

Proakis, J. G.: "Advances in Equalization for Intersymbol Interference," in *Advances in Communication Systems: Theory & Application*, A. V. Balakrishnan and A. J. Viterbi (eds.), Academic Press, New York, N.Y., 1975.

Qureshi, S.: "Adaptive Equalization," *IEEE Commun*., vol. 20, pg. 9, March 1982.

Qureshi, Shahid U. H.: "Adaptive Equalization," *Proceedings of the IEEE*, IEEE, New York, vol. 73, no. 9, pp. 1349–1387, September 1985.

Radio Amateur's Handbook, American Radio Relay League, Newington, Conn., 1983.

"Receiver Planning Factors Applicable to All ATV Systems," Final Report of PS/WP3, Advanced Television Systems Committee, Washington, D.C., Dec. 1, 1994.

Riegler, R. L., and R. T. Compton, Jr.: "An Adaptive Array for Interference Rejection," *Proc. IEEE*, vol. 61, June 1973.

Rohde, Ulrich L.: "Key Components of Modern Receiver Design—Part 1," *QST*, pg. 29, May 1994.

Rohde, Ulrich L. Rohde and David P. Newkirk: *RF/Microwave Circuit Design for Wireless Applications*, John Wiley & Sons, New York, N.Y., 2000.

Rossweiler, G. C., F. Wallace, and C. Ottenhoff: "Analog versus Digital Null-Steering Controllers," *ICC '77 Conf. Rec.*, 1977.

Sgrignoli, Gary: "Preliminary DTV Field Test Results and Their Effects on VSB Receiver Design," ICEE '99.

"Standards Testing: Bit Error Rate," application note 3SW-8136-2, Tektronix, Beaverton, OR, July 1993.

"Television Receivers and Video Products," UL 1410, Sec. 71, Underwriters Laboratories, Inc., New York, 1981.

Ungerboeck, Gottfried: "Fractional Tap-Spacing Equalizer and Consequences for Clock Recovery in Data Modems," *IEEE Transactions on Communications*, IEEE, New York, vol. COM-24, no. 8, pp. 856–864, August 1976.

Using Vector Modulation Analysis in the Integration, Troubleshooting and Design of Digital RF Communications Systems, Product Note HP89400-8, Hewlett-Packard, Palo Alto, Calif., 1994.

Watson, R.: "Receiver Dynamic Range; Pt. 1, Guidelines for Receiver Analysis," *Microwaves & RF*, vol. 25, pg. 113, December 1986.

"Waveform Analysis: Noise and Jitter," application note 3SW8142-2, Tektronix, Beaverton, OR, March 1993.

Whitaker, Jerry C.: *Interactive TV Survival Guide*, McGraw-Hill, New York, N.Y., 2001.

Widrow, B., and J. M. McCool: "A Comparison of Adaptive Algorithms Based on the Methods of Steepest Descent and Random Search," *IEEE Trans.*, vol. AP-24, pg. 615, September 1976.

Widrow, B., P. E. Mantey, L. J. Griffiths, and B. B. Goode: "Adaptive Antenna Systems," *Proc. IEEE*, vol. 55, pg. 2143, December 1967.

Wilson, E.: "Evaluate the Distortion of Modular Cascades," *Microwaves*, vol. 20, March 1981.

Wugofski, T. W.: "A Presentation Engine for Broadcast Digital Television," *International Broadcasting Convention Proceedings*, IBC, London, England, pp. 451–456, 1999.

Yamada and Uematsu, "New Color TV with Composite SAW IF Filter Separating Sound and Picture Signals," *IEEE Trans.*, vol. CE-28, no. 3, p. 193.

Broadcasting, vol. 35, no. 4, pp 362–367, December 1989.

Zborowski, R. W.: "Application of On-Channel Boosters to Fill Gaps in DTV Broadcast Coverage," *NAB Engineering Conference Proceedings*, National Association of Broadcasters, Washington, D.C., 2000.

15.3r Video Signal Measurement and Analysis

ANSI Standard T1.801.03-1996, "Digital Transport of One-Way Video Signals: Parameters for Objective Performance Assessment," ANSI, Washington, D.C., 1996.

ATSC, "Transmission Measurement and Compliance for Digital Television," Advanced Television Systems Committee, Washington, D.C., Doc. A/64, Nov. 17, 1997.

ATSC, "Transmission Measurement and Compliance for Digital Television," Advanced Television Systems Committee, Washington, D.C., Doc. A/64-Rev. ZA, May 30, 2000.

Bender, Walter, and Alan Blount: "The Role of Colorimetry and Context in Color Displays," *Human Vision, Visual Processing, and Digital Display III*, Bernice E. Rogowitz (ed.), Proc. SPIE 1666, SPIE, Bellingham, Wash., pp. 343–348, 1992.

Bentz, Carl, and Jerry C. Whitaker: "Video Transmission Measurements," in *Maintaining Electronic Systems*, CRC Press, Boca Raton, Fla., pp. 328–346, 1991.

Bentz, Carl: "Inside the Visual PA, Part 2," *Broadcast Engineering*, Intertec Publishing, Overland Park, Kan., November 1988.

Bentz, Carl: "Inside the Visual PA, Part 3," *Broadcast Engineering*, Intertec Publishing, Overland Park, Kan., December 1988.

Bishop, Donald M.: "Practical Applications of Picture Quality Measurements," *Proceedings of Digital Television '98*, Intertec Publishing, Overland Park, Kan., 1998.

Boston, J., and J. Kraenzel: "SDI Headroom and the Digital Cliff," *Broadcast Engineering*, Intertec Publishing, Overland Park, Kan., pg. 80, February 1997.

"Broadening the Applications of Zone Plate Generators," Application Note 20W7056, Tektronix, Beaverton, Oreg., 1992.

DTV Express Training Manual on Terrestrial DTV Broadcasting, Harris Corporation, Quincy, Ill., September 1998.

"Eye Diagrams and Sampling Oscilloscopes," *Hewlett-Packard Journal*, Hewlett-Packard, Palo Alto, Calif., pp. 8–9, December 1996.

Fibush, David K.: "Error Detection in Serial Digital Systems," *NAB Broadcast Engineering Conference Proceedings*, National Association of Broadcasters, Washington, D.C., pp. 346-354, 1993.

Fibush, David K.: "Picture Quality Measurements for Digital Television," *Proceedings of the Digital Television '97 Summit*, Intertec Publishing, Overland Park, Kan., December 1997.

Fibush, David K.: "Practical Application of Objective Picture Quality Measurements," *Proceedings IBC '97*, IEE, pp. 123–135, Sept. 16, 1997.

Finck, Konrad: "Digital Video Signal Analysis for Real-World Problems," in *NAB 1994 Broadcast Engineering Conference Proceedings*, National Association of Broadcasters, Washington, D.C., pg. 257, 1994.

Gloeggler, Peter: "Video Pickup Devices and Systems," in *NAB Engineering Handbook*," 9th Ed., Jerry C. Whitaker (ed.), National Association of Broadcasters, Washington, D.C., 1999.

Haines, Steve: "Serial Digital: The Networking Solution?," in *NAB 1994 Broadcast Engineering Conference Proceedings*, National Association of Broadcasters, Washington, D.C., pg. 270, 1994.

Hamada, T., S. Miyaji, and S. Matsumoto: "Picture Quality Assessment System by Three-Layered Bottom-Up Noise Weighting Considering Human Visual Perception," *SMPTE Journal*, SMPTE, White Plains, N.Y., pp. 20–26, January 1999.

MacAdam, D. L.: "Visual Sensitivities to Color Differences in Daylight," *J. Opt. Soc. Am.*, vol. 32, pp. 247–274, 1942.

Mertz, P.: "Television and the Scanning Process," *Proc. IRE*, vol. 29, pp. 529–537, October 1941.

Pank, Bob (ed.): *The Digital Fact Book*, 9th ed., Quantel Ltd, Newbury, England, 1998.

Quinn, S. F., and C. A. Siocos: "PLUGE Method of Adjusting Picture Monitors in Television Studios—A Technical Note," *SMPTE Journal*, SMPTE, White Plains, N.Y., vol. 76, pg. 925, September 1967.

Reed-Nickerson, Linc: "Understanding and Testing the 8-VSB Signal," *Broadcast Engineering*, Intertec Publishing, Overland Park, Kan., pp. 62–69, November 1997.

SMPTE Engineering Guideline EG 1-1990, "Alignment Color Bar Test Signal for Television Picture Monitors," SMPTE, White Plains, N.Y., 1990.

SMPTE Recommended Practice: RP 166-1995, "Critical Viewing Conditions for Evaluation of Color Television Pictures," SMPTE, White Plains, N.Y., 1995.

SMPTE Recommended Practice: RP 167-1995, "Alignment of NTSC Color Picture Monitors," SMPTE, White Plains, N.Y., 1995.

SMPTE Recommended Practice: SMPTE RP 192-1996, "Jitter Measurement Procedures in Bit-Serial Digital Interfaces," SMPTE, White Plains, N.Y., 1996.

SMPTE Standard: SMPTE 259M-1997, "Serial Digital Interface for 10-bit 4:2:2 Components and $4F_{sc}$ NTSC Composite Digital Signals," SMPTE, White Plains, N.Y., 1997.

SMPTE Standard: SMPTE 303M, "Color Reference Pattern," SMPTE, White Plains, N.Y., 1999.

Stremler, Ferrel G.: "Introduction to Communications Systems," Addison-Wesley Series in Electrical Engineering, Addison-Wesley, New York, December 1982.

Tannas, Lawrence E., Jr.: *Flat Panel Displays and CRTs*, Van Nostrand Reinhold, New York, pg. 18, 1985.

Uchida, Tadayuki, Yasuaki Nishida, and Yukihiro Nishida: "Picture Quality in Cascaded Video-Compression Systems for Digital Broadcasting," *SMPTE Journal*, SMPTE, White Plains, N.Y., pp. 27–38, January 1999.

Verona, Robert: "Comparison of CRT Display Measurement Techniques," *Helmet-Mounted Displays III*, Thomas M. Lippert (ed.), Proc. SPIE 1695, SPIE, Bellingham, Wash., pp. 117–127, 1992.

15.3s Standards and Practices

"Advanced Television Enhancement Forum Specification," Draft, Version 1.1r26 updated 2/2/99, ATVEF, Portland, Ore., 1999.

Appelquist, P.: "The HD-Divine Project: A Scandinavian Terrestrial HDTV System," *1993 NAB HDTV World Conference Proceedings*, National Association of Broadcasters, Washington, D.C., pg. 118, 1993.

ATSC Digital Television Standard, Doc. A/53, Advanced Television Systems Committee, Washington, D.C., 1996.

ATSC: "Comments of The Advanced Television Systems Committee, MM Docket No. 00-39," ATSC, Washington, D.C., May, 2000.

"ATV System Recommendation," *1993 NAB HDTV World Conference Proceedings*, National Association of Broadcasters, Washington, D.C., pp. 253–258, 1993.

Baron, Stanley: "International Standards for Digital Terrestrial Television Broadcast: How the ITU Achieved a Single-Decoder World," *Proceedings of the 1997 BEC*, National Association of Broadcasters, Washington, D.C., pp. 150–161, 1997.

Battison, John: "Making History," *Broadcast Engineering*, Intertec Publishing, Overland Park, Kan., June 1986.

Baumgartner, Fred, and Terrence Baun: "Broadcast Engineering Documentation," in *NAB Engineering Handbook*, 9th ed., Jerry C. Whitaker (ed.), National Association of Broadcasters, Washington, D.C., 1999.

Baumgartner, Fred, and Terrence Baun: "Engineering Documentation," in *The Electronics Handbook*, Jerry C. Whitaker (ed.), CRC Press, Boca Raton, Fla., 1996.

Benson, K. B., and D. G. Fink: *HDTV: Advanced Television for the 1990s*, McGraw-Hill, New York, N.Y., 1990.

Benson, K. B., and J. C. Whitaker (eds.): *Television and Audio Handbook for Engineers and Technicians*, McGraw-Hill, New York, N.Y., 1989.

Benson, K. B., and Jerry C. Whitaker (eds.): *Television Engineering Handbook*, rev. ed., McGraw-Hill, New York, N.Y., 1992.

CCIR Document PLEN/69-E (Rev. 1): "Minutes of the Third Plenary Meeting," pp. 2–4, May 29, 1990.

CCIR Report 801-3: "The Present State of High-Definition Television," pg. 37, June 1989.

CCIR Report 801-3: "The Present State of High-Definition Television," pg. 46, June 1989.

Code of Federal Regulations, 40, Part 761.

"Current Intelligence Bulletin #45," National Institute for Occupational Safety and Health, Division of Standards Development and Technology Transfer, February 24, 1986.

Delatore, J. P., E. M. Prell, and M. K. Vora: "Translating Customer Needs Into Product Specifications", Quality Progress, January 1989.

DeSantis, Gene: "Systems Engineering Concepts," in *NAB Engineering Handbook*, 9th ed., Jerry C. Whitaker (ed.), National Association of Broadcasters, Washington, D.C., 1999.

DeSantis, Gene: "Systems Engineering," in *The Electronics Handbook*, Jerry C. Whitaker (ed.), CRC Press, Boca Raton, Fla., 1996.

Digital Audio Compression (AC-3) Standard, Doc. A/52, Advanced Television Systems Committee, Washington, D.C., 1996.

"Dr. Vladimir K. Zworkin: 1889–1982," *Electronic Servicing and Technology*, Intertec Publishing, Overland Park, Kan., October 1982.

"Electrical Standards Reference Manual," U.S. Department of Labor, Washington, D.C.

Federal Communications Commission: Notice of Proposed Rule Making 00-83, FCC, Washington, D.C., March 8, 2000.

Finkelstein, L.: "Systems Theory", *IEE Proceedings*, vol. 135, Part A, no. 6, July 1988.

General Services Administration, Information Technology Service, National Communications System: *Glossary of Telecommunication Terms*, Technology and Standards Division, Federal Standard 1037C, General Services Administration, Washington, D.C., August, 7, 1996.

Gilder, George: "IBM-TV?," *Forbes*, Feb. 20, 1989.

Guide to the Use of the ATSC Digital Television Standard, Doc. A/54, Advanced Television Systems Committee, Washington, D.C., 1996.

Hammar, Willie: *Occupational Safety Management and Engineering*, Prentice Hall, New York, N.Y.

Hammett, William F.: "Meeting IEEE C95.1-1991 Requirements," *NAB 1993 Broadcast Engineering Conference Proceedings*, National Association of Broadcasters, Washington, D.C., pp. 471–476, April 1993.

Hoban, F. T., and W. M. Lawbaugh: *Readings In Systems Engineering*, NASA, Washington, D.C., 1993.

Hopkins, R.: "Advanced Television Systems," *IEEE Transactions on Consumer Electronics*, vol. 34, pp. 1–15, February 1988.

Krivocheev, Mark I., and S. N. Baron: "The First Twenty Years of HDTV: 1972–1992," *SMPTE Journal*, SMPTE, White Plains, N.Y., pg. 913, October 1993.

Lincoln, Donald: "TV in the Bay Area as Viewed from KPIX," *Broadcast Engineering*, Intertec Publishing, Overland Park, Kan., May 1979.

Markley, Donald: "Complying with RF Emission Standards," *Broadcast Engineering*, Intertec Publishing, Overland Park, Kan., May 1986.

McCroskey, Donald C.: "Standardization: History and Purpose," in *The Electronics Handbook*, Jerry C. Whitaker (ed.), CRC Press, Boca Raton, Fla., 1996.

McCroskey, Donald: "Setting Standards for the Future," *Broadcast Engineering*, Intertec Publishing, Overland Park, Kan., May 1989.

National Electrical Code, NFPA #70.

"Occupational Injuries and Illnesses in the United States by Industry," OSHA Bulletin 2278, U.S. Department of Labor, Washington, D.C, 1985.

OSHA, "Electrical Hazard Fact Sheets," U.S. Department of Labor, Washington, D.C, January 1987.

OSHA, "Handbook for Small Business," U.S. Department of Labor, Washington, D.C.

Pank, Bob (ed.): *The Digital Fact Book*, 9th ed., Quantel Ltd, Newbury, England, 1998.

Pfrimmer, Jack, "Identifying and Managing PCBs in Broadcast Facilities," *1987 NAB Engineering Conference Proceedings*, National Association of Broadcasters, Washington, D.C, 1987.

Program Guide for Digital Television, Doc. A/55, Advanced Television Systems Committee, Washington, D.C., 1996.

Reimers, U. H.: "The European Perspective for Digital Terrestrial Television, Part 1: Conclusions of the Working Group on Digital Terrestrial Television Broadcasting," *1993 NAB HDTV World Conference Proceedings*, National Association of Broadcasters, Washington, D.C., p. 117, 1993.

"Safety Precautions," Publication no. 3386A, Varian Associates, Palo Alto, Calif., March 1985.

Schow, Edison: "A Review of Television Systems and the Systems for Recording Television," *Sound and Video Contractor*, Intertec Publishing, Overland Park, Kan., May 1989.

Schreiber, W. F., A. B. Lippman, A. N. Netravali, E. H. Adelson, and D. H. Steelin: "Channel-Compatible 6-MHz HDTV Distribution Systems," *SMPTE Journal*, SMPTE, White Plains, N.Y., vol. 98, pp. 5-13, January 1989.

Schreiber, W. F., and A. B. Lippman: "Single-Channel HDTV Systems—Compatible and Noncompatible," Report ATRP-T-82, Advanced Television Research Program, MIT Media Laboratory, Cambridge, Mass., March 1988.

Schubin, Mark: "From Tiny Tubes to Giant Screens," *Video Review*, April 1989.

Shinners, S. M.: *A Guide to Systems Engineering and Management*, Lexington, 1976.

"Sinclair Seeks a Second Method to Transmit Digital-TV Signals," *Wall Street Journal*, Dow Jones, New York, N.Y., October 7, 1999.

Smith, Milford K., Jr.: "RF Radiation Compliance," *Proceedings of the Broadcast Engineering Conference*, Society of Broadcast Engineers, Indianapolis, IN, 1989.

SMPTE and EBU: "Task Force for Harmonized Standards for the Exchange of Program Material as Bitstreams," *SMPTE Journal*, SMPTE, White Plains, N.Y., pp. 605–815, July 1998.

System Engineering Management Guide, Defense Systems Management College, Virginia, 1983.

System Information for Digital Television, Doc. A/56, Advanced Television Systems Committee, Washington, D.C., 1996.

"Television Pioneering," *Broadcast Engineering*, Intertec Publishing, Overland Park, Kan., May 1979.

"Toxics Information Series," Office of Toxic Substances, July 1983.

Tuxal, J. G.: *Introductory System Engineering*, McGraw-Hill, New York, N.Y., 1972.

"Varian Associates: An Early History," Varian publication, Varian Associates, Palo Alto, Calif.

Whitaker, Jerry C.: *AC Power Systems*, 2nd Ed., CRC Press, Boca Raton, Fla., 1998.

Whitaker, Jerry C.: *Electronic Displays: Technology, Design, and Applications*, McGraw-Hill, New York, N.Y., 1994.

Whitaker, Jerry C.: G. DeSantis, and C. Paulson*: Interconnecting Electronic Systems*, CRC Press, Boca Raton, Fla., 1993.

Whitaker, Jerry C.: *Maintaining Electronic Systems*, CRC Press, Boca Raton, Fla. 1991.

Whitaker, Jerry C.: *Power Vacuum Tubes Handbook*, 2nd Ed., CRC Press, Boca Raton, Fla., 1999.

Whitaker, Jerry C.: *Radio Frequency Transmission Systems: Design and Operation*, McGraw-Hill, New York, N.Y., 1990.

Appendix

A

A Brief History of Radio

Jerry C. Whitaker, editor

A.1 Introduction

From 1887, when Heinrich Hertz first sent and received radio waves, to the present, an amazing amount of progress has been made by radio engineers and scientists. We take for granted today what was considered science fiction just a decade or two ago. The route from the primitive spark-gap transmitters to the present state-of-the-art has been charted by the pioneering efforts of many.

Broadcasting (along with telephone technology) has brought the world closer together than the early pioneers of the art could have imagined. Eighty years have passed since Charles D. (Doc) Herrold founded a voice station (as it then was known) at San Jose, CA. Developments between then and now have been marked by many inspired breakthroughs, and many years of plain hard work.

A.2 In the Beginning

In 1895, at age 21, Guglielmo Marconi and his brother Alfonso first transmitted radio signals across the hills behind their home in Bologna, Italy. Unable to interest the Italian government in his invention, Marconi took his crude transmitter and receiver to England, where the British navy quickly realized the maritime potential of radio. Within two years, the Marconi Wireless Telegraph Company had been founded.

The invention of the vacuum tube diode by J. Ambrose Fleming in 1904 and the triode vacuum tube amplifier by Lee DeForest in 1906 launched broadcasting as we know it. Early experimental stations took this new technology and began developing their own tubes using in-house capabilities, including glass blowing. As the young electronics industry began to grow, vacuum tubes were produced in great quantity and standardized (to a point), making it possible to share new developments and applications.

It is difficult to answer the question, "Who was the first broadcaster?" Much depends on what is defined as broadcasting. As far as AM radio is concerned, the grandfather of the broadcast industry, there were five stations that exhibited a rich tradition of being first in broadcasting:

- KDKA, Pittsburgh. Dr. Frank Conrad conducted the experimental work that led to the establishment of KDKA, which made its formal debut on November 2, 1920. Conrad was apparently the first to use the term "broadcast" to describe a radio service.
- WWJ, Detroit. The birthplace of broadcasting at WWJ was the Detroit News. The station signed on the air August 20, 1920. It was the first station to be operated by a newspaper, and the first commercial station to broadcast regularly scheduled daily programs.
- KCBS, San Jose. Doc Herrold's station at San Jose (which eventually became KCBS, San Francisco) began as an experimental operation with the first documented transmissions occurring in 1909. It is said of Herrold that he conceived the idea of broadcasting information and entertainment programs to the public.
- WHA, the University of Wisconsin. 9XM-WHA achieved its first successful transmission of voice and music in 1917 from the University of Wisconsin campus in Madison. Pioneers in the establishment of the station were Malcolm Hansen and Professor Earle Terry.
- WGY, Schnectady, NY. Operated by the General Electric Company, WGY served as the test bed for many experiments in AM radio. Later efforts at the facility were directed toward perfecting FM and television transmission.

Each of these stations was first in its own way, and each played a significant role in establishing the foundation for broadcasting. Contributions included both equipment and technology.

Most stations in the 1910s and '20s built their own gear. For example, at the University of Wisconsin, Madison, special transmitting tubes were built by hand as needed to keep radio station 9XM, which later became WHA, on the air. The tubes were designed, constructed and tested by Professor E. M. Terry and a group of his students in the University laboratories. Some of the tubes were also used in wireless telephonic experiments carried on with the Great Lakes Naval Training Station during 1918, when a war-time ban was imposed on wireless broadcasts.

It took many hours to make each tube. The air was extracted by means of a mercury vapor vacuum pump while the filaments were lighted and the plate voltage was on. As the vacuum increased, the plate current was raised until the plate became red hot. This *out-gassing* process was primitive, but it worked. The students frequently worked through the night to get a tube ready for the next day's broadcasting. When completed, the device might last only a few minutes before burning out.

Plate dissipation on the early tubes, designated #1, #2, and so on, was about 25W. Tube #5 had a power output of about 50 W. Tubes #6–8 were capable of approximately 75 W. Tube #8 was one of the earliest hand-made commercial products.

A.2a Sarnoff: The Visionary

The "broadcasters" of the 1920s and early 30s were experimenters, not businessmen. Some businesses did, however, spring from the makeshift laboratories of the early scientists. David Sarnoff, general manager of the Radio Corporation of America, was quick to capitalize on the new medium.

Sarnoff was a powerful figure in the development of radio broadcasting in the 1920s and '30s, and a key mover in the development of television from its beginning in the late 1920s through maturity in the '50s and beyond. Sarnoff was born in Russia in 1891, and came to the United States at the age of 9. After completing his schooling, Sarnoff secured a job as a telegraph operator with the Marconi Wireless Company and quickly proceeded to make a name for himself. In 1912, at the age of 21, he intercepted distress signals from the doomed ship Titanic. Legend has it that Sarnoff stayed at his post for 72 straight hours to keep the world apprised of the rescue attempts, and the names of survivors. As the story goes, when the disaster was over, Sarnoff was a household word.

Four years later Sarnoff, then a contracts manager at the Marconi Company, sent a memo to his boss suggesting the use of radio for entertainment: "I have in mind a plan of development which would make radio a household utility in the same sense as the piano or phonograph... The receiver can be designed in the form of a simple 'Radio Music Box'...(which) can be placed in the parlor or living room."

Sarnoff's marketing vision and appetite for hard work led him in 1919 to join the Radio Corporation of America (RCA) as general manager. RCA was created by General Electric when the American Marconi Company was returned to private control by the U.S. government following wartime (WW-1) operations. RCA's function was principally to handle the nation's overseas communications. RCA later became a stand-alone company, independent of GE. Sarnoff was best known at RCA in the years following World War II as "The General." (In 1944 he was awarded the rank of Brigadier General.)

Radio Central

Sarnoff and RCA's first major project was the construction of a huge radio transmitting station at Rocky Point, N.Y. The facility, completed in 1921, was hailed by President Harding as a milestone in wireless progress. The President, in fact, put the station into operation by throwing a switch that had been rigged-up at the White House. Wireless stations around the globe had been alerted to tune in for a congratulatory statement by the President.

For a decade, Radio Central—as it was known—was the only means of direct communications with Europe. It was also the "hopping off" point for messages transmitted by RCA to Central and South America.

The Rocky Point site was not only famous for its role in communications, but also for the pioneers of the radio age who regularly visited there. The guest book lists such men as Guglielmo Marconi, Lee DeForest, Charles Steinmetz, Nikola Tesla, David Sarnoff, and many others. Radio Central was a milestone in transatlantic communications.

There were originally two antenna structures at the Rocky Point site, each with six 410 foot towers. The towers stretched over a 3-mile area on the eastern end of Long Island.

The facility long outlived its usefulness. RCA demolished one group of 6 towers in the 1950s; five more were destroyed in early 1960. The last tower of the once mighty Radio Central was taken down on December 13, 1977.

Radio City at Rockefeller Center

NBC was created in 1926 through the acquisition of two early experimental radio stations that had been founded by Westinghouse and American Telephone and Telegraph. The network's commercial broadcast operations began in October 1927 at 711 Fifth Avenue in New York. The facilities, then viewed as spacious, were designed by venerated chief engineer O. B. Hanson, who would later go on to design the network's 30 Rockefeller Center headquarters.

The first station, WJZ, had been established in 1921 by Westinghouse Electric and Manufacturing Company. A Newark, N.J., factory cloakroom was the original studio, equipped with a rented piano and an acoustic phonograph. After the station was purchased by RCA, the studios and transmitter were moved to Aeolian Hall in New York.

The second station, WEAF, on New York's Walker Street, was put on the air by AT&T, which was—in the words of the company—"anxious to study the possibilities of radio broadcasting."

In just six years, the pioneer network outgrew the Fifth Avenue quarters. On November 11, 1933, NBC began network operations at 30 Rockefeller Center. The 11-story building, just to the west of the 70-story RCA Building Tower, simultaneously fed two networks and the two local stations. The building contained 22 studios and five audition rooms, with associated client and observation rooms, switching booths, a master control center, and other technical facilities. The new radio complex occupied almost 280,000 square feet.

Hanson's Vision

Hanson's master plan for radio broadcasting from 30 Rock evolved from his radio communication experience, which began in 1912 with his education in wireless at the Marconi Company (later the RCA Institute). He went to sea twice, and between voyages, worked for Marconi's testing department. Hanson eventually became chief engineer of the department and helped put WAAM radio in Newark on the air. Hanson moved to WEAF in 1922, becoming the assistant to the plant engineer. When NBC bought the station in 1926, he became the chief engineer.

Hanson's vision in broadcast facilities design was implicit in the radio network's decades-long occupancy of 30 Rock, with the continually expanding TV network as a co-occupant since the late 1930s. His accomplishments lend credence to the old saw that experience is the best teacher. His design knowledge was the amalgamation of lessons learned at sea, at Marconi, at the experimental facilities of WAAM and WEAF and, later, in the installation and operation of the network's Fifth Avenue facilities.

It is particularly interesting that Hanson designed 30 Rock for both radio network broadcasting, and for what he described as "elementary visual broadcasting, or television, as it is more popularly known."

His foresight was evident. "Anticipating the advent of television," he wrote, "the entire lighting system in the studio section was designed to operate on direct current to obviate the possibility of stroboscopic interference from alternating current lighting. Five 750 kW motor generator sets in the basement (have) a considerable portion of capacity intended for flood lighting, a requirement for television."

The design of a ninth floor, 4-studio group anticipated special needs for complex live radio production as well as live television. The four 30-ft, almost-square studios were arranged around a single central control room. Hanson installed sound-insulated windows from this control room into each of the four studios. The floor of the control room was designed so that it could be converted into a turntable that revolved via an electric motor.

"One purpose of this studio group is to allow, in sound broadcasting, for the separation of various performers," Hanson wrote. "For example, the orchestra might be in one studio, the principal players in another and the remainder of the cast in still another. The outputs from the three studios would then be electrically combined or mixed. The arrangement allows the producer a more accurate control over the balances of the units involved."

"Another purpose of these studios is to allow for rapid scene shifts in television. This would be accomplished by simply revolving the control-room floor so that the scanning equipment, which would be mounted thereon, would face the desired studio," he said. (What was Hanson's

1930 vision as to the form and size of the TV camera?) "Thus, four scene changes could be effected, if necessary, within a very short time."

According to Hanson, "The maximum practicable degree of centralization (is needed) to simplify the problem of routine inspection and maintenance, and to facilitate the location or correction of trouble in the minimum possible time."

All of the centralized switching, master control and input/output equipment was situated originally on the fifth floor, along with a power room and battery room (providing 14 Vdc filament power at 250 A, and 400 Vdc plate B+ voltage at 2.5 A), and a main equipment room filled with 330 racks including 300 audio amplifiers. On the periphery of this floor (one of two with windows) were operating staff offices, recording rooms (for electrical transcriptions), a laboratory (specific capabilities or purposes not identified), and operating and maintenance shops.

With the sale of NBC radio in 1988, the 11-story studio core was transformed into an all-TV plant.

A.3 WLW: The Nation's Station

Radio station WLW has a history as colorful and varied as any in the United States. It is unique in that it was the only station ever granted authority to broadcast with 500 kW.

The station actually began with 20 watts of power as a hobby of Powel Crosley, Jr. The first license for WLW was granted by the Department of Commerce in 1922. Crosley was authorized to broadcast on a wavelength of 360 m with a power of 50 W, three evenings a week.

Growth of the station was continuous. It operated at various frequencies and power levels until, in 1927, WLW was assigned to 700 kHz at 50 kW and remained there. Operation at 50 kW commenced on October 4, 1928. The transmitter was located in Mason, OH. The station could be heard as far away as Jacksonville, FL and Washington, D.C.

The super-power era of WLW began in 1934. The contract for construction of the enormous transmitter was awarded to RCA in February 1933. Tests on the unit began on January 15, 1934. The cost of the transmitter and associated equipment was approximately $400,000, not much today, but a staggering sum in the middle of the Great Depression.

At 9:02 p.m. on May second, programming was commenced with full 500 kW of power. The super-power operation was designed to be experimental, but Crosley managed to renew the license every 6 months until 1939. The call sign W8XO was occasionally used during test periods, but the regular call sign of WLW was used for programming.

"Immense" is the only way to describe the WLW facility. The antenna reached a height (including the flagpole at the top) of 831 ft. The antenna rested on a single ceramic insulator that supported the combined force of 135 tons of steel and 400 tons exerted by the guys. The tower was guyed with eight 1 7/8-inch cables anchored 375 ft from the base of the antenna.

The main antenna was augmented by a directional tower designed to protect CFRB, Toronto, when the station was using 500 kW at night. The directional system was unique in that it was the first designed to achieve both horizontal directivity and vertical-angle suppression.

A spray pond in front of the building provided cooling for the system, moving 512 gallons of water per minute. Through a heat exchanger, the water then cooled 200 gallons of distilled water in a closed system that cooled the transmitting tubes.

The transmitter consumed an entire building. Modulation transformers weighing 37,000 lbs each were installed in the basement. Three plate transformers, a rectifier filter reactor, and a modulation reactor were installed outside the building. The exciter for the transmitter produced 50 kW of RF power. A motor-generator was used to provide 125 Vdc for control circuits.

The station had its own power substation. While operating at 500 kW, the transmitter consumed 15,450,000 kWh. The facility was equipped with a complete machine shop because station personnel had to build much of the ancillary hardware they needed. Equipment included gas, arc and spot welders, a metal lathe, milling machine, engraving machine, sander, drill press, metal brake, table saw, and other equipment. A wide variety of electrical components were also on hand.

WLW operated at 500 kW until March 1, 1939, when the FCC ordered the station to reduce power to 50 kW. The station returned to super-power operation a few times during World War II for government research. The days when WLW could boast to being "the nation's station," were, however, in the past.

A.4 The FCC Enters the Picture

As broadcasting began to develop, it became obvious to lawmakers that some type of regulation was needed to provide for orderly use of the airwaves. To fill this need, the Federal Communications Commission was formed in 1934, the end result of the Communications Act of the same year. The FCC was officially established on June 19, 1934, when President Franklin Roosevelt signed the enabling legislation. Some noteworthy actions during the first year of operation included:

- First broadcast license denied (KGIX, Las Vegas) for failure to complete its construction as required.
- Current ownership information required from broadcast stations.
- Hearings held on non-profit education broadcasting allocations.
- Allocations for clear and other channels (with several stations operating simultaneously at night) to be surveyed.
- First amateur license revoked.

A.5 The Golden Era of Radio

Radio of the pre-World War II days was more than an entertainment and information medium. It was a friend; a connection with the rest of the world; an escape from the hardships of the Great Depression. Radio was special. People involved in it were special. This, legend has it, was the "golden age" of radio.

If you were to step back 50 years into the studios of a local station you would find a much different world than the broadcast station of today: different equipment, different types of people, and different business goals. Broadcasting of this era was more than just a business or a job. It was the profession of magic.

Radio had a distinctly formal air about it. Announcers and musicians dressed in tuxedos. Female performers were elegantly attired as well, even when there was no studio audience. Announcing was formal. Broadcasting was regarded as a grand production, almost theatrical in nature. Enunciation and vocal clarity were essential, partly due to limitations of the equipment, but also because of the tradition of the theater. This formality of attitude and style would remain a part of radio well into the 1940s.

Many local stations had a staff orchestra, some for playing jazz, others for symphonic programs. A few stations had their own dramatic groups. Each station had its following of loyal fans who would structure their days around their favorite radio programs. Those were the days that brought radio to its peak of popularity and influence.

Nearly every station had several studios of varying sizes. In New York and Hollywood, the networks used theaters for programs presented before audiences. Because audio consoles of stock design did not appear until the late 1930s, each station assembled its own facilities. Mixers were put together from individual faders. Amplifiers were stock items consisting of several basic types. Monitor loudspeakers were usually electro-dynamics mounted in a baffle of no particular design. Performance left a lot to be desired.

The standard volume unit indicator (VU meter) was adopted by the industry in 1939. Prior to that time many different instruments were used to adjust program levels. Ballistic characteristics were far from standard, and the rectifier configurations inside the meters varied from model to model.

The mainstay of the studio was the venerable RCA 44 series ribbon microphone. The mic provided fairly good directional characteristics, but its size and vulnerability to wind noise limited its use mainly to inside pickups. Condenser microphones were used to some extent in the 1930s. Compared with today's versions, the early condensers were large, heavy, and prone to produce noise under humid conditions. Dynamic microphones also enjoyed popularity, chiefly for remotes because of their ruggedness.

For many years, phonograph records were frowned upon for actual broadcast use, due in part to questions regarding license and royalty matters. The main source of recorded music was transcription libraries. These libraries were leased to stations for use on the air. The 16-inch discs utilized various forms of modulation. Some used a vertical (hill and dale) cut while others used a laterally modulated groove similar to modern discs. Turntables had to be large to accommodate the 16-inch discs; the machines were heavy for speed stability.

Remote amplifiers were back-breakers in the literal sense. These so-called portable units weighed in at 35 to 40 pounds. Add to this several microphones, stands, cables, and headphones, and you can see why only the bravest souls liked to do remotes.

The use of radio relay equipment was limited to frequencies between 1.6 and 3 MHz. This required relatively long antennas, even when loaded with an inductance. Skywave effects were also a problem, and frequently caused interference. Because AM was the only method available, the systems were highly susceptible to noise.

A.6 The Origins of the Networks

The formation and growth of the national radio networks is an exciting chapter in the history of broadcasting. The growth and prosperity of networks was linked tightly with the number of potential affiliates.

The giant RCA organization was the first company to recognize that the development of broadcast technology and management of broadcast services could best be performed by an independent organization. Accordingly, RCA under the guidance of David Sarnoff set the trend in network formation by creating the National Broadcasting Company in 1926. By the following year it had two divisions, the Red and Blue networks.

The origins of the Columbia Broadcasting System (CBS) can be traced back to 1926 and a company called United Independent Broadcasters (UIB). The firm came close to bankruptcy several times during its first year or so of operation. One of the organizations that kept UIB afloat for awhile was the Columbia Phonograph Company. When the record company sold back to the original investors their interest, they retained the name Columbia Broadcasting System. In 1927 William Paley became interested in the fledgling company, and decided to invest in it and help run it. The rest, as they say, is history. Incidentally, in 1938 CBS brought the Columbia Phonograph company (renamed Columbia Records) and held it until the late 1980s.

In 1934, the Mutual Broadcasting System was created to serve the increasing number of radio stations on the air at that time. During the 1940s, two additional radio networks were founded, the DuMont Network (1946) and the Liberty Broadcasting System (1949). They played an important role in the broadcast industry of their time, but later bowed to the giant networks and their well-established affiliates.

In 1943, Edward J. Nobel bought the NBC Blue Network (operated by NBC) and renamed it the American Broadcasting Company. The sale to Nobel was prompted by a federal antitrust ruling. The sale price to the Lifesaver manufacturer was $8-million. In 1953, the company was merged with Paramount, providing a valuable entry for the network into the telefilm business.

A.7 FM Grows Up

Although FM radio had demonstrated impressive audio fidelity from the time it was first demonstrated for broadcast applications by inventor Major Edwin Armstrong, FM remained the forgotten stepchild of radio—at least until the early 1960s. The awakening came as the result of FM license allocation changes and stereo transmission. In 1962 the FCC revised its commercial FM rules to divide the United States into 3 zones (instead of the previous 2). Three classes of commercial FM stations were also created. Until the '62 decision, FM stations were authorized on the basis of protecting the predicted service contours of existing stations. The new rules, however, changed the FM assignment scheme to one requiring minimum mileage separations between stations.

In 1963 the table of assignments for commercial FM stations was created, and nearly 3,000 assignments were made to nearly 2,000 mainland communities. Assignments in Alaska, Hawaii, Puerto Rico, and the Virgin Islands were added in 1964.

The new scheme enacted by the commission was not universally popular, however. Some FM stations on the air at the time were faced with having to reduce their power level and/or antenna height to meet the new guidelines. Stations operating at high powers—125 kW ERP was not uncommon—protested and sought to gain public support. In Northern California, for example, stations that were to be impacted by the new rules organized an appeal to listeners—broadcast simultaneously on the stations. All of the stations to be affected reduced their power during the broadcast to show the public the negative impact such a reduction would have on service.

The effort was copied in Los Angeles and other markets. Finally, in 1963, the grandfather rights of stations that had power in excess of Class B limits were granted.

The second spark that made FM come alive was stereo. Although stereo audio dates back to experiments performed over wire lines by telephone engineers in the 1880s, real development came only with post-World War II technology. In 1959, the National Stereophonic Radio Committee was created to examine the many proposed systems of transmitting FM stereo and submit a final recommendation to the FCC. In the summer of 1960, six systems were field tested over KDKA-FM in Pittsburgh, with receivers set up at Uniontown, PA. The system proposed by General Electric and Zenith was adopted, with broadcasting authorized to start on June 1, 1961.

The first stations to begin stereophonic programing under the new rules were, quite appropriately, WGHM, Schenectady, N.Y. (owned by General Electric), and WEFM, Chicago, IL, (owned by Zenith).

Circular polarization of the transmitted signal was another major step for FM radio. One of the early proponents was KPEN-FM, Atherton, Calif. (on the San Francisco Peninsula), which later would become KIOI (better known as K-101), San Francisco. KPEN received a special temporary authorization from the FCC in the later part of 1963 to start testing the effects of adding a vertical component to the existing horizontal signal.

A second Western Electric 10 kW transmitter was purchased and modified to provide the needed power. Separate vertical dipoles were manufactured and installed on the station's tower. With this setup, engineers were able to vary the phase relationship and amplitude so that the station could switch from a horizontally polarized signal to a circular pattern. Monitoring points were established in rugged areas of San Francisco to observe the results. It was found that as the vertical component of the transmitted signal was increased, reception of stereo signals improved significantly.

At the same time, Lew Wetzel of WFIL-FM was proving to the commission that the vertical component of a circularly polarized transmission did not extend the 1 mV contour. With these two reports, the FCC decided that it would indeed be in the public interest for FM stations to transmit with circular polarization.

A.8 Bibliography

Burke, William: "WLW: The Nation's Station," *Broadcast Engineering*, Intertec Publishing, Overland Park, KS, November 1967.

Dorschug, Harold: "The Good Old Days of Radio," *Broadcast Engineering*, Intertec Publishing, Overland Park, KS, May 1971.

Gabbert, Jim: "Case Study: K101, an FM Pioneer," *Broadcast Engineering*, Intertec Publishing, Overland Park, KS, May 1979.

Nelson, Cindy: "RCA Demolishes Last Antenna Tower at Historic Radio Central," *Broadcast Engineering*, Intertec Publishing, Overland Park, KS, February 1978.

Paulson, Robert: "The House that Radio Built," *Broadcast Engineering*, Intertec Publishing, Overland Park, KS, April 1989.

"Radio Pioneers," *Broadcast Engineering*, Intertec Publishing, Overland Park, KS, May 1979.

Riggins, George: "The Real Story on WLW's Long History," *Radio World*, IMAS Publishing, Falls Church, VA, March 8, 1989.

Schubin, Mark: "From Tiny Tubes to Giant Screens," *Video Review*, April 1989.

"The Last 20 Years at the FCC," *Broadcast Engineering*, Intertec Publishing, Overland Park, KS, May 1979.

Wallechinsky and Wallace: *The People's Almanac*, #2, Bantam Books, 1978.

WTIC: Radio to Remember, WTIC 60th anniversary publication.

Appendix

B

A Brief History of Television

Jerry C. Whitaker, editor

B.1 Introduction

From humble beginnings, television has advanced to become the most effective communications medium in the history of this planet.

"Standardization at the present stage is dangerous. It is extremely difficult to change a standard, however undesirable it may prove, after the public has invested thousands of dollars of equipment. The development goes on, and will go on. There is no question that the technical difficulties will be overcome."

The writer is not addressing the problems faced by high-definition television or fiber optic delivery of video to the home. The writer is addressing the problems faced by *television*. The book containing this passage was published in April 1929. Technology changes, but the problems faced by the industry do not.

B.2 Television: A Revolution in Communications

The mass communications media of television is one of the most significant technical accomplishments of the 20th century. The ability of persons across the country and around the world to *see* each other, to communicate with each other, and experience each other's cultures and ideas is a monumental development. Most of us have difficulty conceiving of a world without instant visual communication to virtually any spot on earth. The technology that we enjoy today, however, required many decades to mature.

B.2a The Nipkow Disc

The first working device for analyzing a scene to generate electrical signals suitable for transmission was a scanning system proposed and built by Paul Nipkow in 1884. The scanner consisted of a rotating disc with a number of small holes (or *apertures*) arranged in a spiral, in front of a photo-electric cell. As the disc rotated, the spiral of 18 holes swept across the image of the scene from top to bottom in a pattern of 18 parallel horizontal lines.

The Nipkow disc was capable of about 4,000 picture "dots" (or pixels) per second. The scanning process analyzed the scene by dissecting it into picture elements. The fineness of picture detail that the system was capable of resolving was limited in the vertical and horizontal axes by the diameter of the area covered by the aperture in the disc. For reproduction of the scene, a light source controlled in intensity by the detected electrical signal was projected on a screen through a similar Nipkow disc rotated in synchronism with the pickup disc.

Despite subsequent improvements by other scientists (J. L. Baird in England and C. F. Jenkins in the United States) and in 1907 the use of Lee De Forest's vacuum-tube amplifier, the serious limitations of the mechanical approach discouraged any practical application of the Nipkow disc. The principle shortcomings were:

- Inefficiency of the optical system
- Use of rotating mechanical components
- Lack of a high-intensity light source capable of being modulated by an electrical signal at the higher frequencies required for video signal reproduction.

Nevertheless, Nipkow demonstrated a scanning process for the for the analysis of images by dissecting a complete scene into an orderly pattern of picture elements that could be transmitted by an electrical signal and reproduced as a visual image. This approach is—of course—the basis for present-day television.

Nipkow lived in Berlin, although he was of Russian birth. The U.S.S.R. claims a Russian invented television, not because of Nipkow, but another man who experimented with the Nipkow disc in 1905 in Moscow. The Germans, English and Japanese also claim their share of the fame for inventing television.

No one argues, however, that credit for the development of modern *electronic* television belongs to two men: Philo T. Farnsworth and Vladimir Zworykin. Each spent their lives perfecting this new technology.

B.2b Zworykin: The Brains of RCA

A Russian immigrant, Vladimir Zworykin came to the United States after World War I and went to work for Westinghouse in Pittsburgh. During his stay at the company, 1920 until 1929, Zworykin performed some of his early experiments in television. Zworykin had left Russia for America to develop his dream: television. His conception of the first practical TV camera tube, the *Iconoscope* (1923), and his development of the *kinescope* picture tube formed the basis for subsequent advances in the field. Zworykin is credited by most historians as *the father of television*.

Zworkin's Iconoscope (from Greek for "image" and "to see") consisted of a thin aluminum-oxide film supported by a thin aluminum film and coated with a photosensitive layer of potassium hydride. With this crude camera tube and a CRT as the picture reproducer, he had the essential elements for electronic television.

Continuing his pioneering work, Zworykin developed an improved Iconoscope six years later that employed a relatively thick, 1-sided target area. He had, in the meantime, continued work on improving the quality of the CRT and presented a paper on his efforts to the Eastern Great Lakes District Convention of the Institute of Radio Engineers (IRE) on November 18, 1929. The presentation attracted the attention of another former Russian immigrant, David Sarnoff, then vice president and general manager of RCA. Sarnoff persuaded Zworykin to join RCA Victor in Camden, NJ, where he was made director of RCA's electronics research laboratory. The company provided the management and financial backing that enabled Zworykin and the RCA scientists working with him to develop television into a practical system.

Both men never forgot their first meeting. In response to Sarnoff's question, Zworykin—thinking solely in research terms—estimated that the development of television would cost $100,000. Years later, Sarnoff delighted in teasing Zworykin by telling audiences what a great salesman the inventor was. "I asked him how much it would cost to develop TV. He told me $100,000, but we spent $50 million before we got a penny back from it."

By 1931, with the Iconoscope and CRT well-developed, electronic television was ready to be launched—and Sarnoff and RCA were ready for the new industry of television.

B.2c Farnsworth: The Boy Wonder

Legend has it that Philo Farnsworth conceived of electronic television when he was a 15 year old high school sophomore in Rigby, Idaho, a small town about 200 miles north of Salt Lake City. Farnsworth met a financial expert by the name of George Everson in Salt Lake City when he was 19 years old and persuaded him to try and secure venture capital for an all-electronic television system.

The main concern of the financial investors whom Everson was able to persuade to put up money for this unproven young man with unorthodox ideas, was that no one else was investigating an electronic method of television. Obviously, many people were interested in capturing the control over patents of a vast new field for profit. If no one was working on this method, then Farnsworth had a clear field. If, on the other hand, other companies were working in secret without publishing their results, then Farnsworth would have little chance of receiving the patent awards and the royalty income that would surely result. Farnsworth and Everson were able to convince the financial backers that they alone were on the trail of a total electronic television system.

Farnsworth established his laboratory first in Los Angeles, and later in San Francisco at the foot of Telegraph Hill. Farnsworth was the proverbial lone basement experimenter. It was at his Green Street (San Francisco) laboratory that Farnsworth gave the first public demonstration in 1927 of the television system he had dreamed of for six years.

He was not yet 21 years of age!

Farnsworth was quick to developed the basic concepts of an electronic television system, giving him an edge on most other inventors in the race for patents. His patents included the principle of blacker-than-black synchronizing signals, linear sweep and the ratio of forward sweep to retrace time. Zworykin won a patent for the principle of field interlace.

In 1928 Farnsworth demonstrated a non-storage electronic pickup and image scanning device he called the *Image Dissector*. The detected image was generated by electrons emitted from a photocathode surface and deflected by horizontal and vertical scanning fields (applied by coils surrounding the tube) so as to cause the image to scan a small aperture. In other words, rather than an aperture or electron beam scanning the image, the aperture was stationary and the electron image was moved across the aperture. The electrons passing through the aperture were collected to produce a signal corresponding to the charge at an element of the photocathode at a given instant.

The limitation of this invention was the extremely high light level required because of the lack of storage capability. Consequently, the Image Dissector found little use other than as a laboratory signal source. Still, in 1930, the 24-year-old Farnsworth received a patent for his Image Dissector, and in the following year entertained Zworykin at his San Francisco laboratory.

Farnsworth's original "broadcast" included the transmission of graphic images, film clips of a Dempsey/Tunney fight and scenes of Mary Pickford combing her hair (from her role in the *Taming of the Shrew*). In his early systems, Farnsworth could transmit pictures with 100- to 150-line definition at a repetition rate of 30 lines per second. This pioneering demonstration set in motion the progression of technology that would lead to commercial broadcast television a decade later.

Farnsworth held many patents for television, and through the mid-1930s remained RCA's fiercest competitor in developing new technology. Indeed, Farnsworth's thoughts seemed to be directed toward cornering patents for the field of television and protecting his ideas. In the late 1930s, fierce patent conflicts between RCA and Farnsworth flourished. They were settled in September 1939 when RCA capitulated and agreed to pay continuing royalties to Farnsworth for the use of his patents. The action ended a long period of litigation. By that time Farnsworth held an impressive list of key patents for electronic television.

Farnsworth died in 1971 and is credited only slightly for the giant television industry that he helped create.

B.2d Other Experimenters

Unsuccessful attempts were made to use pickup devices without storage capability, such as the Farnsworth Image Dissector, for studio applications. The most ambitious was the Allen B. DuMont Laboratories' experiments in the 1940s with an electronic *flying-spot* camera. The set in the studio was illuminated with a projected raster frame of scanning lines from a cathode-ray tube. The light from the scene was gathered by a single photo-cell to produce a video signal.

The artistic and staging limitations of the dimly-lit studio are all too obvious. Nevertheless, while useless for live pickups, it demonstrated the flying-spot principle, a technology that is widely used today for television transmission of motion picture film.

General Electric also played an early role in the development of television. In 1926, Ernst Alexanderson, an engineer at the company, developed a mechanical scanning disc for video transmission. He gave a public demonstration of the system two years later. Coupled with the GE experimental TV station, WGY (Schenectady, N.Y.), Alexanderson's system made history on September 11, 1928, by broadcasting the first dramatic program on television. It was a 40-minute play titled, *The Queen's Messenger*. The program consisted of two characters performing before three simple cameras.

There was a spirited race to see who could begin bringing television programs to the public first. In fact, the 525 line 60Hz standards promoted in 1940 and 1941 were known as "high-definition television," as compared with some of the experimental systems of the 1930s. The original reason for the 30 frame per second rate was the simplified receiver design that it afforded. With the field scan rate the same as the power system frequency, ac line interference effects were minimized in the reproduced picture. Both Zworykin and Farnsworth were members of the committee that came up with proposed standards for a national (U.S.) system. The standard was to be in force before any receiving sets could be sold to the public.

The two men knew that to avoid flicker, it would be necessary to have a minimum of 40 complete pictures per second; this was known from the motion picture industry. Although film is exposed at 24 frames per second, the projection shutter is opened twice for each frame, giving a net effect of 48 frames per second. If 40 complete pictures per second were transmitted, even with 441 lines of horizontal segmentation (which was high-definition TV prior to WW II), the required bandwidth of the transmitted signal would have been greater than the technology of the day could handle. The *interlace* scheme was developed to overcome the technical limitations faced by 1940s technology.

B.2e Pickup Tubes

Zworykin's Iconoscope television camera tube, first patented in 1923, stored an image of a scene as a mosaic pattern of electrical charges. A scanning electron-beam then released secondary electrons from the photosensitive mosaic to be read out sequentially as a video signal.

Although the Iconoscope provided good resolution, relatively high light levels (studio illumination of 500 or more foot-candles) was necessary. In addition, picture quality was degraded by spurious flare. This was caused by photoelectrons and secondaries, resulting from the high potential of the scanning beam, falling back at random on the storage surface. The presence of flare, and the lack of a reference black signal (because of capacitive coupling through the signal plate) resulted in a gray scale that varied with scene content. The pickup system thus required virtually continuous manual correction of video-gain and blanking levels.

Furthermore, because the light image was focused on the same side of the signal plate as the charge image, it was necessary to locate the electron gun and deflection coils off the optical axis in order to avoid obstructing the light path. Because the scanning beam was directed at the signal plate at an average angle of 45 degrees, vertical keystone correction of the horizontal scan was needed.

With all of its shortcomings, the Iconoscope was the key to the introduction of the first practical all-electronic television system. Because a cathode-ray picture-display tube (necessary to supplant the slowly-reacting modulated light source and cumbersome rotating disc of the Nipkow display system) had been demonstrated as early as 1905, a television system composed entirely of electronic components was then feasible.

Zworykin continued to refine the Iconoscope, demonstrating improved tubes in 1929 and 1935. His work culminated in the development of the *Image Iconoscope* in 1939, which offered greater sensitivity and overcame some of the inherent problems of the earlier devices. In the Image Iconoscope, a thin-film transparent photocathode was deposited on the inside of the faceplate. Electrons emitted and accelerated from this surface at a potential several hundred electron-volts were directed and focused on a target storage plate by externally-applied magnetic fields. A positive charge image was formed on the storage plate, this being the equivalent of the storage mosaic in the Iconoscope. A video signal was generated by scanning the positive charge image on the storage plate with a high-velocity beam in exactly the same manner as the Iconoscope.

Both types of Iconoscopes had a light-input/video-output characteristic that compressed highlights and stretched lowlights. This less-than-unity relationship produced signals that closely matched the exponential input-voltage/output-brightness characteristics of picture display tubes, thus producing a pleasing gray-scale of photographic quality.

The *Orthicon* camera tube, introduced in 1943, was the next major development in tube technology. It eliminated many of the shortcomings of the Iconoscope through the use of low-velocity scanning. The Orthicon, so named because the scanning beam landed on the target at right angles to the charge surface, used a photoemitter composed of isolated light-sensitive granules deposited on an insulator. A similar tube, the *CPS Emitron* (so named for its cathode-potential stabilized target scanning), was developed in England. The CPS Emitron target was made up of precise squares of semitransparent photoemissive material deposited on the target insulator through a fine mesh. Both of these tubes produced high-resolution pictures with precise gray-scales.

In 1943, an improved Orthicon pickup device, the *Image Orthicon*, was introduced. The tube incorporated 3 important technologies to make possible studio and field operations under reduced and varied lighting conditions. The technologies involved were:

- Imaging the charge pattern from a photosensitive surface on an electron storage target
- Modulation of the scanning beam by the image charge of the target
- Amplification of the scanning-beam modulation signal by secondary-electron emission in a multistage multiplier.

The image of the scene being televised was focused on a transparent photocathode on the inside of the faceplate of the tube. The diameter of the photocathode on a 3-inch diameter Image Orthicon was 1.6 in. (41 mm), the same as double-frame 35 mm film, a fortunate choice because it permitted the use of already-developed conventional lenses. Light from the scene caused a charge pattern of the image to be set up. Because the faceplate was at a negative voltage (approximately –450V), electrons were emitted in proportion to scene illumination and accelerated to the target-mesh surface, which was at (nearly) zero potential. The fields from the accelerator grid and focusing coil focused the electrons on the target.

The *Vidicon* was introduced to the broadcast industry in 1950. It was the first successful television camera tube to use a photoconductive surface to derive a video signal. The photosensitive target material of the Vidicon consisted of a continuous light-sensitive film deposited on a transparent signal electrode. An antimony trisulphide photoconductor target was scanned by a low-velocity electron beam to provide an output signal directly. No intermediate electron-imaging or electron-emission processes, as in the Image Orthicon or Iconoscope, were employed.

While a variety of tubes have since been developed and identified under different trade names (or by the type of photoconductor used), the name *Vidicon* has become the generic classification for all such photoconductive devices. Milestones in this developmental process include introduction of the 1-inch Plumbicon in 1968, the 2/3-inch Plumbicon in 1974, the 1/2-inch Plumbicon in 1981, and the 1/2-inch Saticon also in 1981.

B.2f Sold State Imaging

Solid-state imaging devices using a flat array of photosensitive diodes were proposed as early as 1964 and demonstrated publicly in 1967. The charge voltage of each sensor element was sampled in a horizontal and vertical, or *X-Y*, addressing pattern to produce an output voltage corresponding to a readout of the image pixels. The resolution capability of these first laboratory models did not exceed 180 × 180 pixels, a tenth of that required for television broadcasting applications. Nevertheless, the practicability of solid-state technology was demonstrated.

In the first solid state camera system, a video signal was generated by sampling the charge voltages of the elements of the array directly in an *X* and *Y* (horizontal and vertical) scanning pattern. In the early 1970s a major improvement was achieved with the development of the *charge-coupled device* (CCD), in operation a charge-transfer system. The photosensitive action of a simple photodiode was combined in one component with the charge-transfer function and metal-oxide capacitor storage capability the CCD. The photo-generated charges were transferred to metal-oxide semiconductor (MOS) capacitors in the CCD and stored for subsequent readout as signals corresponding to pixels.

Thus, rather than sampling directly the instantaneous charge on each photosensitive picture element, the charges were stored for readout either as a series of picture scanning lines in the *interline-transfer* system, or as image fields in the *frame-transfer* system.

The early CCD chips were interline-transfer devices in which vertical columns of photosensitive picture elements were alternated with vertical columns of sampling gates. The gates in turn fed registers to store the individual pixel charges. The vertical storage registers were then sampled one line at a time in a horizontal and vertical scanning pattern to provide an output video signal. This approach was used in early monochrome cameras and in three-sensor color cameras. It was also used with limited success in a single-tube color camera wherein cyan, green, and yellow stripe filters provided three component color signals for encoding as a composite signal. The interline system is of only historical interest. Frame-transfer technology is now used in all professional-quality cameras.

Milestones in the development of CCD devices for professional applications include the introduction in 1979 by Bosch of the FDL-60 CCD-based telecine, the NEC SPC-3 CCD camera in 1983, and the RCA CCD-1 camera in 1984.

B.2g Image Reproduction

From the start of commercial television in the 1940s until the emergence of color as the dominant programming medium in the mid-1960s, virtually all receivers were the direct-view monochrome type. A few large-screen projection receivers were produced, primarily for viewing in public places by small audiences. Initially the screen sizes were 10 to 12-in. diagonal.

The horizontal lines of the two fields on a receiver or monitor screen are produced by a scanning electron beam which, upon striking the back of the picture tube screen, causes the phosphor to glow. The density of the beam, and the resultant brightness of the screen, is controlled by the voltage level of a video signal applied between the controlling aperture and the cathode in the electron gun.

In the old days, viewers were advised to sit at least one foot away from the screen for every inch of screen size as measured diagonally. Thus, if you had a 25-inch screen TV set, you were supposed to sit 25-feet away. In those early days the electron beam scan of the CRT phosphor revealed with crisp sharpness the individual scanning lines in the raster. In fact, the focus of the electron beam was sometimes purposely set for a soft focus so the scan lines were not as easily seen.

All color television picture displays synthesize the reproduction of a color picture by generating light, point by point, from three fluorescent phosphors, each of a different color. This is called an *additive* system. The chroma characteristic, or hue, of each of color light source is defined as a primary color. The most useful range of reproduced colors is obtained from the use of three primaries with hues of red, green, and blue. A combination of the proper intensities of red, green and blue light will be perceived by an observer as white.

Utilizing this phenomenon of physics, color television signals were first produced by optically combining the images from three color tubes, one for each of the red, green and blue primary transmitted colors. This early *Trinescope*, as it was called by RCA, demonstrated the feasibility of color television. The approach was, however, too cumbersome and costly to be a practical solution for viewing in the home.

The problem was solved by the invention of the shadow-mask picture tube in 1953. The first successful tube used a triad assembly of electron guns to produce three beams that scanned a screen composed of groups of red, green, and blue phosphor dots. The dots were small enough not to be perceived as individual light sources at normal viewing distances. Directly behind the screen, a metal mask perforated with small holes approximately the size of each dot triad, was aligned so that each hole was behind an R-G-B dot cluster.

The three beams were aligned by *purity* magnetic fields so that the mask *shadowed* the green and blue dots from the beam driven by the red signal. Similarly, the mask shadowed the red and blue dots from the from green beam, and the red and green dots from the blue beam.

B.2h Who was First?

The technology of television actually was the creation of many people in many countries over many years. It is correctly viewed as an international invention.

British developments in television saw great advances during the thirties. While Baird was conducting daily nighttime 30 line broadcasts from 1929 on two BBC MW transmitters (video and audio), EMI Laboratories in Hayes (Middlesex, England) was developing electronic television of their own. Patents were freely shared, but much development took place in the U.K. (EMI was linked to RCA from the days when the Victor Talking Machine Company of Camden and Gramophone Company of Hayes were related.)

EMI and Marconi in 1934 proposed to the Selsdon Committee on the future of U.K. Television, a system comprised of 405 line interlaced scanning; the 405 waveform included the serrated vertical sync pulse but did not include "equalizing" pulses to improve interlace. Baird proposed 240 lines 25 frames progressive scanning. On November 2, 1936, daily programming was inaugurated, with the BBC alternating weekly each transmission system. Programming began with two hours a day (not including test transmissions). The first sets were made switchable to both systems. (Fortunately each system shared the same video/audio carrier frequencies of 45 MHz and 41.5 MHz, respectively, and each used AM sound and positive modulation of the video.) EMI used iconoscope (Zworykin) camera tubes, whereas Baird, in 1936, used a flying-spot studio camera (four years earlier than DuMont), an "Intermediate Film/Flying Spot" camera of German origin, and an imported Farnsworth Image Dissector camera.

The Baird System was ordered to shut down on February, 8, 1937, and the 405 line system remained. It is estimated that 20,000 TV sets were sold to the public by September, 1939, when the BBC Television service was ordered to close for the duration of WW-II. Prices of receivers plummeted. Many of the prewar sets still exist.

The Germans claimed to have begun broadcasting television to public audiences in 1935. The Farnsworth system was used in Germany to broadcast the 1936 Olympics at Berlin.

The U.S. did not come on-line with a broadcast system that sought to inform or entertain audiences until shortly before WW II. While both Farnsworth and Zworykin had transmitters in place and operational early in their experiments, whatever programing present was incidental. The main purpose was to experiment with the new communication medium. The goal at the time was to improve the picture being transmitted until it compared reasonably well with the 35 mm photographic images available in motion picture theaters. (We are still trying accomplish that task today with HDTV!) There were, however, some pioneer TV broadcasters during the 1930s that offered entertainment and information programs to the few people who had a television receiver.

In 1933, station W9XK and radio station WSUI, broadcasting from the campus of the State University of Iowa, thrilled select midwesterners with a regular evening program of television. WSUI broadcast the audio portion on its assigned frequency of 880 kHz, and W9XK transmitted the video at 2.05 MHz with a power of 100 W. The twice-per-week program, initiating educational television, included performances by students and faculty with brief skits, lectures, and musical selections. During the early 1930s, W9XK was the only television station in the world located on a university campus, transmitting simultaneous video and audio programs.

B.2i TV Grows Up

Both NBC and CBS took early leads in paving the way for commercial television. NBC, through the visionary eyes of David Sarnoff and the resources of RCA, stood ready to undertake pioneering efforts to advance the new technology. Sarnoff accurately reasoned that TV could establish an industry-wide dominance only if television set manufacturers and broadcasters were using the same standards. He knew this would only occur if the FCC adopted suitable standards and allocated the needed frequency spectrum. Toward this end, in April 1935, Sarnoff made a dramatic announcement that RCA would put millions of dollars into television development. One year later, RCA began field testing television transmission methods from a transmitter atop the Empire State Building.

In a parallel move, CBS (after several years of deliberation) was ready to make its views public. In 1937, the company announced a $2 million experimental program that consisted of field testing various TV systems. It is interesting to note that many years earlier, in 1931, CBS put an experimental TV station on the air in New York City and transmitted programs for more than a year before becoming disillusioned with the commercial aspects of the new medium.

The Allen B. DuMont Laboratories also made significant contributions to early television. While DuMont is best known for CRT development and synchronization techniques, the company's major historical contribution was its production of early electronic TV sets for the public beginning in 1939.

It was during the 1939 World's Fair in New York and the Golden Gate International Exposition in San Francisco the same year that exhibits of live and filmed television were demonstrated on a large scale for the first time. Franklin Roosevelt's World's Fair speech (April 30, 1939) marked the first use of television by a U.S. president. The public was fascinated by the new technology.

Television sets were available for sale at the New York Fair's RCA pavilion. Prices ranged from $200 to $600, a princely sum for that time. Screen sizes ranged from 5 in. to 12 in. (the "big screen" model). Because CRT technology at that time did not permit wide deflection angles, the pictures tubes were long. So long, in fact, that the devices were mounted (in the larger-sized models) vertically. A hinge-mounted mirror at the top of the receiver cabinet permitted viewing.

At the San Francisco Exposition, RCA had another large exhibit that featured live television. The models used in the demonstrations could stand the hot lights for only a limited period. The studio areas were small, hot, and suitable only for interviews and commentary. People were allowed to walk through the TV studio and stand in front of the camera for a few seconds. Friends and family members were able to watch on monitors outside the booth. It was great fun, the lines were always long, and the crowds enthusiastic. The interest caused by these first mass demonstrations of television sparked a keen interest in the commercial potential of television broadcasting. Both expositions ran for a second season in 1940, but the war had started in Europe and television development was about to grind to a halt.

B.2j Transmission Standard Developed

In 1936 the Radio Manufacturers Association (RMA), the forerunner of today's EIA, set up a committee to recommend standards for a commercial TV broadcasting service. In December 1937 the committee advised the FCC to adopt the RCA 343-line/30-frame system that had been undergoing intensive development since 1931. The RCA system was the only one tested under both laboratory and field conditions. A majority of the RMA membership objected to the RCA system because of the belief that rapidly advancing technology would soon render this marginal system obsolete, and perhaps more importantly, would place them at a competitive disadvantage (RCA was prepared to immediately start manufacturing TV equipment and sets). Commercial development of television was put on hold.

At an FCC hearing in January 1940, a majority of the RMA was willing to embrace the RCA system, now improved to 441 lines. However, a strong dissenting minority (Zenith, Philco, and DuMont) was still able to block any action.

The result was that the National Television Systems Committee functioned essentially as a forum to investigate various options. DuMont proposed a 625-line/15-frame/4-field interlaced system. Philco advocated a 605-line/24-frame system. Zenith took the stance that it was still premature to adopt a national standard. Not until June 1941 did the FCC accept the consensus of a 525-line/30-frame (60 Hz) black and white system, which still exists today with minor modifications.

Television was formally launched in July 1941 when the FCC authorized the first two commercial TV stations to be constructed in the U.S. However, the growth of early television was ended by the licensing freeze that accompanied World War II. By the end of 1945 there were just nine commercial TV stations authorized, with six of them on the air. The first post-war full-service commercial license was issued to WNBW, the NBC-owned station in Washington, D.C.

As the number of TV stations on the air began to grow, the newest status symbol became a TV antenna on the roof of your home. Sets were expensive and not always reliable. Sometimes there was a waiting list to get one. Nobody cared, it was all very exciting—pictures through the air. People would stand in front of a department store window just to watch a test pattern.

B.2k Color Standard

During early development of commercial television systems—even as early as the 1920s—it was assumed that color would be demanded by the public. Primitive field sequential systems were demonstrated in 1929. Peter Goldmark of CBS showed a field sequential (color filter wheel) system in the early 1940s and promoted it vigorously during the post war years. Despite the fact that it was incompatible with existing receivers, had limited picture size possibilities, and was mechanically noisy, the CBS system was adopted by the FCC as the national color television standard in October 1950.

The engineering community felt betrayed (CBS excepted). Monochrome TV was little more than 3 years old with a base of 10 to 15 million receivers; broadcasters and the public were faced with the specter of having much of their new, expensive equipment become obsolete. The general wisdom was that color must be an adjunct to the 525/30 monochrome system so that existing terminal equipment and receivers could accept color transmissions.

Was the decision to accept the CBS color wheel approach a political one? Not entirely, because it was based on engineering tests presented to the FCC in early 1950. Contenders were the RCA dot sequential, the CTI (Color Television Incorporated) line sequential, and the CBS field sequential systems. The all-electronic compatible approach was in its infancy and there were no suitable display devices. Thus, for a decision made in 1950 based on the available test data, the commission's move to embrace the color wheel system was reasonable. CBS, however, had no support from other broadcasters or manufacturers; indeed, the company had to purchase Hytron-Air King to produce color TV sets (which would also receive black and white NTSC). Two hundred sets were manufactured for public sale.

Programming commenced on a 5 station East Coast network on June 21, 1951, presumably to gain experience. Color receivers went on sale in September, but only 100 total were sold. The last broadcast was on October 21, 1951. The final curtain fell in November when the National Production Authority (an agency created during the Korean war) imposed a prohibition on manufacturing color sets for public use. Some cynics interpreted this action as designed to get CBS off the hook because the production of monochrome sets was not restricted.

The proponents of compatible, all electronic color systems were, meanwhile, making significant advances. RCA had demonstrated a tri-color delta-delta kinescope. Hazeltine demonstrated the constant luminance principle, as well as the "shunted monochrome" idea. GE introduced the frequency interlaced color system. Philco showed a color signal composed of wideband luminance and two color difference signals encoded by a quadrature-modulated subcarrier. These, and other manufacturers, met in June 1951 to reorganize the NTSC for the purpose of pooling their resources in the development of a compatible system. By November a system employing the basic concepts of today's NTSC color system was demonstrated.

Field tests showed certain defects, such as sound interference caused by the choice of color subcarrier. This was easily corrected by the choice of a slightly different frequency, but at the expense of lowering the frame rate to 29.97 Hz. Finally, RCA demonstrated the efficacy of unequal I and Q color difference bandwidths. Following further field tests, the proposal was forwarded to the FCC on July 22, 1953.

A major problem, however, remained: the color kinescope. It was expensive and could only be built to yield a 9 × 12 in. picture. Without the promise of an affordable large screen display, the future of color television would be uncertain. Then came the development of a method of directly applying the phosphor dots on the faceplate together with a curved shadow mask mounted directly on the faceplate. This breakthrough came from the CBS-Hytron company! The commission adopted the color standard on December 17, 1953. It is interesting to note that the *phase alternation line* (PAL) principle was tried, but practical hardware to implement the scheme would not be available until 10 years later.

After nearly 50 years, NTSC still bears a remarkable aptitude for improvement and manipulation. But even with the advantage of compatibility for monochrome viewers, it took 10 years of equipment development and programming support from RCA and NBC before the general public starting buying color receivers in significant numbers.

In France, Germany and other countries, engineers began work aimed at improving upon the U.S. system. France developed the SECAM system and Germany produced the PAL system. Proponents of these three system still debate their advantages today.

B.2l UHF Comes of Age

The early planners of the U.S. television system thought that 13 channels would more than suffice for a given society. The original channel 1 was from 44 MHz to 50 MHz, but was latter dropped prior to any active use because of possible interference with other services. There remained 12 channels for normal use.

Bowing to pressure from various groups, the FCC revised its allocation table in 1952 to permit UHF TV broadcasting for the first time. The new band was not, however, a bed of roses. Many people went bankrupt building UHF stations because there were few receivers available to the public. UHF converters soon became popular. The first converters were so-called *matchbox types* that were good for one channel only. More expensive models mounted on top of the TV receiver and were tunable.

Finally, the commission requested the authority from Congress to require that all TV set manufacturers include UHF tuning in their receivers. President Kennedy signed the bill into law on July 10, 1962, authorizing the FCC to require television receivers "be capable of adequately receiving all frequencies allocated by the commission to television broadcasting." In September of that year, the FCC released a Notice of Proposed Rulemaking requiring any television set manufactured after April 30, 1964, to be an all-channel set. The commission also proposed a maximum noise figure for the tuner. These moves opened the doors for significant market penetration for UHF broadcasters. Without that mandate, UHF broadcasting would have lingered in the dark ages for many years.

The klystron has been the primary means of generating high power UHF-TV signals since the introduction of UHF broadcasting. The device truly revolutionized the modern world when it was quietly developed in 1937. Indeed, the klystron may have helped save the world as we know it. And, more than 50 years after it was first operated in a Stanford University laboratory by Russell Varian and his brother Sigurd, the klystron remains irreplaceable, even in this solid-state electronic age.

B.2m Birth of the Klystron

The Varian brothers were unusually bright and extremely active. The two were mechanically minded, producing one invention after another. Generally, Sigurd would think up an idea, Russell would devise a method for making it work, and Sigurd would then build the device.

Through the influence of William Hansen, a former roommate of Russell and at the time (the mid 1930s) a physics professor at Stanford, the Varians managed to get non-paying jobs as research associates in the Stanford physics lab. They had the right to consult with members of the faculty and were given the use of a small room in the physics building.

Hansen's role, apparently, was to shoot down ideas as fast as the Varians could dream them up. As the story goes, the Varians came up with 36 inventions of varying impracticality. Then they came up with idea #37. This time Hansen's eyes widened.

On June 5, 1937 Russell proposed the concept that eventually became the klystron tube. The device was supposed to amplify microwave signals. With $100 for supplies granted by Stanford, Sigurd built it.

The device was simple. A filament heated by an electric current in turn heated a cathode. A special coating on the cathode gave off electrons when it reached a sufficiently high temperature.

Negatively-charged electrons attracted by a positively-charged anode passed through the first cavity of the klystron tube. Microwaves in the cavity interacted with the electrons and passed through a narrow passage called a *drift tube*. In the drift tube, the electrons tended to bunch up: some speeded up, some slowed down. At the place in the drift tube where the bunching was most pronounced, the electrons entered a second cavity, where the stronger microwaves were excited and amplified in the process.

The first klystron device was lit up on the evening of August 19, 1937. Performance was marginal, but confirmed the theory of the device. An improved klystron was completed and tested on August 30.

The Varians published the results of their discovery in the *Journal of Applied Physics*. For reasons that have never been clear, their announcement immediately impressed British scientists working in the same field, but was almost entirely ignored by the Germans. The development of the klystron allowed British and American researchers to build smaller, more reliable radar systems. Klystron development paralleled work being done in England on the magnetron.

The successful deployment of microwave radar was accomplished by the invention to the cavity magnetron at Manchester University in the late 1930s. It was one of Britain's "Top Secrets" handed over to the Americans early in the war. The cavity magnetron was delivered to the Radiation Laboratory at MIT, where it was incorporated into later wartime radar systems. During the Battle of Britain in May 1940, British defenses depended upon longer wavelength radar (approximately 5 m), which worked but with insufficient resolution. The magnetron provided high-power microwave energy at 10 cm wavelengths, which improved detection resolution enormously. (Incidentally, the mass production of television CRTs prior to WW-II also helped radar development.)

Armed with the magnetron and klystron, British and American scientists perfected radar, a key element in winning the Battle of Britain. So valuable were the secret devices that the British decided not to put radar in planes that flew over occupied Europe lest one of them crash, and the details of the components be discovered.

After the war, the Varians—convinced of the potential for commercial value of the klystron and other devices they had conceived—established their own company. For Stanford University, the klystron represents one of its best investments: $100 in seed money and use of a small laboratory room were turned into $2.56 million in licensing fees before the patents expired in the 1970s, three major campus buildings, and hundreds of thousands of dollars in research funding.

B.3 Early Development of HDTV

The first developmental work on a high-definition television system began in 1968 at the technical research laboratories of Nippon Hoso Kyokai (NHK) in Tokyo [1]. In the initial effort, film transparencies were projected to determine the levels of image quality associated with different numbers of lines, aspect ratios, interlace ratios, and image dimensions. Image diagonals of 5 to 8 ft, aspect ratios of 5:3 to 2:1, and interlace ratios from 1:1 to 5:1 were used. The conclusions reached in the study were that a 1500-line image with 1.7:1 aspect ratio and a display size of 6.5 ft diagonal would have a subjective image quality improvement of more than 3.5 grades on a 7-grade scale, compared to the conventional 525-line, 4:3 image on a 25-in display. To translate these findings to television specifications, the following parameters were thus adopted:

- 2-to-1 interlaced scanning at 60 fields/s
- 0.7 Kell factor

- 0.5 to 0.7 interlace factor
- 5:3 aspect ratio
- Visual acuity of 9.3 cycles per degree of arc

Using these values, the preferred viewing distance for an 1125-line image was found to be 3.3 times the picture height. A display measuring 3 × 1.5 ft was constructed in 1972 using half mirrors to combine the images of three 26-in color tubes. Through the use of improved versions of wide-screen displays, the bandwidth required for luminance was established to be 20 MHz and, for wideband and narrowband chrominance, 7.0 and 5.5 MHz, respectively. The influence of viewing distance on sharpness also was investigated.

Additional tests were conducted to determine the effect of a wide screen on the realism of the display. In an elaborate viewing apparatus, motion picture film was projected on a spherical surface. As the aspect ratio of the image was shifted, the viewers' reactions were noted. The results showed that the realism of the presentation increased when the viewing angle was greater than 20°. The horizontal viewing angles for NTSC and PAL/SECAM were determined to be 11° and 13°, respectively. The horizontal viewing angle of the NHK system was set at 30°.

The so-called provisional standard for the NHK system was published in 1980. Because the NHK HDTV standard of 1125 lines and 60 fields/s was, obviously, incompatible with the conventional (NTSC) service used in Japan, adoption of these parameters raised a number of questions. No explanations appear in the literature, but justification for the values can be found in the situation faced by NHK prior to 1980. At that time, there was widespread support for a single worldwide standard for HDTV service. Indeed, the International Radio Consultative Committee (CCIR) had initiated work toward such a standard as early as 1972. If this were achieved, the NTSC and PAL/SECAM field rates of 59.94 and 50 Hz would prevent one (or all) of these systems from participation in such a standard. The 50 Hz rate was conceded to impose a severe limit on display brightness, and the 59.94 (vs. 60) Hz field rate posed more difficult problems in transcoding. Thus, a 60-field rate was proposed for the world standard. The choice of 1125 lines also was justified by the existing operating standards. The midpoint between 525 and 625 lines is 575 lines. Twice that number would correspond to 1150 lines for a *midpoint* HDTV system. This even number of lines could not produce alternate-line interlacing, then thought to be essential in any scanning standard. The nearest odd number having a common factor with 525 and 625 was the NHK choice: 1125 lines. The common factor—25—would make line-rate transcoding among the NHK, NTSC, and PAL/SECAM systems comparatively simple.

B.3a 1125/60 Equipment Development

The 1970s saw intense development of HDTV equipment at the NHK Laboratories. By 1980, when the NHK system was publicly demonstrated, the necessary camera tubes and cameras, picture tubes and projection displays, telecines, and videotape recorders were available. Also, the choices of transmission systems, signal formats, and modulation/demodulation parameters had been made. Work with digital transmission and fiber optics had begun, and a prototype tape recorder had been designed. Much of the HDTV equipment built to the 1125/60 NHK standard and brought to market by various vendors can be traced to these early developments.

In 1973, the NHK cameras used three 1-½-in. Vidicons, then commercially available. The devices, however, lacked the necessary resolution and sensitivity. To solve this problem, NHK developed the *return-beam Saticon* (RBS), which had adequate resolution and sensitivity, but about 30 percent lag. Cameras using three RBS tubes came into production in 1975 and were used during much of the NHK HDTV system development. By 1980, another device—the *diode-gun impregnated-cathode Saticon* (DIS) tube—was ready for public demonstration. This was a 1-in. tube, having a resolution of 1600 lines (1200 lines outside the 80 percent center circle), lag of less than 1 percent, and 39 dB signal-to-noise ratio (S/N) across a 30 MHz band. Misregistration among the primary color images, about 0.1 percent of the image dimensions in the earlier cameras, was reduced to 0.03 percent in the DIS camera. When used for sporting events, the camera was fitted with a 14-× zoom lens of *f*/2.8 aperture. The performance of this camera established the reputation of the NHK system among industry experts, including those from the motion picture business.

The task of adapting a conventional television display to high definition began in 1973, when NHK developed a 22-in. picture tube with a shadow-mask hole pitch of 310 μm (compared with 660 μm for a standard tube) and an aspect ratio of 4:3. In 1978, a 30-in. tube with hole pitch of 340 μm and 5:3 aspect ratio was produced. This tube had a peak brightness of 30 foot-lamberts (100 cd/m^2). This was followed in 1979 by a 26- in. tube with 370 μm hole pitch, 5:3 aspect ratio, and peak brightness of 45 foot-lamberts (ft-L).

The need for displays larger than those available in picture tubes led NHK to develop projection systems. A system using three CRTs with Schmidt-type focusing optics produced a 55-in. image (diagonal) on a curved screen with a peak brightness of about 30 ft-L. A larger image (67 in.) was produced by a light-valve projector, employing Schlieren optics, with a peak brightness of 100 ft-L at a screen gain of 10.

Development by NHK of telecine equipment was based on 70 mm film to assure a high reserve of resolution in the source material. The first telecine employed three Vidicons, but they had low resolution and high noise. It was decided that these problems could be overcome through the use of highly monochromatic laser beams as light sources: helium-neon at 632.8 nm for red, argon at 514.5 nm for green, and helium-cadmium at 441.6 nm for blue. To avoid variation in the laser output levels, each beam was passed though an acoustic-optical modulator with feedback control. The beams then were combined by dichroic mirrors and scanned mechanically. Horizontal scanning was provided by a 25-sided mirror, rotating at such a high speed (81,000 rpm) as to require aerostatic bearings. This speed was required to scan at 1125 lines, 30 frames/s, with 25 mirror facets. The deflected beam was passed through relay lenses to another mirror mounted on a galvanometer movement, which introduced the vertical scanning. The scanned beam then passed through relay lenses to a second mirror-polygon of 48 sides, rotating at 30 rpm, in accurate synchronism with the continuous movement of the 70 mm film.

The resolution provided by this telecine was limited by the mechanical scanning elements to 35 percent modulation transfer at 1000 lines, 2:1 aspect ratio. This level was achieved only by maintaining high precision in the faces of the horizontal scanning mirror and in the alignment of successive faces. To keep the film motion and the frame-synchronization mirror rotation in precise synchronism, an elaborate electronic feedback control was used between the respective motor drives. In all other respects, the performance was more than adequate.

The processes involved in producing a film version of a video signal were, essentially, the reverse of those employed in the telecine, the end point being exposure of the film by laser beams controlled by the R, G, and B signals. A prototype system was shown in 1971 by CBS Laboratories, Stamford, Conn. The difference lay in the power required in the beams. In the telecine, with highly sensitive phototubes reacting to the beams, power levels of approximately 10 mW were sufficient. To expose 35 mm film, power approaching 100 mW is needed. With the powers available in the mid-1970s, the laser-beam recorder was limited to the smaller area of 16 mm film. A prototype 16 mm version was constructed. In the laser recorder, the R, G, and B video signals were fed to three acoustic-optical modulators, and the resulting modulated beams were combined by dichroic mirrors into a single beam that was mechanically deflected. The scanned beam moved in synchronism with the moving film, which was exposed line by line. Color negative film typically was used in such equipment, but color duplicate film, having higher resolution and finer grain, was preferred for use with the 16 mm recorder. This technique is only of historical significance; the laser system has been discarded for the *electron-beam recording* system.

B.3b The 1125/60 System

In the early 1980s, NHK initiated a development program directed toward providing a high-definition television broadcasting service to the public. The video signal format was based upon the proposed 1125/60 standard published in 1980 by NHK. Experimental broadcasts were transmitted to prototype receivers in Japan over the MUSE (multiple sub-Nyquist encoding) satellite system. The 1984 Olympic Games, held in the United States and televised by NHK for viewers in Japan, was the first event of worldwide interest to be covered using high-definition television. The HDTV signals were *pan-scanned* to a 1.33 aspect ratio and transcoded to 525 lines for terrestrial transmission and reception over regular TV channels. On June 3, 1989, NHK inaugurated regular HDTV program transmissions for about an hour each day using its MS-2 satellite.

Various engineering committees within the Society of Motion Picture and Television Engineers (SMPTE), as well as other engineering committees in the United States, had closely studied the 1125/60 format since the 1980 NHK publication of the system details. Eventually, a production specification describing the 1125/60 format was proposed by SMPTE for adoption as an American National Standard. The document, SMPTE 240M, was published in April 1987 by the Society and forwarded to the American National Standards Institute (ANSI). Because of objections by several organizations, however, the document was not accepted as a national standard. One argument against the proposed standard was that it would be difficult to convert to the NTSC system, particularly if a future HDTV version of NTSC was to be compatible with the existing service.

NBC, for one, recommended that the HDTV studio standard be set at 1050 lines (525 × 2) and that the field rate of 59.94 Hz be retained, rather than the 60 Hz value of SMPTE 240M. Philips Laboratories and the David Sarnoff Laboratories concurred and based their proposed HDTV systems on the 1050-line, 59.94-field specifications.

Despite the rejection as a national standard for production equipment, SMPTE 240M remained a viable *recommended practice* for equipment built to the 1125/60 system.

The first full-scale attempt at international HDTV standardization was made by the CCIR in Dubrovnik, Yugoslavia, in May 1986. Japan and the United States pushed for a 1125-line/60 Hz production standard. The Japanese, of course, already had a working system. The 50 Hz countries, which did not have a working HDTV system of their own, demurred, asking for time to perfect and demonstrate a non-60 Hz (i.e., 50 Hz) system. Because of this objection, the CCIR took no action on the recommendation of its Study Group, voting to delay a decision until 1990, pending an examination of alternative HDTV production standards. There was strong support of the Study Group recommendations in the United States, but not among key European members of the CCIR.

The HDTV standardization fight was by no means simply a matter of North America and Japan vs. Europe. Of the 500 million receivers then in use worldwide, roughly half would feel the effect of any new frame rate. The Dubrovnik meeting focused mainly on a production standard. Television material can, of course, be produced in one standard and readily converted to another for transmission. Still, a single universal standard would avoid both the bother and degradation of the conversion process.

During this developmental period, the commercial implications of HDTV were inextricably intertwined with the technology. Even more in Europe than elsewhere, commercial considerations tended to dominate thinking. The 1125/60 system was, basically, a Japanese system. The U.S. came in late and jumped on Japan's coattails, aided greatly by the fact that the two countries have identical television standards (CCIR System *M*). But the Europeans did not want to have to buy Japanese or U.S. equipment; did not want to pay any Japanese or U. S. royalties; and did not want to swallow their NIH ("not invented here") pride. This feeling also emerged in the United States during the late 1980s and early 1990s for the same reasons—with American manufacturers not wanting to be locked into the Japanese and/or European HDTV systems.

B.3c European HDTV Systems

Early hopes of a single worldwide standard for high-definition television began a slow dissolve to black at the 1988 International Broadcasting Convention (IBC) in Brighton, England. Brighton was the public debut of the HDTV system developed by the European consortium *Eureka EU95*, and supporters made it clear that their system was intended to be a direct competitor of the 1125/60 system developed by NHK.

The Eureka project was launched in October 1986 (5 months after the Dubrovnik CCIR meeting) with the goal of defining a European HDTV standard of 1250-lines/50 Hz that would be compatible with existing 50 Hz receivers. EU95 brought together 30 television-related organizations, including major manufacturers, broadcasters, and universities. The Brighton showing included products and technology necessary for HDTV production, transmission, and reception. HD-MAC (high-definition multiplexed analog component) was the transmission standard developed under the EU95 program. HD-MAC was an extension of the MAC-packet family of transmission standards.

The primary movers in EU95 were the hardware manufacturers Bosch, Philips, and Thomson. The aim of the Eureka project was to define a 50 Hz HDTV standard for submission to the plenary assembly of the CCIR in 1990. The work carried out under this effort involved defining production, transmission, recording, and projection systems that would bring high-definition pictures into viewers' homes.

Supporters of the 1125/60 system also were planning to present their standard to the CCIR in 1990 for endorsement. The entry of EU95 into the HDTV arena significantly changed the complexion of the plenary assembly. For one thing, it guaranteed that no worldwide HDTV production standard—let alone a broadcast transmission standard—would be developed.

The 1250/50 Eureka HDTV format was designed to function as the future direct-broadcast satellite transmission system to the entire Western European Community. Although some interest had been expressed in the former Eastern Bloc countries, work there was slowed by more pressing economic and political concerns.

B.3d A Perspective on HDTV Standardization

HDTV production technology was seen from the very beginning as an opportunity to simplify program exchange, bringing together the production of programs for television and for the cinema [1]. Clearly, the concept of a single production standard that could serve all regions of the world *and* have application in the film community would provide benefits to both broadcasting organizations and program producers. All participants stated their preference for a single worldwide standard for HDTV studio production and international program exchange.

The work conducted in the field of studio standards showed that the task of creating a recommendation for HDTV studio production and international program exchange was made somewhat difficult by the diversity of objectives foreseen for HDTV in different parts of the world. There were differences in approach in terms of technology, support systems, and compatibility. It became clear that, for some participants, the use of HDTV for production of motion pictures and their subsequent distribution via satellites was the most immediate need. For others, there was a greater emphasis on satellite broadcasting, with a diversity of opinion on both the time scale for service introduction and the frequency bands to be used. For still others, the dominant consideration was terrestrial broadcasting services.

The proposal for a draft recommendation for an HDTV studio standard based on a 60 Hz field rate was submitted to the CCIR, initially in 1985. The proposal for a draft recommendation for an HDTV studio standard based on a 50 Hz field rate was submitted to the CCIR in 1987. Unfortunately, neither set of parameters in those drafts brought a consensus within the CCIR as a single worldwide standard. However, both had sufficient support for practical use in specific areas to encourage manufacturers to produce equipment.

Despite the lack of an agreement on HDTV, a great deal of progress was made during this effort in the area of an HDTV source standard. The specific parameters agreed upon included:

- Picture aspect ratio of 16:9
- Color rendition
- Equation for luminance

Thus, for the first time in the history of television, all countries of the world agreed on the technical definition of a basic tristimulus color system for display systems. Also agreed upon, in principle, were the digital HDTV bit-rate values for the studio interface signal, which was important in determining both the interface for HDTV transmission and the use of digital recording. All of these agreements culminated in Recommendation 709, adopted by the XVII Plenary Assembly of the CCIR in 1990 in Dusseldorf [2].

B.3e Digital Systems Emerge

By mid-1991, publications reporting developments in the U.S., the United Kingdom, France, the Nordic countries, and other parts of the world showed that bit-rate reduction schemes on the order of 60:1 could be applied successfully to HDTV source images. The results of this work implied that HDTV image sequences could be transmitted in a relatively narrowband channel in the range of 15 to 25 Mbits/s. Using standard proven modulation technologies, it would therefore be possible to transmit an HDTV program within the existing 6, 7, and/or 8 MHz channel bandwidths provided for in the existing VHF- and UHF-TV bands.

One outgrowth of this development was the initiation of studies into how—if at all—the existing divergent broadcasting systems could be included under a single unifying cover. Thus was born the *HDTV-6-7-8* program. HDTV-6-7-8 was based on the following set of assumptions [1]:

- First, the differences between the bandwidths of the 6, 7, and 8 MHz channels might give rise to the development of three separate HDTV scenarios that would fully utilize the bandwidth of the assigned channels. It was assumed that the 6 MHz implementation would have the potential to provide pictures of sufficiently high quality to satisfy viewers' wishes for a "new viewing experience." The addition of a 1 or 2 MHz increment in the bandwidth, therefore, would not be critical for further improvement for domestic reception. On this basis, there was a possibility of adopting, as a core system, a single 6 MHz scheme to provide a minimum service consisting of video, two multichannel sound services, and appropriate support data channels for conditional access, and—where appropriate—closed captioning, program identification, and other user-oriented services.
- Second, given the previous assumption, the 1 or 2 MHz channel bandwidth that could be saved in a number of countries might be used for transmission of a variety of additional information services, either within a 7 or 8 MHz composite digital signal or on new carriers in the band above the HDTV-6 signal. Such additional information might include narrowband TV signals that provide for an HDTV stereoscopic service, enhanced television services, multiprogram TV broadcasting, additional sound services, and/or additional data services.
- Third, it would be practical to combine audio and video signals, additional information (data), and new control/test signals into a single HDTV-6-7-8 signal, in order to avoid using a secondary audio transmitter. In combination with an appropriate header/descriptor protocol and appropriate signal processing, the number of frequency channels could be increased, and protection ratio requirements would be reduced. In the course of time, this could lead to a review of frequency plans at the national and international levels for terrestrial TV transmission networks and cable television. This scheme, therefore, could go a considerable distance toward meeting the growing demand for frequency assignments.

This digital television system offered the prospect of considerably improved sound and image quality, while appreciably improving spectrum utilization (as compared to the current analog services). It was theorized that one way of exploiting these possibilities would be to use the bit stream available in digital terrestrial or satellite broadcasting to deliver to the public a certain number of digitally compressed conventional television programs instead of a single conventional, enhanced-, or high-definition program. These digitally compressed TV signals would be accompanied by digital high-quality sound, coded conditional access information, and ancillary data channels. Furthermore, the same approach could be implemented in the transmission of multiprogram signals over existing digital satellite or terrestrial links, or cable TV networks.

Despite the intrinsic merits of the HDTV-6-7-8 system, it was quickly superseded by the Digital Video Broadcasting project in Europe and the Grand Alliance project in the United States.

HD-DIVINE

HD-DIVINE began as a developmental project in late 1991. The aim was to prove that a digital HDTV system was possible within a short time frame without an intermediate step based on analog technology. The project was known as "Terrestrial Digital HDTV," but later was renamed "HD-DIVINE" (*digital video narrowband emission*). Less than 10 months after the project started, HD-DIVINE demonstrated a digital terrestrial HDTV system at the 1992 IBC show in Amsterdam. It was a considerable success, triggering discussion in the Nordic countries on cooperation for further development.

HD-DIVINE subsequently was developed to include four conventional digital television channels—as an alternative to HDTV—contained within an 8 MHz channel. Work also was done to adapt the system to distribution via satellite and cable, as demonstrated at the Montreux Television Symposium in 1993.

Meanwhile, in the spring of 1992, a second coordinated effort began with the goal of a common system for digital television broadcasting in Europe. Under the auspices of a European Launching Group composed of members of seven countries representing various organizations involved in the television business, the Working Group for Digital Television Broadcasting (WGDTB) defined approaches for a digital system. Two basic schemes were identified for further study, both multilayer systems that offered multiple service levels, either in a hierarchical or multicast mode. The starting point for this work was the experience with two analog TV standards, D2-MAC and HD-MAC, which had been operational for some time.

Eureka ADTT

On June 16, 1994, the leaders of the Eureka project approved the start of a new research effort targeted for the development of future television systems in Europe based on digital technologies. The Eureka 1187 *Advanced Digital Television Technologies* (ADTT) project was formed with the purpose of building upon the work of the Eureka 95 HDTV program. The effort, which was to run for 2½ years, included as partners Philips, Thomson Consumer Electronics, Nokia, and English and Italian consortia. The initial objectives of Eureka ADTT were to address design issues relating to production, transmission, reception, and display equipment and their key technologies.

A prototype high-definition broadcast system was developed and tested, based on specifications from—and in close consultation with—the European Project for Digital Video Broadcasting. The Eureka ADTT project also was charged with exploring basic technologies and the development of key components for products, such as advanced digital receivers, recording devices, optical systems, and multimedia hardware and software.

The underlying purpose of the Eureka ADTT effort was to translate, insofar as possible, the work done on the analog-based Eureka 95 effort to the all-digital systems that emerged from 1991 to 1993.

B.3f Digital Video Broadcasting

The European DVB project began in the closing months of 1990. Experimental European projects such as SPECTRE showed that the digital video-compression system known as *motion-compensated hybrid discrete cosine transform coding* was highly effective in reducing the transmission capacity required for digital television [3]. Until then, digital TV broadcasting was considered impractical to implement.

In the U.S., the first proposals for digital terrestrial HDTV were made. In Europe, Swedish Television suggested that fellow broadcasters form a concerted pan-European platform to develop digital terrestrial HDTV. During 1991, broadcasters and consumer equipment manufacturers discussed how this could be achieved. Broadcasters, consumer electronics manufacturers, and regulatory bodies agreed to come together to discuss the formation of a pan-European group that would oversee the development of digital television in Europe—the European Launching Group (ELG). Over the course of about a year, the ELG expanded to include the major European media interest groups—both public and private—consumer electronics manufacturers, and common carriers.

The program officially began in September 1993, and the European Launching Group became the DVB (*Digital Video Broadcasting*) Project. Developmental work in digital television, already under way in Europe, then moved forward under this new umbrella. Meanwhile, a parallel activity, the *Working Group on Digital Television*, prepared a study of the prospects and possibilities for digital terrestrial television in Europe.

By 1999, a bit of a watershed year for digital television in general, the Digital Video Broadcasting Project had grown to a consortium of over 200 broadcasters, manufacturers, network operators, and regulatory bodies in more than 30 countries worldwide. Numerous broadcast services using DVB standards were operational in Europe, North and South America, Africa, Asia, and Australia.

At the '99 NAB Convention in Las Vegas, mobile and fixed demonstrations of the DVB system were made using a variety of equipment in various typical situations. Because mobile reception is the most challenging environment for television, the mobile system received a good deal of attention. DVB organizers used the demonstrations to point out the strengths of their chosen modulation method, the multicarrier *coded orthogonal frequency division multiplexing* (COFDM) technique.

In trials held in Germany beginning in 1997, DVB-T, the terrestrial transmission mode, has been tested in slow-moving city trams and at speeds in excess of 170 mph.

B.3g Involvement of the Film Industry

From the moment it was introduced, HDTV was the subject of various aims and claims. Among them was the value of video techniques for motion picture production, making it possible to bypass the use of film in part or completely. Production and editing were said to be enhanced, resulting in reduced costs to the producer. However, the motion picture industry was in no hurry to discard film, the medium that had served it well for the better part of a century. Film quality continues to improve, and film is unquestionably *the* universal production standard. Period. Although HDTV has made inroads in motion picture editing and special effects production, Hollywood has not rushed to hop on board the HDTV express.

Nevertheless, it is certainly true that the film industry has embraced elements of high-definition imaging. As early as 1989, Eastman Kodak unveiled the results of a long-range program to develop an *electronic-intermediate* (EI) digital video postproduction system. The company introduced the concept of an HDTV system intended primarily for use by large-budget feature film producers to provide new, creative dimensions for special effects without incurring the quality compromises of normal edited film masters.

In the system, original camera negative 35 mm film input to the electronic-intermediate system was transferred to a digital frame store at a rate substantially slower than real time, initially about 1 frame/s. Sequences could be displayed a frame at a time for unfettered image manipulation and compositing. This system was an electronic implementation of the time-standing-still format in which film directors and editors have been trained to exercise their creativity.

Kodak established a consortium of manufacturers and software developers to design and produce elements of the EI system. Although the announced application was limited strictly to the creation of artistic high-resolution special effects on film, it was hoped that the EI system eventually would lead to the development of a means for electronic real-time distribution of theatrical films.

B.3h Political Considerations

In an ideal world, the design and direction of a new production or transmission standard would be determined by the technical merits of the proposed system. However, as demonstrated in the past by the controversy surrounding the NTSC monochrome and color standards, technical considerations are not always the first priority. During the late 1980s, when concern over foreign competition was at near-frenzy levels in the U.S. and Europe, the ramifications of HDTV moved beyond just technology and marketing, and into the realm of politics. In fact, the political questions raised by the push for HDTV promised to be far more difficult to resolve than the technical ones.

The most curious development in the battle over HDTV was the interest politicians took in the technology. In early 1989, while chairing the House Telecommunications subcommittee, Representative Ed Markey (D-Massachusetts) invited comments from interested parties on the topic of high-definition television. Markey's subcommittee conducted two days of hearings in February 1989. There was no shortage of sources of input, including:

- The *American Electronics Association*, which suggested revised antitrust laws, patent policy changes, expanded exports of high-tech products, and government funding of research on high-definition television.
- *Citizens for a Sound Economy*, which argued for a relaxation of antitrust laws.
- *Committee to Preserve American Television*, which encouraged strict enforcement of trade laws and consideration of government research and development funds for joint projects involving semiconductors and advanced display devices.
- *Maximum Service Telecasters*, which suggested tax credits, antitrust exemptions, and low-interest loans as ways of encouraging U.S. development of terrestrial HDTV broadcasting.
- *Committee of Corporate Telecommunications Users*, which suggested the creation of a "Technology Corporation of America" to devise an open architecture for the production and transmission of HDTV and other services.

Representative Don Ritter (R-Pennsylvania) expressed serious concern over the role that U.S.-based companies would play—or more to the point, might not play—in the development of high-definition television. Ritter believed it was vital for America to have a piece of the HDTV manufacturing pie. Similar sentiments were echoed by numerous other lawmakers.

The Pentagon, meanwhile, expressed strong interest in high-definition technology for two reasons: direct military applications and the negative effects that the lack of domestic HDTV expertise could have on the American electronics industry. The Department of Defense, accordingly, allocated money for HDTV research.

Not everyone was upset, however, about the perceived technological edge the Japanese (and the Europeans to a lesser extent) had at that time over the U.S. A widely read, widely circulated article in *Forbes* magazine [4] described HDTV as a technology that would be "obsolete" by the time of its introduction. The author argued, "The whole issue is phony," and maintained that

HDTV products would hit the market "precisely at the time when the U.S. computer industry will be able to supply far more powerful video products at a lower price."

Although many of the concerns relating to HDTV that were voiced during the late 1980s now seem rather baseless and even hysterical, this was the atmosphere that drove the pioneering work on the technology. The long, twisting road to HDTV proved once again that the political implications of a new technology may be far more daunting than the engineering issues.

B.3i Terminology

During the development of HDTV, a procession of terms was used to describe performance levels between conventional NTSC and "real" HDTV (1125/60-format quality). Systems were classified in one or more of the following categories:

- *Conventional systems*: The NTSC, PAL, and SECAM systems as standardized prior to the development of advanced systems.
- *Improved-definition television* (IDTV) systems: Conventional systems modified to offer improved vertical and/or horizontal definition, also known as *advanced television* (ATV) or *enhanced-definition television* (EDTV) systems.
- *Advanced systems*: In the broad sense, all systems other than conventional ones. In the narrow sense, all systems other than conventional and "true" HDTV.
- *High-definition television* (HDTV) systems: Systems having vertical and horizontal resolutions approximately twice those of conventional systems.
- *Simulcast systems*: Systems transmitting conventional NTSC, PAL, or SECAM on existing channels and HDTV of the same program on one or more additional channels.
- *Production systems*: Systems intended for use in the production of programs, but not necessarily in their distribution.
- *Distribution systems*: Terrestrial broadcast, cable, satellite, videocassette, and videodisc methods of bringing programs to the viewing audience.

B.4 Digital Television in the U.S.

Although HDTV production equipment had been available since 1984, standardization for broadcast service was slowed by lack of agreement on how the public could best be served. The primary consideration was whether to adopt a system compatible with NTSC or a simulcast system requiring additional transmission spectrum and equipment.

On November 17, 1987, at the request of 58 U.S. broadcasters, the FCC initiated a rulemaking on advanced television (ATV) services and established a blue ribbon Advisory Committee on Advanced Television Service (ACATS) for the purpose of recommending a broadcast standard. Former FCC Chairman Richard E. Wiley was appointed to chair ACATS. At that time, it was generally believed that HDTV could not be broadcast using 6 MHz terrestrial broadcasting channels. Broadcasting organizations were concerned that alternative media would be used to deliver HDTV to the viewing public, placing terrestrial broadcasters at a severe disadvantage. The FCC agreed that this was a subject of importance and initiated a proceeding (MM Docket No. 87-268) to consider the technical and public policy issues of ATV.

The first interim report of the ACATS, filed on June 16, 1988, was based primarily on the work of the Planning Subcommittee. The report noted that proposals to implement improvements in the existing NTSC television standard ranged from simply enhancing the current standard all the way to providing full-quality HDTV. The spectrum requirements for the proposals fell into three categories: 6 MHz, 9 MHz, and 12 MHz. Advocates of a 12 MHz approach suggested using two channels in one of two ways:

- An existing NTSC-compatible channel supplemented by a 6 MHz *augmentation channel* (either contiguous or noncontiguous)
- An existing NTSC-compatible channel, unchanged, and a separate 6 MHz channel containing an independent non-NTSC-compatible HDTV signal

It was pointed out that both of these methods would be "compatible" in the sense that existing TV receivers could continue to be serviced by an NTSC signal.

The first interim report stated that, "based on current bandwidth-compression techniques, it appears that full HDTV will require greater spectrum than 6 MHz." The report went on to say that the Advisory Committee believed that efforts should focus on establishing, at least ultimately, an HDTV standard for terrestrial broadcasting. The report concluded that one advantage to simulcasting was that at some point in the future—after the NTSC standard and NTSC-equipped receivers were retired—part of the spectrum being utilized might be reemployed for other purposes. On the basis of preliminary engineering studies, the Advisory Committee stated that it believed sufficient spectrum capacity in the current television allocations table might be available to allow all existing stations to provide ATV through either an augmentation or simulcast approach.

B.4a The Process Begins

With this launch, the economic, political, and technical implications of HDTV caused a frenzy of activity in technical circles around the world; proponents came forward to offer their ideas. The Advanced Television Test Center (ATTC) was set up to consider the proposals and evaluate their practicality. In the first round of tests, 21 proposed methods of transmitting some form of ATV signals (from 15 different organizations) were considered. The ATTC work was difficult for a number of reasons, but primarily because the 21 systems were in various stages of readiness. Most, if not all, were undergoing continual refinement. Only a few systems existed as real, live black boxes, with "inputs" and "outputs". Computer simulation made up the bulk of what was demonstrated in the first rounds. The ATTC efforts promised, incidentally, to be as much a test of computer simulation as a test of hardware. Of the 21 initial proposals submitted to ACATS in September 1988, only six actually were completed in hardware and tested.

The race begun, engineering teams at various companies began assembling the elements of an ATV service. One of the first was the Advanced Compatible Television (ACTV) system, developed by the Sarnoff Research Center. On April 20, 1989, a short ACTV program segment was transmitted from the center, in New Jersey, to New York for broadcast over a WNBC-TV evening news program. The goal was to demonstrate the NTSC compatibility of ACTV. Consisting of two companion systems, the scheme was developed to comply with the FCC's *tentative decision* of

September 1988, which required an HDTV broadcast standard to be compatible with NTSC receivers.

The basic signal, ACTV-I, was intended to provide a wide-screen picture with improved picture and sound quality on new HDTV receivers, while being compatible with NTSC receivers on a single 6 MHz channel. A second signal, ACTV-II, would provide full HDTV service on a second augmentation channel when such additional spectrum might be available.

In the second interim report of the ACATS (April 26, 1989), the committee suggested that its life be extended from November 1989 to November 1991. It also suggested that the FCC should be in a position to establish a single terrestrial ATV standard sometime in 1992. The Advisory Committee noted that work was ongoing in defining tests to be performed on proponent systems. An issue was raised relating to subjective tests and whether source material required for testing should be produced in only one format and transcoded into the formats used by different systems to be tested, or whether source material should be produced in all required formats. The Advisory Committee also sought guidance from the FCC on the minimum number of audio channels that an ATV system would be expected to provide.

The large number of system proponents, and delays in developing hardware, made it impossible to meet the aggressive timeline set for this process. It was assumed by experts at the time that consumers would be able to purchase HDTV, or at least advanced television, sets for home use by early 1992.

The FCC's tentative decision on compatibility, although not unexpected, laid a false set of ground rules for the early transmission system proponents. The requirement also raised the question of the availability of frequency spectrum to accommodate the added information of the ATV signal. Most if not all of the proposed ATV systems required total bandwidths of one, one and a half, or two standard TV channels (6 MHz, 9 MHz, or 12 MHz). In some cases, the added spectrum space that carried the ATV information beyond the basic 6 MHz did not have to be contiguous with the main channel.

Any additional use of the present VHF- and UHF-TV broadcast bands would have to take into account co-channel and adjacent-channel interference protection. At UHF, an additional important unanswered question was the effect of the UHF "taboo channels" on the availability of extra frequency space for ATV signals. These "taboos" were restrictions on the use of certain UHF channels because of the imperfect response of then-existing TV receivers to unwanted signals, such as those on image frequencies, or those caused by local oscillator radiation and front-end intermodulation.

The mobile radio industry had been a longtime combatant with broadcasters over the limited available spectrum. Land mobile had been asking for additional spectrum for years, saying it was needed for public safety and other worthwhile purposes. At that time, therefore, the chances of the FCC allocating additional spectrum to television broadcasters in the face of land mobile demands were not thought to be very good. Such was the case; the land mobile industry (and other groups) made a better case for the spectrum.

In any event, the FCC informally indicated that it intended to select a simulcast standard for HDTV broadcasting in the United States using existing television band spectrum and would not consider any augmentation-channel proposals.

B.4b System Testing: Round 2

With new groundwork clearly laid by the FCC, the second round of serious system testing was ready to begin. Concurrent with the study of the various system proposals, the ACATS began in late 1990 to evaluate the means for transmission of seven proposed formats for the purpose of determining their suitability as the U.S. standard for VHF and UHF terrestrial broadcasting. The initial round of tests were scheduled for completion by September 30, 1992.

The FCC announced on March 21, 1990 that it favored a technical approach in which high-definition programs would be broadcast on existing 6 MHz VHF and UHF channels separate from the 6 MHz channels used for conventional (NTSC) program transmission, but the commission did not specifically address the expected bandwidth requirements for HDTV. However, the implication was that only a single channel would be allocated for transmission of an HDTV signal. It followed that this limitation to a 6 MHz channel would require the use of video-compression techniques. In addition, it was stated that no authorization would be given for any enhanced TV system, so as not to detract from development of full high-definition television. The spring of 1993 was suggested by the FCC as the time for a final decision on the selection of an HDTV broadcasting system.

Under the simulcast policy, broadcasters would be required to transmit NTSC simultaneously on one channel of the VHF and UHF spectra and the chosen HDTV standard on another 6 MHz TV broadcast channel. This approach was similar to that followed by the British in their introduction of color television, which required monochrome programming to continue on VHF for about 20 years after 625/50 PAL color broadcasting on UHF was introduced. Standards converters working between 625-line PAL color and 405-line monochrome provided the program input for the simultaneous black-and-white network transmitters. The British policy obviously benefited the owners of old monochrome receivers who did not wish to invest in new color receivers; it also permitted program production and receiver sales for the new standard to develop at a rate compatible with industry capabilities. All television transmission in Great Britain now is on UHF, with the VHF channels reassigned to other radio services.

For the development of HDTV, the obvious advantage of simulcasting to viewers with NTSC receivers is that they may continue to receive all television broadcasts in either the current 525-line standard or the new HDTV standard—albeit the latter without the benefit of wide-screen and double resolution—without having to purchase a dual-standard receiver or a new HDTV receiver. Although it was not defined by the FCC, it was presumed that the HDTV channels also would carry the programs available only in the NTSC standard. Ideally for the viewer, these programs would be converted to the HDTV transmission standard from the narrower 1.33 aspect ratio and at the lower resolution of the 525-line format. A less desirable solution would be to carry programs available only in the NTSC standard without conversion to HDTV and require HDTV receivers to be capable of switching automatically between standards. A third choice would be not to carry NTSC-only programs on the HDTV channel and to require HDTV receivers to be both HDTV/NTSC channel and format switchable.

The development of HDTV involved exhaustive study of how to improve the existing NTSC system. It also meant the application of new techniques and the refinement of many others, including:

- Receiver enhancements, such as higher horizontal and vertical resolution, digital processing, and implementation of large displays.

- Transmission enhancements, including new camera technologies, image enhancement, adaptive prefilter encoding, digital recording, and advanced signal distribution.
- Signal compression for direct satellite broadcasting.
- Relay of auxiliary signals within conventional TV channels.
- Allocation and optimization of transmission frequency assignments.

Concurrently, an extensive study was undertaken concerning the different characteristics of the major systems of program distribution: terrestrial broadcasting, cable distribution by wire or fiber optics, satellite broadcasting, and magnetic and optical recorders. The major purposes of this study were to determine how the wide video baseband of HDTV could be accommodated in each system, and whether a single HDTV standard could embrace the needs of all systems. This work not only provided many of the prerequisites of HDTV, but by advancing the state of the conventional art, it established a higher standard against which the HDTV industry must compete.

In the third interim report (March 21, 1990), the Advisory Committee approved the proposed test plans and agreed that complete systems, including audio, would be required for testing. It also was agreed that proposed systems must be precertified by June 1, 1990. That date, naturally, became quite significant to all proponents. The pace of work was accelerated even further.

It is noteworthy that the first all-digital proposal was submitted shortly before the June deadline. The third interim report also stated that psychophysical tests of ATV systems would be conducted and that the Planning Subcommittee, through its Working Party 3 (PS/WP3), would undertake the development of preliminary ATV channel allotment plans and assignment options.

In the fourth interim report (April 1, 1991), the Advisory Committee noted changes in proponents and proposed systems. Most significant was that several all-digital systems had been proposed. Testing of proponent systems was scheduled to begin later that year. Changes had been required in the test procedures because of the introduction of the all-digital systems. It was reported also that the System Standards Working Party had defined a process for recommending an ATV system, and that PS/WP3 was working toward the goal of providing essentially all existing broadcasters with a simulcast channel whose coverage characteristics were equivalent to NTSC service.

By the time the fifth interim report was issued (March 24, 1992), there were five proponent systems, all simulcast—one analog and four all-digital. The Planning Subcommittee reported that it had reconstituted its Working Party 4 to study issues related to harmonizing an ATV broadcast transmission standard with other advanced imaging and transmission schemes that would be used in television and nonbroadcast applications. The Systems Subcommittee reported that its Working Party 2 had developed procedures for field-testing an ATV system. It was noted that the intent of the Advisory Committee was to field-test only the system recommended to the FCC by the Advisory Committee based on the laboratory tests. It also was reported that the Systems Subcommittee Working Party 4 had developed a process for recommending an ATV system and had agreed to a list of 10 selection criteria.

Hundreds of companies and organizations worked together within the numerous sub-committees, working parties, advisory groups, and special panels of ACATS during the 8-year existence of the organization. The ACATS process became a model for international industry-government cooperation. Among its accomplishments was the development of a competitive process by which proponents of systems were required to build prototype hardware that would then be thoroughly tested. This process sparked innovation and entrepreneurial initiative.

B.4c Formation of the Grand Alliance

Although the FCC had said in the spring of 1990 that it would determine whether all-digital technology was feasible for a terrestrial HDTV transmission standard, most observers viewed that technology as being many years in the future. Later the same year, however, General Instrument became the first proponent to announce an all-digital system. Later, all-digital systems were announced by MIT, the Philips-Thomson-Sarnoff consortium, and Zenith-AT&T.

The FCC anticipated the need for interoperability of the HDTV standard with other media. Initially, the focus was on interoperability with cable television and satellite delivery; both were crucial to any broadcast standard. But the value of interoperability with computer and telecommunications applications became increasingly apparent with the advent of all-digital systems.

Proponents later incorporated packetized transmission, headers and descriptors, and composite-coded surround sound in their subsystems. (The Philips-Thomson-Sarnoff consortium was the first to do so.) These features maximized the interoperability of HDTV with computer and telecommunications systems. The introduction of all-digital systems had made such interoperability a reality.

The all-digital systems set the stage for another important step, which was taken in February 1992, when the ACATS recommended that the new standard include a flexible, adaptive data-allocation capability (and that the audio also be upgraded from stereo to surround sound). Following testing, the Advisory Committee decided in February 1993 to limit further consideration only to those proponents that had built all-digital systems: two systems proposed by General Instrument and MIT; one proposed by Zenith and AT&T; and one proposed by Sarnoff, Philips, and Thomson. The Advisory Committee further decided that although all of the digital systems provided impressive results, no single system could be proposed to the FCC as the U.S. HDTV standard at that time. The committee ordered a round of supplementary tests to evaluate improvements to the individual systems.

At its February 1993 meeting, the Advisory Committee also adopted a resolution encouraging the digital HDTV groups to try to find a way to merge the four remaining all-digital systems. The committee recognized the merits of being able to combine the best features of those systems into a single "best of the best" system. With this encouragement, negotiations between the parties heated up, and on May 24, the seven companies involved announced formation of the Digital HDTV Grand Alliance.

By the spring of 1994, significant progress had been made toward the final HDTV system proposal. Teams of engineers and researchers had finished building the subsystems that would be integrated into the complete HDTV prototype system for testing later in the year. The subsystems—scanning formats, digital video compression, packetized data, audio, and modulation—all had been approved by the ACATS. Key features and specifications for the system included:

- Support of two fundamental arrays of pixels (picture elements): 1920 × 1080 and 1280 x 720. Each of these pixel formats supported a wide-screen 16:9 aspect ratio and square pixels, important for computer interoperability. Frame rates of 60, 30, and 24 Hz were supported, yielding a total of six different possible scanning formats—two different pixel arrays, each having three frame rates. The 60 and 30 Hz frame rates were important for video source material and 24 Hz for film. A key feature of the system was the Grand Alliance's commitment to using progressive scanning, also widely used in computer displays. Entertainment television traditionally had used interlaced scanning, which was efficient but subject to various unwanted artifacts. Of the six video formats, progressive scanning was used in all three 720-

line formats and in the 30 and 24 Hz 1080-line formats. The sixth video format was a 60 Hz 1080-line scheme. It was neither technically or economically feasible to initially provide this as a progressive format, although it was a longer-term goal for the Grand Alliance. The 1080-line, 60-Hz format was handled in the initial standard by using interlaced rather than progressive scanning.

- Video compression: Utilizing the MPEG-2 (Moving Picture Experts Group)-proposed international standard allowed HDTV receivers to interoperate with MPEG-2 and MPEG-1 computer, multimedia, and other media applications.
- Packetized data transport: Also based on MPEG-2, this feature provided for the flexible transmission of virtually any combination of video, audio, and data.
- Compact-disc-quality digital audio: This feature was provided in the form of the 5.1-channel Dolby AC-3 surround sound system.
- 8-VSB (8-level vestigial sideband): The modulation system selected for transmission provided maximum coverage area for terrestrial digital broadcasting.

The Grand Alliance format employed principles that made it a highly interoperable system. It was designed with a layered digital architecture that was compatible with the international Open Systems Interconnect (OSI) model of data communications that forms the basis of virtually all modern digital systems. This compatibility allowed the system to interface with other systems at any layer, and it permitted many different applications to make use of various layers of the HDTV architecture. Each individual layer of the system was designed to be interoperable with other systems at corresponding layers.

Because of the interoperability of the system between entertainment television and computer and telecommunications technologies, the Grand Alliance HDTV standard was expected to play a major role in the establishment of the *national information infrastructure* (NII). It was postulated that digital HDTV could be an engine that helped drive deployment of the NII by advancing the development of receivers with high-resolution displays and creating a high-data-rate path to the home for a multitude of entertainment, education, and information services.

B.4d Testing the Grand Alliance System

Field tests of the 8-VSB digital transmission subsystem began on April 11, 1994, under the auspices of the ACATS. The 8-VSB transmission scheme, developed by Zenith, had been selected for use in the Grand Alliance system two months earlier, following comparative laboratory tests at the ATTC. The field tests, held at a site near Charlotte, North Carolina, were conducted on channel 53 at a maximum effective radiated power of 500 kW (peak NTSC visual) and on channel 6 at an ERP of 10 kW (peak NTSC visual).

The tests, developed by the Working Party on System Testing, included measurements at approximately 200 receiving sites. Evaluations were based solely on a pseudorandom data signal as the input source; pictures and audio were not transmitted. The 8-VSB measurements included carrier-to-noise ratio (C/N), error rate, and *margin tests*, performed by adding noise to the received signal until an agreed-upon threshold of performance error rate occurred and noting the difference between the C/N and the C/N without added noise. Testing at the Charlotte facility lasted for about 3 months, under the direction of the Public Broadcasting System (PBS).

In 1995, extensive follow-up tests were conducted, including:

- Laboratory tests at the Advanced Television Test Center in Alexandria, Virginia
- Lab tests at Cable Television Laboratories, Inc. (CableLabs) of Boulder, Colorado
- Subjective viewer testing at the Advanced Television Evaluation Laboratory in Ottawa, Canada
- Continued field testing in Charlotte, North Carolina, by PBS, the Association for Maximum Service Television (MSTV), and CableLabs

The laboratory and field tests evaluated the Grand Alliance system's four principal subsystems: scanning formats, video and audio compression, transport, and transmission. Test results showed that:

- Each of the proposed HDTV scanning formats exceeded targets established for static and dynamic luminance and chrominance resolution.
- Video-compression testing, using 26 different HDTV sequences, showed that the Grand Alliance MPEG-2 compression algorithm was "clearly superior" to the four original ATV systems in both the 1080 interlaced- and 720 progressive-scanning modes. Significantly, the testing also showed little or no deterioration of the image quality while transmitting 3 Mbits/s of ancillary data.
- The 5.1-channel digital surround sound audio subsystem of the Grand Alliance system, known as Dolby AC-3, performed better than specifications in multichannel audio testing and met the expectations in long-form entertainment listening tests.
- The packetized data transport subsystem performed well when tested to evaluate the switching between compressed data streams, robustness of headers and descriptors, and interoperability between the compression and transport layers. Additional testing also demonstrated the feasibility of carrying the ATV transport stream on an *asynchronous transfer mode* (ATM) telecommunications network.

Field and laboratory testing of the 8-VSB digital transmission subsystem reinforced test results achieved in the summer of 1994 in Charlotte. Testing for spectrum utilization and transmission robustness again proved that the Grand Alliance system would provide broadcasters significantly better transmission performance than the current analog transmission system, ensuring HDTV service "in many instances where NTSC service is unacceptable." Extensive testing on cable systems and fiber optic links of the 16-VSB subsystem also showed superior results.

The final technical report, approved on November 28, 1995, by the Advisory Committee, concluded that—based on intensive laboratory and field testing—the Grand Alliance digital television system was superior to any known alternative system in the world, better than any of the four original digital HDTV systems, and had surpassed the performance objectives of the ACATS.

Marking one of the last steps in an 8-year process to establish a U.S. ATV broadcasting standard, the 25-member blue-ribbon ACATS panel recommended the new standard to the FCC on November 28, 1995. Richard E. Wiley, ACATS chairman commented, "This is a landmark day for many communications industries and, especially, for American television viewers."

B.4e The Home Stretch for the Grand Alliance

With the technical aspects of the Grand Alliance HDTV system firmly in place, work proceeded to step through the necessary regulatory issues. Primary among these efforts was the establishment of a table of DTV assignments, a task that brought with it a number of significant concerns on the part of television broadcasters. Questions raised at the time involved whether a station's DTV signal should be equivalent to its present NTSC signal, and if so, how this should be accomplished.

Approval of the DTV standard by the FCC was a 3-step process:

- A *notice of proposed rulemaking* (NPRM) on policy matters, issued on August 9, 1995.
- Official acceptance of the Grand Alliance system. On May 9, 1996, the commission voted to propose that a single digital TV standard be mandated for over-the-air terrestrial broadcast of digital television. The standard chosen was that documented under the auspices of the Advanced Television Systems Committee (ATSC).
- Final acceptance of a table of assignments for DTV service.

During the comment period for the NPRM on Advanced Television Service (MM Docket 87-268), a number of points of view were expressed. Some of the more troublesome—from the standpoint of timely approval of the Grand Alliance standard, at least—came from the computer industry. Among the points raised were:

- Interlaced scanning. Some computer interests wanted to ban the 1920 × 1080 interlaced format.
- Square pixels. Computer interests recommended banning the use of formats that did not incorporate square pixels.
- 60 Hz frame rate. Computer interests recommended a frame rate of 72 Hz and banning of 60 Hz.

Meanwhile, certain interests in the motion picture industry rejected the 16:9 (1.78:1) widescreen aspect ratio in favor of a 2:1 aspect ratio.

One by one, these objections were dealt with. Negotiations between the two primary groups in this battle—broadcasters and the computer industry—resulted in a compromise that urged the FCC to adopt a standard that does not specify a single video format for digital television, but instead lets the various industries and companies choose formats they think will best suit consumers. The lack of a mandated video format set the stage for a lively competition between set makers and personal computer (PC) manufacturers, who were expected to woo consumers by combining sharp pictures with features peculiar to computers.

By early December, a more-or-less unified front had again been forged, clearing the way for final action by the FCC. With approval in hand, broadcasters then faced the demands of the commission's timetable for implementation, which included the following key points:

- By late 1998, 26 TV stations in the country's largest cities—representing about 30 percent of U.S. television households—would begin broadcasting the Grand Alliance DTV system.
- By mid-1999, the initial group would expand to 40; by 2000, it would expand to 120 stations.
- By 2006, every TV station in the country would be broadcasting a digital signal or risk losing its FCC license.

Fierce debates about the wisdom of this plan—and whether such a plan even could be accomplished—then ensued.

B.4f Digital Broadcasting Begins

If HDTV truly was going to be the "next big thing," then it was only fitting to launch it with a bang. The ATSC system received just such a sendoff, playing to excited audiences from coast to coast during the launch of Space Shuttle mission STS-95.

The first nationwide broadcast of a high-definition television program using the ATSC DTV system, complete with promos and commercials, aired on October 29, 1998. The live HDTV broadcast of Senator John Glenn's historic return to space was transmitted by ABC, CBS, Fox, NBC, and PBS affiliates from coast to coast.

The feed was available free for any broadcaster who could receive the signal. The affiliates and other providers transmitted the broadcast to viewing sites in Washington, D.C., New York, Atlanta, Chicago, Los Angeles, and 15 other cities. Audiences in those cities watched the launch on new digital receivers and projectors during special events at museums, retail stores, broadcast stations, and other locations. Many of the stations moved their on-air dates ahead of schedule in order to show the Glenn launch broadcast. The largest scheduled viewing site was the Smithsonian's National Air and Space Museum in Washington, D.C., where viewers watched the launch on an IMAX theatre screen and four new digital receivers.

Beyond the technical details was an even more important story insofar as HDTV production is concerned. All of the cameras used in the coverage provided an HD signal except for one NASA pool feed of the launch control center at the Kennedy Space Center, which was upconverted NTSC. On occasion, the director would switch to the launch center feed, providing a dramatic "A/B" comparison of high-definition versus standard-definition. The difference was startling. It easily convinced the industry observers present at the Air and Space Museum of the compelling power of the HDTV image.

The second production issue to come into focus during the broadcast was the editorial value of the wide aspect ratio of HDTV. At one point in the coverage, the program anchor described to the television audience how the Shuttle was fueled the night before. In describing the process, the camera pulled back from the launch pad shot to reveal a fuel storage tank off to one side of the pad. As the reporter continued to explain the procedure, the camera continued to pull back to reveal a second fuel storage tank on the other side of the launch pad. Thanks in no small part to the increased resolution of HDTV and—of course—the 16:9 aspect ratio, the television audience was able to see the entire launch area in a single image. Such a shot would have been wholly impossible with standard-definition imaging.

The STS-95 mission marked a milestone in space, and a milestone in television.

B.4g Continuing Work on the ATSC Standard

The creation by the Advanced Television Systems Committee of the DTV standard in 1995, and the FCC's subsequent adoption of the major elements of the standard into the FCC Rules in 1996, represented landmark achievements in the history of broadcast television. While these events represented the culmination of the ATSC's work in one sense, they also marked the beginning of a whole new effort to take the DTV standard as developed and turn it into a functioning, and profitable, system that would be embraced by both industry and consumers alike. To that end, the ATSC organized and continues to support the work of three separate technical standards-setting groups, each focusing on a different aspect of DTV deployment. These groups are:

- The *Technology Group on Distribution* (T3), which has as its mission the development and recommendation of voluntary, international technical standards for the distribution of television programs to the public using advanced imaging technology.
- *Technology Group on Production* (T4), established to develop and recommend voluntary, international technical standards for the production of television programs using advanced television technology, encompassing sound, vision, and display sub-systems used for the production of television programming.
- *DTV Implementation Subcommittee* (IS), established to investigate and report on the requirements for implementation of advanced television. The subcommittee evaluates technical requirements, operational impacts, preferred operating methods, time frames, and cost impacts of implementation issues.

The ATSC is composed of more than 160 member corporations, industry associations, standards bodies, research laboratories, and educational institutions. It is an international group whose charter is to develop voluntary standards for the entire range of advanced television systems. Major efforts have included a certification program for television sets, computers, and other consumer video devices in cooperation with the Consumer Electronics Manufacturers Association. Another element of the ATSC work has been to explain and demonstrate the DTV system to international groups, the goal being adoption of the ATSC DTV standard in other countries.

B.4h A Surprising Challenge

As of the 1999 NAB Convention, over 50 stations were transmitting DTV signals. It was no surprise, then, that a proposal by one broadcaster—Sinclair Broadcast Group—that the chosen modulation method for the ATSC system be revaluated, created shock waves throughout the convention, and indeed, throughout the industry. Sinclair's argument was that early production model DTV receivers performed poorly—in their estimation—relative to NTSC, and that the end result would be less reliance by the consumer on over-the-air signals and more reliance on cable television signals, thus putting broadcasters at a disadvantage. Sinclair suggested that COFDM, the method chosen by the European DVB consortium for terrestrial transmission, would perform better than 8-VSB.

Broadcast industry leaders were, generally speaking, in shock over the proposal. Many felt betrayed by one of their own. A number of theories were issued by leaders and observers to explain the Sinclair idea, and few had anything to do with the technical issues of signal coverage. Nevertheless, Sinclair went ahead with plans to conduct side-by-side tests of COFDM and 8-VSB later in the year.

It is important to note that in the early days of the Grand Alliance program, COFDM was tested against 8-VSB. It was, in fact, tested several times. On the basis of technical merit, 8-VSB was chosen by the Grand Alliance, and ultimately endorsed by the FCC.

The Sinclair-sponsored tests did, indeed, take place during the Summer of 1999. The reviews from stations, networks, and industry groups—most notably the NAB—were mixed. In general, observers stated that the tests did show the benefits of COFDM for operation with inside receive antennas in areas that experience considerable multipath. Multipath tolerance, of course, is one of the strengths of the COFDM system. Whether any significant progress was made to change industry minds, however, was unclear.

The ATSC made its position on the Sinclair tests crystal clear, however, in a statement released on July 2, 1999, calling the effort "unwarranted and irresponsible." The statement continued, "It is unwarranted because a growing body of evidence supports the performance of the VSB transmission system, and there is no clear evidence that COFDM is better. It is irresponsible because it seriously understates the impact of a change." The position statement concluded that, "Any decision to revisit the transmission standard would cause years of delay."

There were a number of other organizations and companies that weighed-in on the Sinclair demonstrations. One that received a fair amount of attention was a position paper issued by the Harris Corporation, a major broadcast equipment supplied with perhaps fewer axes to grind than most commenters. The Harris paper, released on August 10, 1999, warned that, "reopening the DTV standard debate would imperil the transition to digital broadcasting." The statement continued, "The burden on those seeking to reopen the DTV standards debate should be extremely high, given the lengthy delay, high cost, and crippling uncertainty necessarily entailed in even considering a switch to COFDM. The related industries have produced a variety of competing studio, transmission, and consumer equipment and 166 stations to date have either initiated DTV broadcasts or purchased transmission equipment. The Sinclair demonstration falls far short of warranting the delay, disruption, and confusion that would accompany reopening this debate."

The Harris position paper raised—rather convincingly—the financial implications inherent in a change to COFDM from 8-VSB, if equivalent NTSC coverage is to be maintained. "The structure of the 8-VSB signal minimizes both co-channel and adjacent channel interference to NTSC signals. By contrast, even at the same power levels, the structure of the COFDM signal will cause perceptibly more interference with existing NTSC services. This interference will be exacerbated if COFDM power levels are increased, as would be required to replicate NTSC coverage. Under any scenario, a change to COFDM would necessitate the Commission adopting a new table of allotments."

The ITU also found itself being drawn into the modulation wars of Summer 1999. A report issued on May 11 by an ITU Radiocommunication Study Group presented an objective, scientifically valid comparison of the modulation schemes under a variety of conditions. The report, titled "Guide for the Use of Digital Television Terrestrial Broadcasting Systems Based on Performance Comparison of ATSC 8-VSB and DVB-T COFDM Transmission Systems," provided considerable detail of the relative merits of each system. The report, interestingly enough, was used by both the COFDM and 8-VSB camps to support their positions.

In essence, the report concluded that the answer to the question of which system is better is: "it depends." According to the document, "Generally speaking, each system has its unique advantages and disadvantages. The ATSC 8-VSB system is more robust in an *additive white Gaussian noise* (AWGN) channel, has a higher spectrum efficiency, a lower peak-to-average power ratio, and is more robust to impulse noise and phase noise. It also has comparable performance to DVB-T on low level ghost ensembles and analog TV interference into DTV. Therefore, the ATSC 8-VSB system could be more advantageous for *multi-frequency network* (MFN) implementation and for providing HDTV service within a 6 MHz channel."

"The DVB-T COFDM system has performance advantages with respect to high level (up to 0 dB), long delay static and dynamic multipath distortion. It could be advantageous for services requiring large scale *single frequency network* (SFN) (8k mode) or mobile reception (2k mode). However, it should be pointed out that large scale SFN, mobile reception and HDTV service cannot be achieved concurrently with any existing DTTB system over any channel spacing, whether 6, 7 or 8MHz."

The ITU report concluded, "DTV implementation is still in its early stage. The first few generations of receivers might not function as well as anticipated. However, with the technical advances, both DTTB systems will accomplish performance improvements and provide a much improved television service."

"The final choice of a DTV modulation system is based on how well the two systems can meet the particular requirements or priorities of each country, as well as other non-technical (but critical) factors, such as geographical, economical, and political connections with surrounding countries and regions. Each country needs to clearly establish their needs, then investigates the available information on the performances of different systems to make the best choice."

Petition Filed

Despite minimal, and sometimes conflicting, data on the relative performance of 8-VSB-versus-COFDM, Sinclair formally put the question before the FCC on October 8, 1999. The petition requested that the Commission allow COFDM transmissions *in addition to* 8-VSB for DTV terrestrial broadcasting. If granted, the petition would, in effect, require receiver manufacturers to develop and market dual mode receivers, something that receiver manufacturers were not predisposed to do. It was estimated that when the Sinclair petition was filed, over 5,000 DTV receivers had been sold to consumers.

Petition Rejected

On February 4, 2000, the FCC released a letter denying a Petition for Expedited Rulemaking, filed by Sinclair requesting that the Commission modify its rules to allow broadcasters to transmit DTV signals using COFDM modulation in addition to the current 8-VSB modulation standard. The Commission said that numerous studies supported the conclusion that NTSC replication is attainable under the 8-VSB standard. It said that the concerns raised in the Sinclair petition had "done no more than to demonstrate a shortcoming of early DTV receiver implementation." The Commission pointed out that receiver manufacturers were aware of problems cited by Sinclair and were aggressively taking steps to resolve multipath problems exhibited in some first-generation TV sets.

The Commission noted that the FCC Office of Engineering and Technology had analyzed the relative merits of the two standards, and concluded that the benefits of changing the DTV transmission standard to COFDM would not outweigh the costs of making such a revision. The Commission reiterated its view that allowing more than one standard could result in compatibility problems that could cause consumers and licensees to postpone purchasing DTV equipment and lead to significant delays in the implementation and provision of DTV services to the public. It said that development of a COFDM standard would result in a multi-year effort, rather than the "unrealistic" 120 days suggested in the Sinclair petition.

At the same time it dismissed the petition, the Commission recognized the importance of the issues raised by Sinclair. The Commission, stated, however, that the issue of the adequacy of the DTV standard is more appropriately addressed in the context of its biennial review of the entire DTV transition effort.

B.4i FCC Reviews the Digital Television Conversion Process

On March 8, 2000, the FCC began its first periodic review of the progress of conversion of the U.S. television system from analog technology to DTV. In a Notice of Proposed Rulemaking (NPRM 00-83), the Commission invited public comment on a number of issues that it said required resolution to insure continued progress in the DTV conversion and to eliminate potential sources of delay. The Commission said its goal was to insure that the DTV transition went smoothly for American consumers, broadcasters, and other interested parties. This periodic review followed through on the conclusion, adopted by the Commission as part of its DTV construction schedule and service rules in the May 1997 5th Report and Order, that it should undertake a periodic review every two years until the cessation of analog service to help the Commission insure that the DTV conversion fully served the public interest.

In the NPRM, the Commission noted that broadcast stations were facing relatively few technical problems in building digital facilities, and that problems encountered by some stations with tower availability and/or local zoning issues did not seem to be widespread. However, it asked for comment on whether broadcasters were able to secure necessary tower locations and construction resources, and to what extent any zoning disputes, private negotiations with tower owners, and the availability of tower construction resources affected the DTV transition.

In the NPRM, the Commission asked for comments on whether to adopt a requirement that DTV licensees replicate their NTSC service area, and whether a replication requirement should be based on the population or the area served by the station. The Commission noted that several licensees had sought authority to move their DTV station to a more central location in their market—or toward a larger market—or had asked to change their DTV allotment—including their assumed transmitter site and/or technical facilities—and it asked for comments on the effect that these situations had on the general replication requirements. In addition, the Commission asked for comments on a proposed requirement that DTV stations' principal community be served by a stronger signal level than that specified for the general DTV service contour.

The Commission asked for comments on what date stations with both their NTSC and DTV channels within the DTV *core* (channels 2–51) would have to choose the channel they intend to keep following the transition. It said that with the target date for the end of DTV transition set for December 31, 2006, it would be reasonable for stations to identify the DTV channels they will be using not later than 2004. It asked for comment on whether this date represented the proper balance between the goals of allowing DTV stations enough time to gain experience with DTV operation, and allowing stations that must move enough time to plan for their DTV channel conversion.

The Commission also invited comments on DTV application processing procedures, including whether to establish DTV application cut-off procedures, how to resolve conflicts between DTV applications to implement "initial" allotments, and the order of priority between DTV and NTSC applications. The Commission said it was seeking comment on whether to adopt a cut-off procedure for DTV area-expansion applications to minimize the number of mutually exclusive applications and to facilitate applicants' planning, and how to resolve any mutual exclusive applications that arise.

In the NPRM, the Commission noted that concerns had arisen regarding the 8-VSB DTV transmission standard. It invited comment on the status of this standard, including information on any additional studies conducted regarding NTSC replication using the 8-VSB standard. It specifically asked for comments on progress that was being made to improve indoor DTV reception, and manufacturers' efforts to implement DTV design or chip improvements.

The Commission noted certain industry agreements relating to cable compatibility, but said these agreements did not cover labeling of digital receivers, and asked whether a failure to reach agreement on the labeling issue would hinder the DTV transition. The Commission also asked for comments on the extent to which a lack of agreement on copy protection technology licensing and related issues would also hinder the DTV transition. Noting that some broadcasters had recommended that the Commission address over-the-air signal reception by setting receiver standards, the Commission asked for comments on whether the FCC had authority to set minimum performance levels for DTV receivers, whether it would be desirable to do so, and, if so, how such requirements should be structured.

In this "periodic review," the Commission said it was not asking for comment on issues that were the subject of separate proceedings—such as the issue of digital broadcast signal carriage on cable systems—or requests for reconsideration of already-decided issues—such as eligibility issues, certain issues relating to public television, and channel allotment or change requests. The Commission also said it is too early in the transition process to address other issues, including reconsidering the flexible approach to ancillary or supplementary services and the application of the simulcast requirement.

Perhaps most importantly—from the standpoint of broadcasters at least—the Commission said it would be inappropriate to review the 2006 target date for complete conversion to DTV because of Congressional action in the Balanced Budget Act of 1997, which confirmed December 31, 2006 as the transition completion date and established procedures and standards for stations seeking an extension of that date.

ATSC Comments Filed

The Advanced Television Systems Committee filed comments on the FCC's DTV review, as expected. What surprised many, however, was the dramatic change of public posture with respect to the 8-VSB-vs.-COFDM issue. The ATSC filing stated, "The ATSC fully endorses and supports the Commission's DTV transmission standard, based on the ATSC DTV standard, and we urge the Commission to take all appropriate action to support and promote the rapid transition to digital television. Even so, the ATSC continues to seek ways to improve its standards and the implementation thereof in order to better meet the existing and evolving service requirements of broadcasters. To this end, and also in response to the specific concerns of some broadcasters regarding RF system performance (modulation, transmission and reception), the Executive Committee of the ATSC (has) formed a task force on RF system performance."

This task force was charged with examining a variety of technical issues that had been raised regarding the theoretical and realized performance of the DTV RF system—including receivers—and based on its findings, to make recommendations to the ATSC Executive Committee as to what, if any, technical initiatives should be undertaken by the ATSC.

At the time of the filing (May, 2000), the task force had identified three areas of focus:

- 8-VSB performance
- Broadcaster requirements
- Field test methodology

The task force was charged with evaluating the extent to which 8-VSB enables broadcasters to achieve reliable DTV reception within their current NTSC service areas. The group then was to assess the current performance of 8-VSB, including 8-VSB receiving equipment, as well as expected improvements. In addition, the task force was to look into potential backward-compatible enhancements to VSB, i.e., enhancements that could be implemented without affecting the operation of existing receivers that are not capable of delivering the enhanced features.

The task force also hoped to recommend potential actions that could be taken by the ATSC and its member organizations that might hasten improvements in the performance of the DTV RF system.

Until the RF task force was established, the ATSC had—publicly at least—refused to even consider the 8-VSB vs. COFDM matter. It is fair to point out, of course, that in this particular case any change in the RF modulation system could have a dramatic impact on the roll-out of DTV. Indeed, within a month of the ATSC filing, Sony became the first major consumer receiver manufacturer to delay introduction of a new line of HDTV sets, citing the uncertainty over the modulation issue (and certain technical obstacles).

B.4j A Turning Point is Reached

January 2001 marked a turning point for the ATSC DTV System. The FCC affirmed its support—for the third time—of the 8-VSB modulation system and a closely watched comparative study of 8-VSB and COFDM concluded that, given the character of the U.S. broadcasting system, 8-VSB was superior to COFDM. The study, managed jointly by the National Association of Broadcasters (NAB) and Maximum Service Television (MSTV), was broad in scope and technical depth. With the support of 30 major broadcast organizations and the oversight of technical committees consisting of some 25 engineers representing all major technical viewpoints, the broadcasting industry conducted a comprehensive, objective, and expedited series of studies and tests to determine whether COFDM should be added to the current 8-VSB standard. Conclusions of the NAB/MSTV study, released on January 15, 2001, were as follows: "We conclude that there is insufficient evidence to add COFDM and we therefore reaffirm our endorsement of the VSB standard. We also conclude that there is an urgent need for swift and dramatic improvement in the performance of the present U.S. digital television system. We therefore will take all necessary steps to promote the rapid improvement of VSB technologies and other enhancements to digital television..."

A few days later, the FCC issued a *Report and Order and a Further Notice of Proposed Rulemaking* (FNRPM) in its first periodic review of the DTV transition. In the Report and Order, the Commission affirmed the 8-VSB modulation system for the DTV transmission standard, concluding that there was no reason to revisit its decision, thereby denying a request to allow use of an alternative DTV modulation system (COFDM).

Reactions came quickly from industry leaders. Robert Graves, Chairman of the ATSC, stated: "The decisions this week by the NAB and MSTV and by the FCC put an end to the debate surrounding the ATSC/VSB transmission system used for terrestrial DTV broadcasts in the U.S. With the transmission system debate behind us, with the impressive array of DTV and HDTV products now available at ever lower prices, and with the increasing amount of compelling DTV content becoming available from broadcasters, the outlook for DTV services in the U.S. is bright. And, we expect this week's actions by the broadcast industry and the FCC to enhance the prospects for adoption of the ATSC Digital Television Standard in other countries."

The NAB/MSTV report noted the need for improvements in receivers and called for a thorough investigation of ways to enhance the 8-VSB system. Graves addressed that issue, saying: "Significant improvements have already been made in the performance of VSB receivers, and further improvements are in the pipeline from a variety of manufacturers. Moreover, work is under way within the ATSC to further enhance the ATSC/VSB standard by adding more robust transmission modes that address emerging DTV applications…"

The consumer electronics industry also responded to the NAB/MSTV report and the FCC action. Consumer Electronics Association (CEA) President and CEO Gary Shapiro said: "The debate over the modulation standard is over. All broadcasters who are seriously committed to the DTV transition should now recommit to the transition by continuing the buildout and producing more high-definition and digital content. The results of the joint NAB/MSTV testing confirm what we have long believed: the Federal Communications Commission (FCC)-approved 8-VSB modulation system is the best choice for broadcasting digital television in the United States. These results also reaffirm that the FCC made the correct decision in initially selecting, and again re-endorsing 8-VSB last year.

A week after these pronouncements, the ATSC made good on its promise to investigate ways to enhance the 8-VSB transmission system to improve reception in fixed indoor, portable, and mobile applications. The ATSC Specialist Group on RF Transmission (T3/S9) issued a Request for Proposal (RFP) for enhancements aimed at addressing emerging DTV applications. The RFP asked for proposals for complementary specifications that will be considered as potential enhancements of the ATSC DTV Standard. Laboratory and field tests of the proposals were expected to begin in the Winter of 2001. Prominent among the specialists participating in the work of T3/S9 were engineers from Sinclair.

With this flurry of activity, the momentum seemed to return to the DTV transition. As of November 2001, there were 215 stations transmitting DTV signals in 75 markets (serving nearly 70 percent of U.S. TV households), according to an NAB survey.

Receiver Performance Issues

In the January 2001 Report and Order, the FCC denied requests to set performance standards for digital receivers, expressing concern that the effect of setting such standards would be to stifle innovation and limit performance to current capabilities. The Commission said, however, it would continue to monitor receiver issues.

The Commission also issued a Further Notice of Proposed Rulemaking to consider whether to require some TV sets to have the capability to demodulate and decode over-the-air DTV signals in addition to displaying the existing analog TV signals. In raising this issue, the Commission recognized broadcasters' concerns that DTV receivers are not yet available in sufficient volume to support a rapid transition to an all-digital broadcast television service. It asked whether a requirement to include DTV reception capability in certain television sets could help to develop the production volumes needed to bring DTV receiver prices down quickly to where they would be more attractive to consumers and could help to promote a more rapid development of high DTV set penetration.

The Commission asked for comments on how best to implement DTV reception capability requirements, if it were to decide to adopt them, adding that it recognized the cost considerations associated with such requirements. The Commission suggested that one approach to minimizing the impact on both consumers and manufacturers would be to impose any requirement first on a percentage of large screen televisions, such as 32 in and larger, because these are typically higher priced units where the cost of DTV components would be a smaller percentage of the cost of the receiver. In addition, it asked whether any requirement should be phased in over time such that manufacturers would increase each year the percentage of units of a designated screen size or larger that are manufactured with DTV receive capability. The FCC noted that separate set-top DTV receivers could also be included in meeting the reception capability requirements.

The Commission additionally requested comment on whether it should require any digital television receivers that cannot receive over-the-air digital broadcast signals to carry a label informing consumers of this limitation. This issue concerns receivers that are intended for use only with cable television or broadcast satellite service. The Commission indicated that while it expects consumers will continue to expect all digital television receivers to be able to receive over-the-air broadcast signals, it suggested that where receivers not able to receive such signals are marketed, consumers should be so notified prior to purchase.

B.5 References

1. Krivocheev, Mark I., and S. N. Baron: "The First Twenty Years of HDTV: 1972–1992," *SMPTE Journal*, SMPTE, White Plains, N.Y., pg. 913, October 1993.
2. CCIR Document PLEN/69-E (Rev. 1), "Minutes of the Third Plenary Meeting," pp. 2–4, May 29, 1990.
3. Based on information supplied by the DVB Project on its Web site: http://www.dvb.com.
4. Gilder, George: "IBM-TV?," *Forbes*, Feb. 20, 1989.

B.6 Bibliography

ATSC: "Comments of The Advanced Television Systems Committee, MM Docket No. 00-39," ATSC, Washington, D.C., May, 2000.

"ATV System Recommendation," *1993 NAB HDTV World Conference Proceedings*, National Association of Broadcasters, Washington, D.C., pp. 253–258, 1993.

Appelquist, P.: "The HD-Divine Project: A Scandinavian Terrestrial HDTV System," *1993 NAB HDTV World Conference Proceedings*, National Association of Broadcasters, Washington, D.C., pg. 118, 1993.

Baron, Stanley: "International Standards for Digital Terrestrial Television Broadcast: How the ITU Achieved a Single-Decoder World," *Proceedings of the 1997 BEC*, National Association of Broadcasters, Washington, D.C., pp. 150–161, 1997.

Battison, John: "Making History," *Broadcast Engineering*, Intertec Publishing, Overland Park, Kan., June 1986.

Benson, K. B., and D. G. Fink: *HDTV: Advanced Television for the 1990s*, McGraw-Hill, New York, 1990.

Benson, K. B., and Jerry C. Whitaker (eds.): *Television Engineering Handbook*, rev. ed., McGraw-Hill, New York, 1992.

Benson, K. B., and J. C. Whitaker (eds.): *Television and Audio Handbook for Engineers and Technicians*, McGraw-Hill, New York, 1989.

CCIR Report 801-3: "The Present State of High-Definition Television," pg 37, June 1989.

CCIR Report 801-3: "The Present State of High-Definition Television," pg. 46, June 1989.

"Dr. Vladimir K. Zworkin: 1889–1982," *Electronic Servicing and Technology*, Intertec Publishing, Overland Park, Kan., October 1982.

Federal Communications Commission: Notice of Proposed Rule Making 00-83, FCC, Washington, D.C., March 8, 2000.

Hopkins, R.: "Advanced Television Systems," *IEEE Transactions on Consumer Electronics*, vol. 34, pp. 1–15, February 1988.

Lincoln, Donald: "TV in the Bay Area as Viewed from KPIX," *Broadcast Engineering*, Intertec Publishing, Overland Park, Kan., May 1979.

McCroskey, Donald: "Setting Standards for the Future," *Broadcast Engineering*, Intertec Publishing, Overland Park, Kan., May 1989.

Pank, Bob (ed.): *The Digital Fact Book*, 9th ed., Quantel Ltd, Newbury, England, 1998.

Reimers, U. H.: "The European Perspective for Digital Terrestrial Television, Part 1: Conclusions of the Working Group on Digital Terrestrial Television Broadcasting," *1993 NAB HDTV World Conference Proceedings*, National Association of Broadcasters, Washington, D.C., p. 117, 1993.

Schow, Edison: "A Review of Television Systems and the Systems for Recording Television," *Sound and Video Contractor*, Intertec Publishing, Overland Park, Kan., May 1989.

Schreiber, W. F., A. B. Lippman, A. N. Netravali, E. H. Adelson, and D. H. Steelin: "Channel-Compatible 6-MHz HDTV Distribution Systems," *SMPTE Journal*, SMPTE, White Plains, N.Y., vol. 98, pp. 5-13, January 1989.

Schreiber, W. F., and A. B. Lippman: "Single-Channel HDTV Systems—Compatible and Noncompatible," Report ATRP-T-82, Advanced Television Research Program, MIT Media Laboratory, Cambridge, Mass., March 1988.

Schubin, Mark, "From Tiny Tubes to Giant Screens," *Video Review*, April 1989.

"Sinclair Seeks a Second Method to Transmit Digital-TV Signals," *Wall Street Journal*, Dow Jones, New York, N.Y., October 7, 1999.

SMPTE and EBU, "Task Force for Harmonized Standards for the Exchange of Program Material as Bitstreams," *SMPTE Journal*, SMPTE, White Plains, N.Y., pp. 605–815, July 1998.

"Television Pioneering," *Broadcast Engineering*, Intertec Publishing, Overland Park, Kan., May 1979.

"Varian Associates: An Early History," Varian publication, Varian Associates, Palo Alto, Calif.

Whitaker, Jerry C.: *Electronic Displays: Technology, Design, and Applications*, McGraw-Hill, New York, N.Y., 1994.

Subject Index

Numerics

1125/60 NHK standard 556
1250/50 Eureka HDTV 559
16 VSB 309
8 VSB 309

A

access unit 309
acquisition phase 5
additive white Gaussian noise 576
advanced systems 565
Advanced Television Systems Committee 57
AES 57
Allen B. DuMont 546
American Broadcasting Company 540
American Electronics Association 564
American National Standards Institute (ANSI) 558
anchor frame 309
ANSI 54
ANSI C95.1-1982 43
ascension numbering 26
Askarel 47
aspect ratio 560
asynchronous transfer mode 309
ATSC 57

B

backward-compatible 580
BeO ceramic 38, 40
beryllium oxide 38, 40
bidirectional pictures 309
bioaccumulation 47
bit rate 309
black body radiation 203
block 309
breaking radiation 208
byte-aligned 309

C

cable schedule 12
capital project budget request 14
centimeter waves 205
change control 8
change order request 15
channel 309
chlorinated hydrocarbons 47
chlorobenzenes 47
Coast Guard National Spill Response Center 50
coded orthogonal frequency division multiplexing 563
coded representation 309
Columbia Broadcasting System 540
Communications Act of 1934 213
compression 310
computer aided design 27
constant bit rate 310
Consumer Electronics Association 56
conventional definition television 310
conventional systems 565
conventions document 25
cosmic rays 208
CRC 310
Critical Path Method 14

D

database documentation 26
David Sarnoff 534
dc to light spectrum 202
decoded stream 310
decoder 310
decoding 310
decoding time-stamp 310
design and development phase 9
D-frame 310
DHTML 310
digital storage media 310
Digital Video Broadcasting project 561
diode-gun impregnated-cathode Saticon (DIS) 557
discrete cosine transform 310
distribution systems 565
documentation 3, 23
documentation system 25
DOM 310

DVB project 562

E

editing 310
Edwin Armstrong 540
electric shock 31
electrocution 32
electromagnetic spectrum 201
Electronic Industries Alliance 56
electronic-intermediate (EI) digital video system 563
Electronics Industries Association 56
elementary stream 310
elementary stream clock reference 310
EM spectrum 201
emergent properties 2
encoder 310
encoding 310
entitlement control message 310
entitlement management message 310
entropy coding 311
entry point 311
Environmental Protection Agency 48
equipment file 28
equipment grounding 39
Eureka 1187 Advanced Digital Television Technologies (ADTT) 562
Eureka EU95 559
European HDTV systems 559
event 311

F

FC-75 42
feasibility study 14
Federal Communications Commission 215, 538
field 311
first aid 36
flow diagram 12
FM radio 540
frame 311
frequency bands 201
frequency segments 201
frequency spectrums 201
from-to coding 26
functional flow block diagram 4
functional needs 4

G

gamma ray band 210
Gantt Chart 14
General Secretariat 215
Grand Alliance 561
ground-fault current interrupter 33
grounding 39
group of pictures 311
Guglielmo Marconi 533

H

hard gamma rays 210
hard X rays 208
HD-DIVINE 561
HD-MAC (high-definition multiplexed analog component) 559
HDTV standardization 560
HDTV-6-7-8 system 561
Heinrich Hertz 533
high level 311
high-definition television 311, 565
HTML 311
HTTP 311
Huffman coding 311

I

Iconoscope 544
IEEE 56
Image Iconoscope 547
Image Orthicon 548
improved-definition television (IDTV) 565
infrared band 202
Institute of Radio Engineers 56
International Radio Consultative Committee 54, 556
International Standards Organization 54
International Telecommunication Convention 213
International Telecommunications Union 54, 214
International Telephone Consultative Committee 214
intra-coded pictures 311
IR band 203
ISO 54
isomorphism 2
ITU 54
ITU Table of Frequency Allocations 217

J

J. Ambrose Fleming 533
Joint Committee for Intersociety Coordination 55

K

kinescope 544
klystron 555

L

layer 311
Lee DeForest 533
level 311
light to gamma ray spectrum 202

M

macroblock 311
main level 311
main profile 311
Maximum Service Telecasters 564
microwave band 204
millimeter waves 205
MIL-STD-499 1
MIME 312
modeling 5
motion vector 312
motion-compensated hybrid discrete cosine transform coding 562
MPEG 312
MPEG-1 312
MPEG-2 312
multi-frequency network 576
MUSE (multiple sub-Nyquist encoding) 558
Mutual Broadcasting System 540

N

National Broadcasting Company 540
National Electrical Code 40
National Environmental Policy Act 43
National Stereophonic Radio Committee 541
National Table of Frequency Allocations 216
National Television Systems Committee 552
Nipkow disc 544
Nippon Hoso Kyokai (NHK) 555
nonionizing radiation 43
NPRM 00-83 578

O

O. B. Hanson 535
Occupational Safety and Health Administration 38
optical spectrum 202
Orthicon camera tube 547
OSHA 38

P

packet 312
packet data 312
packet identifier 312
padding 312
Paul Nipkow 543
payload 312
PCB large capacitor 50
PCB large high-voltage capacitor 50
PCB small capacitor 50
PCB transformer 48
PCBs 47
permitted services 218
PES packet 312
PES stream 312
phase alternation line (PAL) 553
Philo Farnsworth 545
picture 312
pixel 312
Plenipotentiary Conference 215
Plumbicon 548
polychlorinated dibenzofurans 47
polychlorinated dibenzo-p-dioxins 47
predicted pictures 312
presentation time-stamp 312
presentation unit 312
primary gamma rays 210
primary services 218
production systems 565
profile 312
program 313
program clock reference 313
program element 313
program specific information 313
Project Evaluation and Review 14
project management 17
purpose codes 26

Q

quantizer 313

R

radio frequency band 204
radio frequency radiation 43
Radio Manufacturers Association 56
Radiocommunication Sector 215
random access 313
RCA 539
RETMA 56
return-beam Saticon 556
RFR 43

S

Saticon 548
scrambling 313
secondary gamma rays 208, 210
secondary services 218
simulcast systems 565
single frequency network 576
slice 313
SMPTE 56
SMPTE 240M 558
Society of Motion Picture and Television Engineers (SMPTE) 558
soft gamma rays 210
soft X rays 208
source stream 313
spectral lines 205
SPECTRE 562
splicing 313
standard definition television 313
standardization xvii, 53
standardization bodies 55
start codes 313
STD input buffer 313
still picture 313
synchrotron radiation 205, 208
synthesis phase 4
system clock reference 313
system grounding 39
system header 313
system target decoder 313
systematic approach 4
systems engineer 20
systems engineering 1
systems maintenance 23

T

Telecommunication Development Sector 215
Telecommunication Standardization Sector 215
Telecommunications Industry Association 56
television history 543
thermal radiation 205
time-stamp 314
Toxic Substances Control Act 48
toxic waste 43
trade study process 6
trade table 8
trade tree 6
transport stream packet header 314
Trinescope 549

U

U.S. Government Table of Frequency Allocations 219
UHTTP 314
ultraviolet band 202
user documentation 28
utility function graph 7
UUID 314
UV band 204

V

vacuum UV band 204
variable bit rate 314
ventricular fibrillation 32
video buffering verifier 314
video sequence 314
Vidicon 548
visible light band 202
Vladimir Zworykin 544

W

William Paley 540
World Administrative Radio Conference 215

X

X rays 46

About the Editor

Jerry C. Whitaker is Technical Director of the Advanced Television Systems Committee (ATSC), Washington, D.C. He was previously President of Technical Press, a consulting company based in the San Jose (CA) area. Mr. Whitaker has been involved in various aspects of the electronics industry for over 25 years, with specialization in communications. Current book titles include the following:

- Editor-in-chief, *Standard Handbook of Video and Television Engineering*, 3rd ed., McGraw-Hill, 2000
- Editor-in-chief, *Standard Handbook of Audio and Radio Engineering*, 2nd ed., McGraw-Hill, 2001
- *DTV Handbook*, 3rd ed., McGraw-Hill, 2001
- *Video Displays*, McGraw-Hill, 2000
- Editor, *Interactive Television Demystified*, McGraw-Hill, 2001
- Editor, *Audio and Video Professional's Field Manual*, McGraw-Hill, 2001
- Editor, *Video and Television Engineers' Field Manual*, McGraw-Hill, 2000
- *Radio Frequency Transmission Systems: Design and Operation*, McGraw-Hill, 1990
- *Maintaining Electronic Systems*, 2nd ed., CRC Press, 2001
- Editor-in-chief, *The Electronics Handbook*, CRC Press, 1996
- *AC Power Systems Handbook*, 2nd ed., CRC Press, 1999
- *Power Vacuum Tubes Handbook*, 2nd ed., CRC Press, 1999
- *The Communications Facility Design Handbook*, CRC Press, 2000
- *The Resource Handbook of Electronics*, CRC Press, 2000
- Co-author, *Communications Receivers*, 3rd ed., McGraw-Hill, 2000

Mr. Whitaker has lectured extensively on the topic of electronic systems design, installation, and maintenance. He is the former editorial director and associate publisher of *Broadcast Engineering* and *Video Systems* magazines, and a former radio station chief engineer and television news producer.

Mr. Whitaker is a Fellow of the Society of Broadcast Engineers and an SBE-certified professional broadcast engineer. He is also a fellow of the Society of Motion Picture and Television Engineers, and a member of the Institute of Electrical and Electronics Engineers.

Mr. Whitaker has twice received a Jesse H. Neal Award *Certificate of Merit* from the Association of Business Publishers for editorial excellence. He has also been recognized as *Educator of the Year* by the Society of Broadcast Engineers.

Mr. Whitaker resides in Morgan Hill, California.

On-Line Updates

Additional updates relating to audio/video engineering in general, and this book in particular, can be found at the *Standard Handbook of Video and Television Engineering* web site:

www.tvhandbook.com

The tvhandbook.com web site supports the professional audio/video community with news, updates, and product information relating to the broadcast, post production, and business/industrial applications of digital audio and video.

Check the site regularly for news, updated chapters, and special events related to audio and video engineering. The technologies encompassed by *Audio/Video Protocol Handbook* are changing rapidly, with new developments each month. Changing market conditions and regulatory issues are adding to the rapid flow of news and information in this area.

Specific services found at **www.tvhandbook.com** include:

- **Audio/Video Technology News**. News reports and technical articles on the latest developments in digital radio and television, both in the U.S. and around the world. Check in at least once a month to see what's happening in the fast-moving area of digital broadcasting.
- **Resource Center**. Check for the latest information on professional and broadcast audio/video systems. The Resource Center provides updates on implementation and standardization efforts, plus links to related web sites.
- **tvhandbook.com Update Port**. Updated material for books in the McGraw-Hill Audio/Video Series is posted on the site regularly. Material available includes updated sections and chapters in areas of rapidly advancing technologies.
- **Book Store**. Check to find the latest books on digital audio and video technologies. Direct links to authors and publishers are provided. You can also place secure orders from our on-line bookstore.

In addition to the resources outlined above, detailed information is available on other books in the McGraw-Hill Video/Audio Series.

SOFTWARE AND INFORMATION LICENSE

The software and information on this diskette (collectively referred to as the "Product") are the property of The McGraw-Hill Companies, Inc. ("McGraw-Hill") and are protected by both United States copyright law and international copyright treaty provision. You must treat this Product just like a book, except that you may copy it into a computer to be used and you may make archival copies of the Products for the sole purpose of backing up our software and protecting your investment from loss.

By saying "just like a book," McGraw-Hill means, for example, that the Product may be used by any number of people and may be freely moved from one computer location to another, so long as there is no possibility of the Product (or any part of the Product) being used at one location or on one computer while it is being used at another. Just as a book cannot be read by two different people in two different places at the same time, neither can the Product be used by two different people in two different places at the same time (unless, of course, McGraw-Hill's rights are being violated).

McGraw-Hill reserves the right to alter or modify the contents of the Product at any time.

This agreement is effective until terminated. The Agreement will terminate automatically without notice if you fail to comply with any provisions of this Agreement. In the event of termination by reason of your breach, you will destroy or erase all copies of the Product installed on any computer system or made for backup purposes and shall expunge the Product from your data storage facilities.

LIMITED WARRANTY

McGraw-Hill warrants the physical diskette(s) enclosed herein to be free of defects in materials and workmanship for a period of sixty days from the purchase date. If McGraw-Hill receives written notification within the warranty period of defects in materials or workmanship, and such notification is determined by McGraw-Hill to be correct, McGraw-Hill will replace the defective diskette(s). Send request to:

Customer Service
McGraw-Hill
Gahanna Industrial Park
860 Taylor Station Road
Blacklick, OH 43004-9615

The entire and exclusive liability and remedy for breach of this Limited Warranty shall be limited to replacement of defective diskette(s) and shall not include or extend to any claim for or right to cover any other damages, including but not limited to, loss of profit, data, or use of the software, or special, incidental, or consequential damages or other similar claims, even if McGraw-Hill has been specifically advised as to the possibility of such damages. In no event will McGraw-Hill's liability for any damages to you or any other person ever exceed the lower of suggested list price or actual price paid for the license to use the Product, regardless of any form of the claim.

THE McGRAW-HILL COMPANIES, INC. SPECIFICALLY DISCLAIMS ALL OTHER WARRANTIES, EXPRESS OR IMPLIED, INCLUDING BUT NOT LIMITED TO, ANY IMPLIED WARRANTY OF MERCHANTABILITY OR FITNESS FOR A PARTICULAR PURPOSE. Specifically, McGraw-Hill makes no representation or warranty that the Product is fit for any particular purpose and any implied warranty of merchantability is limited to the sixty day duration of the Limited Warranty covering the physical diskette(s) only (and not the software or in-formation) and is otherwise expressly and specifically disclaimed.

This Limited Warranty gives you specific legal rights; you may have others which may vary from state to state. Some states do not allow the exclusion of incidental or consequential damages, or the limitation on how long an implied warranty lasts, so some of the above may not apply to you.

This Agreement constitutes the entire agreement between the parties relating to use of the Product. The terms of any purchase order shall have no effect on the terms of this Agreement. Failure of McGraw-Hill to insist at any time on strict compliance with this Agreement shall not constitute a waiver of any rights under this Agreement. This Agreement shall be construed and governed in accordance with the laws of New York. If any provision of this Agreement is held to be contrary to law, that provision will be enforced to the maximum extent permissible and the remaining provisions will remain in force and effect.